CONTENTS

TABLE OF CASES

Titles of cases shown in bold type
Names of parties reversed shown in ordinary type

INDEX OF SUBJECT-MATTER

AGRICULTURE

A

HOUSE OF LORDS

November 13 1974

(Before Lord DIPLOCK, Viscount DILHORNE, and Lord SIMON OF GLAISDALE)

THREADGOLD v BAHAMAS INTERNATIONAL TRUST CO LTD AND ANOTHER

Estates Gazette January 4 1975

(1974) 233 EG 47

Agricultural holding—" Unique " agreement held not to confer a right of exclusive occupation on licensee—Not, therefore, B **to take effect under Agricultural Holdings Act 1948, section 2 (1), as a tenancy from year to year—Licensors awarded possession of small dairy farm in temporary charge of deceased tenant's son**

This was an appeal by Mr Joel Threadgold, of Sycamore Farm, Bickerton, Malpas, Cheshire, from a decision of the Court of Appeal dated July 10 1974 awarding the respondent landlords, Bahamas International Trust Co Ltd, an order for possession of the property.

Mr M C Nourse QC and Mr J A D Gilliland (instructed by Riders, agents for Bellyse & Eric Smith, of Nantwich) appeared for the appellant, and Mr O M W Swingland QC and Mr G C Nurse (instructed by Mackrell & Co, agents for Bartlett & Son, of Chester) represented the respondents.

In his speech, LORD DIPLOCK said: This appeal is about the right to possession of a small dairy farm in Cheshire, known as Sycamore Farm, Bickerton. It consists of a farmhouse and some 11 acres of grazing land. The respondents (" the freeholders ") are the owners of the freehold. Before her death intestate in January 1968, the farm was held by the appellant's mother, Mrs Threadgold, upon an agricultural tenancy from year to year. The appellant (Mr Threadgold) helped her in the management of the farm. He was her sole next-of-kin, and after her death he would, upon obtaining letters of administration, have D become entitled to the residue of the term. The freeholders exercised their rights under the Agricultural Holdings Act 1948 to terminate the tenancy by notice to quit expiring on March 25 1969. Upon the expiry of this notice, Mr Threadgold was permitted by the freeholders to remain on the farm under the terms of an agreement dated March 27 1969. That agreement reads as follows:

" WHEREAS JOEL THREADGOLD agrees: 1. That the tenancy of Sycamore Farm and also the tenancy of two adjoining blocks of land in the name of Effie Threadgold terminated on March 25 1969 and that he will pay the half-year's rent due on that date.

" THE COMPANY agrees: 1. To allow Joel Threadgold to live in E the farmhouse rent and rates free until February 28 1970. The company reserves the right to terminate this occupation by fifty-six days' notice at any time. 2. To pay Joel Threadgold the sum of ten shillings (10s.) per week as caretaker of Sycamore Farm and the adjoining block of land until his occupation of the farmhouse ceases on February 28, 1970. These payments will automatically end if Joel Threadgold ceases to live in the farmhouse or if his occupation of the farmhouse is brought to an end by fifty-six days' notice before February 28, 1970, in which case payments will cease at the expiry of fifty-six days. 3. To allow Joel Threadgold to retain his cows and other livestock and to sell milk until October 31, 1969, subject always to the provisions of the Milk and Dairies Regulations. The company will be under no liability to remedy any defects which might exist F under these regulations."

Mr Threadgold did not comply with the terms of the agreement. He kept his cattle on the farm after October 31 1969, added to their number, and refused to give up occupation of the farmhouse after February 28 1970. The freeholders, after considerable forbearance, ultimately took proceedings in the Crewe County Court to obtain possession of the farm. Their claim to possession was resisted by Mr Threadgold on the ground that under section 2 (1) of the Agricultural Holdings Act 1948 the agreement took effect as a tenancy of an agricultural holding from year to year, and so gave him the security of tenure provided by that Act. On October 31 1973 the county G court judge upheld this contention. He refused to make an order for possession. On July 10 1974 his judgment was reversed by the Court of Appeal. That court made an order for possession, but gave leave to Mr Threadgold to appeal to your Lordships' House. Despite the fact that they had given leave to appeal, the Court of Appeal refused a stay of execution beyond a period of 56 days, to which the freeholders were willing to consent. The effect of this refusal of a stay of execution would have been that Mr Threadgold and his cattle would have been evicted from the farm long before the appeal could, in the ordinary course of events, have been heard by your Lordships' House; the freeholders had however undertaken not to let the farm to H any other person until the appeal had been determined. It was not brought to the attention of the Court of Appeal that in the present state of the farming industry in Cheshire, the practical result of his eviction would have been that it would have been impossible for Mr Threadgold to find anywhere else to keep his cattle. He would have been compelled to sell them at a sacrificial price and to abandon his milk business. So even if he succeeded at the hearing of the appeal, he would have sustained so serious a financial loss that he would have been unable to re-stock the farm so as to enable him to resume his milk business. In these circumstances, an application was made to this House for a stay of execution until after the hearing of this J appeal. In the ordinary course of events, this would not have taken place until the Easter Term of 1975. In the meantime, the freeholders, in the event of the appeal's failing, would have been kept out of possession for another six months. In these circumstances, to avoid inevitable injustice to one or other of the parties, it was found possible to make arrangements to hear the appeal forthwith.

The agreement of March 27 1969 is idiosyncratic in its terms. The only question in this appeal is whether, upon its true construction, it is caught by the provisions of section 2 (1) of the Agricultural Holdings Act 1948 and is accordingly to " take K effect, with the necessary modifications, as if it were an agreement for the letting of the land for a tenancy from year to year." Before the Court of Appeal the case was argued by counsel for the freeholders on the basis that clause 3 of the agreement gave Mr Threadgold an exclusive licence to occupy the land. The Court of Appeal, without further analysis, accepted this as being the true construction of the clause. Upon this construction, the case did raise questions of law of some difficulty and some general importance as to the construction of section 2 (1) of the Act and the correctness of the interpretation of it which had been suggested in two previous decisions of the Court of Appeal, viz *Goldsack v Shore* [1950] 1 KB 708 and *Harrison-Broadley v Smith* [1964] 1 WLR 456. It was for this reason that leave to appeal to this House was granted.

In a case which turns, as this one does, upon the construction to be given to a written document, a court called upon to construe the document, in the absence of any claim for rectification, cannot be bound by any concession made by any of the parties as to what its language means. This is so even in the court before which the concession is made; a *fortiori* in the court to which an appeal from the judgment of that court is brought. The reason is that the construction of a written document is a question of law. It is for the judge to decide for himself what the law is, not to accept it from any or even all of the parties to the suit; having so decided, it is his duty to apply it to the facts of the cases. He would be acting contrary to his judicial oath if he were to determine the case by applying what the parties conceived to be the law, if in his own opinion it was erroneous. Your Lordships must accordingly approach the construction of the agreement of March 27 1969 untrammelled by any concession as to its meaning that may have been made in the course of the argument before the Court of Appeal. This is a necessary preliminary step to the determination whether it is such an agreement as falls within section 2 (1) of the Agricultural Holdings Act 1948.

The first clause, numbered 1 in the agreement, is in part a recital and in part a covenant by Mr Threadgold. It contains an acknowledgement by him that the previously-existing tenancy of Sycamore Farm and two other blocks of land terminated two days earlier, on March 25 1969, and a personal covenant by him to pay the half-year's rent due on that date. Its significance for present purposes is that it acknowledges that any actual or potential legal relationship between the freeholders and Mr Threadgold arising out of the previous agricultural tenancy of Sycamore Farm had come to an end. The subsequent clauses, of which the first is also numbered 1, which I will call clauses 1, 2 and 3, regulate what is to be the new relationship between the freeholders and Mr Threadgold as respects Sycamore Farm. Clause 1 grants to Mr Threadgold a licence to live rent and rate free in the farmhouse until February 28 1970 unless the licence is terminated earlier by 56 days' notice. His living there is described in this clause and in clause 2 as " occupation " of the farmhouse. The capacity in which he is to live there is indicated in clause 2. He is to be employed as caretaker of Sycamore Farm and the adjoining block of land for a remuneration of 10s. per week, and his continuing to live in the farmhouse is made a condition of this employment. As respects the remainder of Sycamore Farm apart from the farmhouse, and the adjoining block of land, clearly no right to occupy any of the land or the farmbuildings was granted by clause 2. The most that this clause gave to Mr Threadgold was a personal right to enter upon the land and buildings in the course of his employment by the freeholders as caretaker.

Clause 3 is the crucial clause in this appeal. Whatever licence it grants is independent of those granted by clauses 1 and 2. It is for a fixed period until October 31 1969. It would terminate before the licences granted by clauses 1 and 2 if they ran their full course to February 28 1970. It would continue to subsist after they had terminated if notice of termination were given more than 56 days before October 31 1969. The licence granted to Mr Threadgold by clause 3 is expressed to be " to retain his cows and other livestock and to sell milk." It is a necessary implication from the context that he is to be allowed to keep the cattle he retains upon the land comprised in Sycamore Farm and the adjoining block of land referred to in clause 2, and to use the farmbuildings for the purpose of milking them and preparing their milk for sale. The use of the verb " to retain " indicates that the cows and livestock referred to are confined to those which were on the land at the date of the agreement, together with any progeny born to them during the fixed period of the licence. He is not entitled to put on to the land any fresh livestock.

In my view the meaning of this clause is clear and unambiguous. It grants to Mr Threadgold the right for the stipulated period to let any cattle of his which were on the land on March 27 1969 continue to graze on the pastures, and to milk them and prepare their milk for sale in the farmbuildings. In this clause there is no reference to " occupation " by Mr Threadgold of the land or farmbuildings, as there was in clauses 1 and 2 to his " occupation " of the farmhouse. As a matter of construction, I see nothing in the words which either expressly or by necessary implication gives to Mr Threadgold any right to exclude the freeholders from the land or the farmbuildings, or to stop them from making whatever use they please of the land or buildings, so long as that use does not prevent his reasonable exercise of the rights conferred upon him by the licence. The only limitation upon the freeholders' right to enter upon the land and use it for their own purposes is such as would result from the application of the principle that a grantor cannot derogate from his own grant. Whether any particular kind of use of the land by the freeholders would conflict with that principle would depend upon the circumstances existing at the time when that use was sought to be made.

If at the date of the agreement Mr Threadgold had had only two cows on the 11 acres of pasture I do not think that it could have plausibly been argued that the freeholders were debarred by the terms of clause 3 from bringing on to the pasture such number of additional livestock of their own as would be consistent with good farming practice having regard to the total amount of grass available, or from ploughing up and growing cereal crops on such part of the land as was not needed for the keep of Mr Threadgold's two cows. According to the evidence, he had 11 cows on the farm at the date of the agreement. There was no evidence as to whether or not this was the maximum number of livestock of all kinds that the pasture could support consistently with good farming practice throughout the period until October 31 1969. Such evidence, however, would be relevant to the construction of clause 3 only if Mr Threadgold had been bound to keep up the number of cattle on the land throughout the period of the licence; whereas, though he could not increase their number, he was entitled to reduce them to as few as he wished or to none at all if that suited his convenience. So even if at the beginning of the period of the licence it would have involved a derogation from the grant for the freeholders to put any livestock of their own upon the land, this would not necessarily remain the case throughout the period of the licence.

Clause 3 is the only clause of this unique agreement which allows Mr Threadgold to make any use of Sycamore Farm and the adjoining block of land for agricultural purposes. Upon its true construction, the licence thereby granted is non-exclusive. It gives him no right to prevent the freeholders themselves, or other licensees of theirs, from making whatever contemporaneous use of the land for agricultural purposes they think fit, so long as this would leave enough grazing for whatever numbers of Mr Threadgold's original dairy herd were still remaining on the land as the licence ran its course until October 31 1969. To come within section 2 (1) of the Agricultural Holdings Act 1948, the licence granted under an agreement must be a " licence to *occupy* land for use as agricultural land." To satisfy this requirement the right of occupation for agricultural purposes must be an exclusive right under which the grantee is entitled to prevent the grantor and any other person authorised by the grantor from making any use of the land, at any rate for agricultural purposes, during the period of the grant. The application of section 2 (1) of the Act to licences to occupy land for use as agricultural land was in my view correctly stated by Davies LJ in *Harrison-Broadley v Smith* [1964] 1 WLR 456 at 470, where he said: " There cannot be such a licence without a right of exclusive occupation during the currency of the licence in the licensee as against the licensor for that purpose." I would accordingly dismiss this appeal upon the ground that the agreement of March 27 1969, the terms of which must be unique, did not upon its true construction grant to the appellant the right to use the land for agricultural purposes to the exclusion of its use for agricultural purposes by the respondents. The question of law of more general application discussed in the judgment of the Court of Appeal accordingly does not arise. The instant appeal does not

A provide an appropriate occasion for considering it.

LORD DILHORNE and LORD SIMON agreed, and Her Majesty was advised accordingly.

COURT OF APPEAL

November 8 1974

(Before Lord Justice BUCKLEY, Lord Justice BROWNE and Mr Justice MACKENNA)

WYKES AND OTHERS v DAVIS AND ANOTHER

Estates Gazette January 18 1975

(1974) 233 EG 265 B

Agricultural holding—Notice to remedy numerous breaches must specify a reasonable time for all to be remedied, or it will be valid in respect of none—Shepherd v Lomas (1963) interpreted in that sense—Decision and dicta of practical importance to landlords and their agents

This was an appeal by Mr and Mrs Eric Albert John Davis, from a judgment of Judge Pratt at Barnstaple County Court on July 11 1974 upholding the validity of notices to remedy dilapidations at Saunton Barton Farm and East Saunton Farm served on the appellants by their landlords, Norman Gordon Wykes, William Lloyd Baxendale, Peter Baring and Ronald George Price, C on January 26 1973.

Mr A Fletcher (instructed by Parker, Garrett & Co) appeared for the appellants, and Mr D Wood (instructed by Freshfields) represented the respondents.

Giving judgment, BUCKLEY LJ said: This is an appeal from a judgment of His Honour Judge Pratt of July 11 1974 on a case stated by an arbitrator for the opinion of the court under the Agricultural Holdings Act 1968, as amended by the Agriculture (Miscellaneous Provisions) Act 1963. The arbitrator submitted two questions of law for the opinion of the court, with only one of which we are concerned. It is this: on a true construction of section 24 (2) (d) of the Agricultural Holdings Act 1948 and D section 19 (1) of the Agriculture (Miscellaneous Provisions) Act 1963, was a notice to remedy given on January 26 1973 invalid on the ground that the period specified in the notice was not a reasonable period in which to remedy all of the breaches specified in the notice, although it was a reasonable period in which to remedy some of the breaches? The learned judge answered that question in the negative, holding in effect that the notice was valid in part, although invalid as to part.

The appellants are yearly tenants of two farms in Devon known as Saunton Barton and East Saunton Farms, comprising some 500 acres, under an agreement of May 29 1969. The year of the tenancy ends on September 29. The respondents are the E landlords. I shall refer to the parties as " the tenants " and " the landlords " respectively. On January 26 1973 the landlords by their agent gave the tenants notice in the prescribed form required by section 19 (1) (a) of the Act of 1963, requiring them to remedy within seven months from the date of service of that notice the breaches of which particulars were set out in the notice. The terminal date of that notice, which was August 26 1973 or thereabout, was such as to allow the landlords time to serve notice to quit on the tenants before September 29 1973, so as to terminate the tenancy at September 29 1974. On August 29 1973 the landlords served the tenants with notice to quit the farms on September 29 1974 on the ground of their failure to comply with the notice to remedy.

F The tenants did not dispute their liability to carry out any of the remedial works specified in the notice to remedy. These works were listed in the schedule to the notice to remedy, which is headed in accordance with the prescribed form, " Particulars of breaches of terms or conditions of tenancy." They were grouped under three headings: first, " Dilapidations to hedges and banks," under which heading work was required to be carried out upon nine separate hedges or banks; secondly, " Dilapidations to field surfaces," under which heading work was required to be done in five fields; thirdly, " Dilapidations to gates," under which heading work was required upon gates in nine fields. By August 26 1973 the tenants had done all the work required by the notice except that (1) they had not repaired all G the " growth banks " mentioned under the first heading, and (2) they had not fully cleaned one field of couch.

The arbitrator made two important findings of fact. He found that the period of seven months specified in the notice to remedy was not a reasonable time in which to repair all the growth banks, because according to the rules of good husbandry this type of work was to be carried out during the winter months; and he found that the period of seven months specified in the notice to remedy was a reasonable time within which to eradicate the couch from the field, which was still to some extent infested at August 26 1973. The tenants contend that in these circumstances the notice to remedy was wholly invalid, because H seven months was not a reasonable period in which to carry out all the work required by it. The landlords contend that the notice was valid, notwithstanding the first of the two findings which I have mentioned, to the extent of all the works required by the notice with the exception of the repair of the growth banks.

The question to be determined is rightly formulated as a question of construction of section 24 (2) (d) of the Act of 1948 and of section 19 (1) of the Act of 1963. Section 24 (1) of the 1948 Act provides that where a tenant is given notice to quit an agricultural holding he may give a counternotice to the landlord, whereupon, subject to the provisions of subsection (2) the notice to quit shall not have effect unless the Agricultural Land Tribunal consents to the operation thereof. Subsection (2), so J far as material for present purposes, was originally enacted in the following terms:

" The foregoing subsection shall not apply where . . . (d) at the date of the giving of the notice to quit the tenant had failed to comply with a notice in writing served on him by the landlord requiring him . . . within a reasonable time or within such reasonable period as was specified in the notice to remedy any breach by the tenant that was capable of being remedied of any term or condition of his tenancy which was not inconsistent with the fulfilment of his responsibilities to farm in accordance with the rules of good husbandry, and it is stated in the notice to quit that it is given by reason of the matter aforesaid." K

By section 19 (6) of the 1963 Act, section 24 (2) (d) of the 1948 Act was amended by deleting the words " within a reasonable time or." Section 19 (1) provides:

" for the purposes of paragraph (d) of section 24 (2) of the Agricultural Holdings Act 1948 . . . (a) a notice requiring the tenant to remedy a breach of any term or condition must be in the prescribed form and must specify the period within which the breach is to be remedied; (b) where such a notice in the prescribed form requires the doing of any work of repair, maintenance or replacement, any further notice requiring the doing of any such work and served on the tenant less than twelve months after the earlier notice shall be disregarded, unless the earlier notice was withdrawn with his agreement in writing; (c) a period of less than six months shall not be L treated as a reasonable period within which to do any such work."

Subsections (2), (3) and (4) of that section enable the Lord Chancellor to make regulations and to prescribe forms by statutory instrument, subject to annulment by resolution of either House of Parliament. This power includes power to prescribe different forms for the purposes of section 19 (1) (a). The form appropriate to the present case was prescribed by the Agriculture (Forms of Notices to Remedy) Regulations 1964. It was adopted in the present case. It is headed: " Notice to tenant to remedy breach of tenancy agreement by doing work of repair, maintenance or replacement." It provides for identifying the holding in respect of which the notice is given and the tenants to whom it is addressed, and then proceeds in the following terms: M

" 1. I hereby give you notice that I require you to remedy within (blank) months from the date of service of this notice the breaches whereof particulars are given below of the terms or conditions of your tenancy, being breaches which are capable of being remedied of terms or conditions which are not inconsistent with the fulfilment of your responsibilities to farm the holding in accordance with the rules of good husbandry. 2. This notice requires the doing of the work of repair, maintenance or replacement specified below. 3. This notice is given in accordance with section 24 (2) (d) of the Agricultural Holdings Act 1948 and section 19 (1) of the Agriculture (Miscellaneous Provisions) Act 1963. Failure to comply with it within the period specified above may be relied on as a reason for a notice to quit

under section 24 (2) (d) of the Agricultural Holdings Act 1948.
4. Your attention is drawn to the notes following the signature to this notice."

There then follows the schedule, headed in the manner I have already mentioned, which contains spaces for indicating the term or condition of tenancy alleged to have been contravened in each case and particulars of the breach alleged and the work required to remedy it. A marginal note is printed opposite to paragraph 1 of the notice relating to the words "within (blank) months." The note is in these terms: "This period must be a reasonable period for the tenant to remedy the breaches and must in any event be not less than six months." At the foot of the space left to contain the schedule there are some explanatory notes which I do not think it necessary for me to read.

The tenants contend that the natural meaning of the statutory language is such that the landlord must serve a notice allowing a reasonable period for all the work specified in it to be completed. In this connection they rely by way of analogy on decisions under the Conveyancing Act 1881, section 14, now superseded by the Law of Property Act 1925, section 146. They contend that the language of those sections is very similar to the language of section 24 (2) (d) of the 1948 Act as originally enacted, and they say that if a landlord had served a notice under that section requiring works to be done within a reasonable time without specifying any particular period, a reasonable time would not have been held to have elapsed until the expiration of a sufficient period to allow of the completion of all the works required by the notice. They say that it would be illogical if a different principle or construction were to be adopted in a case in which the landlord chose to specify the period. They contend that the 1963 amendment cannot have been intended to alter the effect of the words "within such reasonable period as was specified in the notice" drastically to the advantage of the landlord. They say that it is for the landlord to decide what works he shall require by the notice, and that it is for him to determine what period he thinks will be sufficient to enable the tenant to complete all those works. In this situation, Mr Fletcher says, the landlord is in control of the position and is not exposed to any difficulty, embarrassment or hardship, while the tenant is free, as he should be, to decide how and in what order he will tackle the works which he is required to carry out.

Mr Wood, for the landlords, has placed much emphasis upon the inconvenience which he submits would arise from adopting the tenants' argument. He points out that under the Agriculture (Notices to Remedy and Notices to Quit) Order 1964, article 5 (3), a tenant who does not contest his liability to do the required works need not dispute the validity of the notice to remedy on any other ground until after a notice to quit has been served. Suppose, in the case of a yearly tenancy expiring on the September quarterday in each year, a landlord were to serve a six months' notice to remedy early in March to expire in the following September before the end of the current tenancy year, the tenant, if he did not dispute liability, would be under no obligation to assert the invalidity of the notice on the ground that the period allowed was unreasonable until after the notice to quit had been served. In the following September the landlord serves a notice to quit and the tenant thereupon disputes the validity of the notice to remedy. If the tenants' contention here succeeds, upwards of six months have been wasted and the landlord would be precluded by section 19 (1) (b) of the 1963 Act from serving a new notice to remedy until the following March. The notice to remedy which he would then serve would have to allow a reasonable period for the works to be done, which *ex hypothesi* would be more than six months. He would consequently be unable to serve a notice to quit before the next September quarterday. Instead of being able to terminate the tenancy at about 18 months after his original notice to remedy, the landlord would be unable to recover possession of the holding until some 42 months from that date. The landlord cannot himself initiate arbitration upon the reasonableness of his notice to remedy. Mr Wood says that this state of affairs would enable a tenant to continue in default under his obligations with impunity.

In answer to this Mr Fletcher says that in any ordinary case the alternative remedy of forfeiture would at all times be available to the landlord, subject to the possibility of the tenant obtaining relief from forfeiture. Then Mr Wood goes on to contend that where a notice to remedy specifies a period which is reasonable as regards some of the works comprised in the notice but unreasonable as regards others, the notice is invalid in respect of the latter works but valid in respect of the former, and can be severed so that the bad does not infect the good. The tenant, says counsel, can at any time require arbitration of the question to what extent the notice is a good one: he ought not to be allowed to remain in default in respect of obligations in relation to which a good notice has been served merely because the notice has also included works for which no reasonable time has been allowed. Upon the language of section 24 (2) (d), Mr Wood contends that the words "any breach" should be construed, as he says, distributively; so that the question to be asked in any case would be whether the notice requires the tenant to remedy any breach within a reasonable period, and, if so, whether the tenant has in fact remedied that breach within that period. If the tenant has not done so, Mr Wood says that the landlord is entitled to serve a notice to quit based upon that default, whatever may be the position with regard to any other works specified in the notice to remedy.

The learned judge approached the construction of the subsection in this way: "On a strict construction of the words quoted the period stated in the notice for remedying the breaches (assuming there is more than one breach) is to be a reasonable period for remedying all the breaches. Therefore a notice specifying a period reasonable for remedying some but not all of the breaches is not a notice in accordance with the section. But there is a less strict construction of these quoted words, namely, that the period stated in the notice is to be a reasonable period for remedying each of the breaches. Therefore, a notice specifying a period reasonable for remedying some but not all of the breaches is valid as regards the breaches for which a reasonable time has been given and invalid as regards the others, ie the notice is not void in toto." Before discussing any authorities, I will consider the language of the subsection. It is to be observed that the condition precedent of an effective notice to quit under section 24 (2) (d) is that the tenant shall have failed to comply with a notice. The words which follow are words describing the notice, not words defining the condition. The notice is to be one by which the landlord requires the tenant within a reasonable period to remedy any breach by the tenant of any term or condition of his tenancy. It is common ground that it is open to a landlord to include more than one breach in one notice. So the words "any breach" must be read as equivalent to "any breach or breaches." It may be that the words "reasonable period" should also be read as equivalent to "reasonable period or periods," but that does not arise for determination in this case. It seems to me that the use of the word "any" in this context merely signifies that the landlord may select what breach or breaches he chooses to specify in the notice. By what, then, is the reasonableness of the period to be measured? It must surely be by reference to the length of time required to remedy the breach or breaches in question. If more than one breach is required to be remedied, it seems to me that the period must be such as will reasonably allow time to remedy all the specified breaches. If this is right, a notice comprising a number of breaches will not be such a notice as is described in the subsection unless it allows a reasonable period within which to remedy all of them. If the notice is not such as to satisfy the description in the subsection, failure to comply with it cannot satisfy the condition for an effective notice to quit.

That is the construction which the learned judge described as a "strict construction." In arriving at that conclusion the learned judge was largely influenced by the circumstance that he thought that the strict construction would, for reasons which I have already indicated, throw a burden upon a landlord and give a bonus to a tenant which seemed to him unreasonable and unfair. The effect of these two constructions can perhaps best be demonstrated by an illustration. Suppose that a landlord serves a notice to remedy in which he complains of nine breaches which he requires the tenant to remedy. Suppose that eight of those breaches could all be remedied within six months, but that one could not be remedied in less than 12 months. If the notice were to specify seven months as the period within which the necessary works were to be done, would the notice be entirely ineffective, or would it be effective in relation to the eight breaches but ineffective in relation to the ninth? Upon the strict construction the notice would be wholly bad. Upon the less strict construction it would be bad only as regards the ninth breach. But now suppose that in a similar case any one of the nine breaches could reasonably be remedied within two months, but circumstances were such that all nine could not reasonably be remedied in less than 12 months. In these circumstances, could a notice requiring the tenant to remedy the breaches within seven months be good? Upon the strict construction it would not, but upon the less strict construction it would seem that it would be wholly good, notwithstanding that the tenant would almost certainly be unable to comply with it. The learned judge seems to have considered that this would have occasioned a difficulty for the tenant of a theoretical rather than a practical kind, because he could at once apply for arbitration under article 5 of the order to have the question of the reasonableness of the period determined; but if the reasonableness of the period is to be measured in relation to individual breaches and not in relation to all the specified breaches taken together, it is difficult to see how the arbitrator could in such a case decide in relation to which breaches the notice should be treated as valid and in relation to which it should be treated as invalid.

We were referred to cases in which it has been decided under the Conveyancing Act 1881, section 14, that a notice requiring true breaches to be remedied, and also requiring the tenant to remedy a matter in respect of which he was not in fact in breach, was not invalidated by the latter circumstance (*Pannell v City of London Brewery Co* [1900] 1 Ch 496), and that a notice which specified certain breaches with sufficient particularity would not be invalidated by the fact that other alleged breaches were referred to in the notice with insufficient particularity (*Fox v Jolly* [1916] 1 AC 1). I, for my part, do not find these authorities of great assistance in the present case. The landlords here did not require the tenant to carry out any work which he was not liable to do. It may be said that the landlords failed to give any effective notice in respect of the dilapidations to hedges and banks because seven months was not a reasonable period within which those works could be completed, but this is not a defect of the same kind as reliance upon a non-existent obligation, nor, for reasons which I will shortly explain, do I think that it is a defect of the same kind as failure to specify alleged breaches with sufficient particularity. There is in my opinion an important difference between the Conveyancing Act 1881, section 14 (1), and the subsection with which we are directly concerned in this case. As I have already pointed out, the condition for an effective notice to quit under the latter subsection is a failure on the part of the tenant to comply with a notice. Under section 14 (1) of the Act of 1881 the condition precedent for forfeiture of a lease is the failure of the lessee within a reasonable time after a notice to remedy a breach complained of in the notice. Under that subsection the question to be asked is whether a lessee has failed to remedy a breach and has been required to remedy that breach by a notice given at a sufficient interval before. What else may have been contained in the notice, whether justifiably or unjustifiably, is irrelevant to the answer to that question. It follows, I think, that a notice given under that section in respect of a number of breaches is capable of operating as a separate notice in respect of each of those breaches. This is not, however, in my opinion the case under section 24 (2) (d) of the 1948 Act.

In *Price v Romilly* [1960] 1 WLR 1360, Diplock J (as he then was) had to consider what amounted to compliance with a notice served under section 24 (2) (d). The notice to remedy in that case had specified seven distinct works required to be done. At the expiry of the period limited by the notice the tenant had not done the work comprised in the first of these seven items. The tenant contended that a partial failure to comply with the notice was insufficient to satisfy the subsection. Diplock J disposed of this argument by saying: "It seems to me that if he had not remedied it and remedied it completely, he has failed to comply with the notice." In some closing observations the learned judge referred to the fact that the arbitrator had found as a fact that one of the seven breaches was not capable of being remedied within the time stated in the notice. He mentioned this only to negative an argument that it was not apparent whether the arbitrator had considered whether or not the notice gave a reasonable time for doing the work. It does not appear that the question whether this circumstance might have invalidated the entire notice was canvassed. In *Shepherd v Lomas* [1963] 1 WLR 962, Harman LJ at page 974 referred to this part of Diplock J's judgment, pointing out that if the argument of the tenant in *Shepherd v Lomas* was right the notice in *Price v Romilly* would have been invalid; but Harman LJ pointed out that the point was not taken before Diplock J.

In *Shepherd v Lomas* the court was concerned with a case in which the landlord served a notice to remedy in respect of a number of breaches of covenant, in connection with some of which the landlord was under an obligation to provide materials for carrying out the repairs, while with others he was under no such obligation. The landlord did not deliver the necessary materials until too late to make it possible for the tenant to carry out the required repairs within the period limited by the notice. When the time expired, the tenant had carried out none of the repairs for which the landlord was obliged to provide materials; he had also failed to carry out much of the work for which the landlord was not bound to provide materials. The landlord served a notice to quit and the case went to arbitration under the Act. The arbitrator made this finding, viz: "That the period of time specified in the notice was reasonable in respect of the breaches for the remedy of which the landlords were not required to supply any materials and was unreasonable with regard to those for the remedy of which the landlords were liable to provide materials." The Court of Appeal reached the conclusion that, properly interpreted, this finding meant that the period was reasonable in respect of the first group (where no materials were required), and that it would also have been reasonable in respect of the second group (where materials were required), provided always that the landlord did his part and supplied the materials, but that in the events that happened the period turned out to be unreasonable for the second group because of the landlord's failure to provide the materials. Lord Denning MR said at page 970:

> "So interpreting the finding, I do not think that the notice to remedy breaches was void *ab initio*. It was good in the beginning because six months was reasonable for all the breaches, not only the first group but also the second group, seeing that it could fairly be assumed that the landlord would do his part and fulfil his covenant to provide the materials."

The Master of the Rolls then made a passing reference to *Pannell v City of London Brewery Co* and *Fox v Jolly*, and proceeded as follows:

> "So also it seems to me that under the Agricultural Holdings Act, even though some of the matters that are specified in the notice turn out not to be breaches, or the tenant afterwards is excused from performing some of them, the landlord is still entitled to rely on the others. It follows that in this case, although the tenant was excused from remedying the second group of breaches (because the landlords did not provide the materials) nevertheless he ought to have remedied the first group. He was not entitled to ignore the whole notice and do nothing."

Towards the end of his judgment, the Master of the Rolls said:

> "I would only add this. In considering what is a reasonable time for any of the breaches the arbitrator should always take into account the rest of the breaches specified in the notice. It may often be quite reasonable for the tenant to postpone some work until other work has been done: or to wait until he has men to spare for the task; or to abide a suitable season. The time specified ought to be time to do them all."

Harman and Pearson LJJ expressed similar opinions, the latter saying at page 974:

> "The effect of that notice was not to impose any new obligation to remedy breaches, but to prepare the way for the landlord to serve an effective and operative notice to quit if the tenant failed

to comply with the notice which required him to remedy the breaches—all of them—within a period of six months. That notice when it was served was a good and valid notice. It was at that time reasonable to specify six months as the period within which all those breaches had to be rectified."

Having reached the conclusion which they did about the arbitrator's finding, the court did not have to consider in that case whether, if the time specified in the notice had not been a reasonable time for remedying some one breach referred to in the notice, the notice to remedy would have been wholly ineffective; but, in my opinion, it is an inescapable inference from their judgments that if they had not interpreted the finding in the way in which they did they would have held the notice to be bad and wholly ineffective. I am inclined to think that we are bound by the decision in *Shepherd v Lomas* to adopt what has been called the strict construction, but however that may be, I am of opinion that that is the true meaning and effect of the subsection. On a proper reading of the language the condition of a valid notice to quit is not that the tenant shall have failed to remedy any breach alleged in a notice to remedy within the period specified in that notice, but that the tenant shall have failed to comply with the notice. Compliance with the notice to remedy must involve compliance with all the legitimate requirements of it. It may be that if some alleged breach was not in truth a breach of the tenant's obligations, the fact that the time for compliance specified in the notice did not allow a reasonable period for the work required to remedy that alleged breach would not invalidate the notice; but nothing of that kind arises here, where the tenant has not disputed liability to do any of the required works. If that is the right view of what is involved in compliance and failure to comply, it necessarily follows, in my opinion, that the reasonableness of the period within which the tenant is required to remedy the breach or breaches alleged in the notice must be judged in relation to all the work required to be done. The same reasoning would, I think, have applied before the 1963 amendment to a case in which the landlord by a notice to remedy required a tenant to remedy a number of breaches within a reasonable time without specifying any particular period for compliance. In such a case the landlord could not, in my opinion, have served an effective notice to quit on the ground of non-compliance until a reasonable period within which to remedy all the breaches had elapsed or, perhaps, until it had become manifest that the tenant would not be able to remedy all the breaches within such a period from the date of the notice to remedy. It does not seem to me, therefore, that the amendment can be said to have had any effect upon the true interpretation of the words " within such reasonable period as was specified."

Mr Fletcher presented an argument to us in support of the strict construction based upon the terms of the prescribed form of notice to remedy, and in particular upon the marginal note to paragraph 1 of that form which I have read. I do not find it necessary to deal with that argument, because in my judgment the true interpretation of the subsection is clear without any assistance which might otherwise have been available from the form or the note. So I need not consider whether any such assistance could be invoked in this case. The subsection is not in my judgment susceptible of the construction contended for by the landlords without some violence to the language. Since the so-called strict construction does not, in my opinion, do any violence to the language, and since it leads, as it seems to me, to a reasonable interpretation and result, it must, in my judgment, be preferred. For these reasons, I would hold that this appeal succeeds. I would amend the learned judge's order to declare that the question referred to at the outset of this judgment should be answered affirmatively, to the effect that the notice to remedy dated January 26 1973 was invalid on the ground that the period specified in it was not a reasonable period in which to remedy all the breaches specified in the notice.

Browne LJ delivered a short judgment in which he stated his agreement with the reasoning of both Buckley LJ and MacKenna J, remarking that in his opinion the court was bound by the decision in *Shepherd v Lomas* to come to its conclusion, and adding: I have a good deal of sympathy with Mr Wood's point that if a tenant is guilty of two breaches, one of which could be remedied in six months, but the other of which could not be remedied in less than twelve months, it is unreasonable that the tenant should be allowed twelve months to remedy the former. Mr Wood suggested that by a properly-drafted notice or notices different periods could validly be specified in respect of different breaches, but that point does not arise in this case, and I express no opinion about it.

MacKenna J referred to section 24 (2) (d), observed that notice must be in writing and must have been served by the landlord on the tenant, and continued: The notice must state the landlord's complaint. He is not limited to complaining of a single breach. He may complain of breaches in the plural. The breaches must be of terms and conditions of the tenancy, and the terms must be such that the tenant's observance of them would have been consistent with the rules of good husbandry. And lastly, the notice must specify a period which is reasonable for some purpose. The question is, for what purpose? Must it be, as the tenant in our case contends, a reasonable period in which to do the whole of the remedial work which the notice properly requires him to do, so that if it is too short for that purpose the notice is bad? Or will the notice be *pro tanto* a good one, if the landlord can prove that the period was long enough for the tenant to remedy at least one of the stated breaches, and that the tenant failed to remedy that breach, which is the landlord's contention?

During the argument the meaning of the word " any " in the expression " any breach " was discussed. I do not think that there is any doubt about its meaning. In this context it means " every." Where the notice mentions more than one breach, the tenant will be required to remedy every one of those mentioned, and if he omits to remedy any of them within the prescribed reasonable period, he will have failed to comply with the notice. As the notice will require the tenant to remedy all the breaches complained of, it would seem that the period specified must be reasonable for the whole of the work. There are two different ways in which a period might be unreasonable. First, it might not be sufficient to remedy all the breaches taken together, though each breach could be separately remedied within the period. For example, the notice might require a tenant to repair gates and to eradicate weeds—two separate obligations—within a period of six months. If that period were long enough to do either job separately, but not long enough to do them both together, it would be unreasonable. Secondly, the period might not be sufficient to remedy one breach taken by itself, though long enough to remedy the other, as, in the case I have supposed, if the six months were enough for the gates but not for the weeds. In this case, too, the specified period would be an unreasonable one.

What then is the effect of the landlord's specifying an unreasonable period? The effect, in my opinion, is to make his notice invalid. If the period is, in either of the ways I have described, unreasonably short, the notice imposes no obligation on the tenant and he may disregard it with impunity. The landlord argues that a notice stating several breaches and specifying a single period for remedying all of them can be treated as if it were a series of notices each stating a single breach and specifying the same period for remedying it. If the period is reasonable for remedying any one of the breaches, the composite notice is, it is argued, *pro tanto* valid, and the tenant who has failed to comply with the valid part loses the protection of the Act. I find this argument unacceptable. The notice requires the tenant to remedy *all* the breaches. To treat it as if it required him to remedy just one of them, or even some of them, is to change its character. Answering the argument in other words, I would say that in deciding whether the specified period is reasonable, it is impossible to disregard any of the breaches which the tenant is required to remedy within that time. It is impossible to treat the notice as if it only required him to remedy such breaches as he reasonably can within the specified period.

This conclusion is, I believe, supported by the judgments of Lord Denning MR and Pearson LJ in *Shepherd v Lomas*. I must examine at a little length the facts of that confusing case.

[His Lordship expounded the two judgments concerned in substantially the same terms as those employed by Buckley LJ, concluding:] I have not found it easy to understand the judgment of Harman LJ in this case. Mr Blundell's argument which the Lord Justice describes at p 972 as "formidable" was, I think, the argument that the six months given to do the 30 jobs was insufficient because of their number, and this argument failed because it had not been raised before the arbitrator. What the Lord Justice describes at p 974 as "Mr Blundell's point" must, I think, have been a different argument from the earlier one, and was presumably the argument that the time was too short for doing that part of the work for which the landlord should have provided the materials. The Lord Justice in rejecting this argument pointed out that in the earlier case of *Price v Romilly,* where the point we are considering was not argued or discussed by the judge, the landlord had succeeded, even though the time specified by him had been too short for remedying one of the breaches, and it was clearly the Lord Justice's view that the decision was right on those facts. I find it difficult to reconcile the two passages of Harman LJ's judgment unless it was his view that the landlord may fail where the time is unreasonably short for repairing all the breaches together though long enough to remedy each separately (the "formidable argument"), but succeeds where the time is too short for remedying one of them ("Mr Blundell's point"). If that is the explanation of the two passages, I do not think that the distinction is a sound one.

I have considered the cases cited to us under section 14 of the Conveyancing Act 1881, I do not find that they give us any clear guidance in construing the 1948 Act, and I do not refer to them. For these reasons I would allow the appeal and amend the learned judge's order as proposed by Buckley LJ.

The appellants were awarded their costs of the appeal despite the wording of paragraph 21 of the Sixth Schedule to the Act of 1948, "The costs of, and incidental to, the arbitration and award shall be in the discretion of the arbitrator who may direct to and by whom and in what manner the costs or any part thereof are to be paid." Counsel on both sides referred to the fact that the arbitration was still notionally in progress, the proceedings on appeal being technically the decision of a case stated by the arbitrator; to an agreement by which the costs in the county court were to abide the decision of the main issue posed in the case; and to the observation of the learned county court judge that in view of the provisions in paragraph 21, that agreement would have to be mentioned to the arbitrator. Buckley LJ stated, "As nobody asks for any order as to the costs in the county court, we leave that to be dealt with by agreement between the parties or as the arbitrator shall award." No order was made as to the costs of an interlocutory application by the landlords for £750 to be paid into court as security for costs, which resulted in a consent order for payment-in of £300, the costs of the application being reserved. The landlords were given leave to appeal to the House of Lords.

HOUSE OF LORDS

November 27 1974

(Before Lord REID, Viscount DILHORNE, Lord WILBERFORCE, Lord DIPLOCK and Lord SIMON OF GLAISDALE)

MAUNSELL v OLINS AND ANOTHER

Estates Gazette February 8 1975

(1974) 233 EG 591

Subtenant of farm cottage is not protected against superior landlord though not a service tenant and not subject to eviction by his immediate landlord—Decision of importance both inside and outside the sphere of agricultural holdings—Meaning of "premises" in section 18 (5) of Rent Act 1968

This was an appeal by Mr and Mrs Ilmar Olins, of 2 Hallsannery Cottages, Hallsannery Farm, Bideford, Devon, from a decision of the Court of Appeal (Edmund Davies, Cairns and Lawton LJJ) on February 21 1974 allowing an appeal by Mrs Nesta Gwendoline Maunsell from a decision of Judge Stansfield at Bideford County Court on July 26 1973 and awarding her possession of 2 Hallsannery Cottages.

Mr R H Bernstein QC and Mr S L Tuckey (instructed by Cripps, Harries, Willis & Carter) appeared for the appellants, and Mr D A Wood (instructed by Macfarlanes) represented the respondent.

In a speech with which Lord Reid and Lord Dilhorne in effect concurred, LORD WILBERFORCE said: The appellants claim the protection of the Rent Act 1968 in respect of their occupation of a cottage situated on a 106-acre farm in Devonshire. They are not employed on the farm. The question raised is of general importance both to owners of farms and to cottage occupiers. As will be seen, it is not easy to decide. The respondent is the freehold owner of the farm. The main buildings are approximately in the centre of the holding and include a farmhouse and two cottages. The cottage occupied by the appellants forms, together with the farmhouse itself, a semi-detached pair of dwelling-houses. The other cottage is separate, but nothing in this appeal depends upon any distinction between the physical situation of the one cottage or the other. The farm was let for many years to Mr Ernest Beer on a tenancy protected by the Agricultural Holdings Act 1948. Mr Olins was employed on the farm for some time and lived in the other cottage, but in 1955 he took other employment and in 1959 Mr Beer let the cottage in question to both appellants on an unfurnished tenancy for £1 a week. On April 30 1963 the tenancy agreement between the respondent and Mr Beer, which had previously been oral, was replaced by a written agreement in the standard "short form." Under this agreement the farm was let to Mr Beer from March 25 1963 for one year certain and thereafter from year to year at a rent of £400 pa. Clause 4 (12) of the tenancy agreement provided that the tenant "will reside in the farmhouse and will use the farm for agricultural purposes only and will not assign underlet or part with possession of the farm or any part thereof except the two cottages which may be sublet on monthly tenancies." The appellants were, of course, already in occupation of their cottage under a subtenancy. They continued to live in the cottage and it is not disputed that they were protected tenants *vis-à-vis* Mr Beer. He died on March 15 1971, and the respondent served a notice to quit upon his legal personal representatives terminating the head tenancy on March 25 1973. On June 15 1973 she started these proceedings against the appellants for possession of the cottage. There is no doubt that at common law this claim must succeed, since the appellants' subtenancy could not survive the termination of the tenancy. But the appellants claimed the protection of the Rent Act 1968, section 18 (5). The claim succeeded in the county court on the ground that the cottage formed physically part of the farm buildings, a circumstance supposed to distinguish the case from *Hobhouse v Wall* [1963] 2 QB 124 CA. On appeal to the Court of Appeal, this distinction was rejected and the court followed its earlier decision. The appellants in this House abandoned any attempt to distinguish this case from *Hobhouse v Wall* on the facts. So they are faced with the task of contending that two decisions of the Court of Appeal are wrong. It is fair to say that more than one of the learned judges whose judgments are under review have found difficulties in interpreting the statutory provisions, and the section is certainly one which admits, almost invites, opposing constructions.

As I have stated, the provision actually relied on is section 18 (5) of the Rent Act 1968. This Act is a Consolidation Act, so that it is legitimate to look to the origin of the subsection in question. That is to be found in section 41 of the Housing Repairs and Rents Act 1954. There are some differences in the language of these two provisions, but, in agreement with the Court of Appeal, I do not find them significant. I shall quote, at the present stage, the earlier section—I quote the present section later—because the next step will be to ascertain what was its legal background at the time of its enactment: it was also the section interpreted in *Hobhouse v Wall.* It read as follows:

Where a dwelling-house to which the Act of 1920 applies (hereinafter referred to as "the sublet part") forms parts of premises, not being such a dwelling-house, which have been let as a whole on a superior letting, then from the coming to an end of the superior letting the operation of the Rent Acts in relation to the sublet part shall be the same as if in lieu of the superior letting there had been separate lettings of the sublet part and the remainder of the premises, for the like purposes as under the superior letting, and at rents equal to the just proportion of the rent under the superior letting.

Before we try to interpret this we must look at its context. The Act in which it appears deals with two separate subjects. Part I is essentially a housing enactment dealing with slum clearance and fitness for habitation, and contains a number of amendments of Housing Acts. Part II, in which section 41 appears, is essentially rent legislation. After 10 sections (23-32) dealing with the increase of rents on account of repairs, it proceeds from section 33 onwards to introduce some miscellaneous amendments to previous Rent Acts and also to Housing Acts. Section 41 is one of these amendments; it seems to stand by itself as an independent provision not reflecting or forming part of some wider change in policy. Rather it seems, as we know is common in rent legislation, a section of a piecemeal or patching character. A first reading of the provision does not, to my mind, convey an impression of conspicuous clarity, particularly as to its general scope. I think I would not be alone in finding that the key word "premises" invites reflection. It is true that it is a general word, or rather, a word of some generality, but I know of no rule of construction which requires general words to be interpreted literally regardless of their context. If appeal is made to the principle that the plain meaning of a word should be taken, unless at least some other indication appears, it must be said that a word does not necessarily have a plain meaning just because it appears to be general—certainly not such a word as "premises." We know that this is a word of conveyancing jargon, meaning, strictly (and *pace* Viscount Hailsham in *Whitley v Stumbles* [1930] AC 544, 546) everything in a deed which precedes the *habendum*. From this it has passed into the vernacular, at least a quasi-legal vernacular, as referring to some sort of property, but not with any precise connotation. A reference to Stroud's Judicial Dictionary shows that a number of different meanings have been acquired of which the most central appears to be buildings or some kinds of buildings, but it would be far too much to say that there is any *prima facie*, still less any grammatical meaning from which one should start.

The appellants' main argument was that if the word is not itself a general word equivalent to property, it is given the general meaning in this subsection of any property which can be or has been let. This is said to follow from the phrase "part of premises . . . which have been let as a whole. . . ." But this is a confusion of thought. A relative clause merely qualifies that to which it is relative, and does not define it or help in its definition. What kind of property (which has been let) is meant must be ascertained *aliunde*. I am not, myself, able to solve this problem by a simplistic resort to plain meaning. Most language, and particularly all language used in rents legislation, is opaque: all general words are open to inspection, many general words demand inspection, to see whether they really bear their widest possible meaning. As Viscount Simonds said in *A-G v Prince Ernest Augustus of Hanover* [1957] AC 436, 461:

For words, and particularly general words, cannot be read in isolation: their colour and content are derived from their context. So it is that I conceive it to be my right and duty to examine every word of a statute in its context, and I use "context" in its widest sense, which I have already indicated as including not only other enacting provisions of the same statute, but its preamble, the existing state of the law, other statutes in *pari materia*, and the mischief which I can, by those and other legitimate means, discern the statute was intended to remedy.

A first stage in this inspection might be to ask whether, to an ordinary man, "premises" would be regarded as equivalent to "land" or "property." I doubt very much whether a farmer would accept the word as a description of his farm land, or whether the local hunt would use it when asking for permission to hunt over his land. I note with interest that the tenancy agreement of this farm does not use the word "premises" at all: what is let, and referred to in the covenants, is "the farm." This tenancy agreement, moreover, was made in the standard short form which applies to many holdings all over the country. So the reader of this section is bound to ask himself what the word does include: has it some lesser meaning than property or than leasehold property: is it coloured by the references to dwelling-houses: why is it used in preference to something more precise: is there perhaps a drafting reason for it? It certainly seems to facilitate the later words "the remainder of the premises." And in the same line of thought, what is meant by "forms part of," "for the like purposes as under the superior letting"? In sum, what is the section trying to do?

We do not have to look far for a clue. In 1949 the Court of Appeal decided *Cow v Casey* [1949] 1 KB 474. That case was concerned with a subtenancy of a dwelling-house which was not subject to the Rent Acts. The decision was that the subtenant was not entitled as against the superior landlord to the protection of the Acts. Quite obviously section 41 was intended to deal with that situation and to bring about the opposite result. It frequently happens that legislative changes are made in order to reverse decisions of the courts: sometimes, indeed, the courts themselves invite the change. The decision is then the occasion of the enactment. The question may consequently arise whether the new enactment is confined to dealing with the particular situation with which the court was concerned or whether it goes further and covers a wider field, and if so, how much wider. There is no general rule or presumption as to this. Often Parliament, or its expert advisers, may take the opportunity to review the whole matter in principle and make broad changes (see for example *Smith v Central Asbestos Co Ltd* [1973] AC 518 as to the Limitation Acts). Legislative time is a precious commodity, and it is natural that opportunities, when they arise, will be used. Or, and this happens in the fiscal field, the draftsman, faced with some loophole in a taxing Act which the courts have recognised, will not merely close the particular loophole but will use general language extending much more widely, sometimes so as to sweep the honest and conscientious taxpayer up in the same net as the evader. On the other hand, there may be cases where Parliament takes a narrow and piecemeal view of the matter: time may not admit of an extensive review which may involve wide policy questions, or necessitate consultation with other interests. All these possibilities must be taken into account by courts in assessing the legislative intention. In performing that task here, I think that help is to be gained from setting down the two main elements which the draftsman must have had in mind: the pre-existing law and the decision in *Cow v Casey*. The pre-existing law is section 15 (3) of the Increase of Rent and Mortgage Interest (Restrictions) Act 1920:

Where the interest of a tenant of a dwelling-house to which this Act applies is determined, either as the result of an order or judgment for possession or ejectment, or for any other reason, any subtenant to whom the premises or any part thereof have been lawfully sublet shall, subject to the provisions of this Act, be deemed to become the tenant of the landlord on the same terms as he would have held from the tenant if the tenancy had continued.

The decision in *Cow v Casey*, as reported in the headnote in the Law Reports, reads:

Held, dismissing the appeal, that the dwelling-house, the subject of the original demise, not being a dwelling-house to which the Rent Restriction Acts applied, section 3 of the Rent and Mortgage Interest Restrictions (Amendment) Act 1933 did not apply to the subtenancy of a part of the premises so as to protect a subtenant from ejectment. On the determination of his subtenancy the defendant became at common law a trespasser. Section 15, subsection 3 of the Increase of Rent and Mortgage Interest (Restrictions) Act 1920, under which (where the interest of a tenant of a dwelling-house to which the Act applied was determined) any subtenant to whom any part of the premises had been lawfully sublet was to be deemed to have become the tenant of the head landlord, only applied to the case where the head tenant, whose tenancy had been determined, was tenant of a dwelling-house to which the Acts applied. The defendant accordingly was not entitled to the protection of the Rent Restriction Acts.

The coincidence of language between the holding in *Cow v Casey* and section 41 conveys to my mind a strong impression that section 41 is a section of limited effect, dealing primarily with subleases of dwelling-houses or, at the most, with subleases of property in the same field. And this fits in with the context—a section on the face of it piecemeal, included in what is throughout housing legislation.

A The effect of the change on the previous law can well be seen by looking at the Consolidation Act, which now sets out the old provision and the new as two subsections of the same section:

> 18.—(2) Where a protected or statutory tenancy of a dwelling-house is determined, either as a result of an order for possession or for any other reason, any subtenant to whom the dwelling-house or any part of it has been lawfully sublet shall, subject to the provisions of this Act, be deemed to become the tenant of the landlord on the same terms as he would have held from the tenant if the tenant's protected or statutory tenancy had continued.
>
> . . . (5) Where a dwelling-house:
>
> (*a*) forms part of premises which have been let as a whole on a
> B superior letting but do not constitute a dwelling-house let on a protected tenancy; and
>
> (*b*) is itself let on a protected tenancy, or subject to a statutory tenancy,
>
> then, from the coming to an end of the superior letting, this Act shall apply in relation to the dwelling-house as if, in lieu of the superior letting, there had been separate lettings of the dwelling-house and of the remainder of the premises, for the like purposes as under the superior letting, and at rents equal to the just proportion of the rent under the superior letting.

There seems to be a relation between these two subsections. They cover similar ground—broadly that of tenancies of dwelling-houses. There is no indication that they extend any
C wider. So what should " premises " be taken to mean? One view, the narrowest view, would be that it simply means " dwelling-houses." I find this, linguistically, a perfectly possible construction: the phrase " premises, not being such a dwelling-house " meaning premises being a dwelling-house but not a protected dwelling-house: the exclusion qualifies and colours the preceding generality. A less narrow view would be to say that " premises " includes not only dwelling-houses in the normal popular sense, but premises, which, for the purposes of the Rent Act, are treated as dwelling-houses. Everybody knows, and the draftsman must be taken to have known, that protection under the Rent Acts is given not merely to single, identifiable, pure dwelling-houses or dwelling units, but also to units of a mixed character
D —houses let with a garden or a yard or a garage or a paddock, houses part (even a substantial part) of which is used for business purposes. This is, of course, an untidy situation and it means that no clear definition of a dwelling-house entitled to protection can be given. (We note that a distinction is made between a house let together with land and land let with a house). But it reflects the reality of life, and the county courts are used, and skilful, at solving what are inevitably questions of degree. We should recognise this and, as between the narrow and the less narrow meaning, I would apply to " premises " the latter, which would include any premises which, as a matter of fact, applying accepted principles, would be held to be a dwelling-house for
E the purposes of the Act.

But the alternatives must be considered. One suggested alternative is to give to " premises " a general meaning but to exclude from it any property forming part of an agricultural holding. The basis for this would be that agricultural holdings have a code of their own (see Megarry, the Rent Acts, 10th ed pp 107-8), and should not be brought under the general provisions of the Rent Acts. I see the force of this, but I think that a particular exception limited in this way would have received specific legislative mention, and this is not to be found. The only real alternative seems to me to be to give to " premises " a universal meaning—and that is in effect the appellants' contention. The
F effect of this would be to give protection not only to subtenants of individual dwelling-houses in the middle of agricultural holdings but to subtenants of individual dwelling-houses in other complexes, industrial or business, or in any large estate of any kind. To do this would represent a very great enlargement of the rights of subtenants as compared with the situation pre-1954. No doubt Parliament might think fit to take this course, but it is obvious that a number of problems and policy considerations would arise—quite distinct from those which relate to subtenants of " housing properties " (I use this expression for convenience to refer to what I think is *within* the section). An indication that such problems exist and were recognised is to be found in the Agricultural Holdings Act 1948, section 26 (1) (c) of which confers on the Minister power *by regulation* to safeguard the G
interests of subtenants of agricultural holdings—a power which had been partially, but not relevantly, exercised prior to 1954 (see *Sherwood v Moody* [1952] All ER 389). Nobody can be unaware that the law relating to tenancies of agricultural land is one of considerable political import and delicacy, and I am very reluctant to believe that the particular aspect of this law which relates to subtenancies, treated with evident circumspection in 1948, would have been dealt with in a general provision such as appears in section 41 of the Act of 1954. I cannot finally help noticing that although the Court of Appeal decided in 1963 (in *Hobhouse v Wall*) against protection of subtenants of premises in an agricultural holding, Parliament, normally quick enough
to react in this field, has not disturbed the decision. The changes H
made by the Rent Act 1965 (not directly affecting the point at issue, but relating to agricultural tenancies) cannot affect the construction of a provision enacted in 1954.

It remains to say a few words on the judgments in the Court of Appeal. The leading judgment is that of Upjohn LJ in *Hobhouse v Wall*. The learned Lord Justice clearly felt difficulty about the case, as who does not? Certainly the word " premises " was not clear in its meaning to him; but after a thoughtful consideration of the issues he reached the conclusion with which, broadly, I agree. I do not read him as saying that " premises " means " buildings," though, as I have pointed out, the word has
been given this meaning in other contexts, but as saying that J
his interpretation would in effect cover only buildings. His reference, on the other hand, to " what are essentially dwelling-houses " seems to reflect the conception which I would favour. He did not attempt any definition of " premises," but contented himself with the negative conclusion that the section did not apply to agricultural holdings. I sympathise with his caution, but I think that now in this House we must, unless we are to accept a universal meaning of premises, risk the attempt to draw a dividing line. I have attempted to do so in this opinion. In the present case, Lawton LJ might be taken perhaps as equating " premises " with (conventional) dwelling-houses, but I do not
think that in his reference to large Victorian houses he was using K
more than an illustration. I regret that a matter which is really one of impression should have needed so many words to dispose of, but we have heard some well balanced arguments and the case is not one for quick solution. On the whole I would dismiss the appeal.

Lord Reid said that he agreed with Lord Wilberforce's speech and would add only a few words. Section 41 of the original Act, the Housing Repairs and Rents Act 1954, itself contained a fairly clear pointer to the meaning of " premises." The section directed one to suppose that in lieu of the superior letting there had been two lettings, one of the sublet part and one of the
remainder of the property demised by the superior lease, and L
then to suppose that the property in each of these lettings had been let " for the like purposes as under the superior letting." This caused some difficulty if the purpose of the superior letting was not residential, but was, say, agricultural. Suppose that a farm to be let included several cottages all occupied by farm workers. One would naturally say that the whole original lease, or " superior letting," had only one purpose, which was agricultural. It would at least be unusual to say that it had two purposes, residential as regards the farmer's house and the cottages, and agricultural as regards the land and the other buildings. He (his Lordship) would think that if one of the cottages were sublet
to a person who had no connection with the farm, there was a M
change of purpose. But the section would not work unless it could be said that there were two purposes in the original lease, so that the purpose of letting the house and cottage was residential and not agricultural. While this was not a conclusive argument, it seemed a fairly clear indication that the draftsman responsible for section 41 did not have in mind any " superior letting " other than a letting for residential purposes.

Lord Dilhorne said that he agreed with Lord Wilberforce that the appeal should be dismissed. The Agricultural Holdings Act 1948 contained a separate code for dealing with agricultural holdings. It gave power to the Minister to make regulations dealing with the situation where a tenant of such a holding had

A sublet a dwelling-house on a protected tenancy and the superior tenancy had been lawfully determined. No such regulations had been made, and it would be surprising to him (his Lordship), if Parliament, in an Act dealing with housing and rents, had intended section 41 to apply to agricultural holdings, that no reference should be made to such a holding. In his opinion, the word "premises" in section 41 of the 1954 Act and section 18 (5) of the Act of 1968 meant dwelling-houses, including all properties treated as such for the purposes of the Rent Acts. But if that was not right, then he thought the word should be given the slightly wider meaning of "buildings." It was quite another thing to say that it covered acres of arable or pasture land, and that just by the use of that one word Parliament B intended the section to encroach upon the sphere of the Agricultural Holdings Act 1948.

In a joint speech prepared by LORD DIPLOCK and LORD SIMON OF GLAISDALE in collaboration, Lord Simon said that it was the first and most elementary principle of construction that words and phrases of technical legislation were used in their technical meaning if they had acquired one. If the Rent Act 1968 dealt with legal technicalities of leasehold tenure on which ordinary citizens consulted their lawyers, rather than with everyday affairs, "premises" must be construed as a term of art, ie, as the subject matter of the habendum clause of a lease. In popular parlance, "landlord" could mean merely "inn-keeper," but in ordinary legal parlance, which was the language of legislation C dealing with rent restriction and security of leasehold tenure, "landlord" must be presumed to mean "lessor." Similarly, in popular parlance "premises" could mean "building," but in ordinary legal parlance it meant "the subject matter of a letting." Any presumption there might still be that Parliament did not intend by a Consolidation Act to alter the pre-existing law could have no operation where the words of the Consolidation Act were not, as a matter of legal language, capable of bearing more than one meaning; there was, moreover, no presumption that in reversing decisions of the courts Parliament confined itself to making no more alteration in the law than necessary. To restrict the meaning of "premises" in section 18 (5) to "a building used predominantly for residential purposes," as counsel for the D respondents proposed, was unacceptable. In the first place, this was not a natural meaning of "premises," either in a popular or in a technical context. It therefore involved rewriting the subsection, quite unnecessarily and inadmissibly. In the second place, it produced absurdity and anomaly. Why should Parliament wish to protect a subtenant of a flat in a block containing 40 per cent office accommodation and 60 per cent residential accommodation, but not a subtenant where the percentages were reversed? The case provided a striking example of how disregard of the first and most elementary principle of construction could lead to the ascription of a Parliamentary vagary.

ARBITRATION

A **COURT OF APPEAL**

February 10 1975

(Before Lord Justice BUCKLEY and Lord Justice ORMROD)

MODERN BUILDING WALES LTD v LIMMER & TRINIDAD CO LTD

Estates Gazette April 26 1975

(1975) 234 EG 287

Building contract—Whether dispute should go to arbitration—If there is a question whether an arbitration term exists, that must be determined on the hearing of the
B **application for a stay—Passage in Russell on Arbitration criticised—On facts, an arbitration term did exist—Order made staying action**

This was an appeal by Limmer & Trinidad Co Ltd, of Croydon, Surrey, defendants in an action in which Modern Building Wales Ltd, of Caerphilly, Glamorgan, claimed damages for breach of contract, from an order of Kerr J in chambers on November 20 1974 rejecting an application for a stay of further proceedings in the action pending arbitration.

Mr P Miller (instructed by M E Brown) appeared for the
C appellants, and Mr R Akenhead (instructed by Freedman & Co) represented the respondents.

Giving judgment, BUCKLEY LJ said that the court had been told that the ground upon which the judge made his order was that he considered there was a doubt whether the contractual document on which the respondents were suing, a written order dated December 18 1968 for the supply of certain labour, plant and machinery, imported by reference a form of contract between contractor and subcontractor which had in it an arbitration clause, and possibly that there was a further doubt whether, if that form of contract was imported at all, that part of it which contained the arbitration clause
D was a part which was so imported. The court's attention had been drawn to a passage in *Russell on Arbitration,* 18th ed, p 153, which stated, "Any doubt about the written submission being quite clearly established is . . . a sufficient reason for the judge's exercising his discretion against a stay."

The respondent plaintiffs were head contractors under a contract with the Taf Fechan Water Board for the construction of a building at Nelson, Glamorgan. On September 4 1968 the defendant company, who were nominated subcontractors for the purpose of putting in some suspended ceilings in the building, delivered a quotation to the architects for that work in the sum of £5,018. On December 18 1968 the plaintiff
E company placed an order, referring to that quotation, for the supply of "adequate labour, plant and machinery to carry out complete the ventilated and non-ventilated ceilings at the above contract within the period stipulated in the programme of work and in accordance with the appropriate form for nominated subcontractors (RIBA 1965 edition)." It was common ground that the effect of this document was to constitute a contract between the plaintiff company and the defendant company as contractors and subcontractors respectively in respect of the work detailed in the quotation.

It transpired, however, that in fact the RIBA did not have a F
form of contract between a contractor and a nominated subcontractor, nor indeed did they have any form of contract which existed in a 1965 edition, so that the words in brackets in the order were inappropriate language. The head contract was in a form issued under the sanction of the RIBA and various other bodies. There was in existence a form of contract intended to be used by contractors and nominated subcontractors when contracting *inter se,* but that was a form which was put out not under the authority of the RIBA but under that of the National Federation of Building Trades Employers and the Federation of Associations of Specialists and Subcontractors, also being approved by the Committee G
of Associations of Specialist Engineering Contractors. That form of contract had been referred to in the argument as the "green form."

The plaintiff company issued a writ claiming damages for breach of contract between themselves and the defendant company. The endorsement on the writ had been amended, and the claim for damages was now set out in very considerable detail and amounted in total to £10,284. It was mainly, if not entirely, based on the contention that the defendants failed to do the work which they were obliged to do in due time. The plaintiffs launched an application under order 14
for summary judgment. That was countered by the defendants H
with an application for a stay of the action on the ground that the written order incorporated the green form of contract and that the green form of contract contained an arbitration clause. The application was therefore to stay proceedings in the action pending the dispute being referred to arbitration. The defendants put in an affidavit in support of that application in which they denied that they were in breach of contract. The application was further supported by an affidavit of Mr R C Atkins, contracts administrator of a company which was a wholly-owned subsidiary of Tarmac Ltd, the parent company of the defendant company. His
evidence could be accepted as that of an expert on the subject. J
He said that he had no doubt that anyone in the industry would understand the words "in accordance with the appropriate form for nominated subcontractors (RIBA 1965 edition)" to refer to what was known as the green form. He drew attention to the fact that the green form was headed "for use where the subcontractor is nominated under the 1963 RIBA form of main contract," and said that it was correct that there was no 1965 edition; there was only a 1963 edition of the green form, and he would understand that that was the document intended to be referred to.

The question the court had to decide was whether this was K
a case in which there was an arbitration agreement between the parties, in which event the action should be stayed pending arbitration, or whether it was not. It had been argued by Mr Akenhead on behalf of the plaintiffs that there was a doubt as to whether the green form was or was not imported. Counsel said that where a doubt existed there was a good ground for the judge exercising his discretion under section 4 of the Arbitration Act 1950 in such a way as to allow the action to go on. He relied upon the decision of the Court of

A Appeal in *London Sack & Bag Co Ltd v Dixon & Lugton Ltd* [1943] 2 All ER 763. It was clear that what the court was there deciding was that it was not established that there was any contract containing an arbitration clause between the two parties to the action. In his (Buckley LJ's) judgment, that case was not authority for anything further than this, that in a dispute in which a party who sought a stay failed to prove that there was an arbitration agreement in force between the parties he could not get a stay. That seemed to be a fairly obvious proposition, and he (his Lordship) thought that the way the matter was stated in *Russell on Arbitration* was misleading. It was not sufficient, in his judgment, for a
B party who resisted a stay merely to say, "Well, there is some doubt as to how the contract between the parties should be construed, and therefore it is proper for the court in the exercise of its discretion to refuse a stay." On the contrary, in his judgment, when an application was made by one party on the ground that there was in operation an arbitration agreement, it was incumbent upon the court to discover whether or not there was such an agreement in force, and if that involved determining a question of construction, then such question of construction must be decided there and then. The question clearly had to be decided at an interlocutory stage, because it had to be decided before the defendant took
C any further step in the action.

In the present case, therefore, the court had to ask whether on the language of the order of December 18 the arbitration clause in the green form was or was not imported into the contract. Mr Akenhead had said that there were various ways in which that order could be construed which would result in its being held that the arbitration clause was not so imported. He (counsel) contended first that the words in brackets, "RIBA 1965 edition," were really nonsensical, and that the whole of the importation clause in the order form was therefore of no effect at all. The words were, however, just an added definition which turned out to be factually inaccurate, and in accordance with the ordinary procedure where there was a misdescription of that character, the court D
disregarded what was inaccurate and inapplicable and proceeded upon that which was appropriate and intelligible and on what were evidently intended to be the governing words of definition. The right way to construe this order form was to ignore the words in brackets altogether as being a *falsa demonstratio* and to accept that the reference to the appropriate form for nominated subcontractors was a reference to the green form, that being the only form to which it was suggested those words could apply. Mr Akenhead's next contention was that reading the document as a whole, one could not find that the parties had shown a firm intention that all the terms of the green form should E
be imported. He (counsel) submitted that the language was only appropriate to include those clauses of the green form which could be said to relate to the supply of labour, plant and machinery. But the words "in accordance with the appropriate form for nominated subcontractors" appeared to him (his Lordship) to be fully wide enough to introduce all the terms of the green form. Counsel had suggested that the written contract bore insufficient indication as to how various matters which were left in blank in the green form of contract, and in particular in the appendix to that contract, relating to such matters as the completion period and the F
retention money, should be filled in. But where parties by an agreement imported the terms of some other document as part of their agreement, those terms must be imported in their entirety, subject only to this, that if any of the imported terms in any way conflicted with the expressly-agreed terms the latter must prevail over what would otherwise be imported. It followed that the court must proceed on the basis that there was here a *bona fide* dispute between the parties coming within the terms of the arbitration clause in the green form.

ORMROD LJ agreed, and the appeal was allowed with costs. An order was made that further proceedings in the action be stayed pursuant to section 4 of the Arbitration Act.

AUCTIONS

CHANCERY DIVISION

October 17 1974

(Before Mr Justice TEMPLEMAN)

HARROP v THOMPSON AND ANOTHER

Estates Gazette February 22 1975

(1974) 233 EG 763

Auction of farm—Agreement by purchaser with potential bidder restraining competition, if it could be established, would not invalidate the contract or bar a suit for specific performance—Summary judgment awarded—Further point on sale of house—Triable issue held to arise whether purchaser bought as nominee for trustee vendor

This was an order 86 summons by Mr Ronald Harrop seeking summary judgment in a suit against Mr George Roper Thompson and Mr Malcolm Thompson, personal representatives of the late George Atkinson Thompson, for specific performance of (1) a contract dated March 15 1973 for sale by the defendants to the plaintiff of Dyke Nook Farm and High House Farm, Consett, Co Durham, for £106,000, and (2) a further contract for sale on the same date by the defendants to the plaintiff of Heatherlea, Consett, Co Durham for £10,000.

Mr J F Parker (instructed by Crossman, Block & Keith, agents for Richmonds, of Consett) appeared for the plaintiff; Mr S Goldblatt QC and Mr D G A Jackson (instructed by Robinson & Co, agents for Hewitt, Brown-Humes & Hare, of Bishop Auckland) for the first defendant; and Mr A G Boyle (instructed by Snowball, Tucker & Bibby, of Consett) for the second defendant.

Giving judgment, TEMPLEMAN J said: The question is whether an agreement not to compete invalidates an auction contract. The plaintiff purchased a farm at an auction for £106,000 and now seeks summary judgment under order 86 for specific performance. The vendor defendants resist specific performance on the grounds that prior to the auction the purchaser entered into an agreement with another potential bidder whereby the potential bidder stayed away from the auction and the purchaser acquired the farm cheaply. The purchaser can obtain summary judgment only if there is no triable issue. Although, therefore, the purchaser denies that there was any agreement with a potential bidder, his counsel submits that even if there were such an agreement the vendors cannot resist specific performance.

The point is not devoid of authority. In *Galton v Emuss* (1844) 1 Coll 243, a potential bidder absented himself from the auction on the promise of the successful purchaser to give the potential bidder a right of pre-emption. (A similar promise is alleged in the present case to give an option.) That promise was enforced by Knight Bruce V-C, who said at p 246:

"Two men, severally desirous of effecting a purchase of an estate, become acquainted with each other's intentions, and, with a view to their own benefit, enter into an engagement together that one shall retire leaving the field open to the other. It is not suggested that this arrangement involved any matter of fraud or misrepresentation; it was merely that one should not bid while the other was a competitor. No authority has been cited to show that a contract founded on such a consideration is illegal, and, as the defendants have not in consequence of my intimation, or otherwise, asked an opportunity of bringing the question before a court of law, I assume that it is not their wish to do so."

In *Re Carew's Estate* (1858) 26 Beav 187 on a sale under the court two persons agreed not to bid against each other, but that one should bid up to £1,500 and divide the lot between them. They bought for £650, and Romilly MR refused to avoid the sale, saying at p 189:

"I am not aware of any case, or any principle which establishes that such an agreement is inequitable."

And at p 190:

"All that is proved is that the persons who have sold the estate were not aware of the real value of this lot, and it is to be inferred that if they had known what the court now knows, they would have fixed a much higher reserved price, but this is no reason for opening the sale."

In *Heffer v Martyn* (1867) 36 LJNS Eq 372 the purchaser agreed to pay £500 to a potential bidder if the bidder stayed away and if the purchaser was able to purchase for not more than 6,000 guineas. The purchaser also tried to make arrangements with other persons not to bid. At the auction the purchaser succeeded in a bid of £3,000, and the vendors, discovering the purchaser's artifices, resisted specific performance. Lord Romilly found in favour of the purchaser, and said at p 373:

"This is no doubt very hard upon a vendor that if he combines with others to keep up the price it should be illegal and the sale void but that the purchaser may combine together with others to get the property at less than its value; but the real remedy in such a case as this is in fixing the reserved bidding."

And later:

"It is no doubt extremely irritating to find that the property is sold for one-half of its real value, or less; but still it is the fact that according to the decisions of this court such an agreement is not illegal and the plaintiff is entitled to a decree for specific performance."

These authorities appear to be indistinguishable and to decide the present case in favour of the purchaser. In *Rawlings v General Trading Co* [1920] 3 KB 30 at a sale by public auction of surplus stores belonging to the Ministry of Munitions the plaintiff and the defendant agreed that in order to avoid competition the defendant would bid and then divide the goods with the plaintiff. The defendant bought the goods and was sued by the plaintiff for his share. Shearman J held that the agreement between the plaintiff and the defendant to divide the goods was unenforceable, saying at p 35:

"I am of the opinion that, at any rate in such a case as this where the goods sold are the property of the public, it is against public policy that people should combine at an auction to procure the goods to be sold at a price considerably below the fair value, with the necessary result that the public are defrauded."

This decision was reversed by the Court of Appeal [1921] 1 KB 635. Bankes LJ reviewed the authorities and said at p 641:

"Having regard to the state of the authorities in the Chancery courts for over 70 years I do not think it was open to the learned judge to take the view he did, nor do I think that this court should after this lapse of time overrule those authorities even if this court considered they were wrong, which I am far from suggesting that they are. If there was any conflict between the rules of law and the rules of equity on the point (which I do not consider that there was) the latter must prevail."

Another 50 years has elapsed since the words of Bankes LJ. In the same case, however, Scrutton LJ dissented on the ground that the agreement between the plaintiff and the defendant which was sought to be enforced was, as he says at p 647, "contrary to public policy, as a restraint of trade contrary to the interests of the public." Atkin LJ agreed with Bankes LJ in reversing the decision of Shearman J, and at p 632 he said:

> "I can see no reason for saying that this contract is *ex facie* illegal. It would probably be sufficient to say that for nearly a century courts of equity have held similar agreements legal and enforceable by suit for specific performance. But apart from decided cases as between the parties it appears to be plainly reasonable, and if the defendant wished to establish that the public interest suffered he should have so pleaded as to give the plaintiff notice."

Mr Goldblatt submits that in the light of the observations of Shearman J and Scrutton LJ, it is open to a judge of first instance to hold that a knock-out or similar agreement is against public policy, being in restraint of trade, and that a vendor who complains of such an agreement is not bound by his contract with the purchaser. There is nothing in the *Rawlings* case which overrules the earlier authorities, or enables me to ignore them. Mr Goldblatt relies on the observations in the *Rawlings* case as pointing in the direction in which he submits the law should develop. I am not at liberty to twist the wheel in the direction he wishes. In *Cohen v Roche* [1927] 1 KB 169 at 173, it was accepted by McCardie J that it seems reasonably clear in law that the existence of a knock-out agreement does not of itself afford any answer to this action by the plaintiff. And in *Pallant v Morgan* [1953] Ch 43 there was a formidable array of distinguished counsel. It did not occur to them or to Harman J, who decided the case, that there was anything wrong with an agreement between two bidders not to compete at an auction.

Mr Goldblatt, in the course of an interesting argument, invited me to follow the path indicated by Shearman J and Scrutton LJ. A knock-out or similar agreement, he says, if not proved to be reasonable, is unenforceable and relieves the vendor from his contract. He referred me to *Neugebauer & Co Ltd v Hermann* [1923] App D 564, a South African case, which illustrates that the law of South Africa and the law of some states of America is not always the same as the law of England. He also relied on *Esso Petroleum Co Ltd v Harper's Garage* (*Stourport*) *Ltd* [1968] AC 269 for the proposition that the courts of this country are now astute in applying the doctrine of restraint of trade to strike down any interference with individual liberty. A knock-out agreement, he submitted, is an interference with the liberty of the potential bidder to bid at the auction and ought not to be enforceable if the sole or primary object of the agreement between the potential bidders is to buy the property cheaply. This submission brings most contracts within the grasp of restraint of trade, and therefore unenforceable unless proved to be reasonable, because most contracts interfere with liberty of action one way or another. I am not convinced that the agreement alleged in the present case was in restraint of trade. If it was in restraint of trade, I am not convinced that it invalidated or affected the contract between the vendors and the purchaser. But for the earlier cases which bind me, it might be argued that there was a kind of fraud on the vendors, but restraint of trade in the circumstances I do not understand. Mr Goldblatt's remedy may be worse than any disease, If potential bidders, or a syndicate, enter into some form of agreement which results in some of them keeping quiet at the auction, there is no method open to a judge of first instance of sorting out a good agreement from a bad agreement. The present case may open up a vista of litigation. It is said that an obliging potential bidder confided to the vendors at one stage that he might have bid £130,000, but in the meantime he had confided to the purchaser that he would not bid for a consideration, and then later on confided the whole plot to the vendors. If such allegations and conversations must be investigated before a purchaser is allowed to complete, then only the lawyers may benefit. It may be better that the property should fall with the hammer. A vendor can protect himself by conditions under which he imposes a reserve and keeps the right to bid and to withdraw. A timid vendor need not sell by auction. But these matters can be pursued with the Law Commission, and restraint of trade or fraud may be pursued in the appellate courts. As the law now stands, I am bound by authority to ignore the agreement between the purchaser and the potential bidder which is alleged in the evidence in this case. That being so, there is no triable issue, and the purchaser is entitled to specific performance.

The present application also concerns a second contract comprising the sale of a house by the vendors to the plaintiff. The first defendant has given affidavit evidence that the plaintiff purchased this property as nominee for the first defendant. A sale to the first defendant is void because he is a vendor trustee and cannot purchase trust property. These allegations, if true, debar the purchaser plaintiff from specific performance. Mr Parker, who appeared for the purchaser, submitted that the allegations are inconsistent with the correspondence, but this may not prove to be so at the trial. There is a triable issue, namely whether the plaintiff bought the house for himself or as nominee for the first defendant, and so far as this second contract is concerned, the plaintiff is not in my judgment entitled to summary judgment.

COURT OF APPEAL

January 29 1975

(Before Lord Justice CAIRNS, Lord Justice LAWTON and Mr Justice MacKENNA)

BROWN v DRAPER & CO

Estates Gazette March 8 1975

(1975) 233 EG 929

Auctioneers must either adopt the client's description of goods put up for sale or protect themselves with a suitable condition—Damages awarded against firm which simply substituted its own description without more

This was an appeal by Mr Derek Arthur Russell Brown, of The Firs, Great Plumstead, Norwich, from a judgment of Judge Moylan at Norwich County Court on August 8 1974 dismissing a claim against J R E Draper & Co, auctioneers, of All Saints Green, Norwich, for damages for negligence and/or breach of contract.

The appellant appeared in person. Mr S R Silber (instructed by Mills & Reeve, of Norwich) represented the respondents.

Giving the first judgment, MacKenna J said: The defendants are auctioneers carrying on business in Norfolk. Mr Brown, the plaintiff, is a builder living in Norfolk. The defendants had advertised an auction sale to be held by them on Wednesday November 14 1973 at the Cornhall, Harford. Two days before the sale, on November 12, Mr Brown gave a number of articles to Mr Geoffrey Bullen, one of the defendants' representatives, for sale at the auction. He gave them to Mr Bullen in a box, and handed him at the same time a written list describing each of the articles. Nothing was said either by Mr Brown or by Mr Bullen about the descriptions under which the articles should be auctioned. There were no special terms discussed between them. When auctioning the goods on the 14th, the defendants gave some of the articles different descriptions from those contained in Mr Brown's list and sold them under those different descriptions. One of Mr Brown's articles was described in his list as "14-carat gold bracelet." It was described by the defendants at the sale as "an expanding bracelet," without any mention of its being gold. Another was described in the list as "box silver and marcasite jewellery." It was described at the sale as "miscellaneous jewellery," without any reference to its being silver. Another was described in the list as "1 pair heavy gold cuff links." It was described at the sale as "gold cuff links" without any reference to the weight. Another was described in the list as "an electric watch." It was described at the sale as "gent's wrist-watch and strap," without any reference to its being electric. Another was described in the list as a "silver and gold compact." It was described in the sale as a "silver compact" without any reference to its being gold. Mr Brown was aggrieved, and brought this action, claiming damages against the auctioneers for negligence. He contends that the defendants did wrong in selling the goods, as they did, under descriptions which differed from those he had given to them in his list. He recognises that they were under no obligation to accept the goods for sale, but says in effect that if they did not wish to sell them under his descriptions, they should have given them back to him. At the trial, Mr Bullen gave evidence for the defendants. He said he would not describe anything as gold or silver unless he had satisfied himself, either by marks or tests, that the description was a true one. No evidence was given by him of any custom of the trade in favour of the auctioneers. The judge accepted his evidence and dismissed the action, saying that it was not an implied term of the contract that the goods, if sold, should be sold under the descriptions contained in Mr Brown's list.

In my judgment, the judge misconceived the duty which the defendants owed to Mr Brown. When he gave them his articles for sale with the list describing them, he was, I think, inviting them to sell his goods under those descriptions, and not to sell them under any other description without obtaining his agreement. In these circumstances, it was their duty, if they sold the goods at all, to sell them under the descriptions given to them by Mr Brown, unless they had obtained his consent to sell them under other descriptions. They did sell them under a different description and did not obtain his consent. They acted wrongly. Suppose this case: I send a book to Sotheby's, the auctioneers, under cover of a letter describing the book as a first folio Shakespeare and asking them to sell it by auction. That would mean, in the ordinary way, that I was asking them to offer it for sale as a first folio. It could hardly mean anything else. Suppose that, without any further discussion, Sotheby's sold the book, describing it as an edition of Shakespeare but not mentioning that it was a first folio. Can it be said that I have no right to complain that Sotheby's, my agents, have not acted in accordance with my instructions? Would it be any answer for Sotheby's to say that they were uncertain whether it was a first folio and had had no time to establish its authenticity? I would say not, unless either there is a custom of the trade which entitles the auctioneer to use his own description instead of the customer's, or unless the special terms on which he does business give him that right and he has brought those terms to the customer's notice. I cannot accept Mr Silber's argument that it makes any difference that the sale by Sotheby's would be a sale by catalogue and that this was a non-catalogue sale, the defendants' descriptions being read out to the buyers from a book prepared by the auctioneers shortly before the sale.

The measure of Mr Brown's damage is the difference between the prices which the goods would probably have fetched if they had been sold under his descriptions and the prices which they actually fetched under the defendants' descriptions. If the parties cannot agree on this figure, the case will have to go back to the county court for the registrar to assess the damages.

LAWTON LJ: I agree, and only have this to add: it should not, in my view, be thought by anybody that the decision in this case introduces any new concept of the law relating to auctioneers. It is a case which has been decided on its own facts and, within my limited experience, unusual facts, because at most auction sales there are conditions which are brought to the attention of those who want to put their goods into an auction. If an auctioneer will only accept goods on the terms that he will put his own description on the goods, he is entitled to do so. But in the absence of any express term to that effect, if an auctioneer is offered goods by a prospective seller, it is up to him to decide whether to accept them or not. If the circumstances are such (as they were in this case) that the seller by implication was asking the auctioneer to sell under the seller's description, then the auctioneer must say "Yes, I will" or "No, I won't." But if, as happened in this case, he accepts the goods and then sells them under different descriptions without saying a word to the seller, he has only himself to blame for the kind of claim which has been made against him in this case. I would allow the appeal.

CAIRNS LJ: I agree that this appeal succeeds, for the reasons which my Lords have given, and that the order of this court should be as enunciated by MacKenna J.

The appellant was awarded costs of £12.

COMPULSORY PURCHASE AND COMPENSATION

A **QUEEN'S BENCH DIVISION**

July 31 1974

(Before Mr Justice WILLIS)

SUNLEY HOMES LTD v SECRETARY OF STATE FOR THE ENVIRONMENT AND ANOTHER

Estates Gazette January 4 1975

(1974) 233 EG 57

High-value site in high-value area compulsorily acquired for council housing—Inspector says CPO should not be confirmed, but Secretary of State nevertheless confirms it—Differences between inspector and Secretary of State held to
B **be differences of opinion only—No ground available on which order might be quashed**

This was an application by Sunley Homes Ltd, of Henrietta Street, London WC2, for an interim order suspending the operation of the Westburn Lodge/Sweetmans Compulsory Purchase Order 1972, made by the second respondents, the London Borough of Harrow, and confirmed by the first respondent, the Secretary of State for the Environment.

Mr G Dobry QC and Mr A B Dawson (instructed by E P Rugg & Co) appeared for the applicants; Mr J H R Newey and Mr H K Woolf (instructed by the Treasury Solicitor) for the first respondent; and Mr R P Ground (instructed by the legal
C department, London Borough of Harrow) for the second respondents.

Giving judgment, WILLIS J said that the substantial point of the case was whether the Secretary of State had failed to comply with rule 9 of the Compulsory Purchase by Local Authorities (Inquiries Procedure) Rules 1962. This rule provided that an appointed person, in this case an inspector, should make a report in writing to the Secretary of State with findings of fact and his recommendations and that where the Secretary of State differed from the inspector on a finding of fact or took into consideration any new issue of fact he should take certain steps.

D The council wanted to acquire the site in question for council housing. When they heard it was for sale they negotiated with the vendors, but failed to agree a price. They then heard that tenders were being invited, and they made the compulsory purchase order complained of. Sunley Homes made the successful tender. In view of the compulsory purchase order a public inquiry was held. There was a great deal of evidence on whether the site should be developed as private houses or council houses. The land had been valued at £1.4m. It was argued that this was an exceptional site of high value. Housing in the immediate vicinity was of high value, and the inspector's view was that local authority housing was inappropriate. The quality of
E council housing would have to be of such a high standard as to make council development unacceptable. Mr Dobry had conceded that the Secretary of State was entitled to differ from the inspector, but only if he complied with rule 9. In four instances, said counsel, the Secretary of State had differed from the inspector's findings of fact. Firstly, the inspector concluded that a comparison of the economics of private and council housing made the site too expensive for council housing, and that such use could be justified only by dire necessity. The Secretary of State did not accept that this factor should have great weight attached to it, and he had taken into consideration
F an estimate of what such a council project would cost in comparison with other local authority projects in outer London. Mr Dobry said that the Secretary of State was differing fundamentally here from his inspector, but the defendants argued that the Secretary of State was not differing on a fact but merely expressing a government policy: housing needs must be met at any cost. It seemed to him (his Lordship) that a comparison between what the inspector had said, and his conclusion, and what the Secretary of State had decided revealed no difference of fact. Any difference was one of opinion.

The second difference was on the occupancy rates to be expected. The council maintained that their plan would mean housing for more people. The inspector merely said that he
G had doubts. The Secretary of State differed from that view. This was a conjectural matter and not a finding of fact. The third difference was based on the view that council housing would lead to mixed development in view of the high quality of the private housing in the vicinity, and that this could be harmful socially. The inspector was sympathetic to this view, but this was a matter of judgment and not a finding of fact. The Secretary of State did not accept that mixed development might be harmful. The fourth difference was a comparison between the rival schemes having regard to the extent to which each was worked out. The inspector held the view that the council scheme had not been worked out in sufficient detail, but
H in differing on this matter the Secretary of State had not come to a conclusion of fact. For these reasons the grounds of the application fell away, and there would be judgment, with costs, for the Secretary of State and the Borough of Harrow.

His Lordship granted a stay until the first day of next term provided that the applicants lodged notice of appeal within 28 days.

QUEEN'S BENCH DIVISION J

December 4 1974

(Before Mr Justice CANTLEY)

SAVOURY AND OTHERS v SECRETARY OF STATE FOR WALES AND ANOTHER

Estates Gazette February 22 1975

(1974) 233 EG 766

Cardiff clearance area covering streets with close community spirit—Residents' objections to compulsory purchase order —Council said to have acted without being satisfied, as
required, that "suitable" alternative accommodation was K
available—"Suitable" does not necessarily mean "located so as to preserve the community spirit"—In any event, council's own statements as to satisfaction must be accepted in the absence of evidence to the opposite effect

This was an appeal by Mr Leonard Savoury and other residents of Nora Street and Helen Street, Adamstown, Cardiff, from a decision of the first respondent, the Secretary of State for Wales, dated June 13 1974, confirming, with minor modifications, a compulsory purchase order made by

A the second respondents, Cardiff City Council, in respect of houses in the district.

Lord Gifford (instructed by L Bingham & Co, agents for Anthony W Jeremy & Co, of Cardiff) appeared for the appellants, and Mr H K Woolf (instructed by the Treasury Solicitor and by Collyer-Bristow & Co, agents for the city solicitor) represented both respondents.

Giving a reserved judgment, CANTLEY J said that because of strong objections to the compulsory purchase order in question, the Secretary of State caused a public local inquiry to be held. It was very understandable that the appellants and other residents of the two streets concerned strongly
B objected to the order. It was a very closely-knit community with a strong community spirit. A good percentage of the residents were between 60 and 80 years of age, and some of them had lived in the two streets for the whole of their lives. Many of the elderly residents received practical help and moral support from kind neighbours. In his report, the inspector accepted that a strong case had been presented against the demolition of the houses, based on three aspects: (a) the social consequences of demolition; (b) the economics of preservation as compared with demolition; and (c) the wish of the residents to be allowed to remain undisturbed. The inspector accepted without qualification the case made
C on the basis of community spirit, and reported that while the council hoped to rehouse many of the residents on the same site after redevelopment, the intervening disruption might result in a failure to regain the close community spirit. The Secretary of State nevertheless confirmed the order. He rejected the appellants' argument that the houses should be retained and improved, saying that he accepted that the property was in such a condition that improvement was not practicable. The appellants now moved to quash the order on the grounds that it was not within the powers of the Housing Act 1957 and that they had been substantially
D prejudiced by non-compliance with a requirement of the Act.

The basis of the argument presented on their behalf was the proviso to section 42 of the statute, which stated that before passing a resolution declaring an area to be a clearance area, a local authority must be satisfied, first, that it would be in a position to provide, or secure the provision of, such accommodation as would be necessary for persons displaced, in advance of the displacements which would from time to time become necessary, and secondly, that its resources were sufficient for the purpose of putting the resolution into effect. The appellants said that before the council passed their resolution declaring the area to be a
E clearance area they had not satisfied themselves as to the suitability of alternative accommodation already existing, or secured the provision of accommodation in advance of the displacements. The appellants contended that the council should have satisfied themselves that such accommodation as was, or would be, available would be so located as to preserve the community spirit; at the most, they said, the council had merely satisfied themselves that a sufficient number of suitable houses would be available, and that was not good enough. The effect of a contravention of the second proviso to section 42 (1) had been considered by Lord Parker CJ in *Goddard v Minister of Housing and Local Government*
F [1958] 3 All ER 482, in which he had accepted that compliance with the proviso was a condition precedent when making a valid declaration or clearance order, so that if the requirements set out were not met, the subsequent compulsory purchase order would not be within the powers of the Act. He (his Lordship) respectfully agreed. If, therefore, the present appellants could persuade the court that the council had not satisfied themselves in accordance with the requirements of the first limb of the proviso, the compulsory purchase order must be quashed. It should be noted that this point had not been raised at the inquiry. The objectors had, however, argued strongly that proper regard had not been paid by the local authority to the wishes of the residents to G
remain in the area, and the results of a survey of a local action group, which showed that 60 per cent of the present residents wished to remain, had been put before the inspector. It was also to be observed that it was not suggested that the council had acted otherwise than bona fide in the matter. If, therefore, it appeared that they (the council) had somehow satisfied themselves of the relevant requirements, the court would be in a position to interfere only if it was shown that they had acted capriciously, had not really considered the evidence, or were satisfied on no evidence, or else that such evidence as there was could not possibly have satisfied them.

The council had to be satisfied (1) that accommodation H
was available, or would be provided, in advance of the displacement which would from time to time become necessary, and (2) that the accommodation would be suitable. The statute recognised that displacement of people was inevitable when rehousing took place, and that that meant real inconvenience to everyone. The statute also recognised that demolition might take place in stages, and that everyone in the area would not necessarily be rehoused simultaneously. In the present context, the requirement that the alternative accommodation should be "suitable" meant, as Scarman LJ said in *R v Bristol Corporation* [1974] 1 WLR 498 at 502, "no more and no less than suitable to the requirements of J
the person and his family in the circumstances in which they find themselves." Lord Gifford had adopted that definition on behalf of the appellants, and so did he (his Lordship). He also accepted Lord Gifford's submission that the accommodation to be provided had to be suitable for the requirements of the real people who were being displaced, and not purely hypothetical people. However, "suitable" was not the equivalent of an ideal or perfect alternative, nor was it equivalent to "identical." Regard had to be had to what was reasonably practicable and to what accommodation the actual persons being displaced, temporarily or permanently, could reasonably be asked to accept. Accommodation was K
not unsuitable merely because people displaced by demolition were not going to be rehoused on the old site. The preservation of an established, well-conducted and contented community was a very desirable thing, but it might not always be reasonably practical to achieve it.

He (Cantley J) had been shown an extract from the minutes of the council's estates committee for July 20 1971. This read:

> "Proposed clearance area. *Resolved,* that subject to the finance committee being satisfied that the resources of the council are sufficient for the purpose of carrying this decision into effect, a L
> compulsory purchase order be made under section 43 of the Housing Act 1957 in respect of the land shown on the plan annexed, and in so far as suitable accommodation available for persons who will be displaced by the clearance of this site does not already exist, the council do secure provision of such accommodation in advance of displacement, and do undertake to carry out or secure the carrying-out of such rehousing within such a period of time as the Secretary of State may consider reasonably necessary."

It did, therefore, appear that the council had satisfied themselves of the matters set out in the proviso, and the court could in these circumstances interfere only if it was shown that they had not really considered the question at all, or had M
decided it either upon no evidence, or upon evidence on which they could not reasonably be satisfied. Lord Gifford contended that there was no evidence and that the inspector's report indicated there had been no consultation with residents. But he (his Lordship) did not read the inspector's report as indicating that the council were unaware of the community spirit in the area or of the fact that many residents wanted to stay there. Counsel had gone on to contend that it was plain from the evidence that the council had done no more than make a numerical comparison between houses likely to be available and the number of family units in the

streets. That did not follow from the evidence. In any event, the question with which the court was concerned was whether the council had acted arbitrarily or whether, in declaring themselves satisfied that suitable accommodation was, or would be, available, they had reached a conclusion they could not reasonably have reached. The answer to this must be no.

It was finally contended on behalf of the appellants that the council appeared to have thought that the subsequent rehousing of some of the people on the site would meet the objections raised. That was not an inference which could properly be drawn from the evidence. If the intention was to rehouse 50 per cent on the site, that did not mean that the accommodation for the other 50 per cent would be unsuitable. The difficulty in the way of the appellants was that there was no real evidence as to the information upon which the council, before passing their resolution, had decided that suitable accommodation was, or would be, available. On the face of the resolution and the compulsory purchase order, the council had carried out their duties, and there was no evidence that they had not acted regularly and properly. There was nothing established which would justify the court in quashing the order, and the appeal must be dismissed.

No order was made as to costs.

COURT OF APPEAL

December 5 1974

(Before Lord Justice RUSSELL, Lord Justice ORR and Sir John PENNYCUICK)

THOMAS LANGLEY GROUP v ROYAL LEAMINGTON SPA BOROUGH COUNCIL (now Warwick District Council)

Estates Gazette April 5 1975

(1974) 234 EG 39

Compensation—Purchase notice—Successive planning permissions either inconsistent, or consistent only on the basis that the later one constituted an agreed variation of the rights conferred by the earlier one—By acting on the later permission, developer abandoned rights earlier granted—Lands Tribunal's award of £500 for 2-acre site devoted to open space confirmed by court

This was an appeal by the Thomas Langley Group Ltd from a decision of the Lands Tribunal dated May 10 1973 assessing at £500 the compensation payable by the respondents, the Borough of Royal Leamington Spa (now Warwick District Council), on acquisition, pursuant to a purchase notice, of two acres of land forming part of the Campion Estate, Leamington Spa. The decision of the Lands Tribunal was reported at (1973) 228 EG 339.

Mr P Freeman QC and Mr J M Sullivan (instructed by Southall & Co, agents for Southall & Co, of Birmingham) appeared for the appellants, and Mr R P Ground (instructed by C J George, of Leamington Spa) represented the respondents.

Giving the judgment of the court, ORR LJ said: The appellants in this case referred to the Lands Tribunal the assessment of compensation payable to them by the respondents, the Borough of Royal Leamington Spa (whose functions have since been taken over by the Warwick District Council), as a result of the respondents' compulsory acquisition, under the provisions of Part VIII of the Town and Country Planning Act 1962, of some two acres of land situated on the appellants' Campion Estate in Leamington Spa; and this appeal is now brought, on the ground that it is too low, against the Lands Tribunal's assessment of the compensation in the sum of £500. We do not find it necessary to refer in any detail to the provisions of Part VIII of the Act. It is sufficient for present purposes to say that where, on an application for planning permission to develop any land, permission has been refused or has been granted subject to conditions, and provided that certain requirements are satisfied as to the incapacity of the land in question for reasonably beneficial use, those provisions enable the landowner to serve on the local authority a notice requiring it to purchase his interest, and if it refuses to do so the appropriate Minister, if satisfied as to certain matters, may confirm the notice, and thereupon the local authority is deemed to be authorised to acquire the interest compulsorily under the provisions of Part V of the Act, and to have served for that purpose a notice to treat, with the result that in the absence of agreement the amount of the compensation payable for the acquisition falls to be assessed by the Lands Tribunal. On the reference to assess the compensation in this case the Lands Tribunal were not concerned with any issue of valuation, it being agreed between the parties that if an outline planning permission admittedly granted by the respondents on June 5 1964 in respect of a larger area embracing the two acres remained effective at the time of the compulsory acquisition the compensation should be £5,000, but that if that permission did not remain effective the compensation should be £500. The sole issue before the Lands Tribunal was therefore whether this outline permission remained valid; it being contended for the claimants that it did, and for the local authority that it had been abandoned by the claimants. The conclusion of the Lands Tribunal was that it had been abandoned, and the main issue in this appeal is whether that conclusion was justified on the material before the tribunal, which consisted of an agreed statement of facts together with the plans and other documents therein referred to and may be summarised as follows, the respondents being referred to for this purpose as "the council."

The Campion Estate (which we will call "the estate"), comprising some 13 acres, was formerly a clay works and brickyard, and was shown both on the 1956 Leamington Spa town map and on the amended town map operative from February 1969 as land allocated for surface mineral workings with the notation SM/brickclay. The two acres, situated approximately in the middle of the estate, which became the subject of the compulsory acquisition, and which we shall refer to as "the two-acre site," had been worked as a deep clay pit. On February 5 1964 the claimants applied to the council for outline planning permission for residential development of the estate, enclosing for that purpose only a site plan which gave no indication of the proposed development, and it is common ground that on June 5 1964 the council granted outline planning permission for residential development of the whole of the estate, subject (*inter alia*) to the following conditions:

> 3 No building shall be erected pursuant to this permission until the expiration of a period of ten years after satisfactory completion of filling.
>
> 5 This permission shall cease to have effect at the expiration of a period of three years from the period referred to in condition 3 above unless all necessary detailed drawings and particulars shall previously have been submitted to and approved by the local planning authority.

On February 17 1965 the claimants purchased the estate. On February 22 they applied for permission to carry out development described as "Erection of detached and semi-detached houses (residential)," and appended to the application a plan showing a proposed layout of 147 houses. On March 16 1965 the council refused this permission, giving the following among other reasons for the refusal:

> 1 The outline permission prohibits development until ten years after satisfactory completion of filling the clay workings and submission of details at this time is therefore premature.

A 4 There is no provision of public open space which is considered to be a necessary amenity in an estate of this size.

Following this refusal, the council's engineer wrote to the claimants a letter dated March 24 1965 which referred to the council's first reason for their refusal and continued as follows:

> I appreciate that this may give rise to a conflict between the three and ten year periods in the conditions attached to the outline permission and if you will submit a further outline application I will place it before the council with a view to regularising the matter. I understand that you propose to submit an application for carrying out the development in stages. Such B an application should be for stage 1 only, with details of the further stages shown on the plan. The other stages would be the subject of subsequent applications when you were ready for starting them. To assist you in preparing satisfactory proposals for such an application I give below preliminary observations on the plans which have been refused. . . . 2 An estate accommodating 450/500 persons should include some public open space. I consider that an area of two acres would be reasonable and I see no objection to part of the filled ground being used for this purpose.

In this letter the engineer was clearly under a misapprehension that the 10-year and the three-year periods referred to in conditions three and five of the 1964 permission would C run consecutively and not concurrently, but the importance of the letter lies in the views expressed as to the necessity of allocating part of the estate as a public open space and as to the suitability of the two-acre site for this purpose. On September 17 1965 the claimants made application for permission to carry out development described as "Stage 1 of construction of detached and semi-detached houses (residential)," and appended a plan showing a proposed layout of dwellings to be constructed under stages 1, 2 and 3, but showing no dwellings on the two-acre site and labelling it as "public open space." There accompanied this application D a letter from the claimants' general manager in which he referred to a recent meeting, and the material part of this reads as follows:

> The application, as you will see, is for Stage 1 only, although you will notice that we have sketched on the plan our proposals for Stages 2 and 3. We have also shown on the plan the public open space, which apparently your council require on a development of this size.

On November 9 1965 this application was refused by the council, the reasons given being that the amenity value of the open space was seriously reduced by the rear gardens backing on to it, and the want of satisfactory details as to E its treatment and landscaping. On December 31 1965 the claimants applied for permission to carry out development described as "Layout for erection of 73 dwellings, stage 1, and deletion of condition 3 on outline planning permission for this stage," and appended a plan showing the proposed layout of the 73 houses and of other houses, but leaving the two-acre site blank and also eliminating the rear gardens backing on to that area of which the council had made criticism. On February 1 1966 the council granted permission for the erection of 73 houses and construction of roads subject to the following among other conditions:

> F 5 Before any houses on the site are occupied the deep clay workings shall be filled with materials satisfactory to the local planning authority to levels which will ensure that there shall be no danger to occupiers of the houses and within six months from completion of stage 1 development the public open space shall be satisfactorily landscaped in accordance with the deposited plan.

On June 10 1966 the claimants applied for permission to carry out development described as "Layout consisting of 73 dwellings including estate roads" and appended a plan showing the proposed layout of those houses and also others, but again showing no projected houses on the two-acre site. On July 11 1966 the council approved this plan, and in doing so referred to "the outline permission dated February 1 1966." G The construction of the roads for stage 1 was begun on August 28 1966, and of the houses on November 2 1966. On February 2 1967 the claimants applied for permission to carry out development described as "stage 2: Construction of roads and erection of four detached and 11 pairs of semi-detached houses; stage 3: renewal of outline planning approval for 24 houses," and again appended a plan showing no projected houses on the two-acre site; and on February 21 1967 the council granted permission for "residential development: stage 2 details, stage 3 renewal of outline," subject to, *inter alia*, a condition that no buildings within stage 3 should be erected until the expiry of 10 years from satis- H factory completion of filling of the workings, in explanation of which it was added that this development was sited on made ground which would settle. The erection of the first pair of houses for stage 2 was begun on September 27 1967. Finally, on May 18 1967 the claimants applied for permission to carry out development described as "erection of semi-detached houses," and appended a plan showing 15 projected buildings of which 14 were within the two-acre site. It is common ground that this application was made in the expectation of refusal and with a view to subsequent proceedings under Part VIII of the Act, and it was in fact refused by the council on August 7 1967, J among the reasons given for the refusal being that the proposed development would conflict with conditions imposed on the permission to develop adjoining land in the same ownership and that the proposed development of land which had been shown on the layout plan as open space would raise the overall density above that considered appropriate for the site and type of dwelling. The compulsory purchase proceedings followed and were originally defective but were put right, the Minister having granted an extension of time for that purpose, by a fresh application for planning permission, which was refused on December 19 1968, and a fresh purchase notice with which K the council was unwilling to comply but which the Minister confirmed.

On these facts it was contended for the council before the Lands Tribunal that the case fell directly within the principle applied by this court in *Slough Estates Ltd v Slough Borough Council and Another* [1969] 2 WLR 1157, that "when a man is entitled to one of two inconsistent rights, then if he, with full knowledge, has done an unequivocal act showing that he has chosen the one, he cannot afterwards pursue the other"; that "If he does accept the one right, then by law he waives—he 'abandons'—the other," and that for this purpose abandonment depends on conduct and not upon the existence L of an intention to abandon (per Lord Denning MR at p 1164). For the claimants it was contended that this principle did not apply to the present facts, and that the tribunal should follow an earlier decision of its own in *Richardsons Developments Ltd v Stoke-on-Trent Corporation* (1971) 22 P & CR 958. The tribunal held that the claimants by their conduct subsequent to the permission of February 1 1966 had abandoned the outline planning permission of June 5 1964 within the principle laid down in the *Slough* case, and that the *Richardsons* case was distinguishable on the grounds, first, that no issue of abandonment had arisen in that case, and secondly, that there were important differences between that and the M present case both as to the facts and as to the statutory assumptions to be made in assessing compensation. On this appeal the main argument has been as to the applicability of the *Slough* principle to the agreed facts. The appellants claim that the permission of February 1 1966 ought not to be construed as a fresh outline planning permission, with the result that the facts do not disclose any inconsistency between rights conferred by the two outline permissions. The respondents contend that the permission of February 1 1966 should be construed as a fresh outline permission covering the whole of the estate save two acres in the centre of it to be reserved as a public open space. In support of this argument it is claimed that it was not in law open to the respondents to effect, by a mere approval of plans, an alteration of substance in the terms of a condition attached to an outline planning permission, and in this connection we were referred to the wording of paragraph 5 of the Town and Country Planning General Development Order 1963, and to the statement of Lord Denning (with whose judgment Megaw LJ agreed) in *Lever Finance Ltd v Westminster City Council* [1971] 1 QB 222 that "a planning permission covers works specified in detailed plans and any immaterial variation," which, it is argued, indicate that material variations can only be effected by the grant of a fresh outline planning permission.

We do not, however, find it necessary for the purpose of determining this appeal to decide whether the document of February 1 1966 is to be construed as a fresh outline planning application or only as an approval of detailed plans for stage 1, since in our judgment the outcome of the appeal must in either event be the same, for the following reasons. If on the one hand the document of February 1 1966 is to be construed as a fresh outline planning permission, there can in our judgment be no doubt that that document and the earlier outline planning permission of June 5 1964 gave rise to inconsistent rights with the result that when the appellants began to build, as they were entitled to do under the later but not under the earlier permission, they must be taken to have made an election and to have abandoned the rights under the earlier document. If on the other hand the document of February 1 1966 was not a fresh outline planning permission, we find the conclusion inescapable that it constituted an agreed variation of the rights conferred by the 1964 permission, with the consequence that those rights, to the extent that they were varied, disappeared at an earlier stage. It is clear, in our judgment, from the terms of the appellants' application of December 1965 that they were inviting the council to vary the 1964 permission by deleting the 10-year ban on the commencement of building, and clear also from the whole of the documentary material that in return for the very considerable advantage to be obtained from removal of that ban the appellants were prepared to allocate two acres at the centre of the site as a public open space, and if in law such a variation of the 1964 permission could be achieved otherwise than by way of a fresh outline planning permission we have no doubt that it was achieved by the document of February 1 1966. On this main issue, therefore, the appeal fails; and before leaving it, we would only add that in our judgment the Lands Tribunal was entirely right in the view that the claimants can derive no assistance from the *Richardsons* case, both because that case raised no issue of abandonment and also because the statutory assumption to be made in that case in the assessment of compensation was that planning permission would be granted for residential development, since the relevant development plan showed the premises as primarily designated for such development, whereas in the present case the only use designated for the estate on the town map was for mineral working, and it was common ground that the minerals had been worked out. We cannot therefore accept Mr Freeman's argument that the *Richardsons* case lends assistance to the appellants.

There remains a subsidiary point taken in the third of the grounds of appeal, with which we can deal very shortly. It is that if the appellants abandoned the outline permission of 1964 they did so only in consequence of a known scheme of the respondents to acquire the land for open-space purposes, and that in these circumstances the compensation should not have been diminished by reason of the scheme. In our judgment, for the reasons already given, it is clear that the main reason why the appellants abandoned the permission was that they wished to free themselves from the 10-year ban on the commencement of building, and we can find no substance in this ground. The appeal is, for the foregoing reasons, dismissed, and we do not find it necessary to refer to the additional argument raised in the respondents' notice.

The appeal was dismissed with costs. Leave to appeal to the House of Lords was refused.

COURT OF APPEAL

November 5 1974

(Before Lord DENNING MR, Lord Justice ROSKILL and Sir John PENNYCUICK)

LONDON TRANSPORT EXECUTIVE v PEPYS

Estates Gazette April 19 1975

(1974) 234 EG 195

Claim to compensation for injurious affection by construction of Victoria Line fails before Lands Tribunal, but costs awarded to unsuccessful claimant up to date of substantial sealed offer lodged by acquiring authority—Tribunal has general discretion as to such costs, but this must be judicially exercised—No ground shown for ordering authority to pay any part of claimant's costs in the circumstances—Award set aside

This was an appeal by the London Transport Executive from a decision of the Lands Tribunal on June 28 1973 awarding an unsuccessful claimant, Dr Elizabeth Olga Pepys, costs up to the date of a sealed offer lodged by the appellants. The tribunal's decision was reported at (1973) 227 EG 2068.

Mr K Bagnall QC and Mr J Gaunt (instructed by G S M Birch) appeared for the appellants. The respondent did not appear and was not represented.

Giving judgment, LORD DENNING said that from 1966 to 1968 a new underground line was being constructed as an extension of the Victoria Line. It went 70 ft below the ground in Islington, passing under Dr Pepys's property, 33 Gibson Square, which she bought on February 7 1967 for £14,000. The line was opened for public traffic on December 1 1968. London Transport Executive had acquired the necessary easement from the previous owner of 33 Gibson Square on terms which preserved a right to compensation for injurious affection provided a claim was made within two years of the opening, and in September 1970 Dr Pepys's solicitors put in a claim on the basis that the noise and vibration from trains running underneath had affected the value of the property. About that time, a proposed purchaser, a Mr Matthews, offered as much as £18,000 for the house, subject to contract. But after he had been into it he thought the extra noise was so bad that he declined to take it at that price. Dr Pepys eventually sold the house for £15,850. She said that this loss of £2,150 was due to the injurious affection to the house owing to the noise, and she claimed this amount before the Lands Tribunal. The LTE said that the house had not diminished in value at all, and they declined to pay any compensation. Their opinion was that the noise of the trains was no worse than that of passing lorries and the like. It was fair to say that, according to the evidence, very few people had claimed on account of the railway. Only 10 or 12 claims were received, and only three were still active at the time of the hearing; in other words, most people put up with the noise and made no claim at all. Just before the hearing, the LTE put in a sealed offer of £500, wanting to get rid of the claim. Dr Pepys did not accept that. She went on with the hearing, and lost. The reason seemed to be that she could not prove that the man who eventually bought the house put any less value on it because of the noise and vibration. In other words, there was not sufficient evidence to satisfy the tribunal that there had been any diminution in value because of the railway.

The actual decision was not challenged, and hardly could be, because it was a point of fact and not a point of law. The Lands Tribunal member, who was not a lawyer, said: "... no evidence was adduced that the eventual purchase at £16,000 [£15,850 plus £150 for fixtures and fittings] was a price discounted by reason of noise. . . . It is for the claimant to establish that the market value of the subject house has been depreciated by the running of the trains, and this she has failed to do." But he then said: "Having read the decision in this matter, and having then caused a sealed offer lodged by the acquiring authority to be opened, I find that the claimant had been offered unconditionally an amount of £500; accordingly the acquiring authority will pay the claimant her costs of this reference up to the date of the sealed offer." The LTE were upset. They had won the case, and yet they had been ordered to pay the costs up to the date of the offer having been put in. They wrote a letter to the Lands Tribunal asking whether this was a slip. The answer was that the matter was one in the discretion of the tribunal and the decision was not to be altered. The LTE asked for a case to be stated upon the point of law as to costs, and the question was now brought before the court. The first point was whether the tribunal had any discretion as to the costs incurred *before* the sealed offer. Under section 3 (5) of the Lands Tribunal Act 1949, the tribunal was given a general discretion as to costs, but there was a rule which took away that discretion in cases to which the provisions of section 4 (1), (2) and (3) of the Land Compensation Act 1961 applied. Mr Bagnall had submitted that in these circumstances the tribunal had a discretion as to the costs only *after* the date of the sealed offer, and none as to the costs before that date. Accordingly a claimant was not entitled to any costs before the sealed offer, even though he or she was awarded a sum of compensation. He (counsel) admitted that that was contrary to the practice of the tribunal, which always was that if the claimant was awarded a sum, but a sum which was less than the sealed offer, the claimant got the costs up to the date of the sealed offer and from that date the authority got them. He (his Lordship) thought that present practice well justified by the statutory provisions. The general discretion given in section 3 (5) of the 1949 Act applied to the costs before the sealed offer. If the claimant did recover a sum, even though less than the sealed offer, it was within the discretion of the tribunal to award the claimant the costs up to that date, because section 4 (1), (2) and (3) dealt only with the costs after the sealed offer and not with those before it.

The second point was the way in which the tribunal's discretion should be exercised in a case like the present where the claimant had failed altogether. In those circumstances it seemed to him (Lord Denning) that the practice of the courts in this country—and tribunals and arbitrators equally—was that if a claimant failed altogether no order was made whereby the successful party was to pay any part of his costs. Of course, the successful party might not be entitled to costs, but he was not, in general, to be ordered to pay those incurred by the claimant. The court had been referred to a number of cases starting with ***Wootton v Central Land Board*** [1957] 1 WLR 424 and ending with ***Knight v Clifton*** [1971] 1 Ch 700, and from these it appeared that the discretion of the court was usually exercised in this way: that if the plaintiff failed, he had to bear his own costs and often those of the defendant, but that while each side could be ordered to pay its own costs, it was exceedingly rare for a successful defendant to be required to pay the costs of a plaintiff who had failed. This should never be done except for very special reasons, and then the tribunal or arbitrator or whoever it might be that was dealing with the case ought to state those reasons. As Lord Goddard CJ said in *Lewis v Haverfordwest Rural District Council* [1953] 1 WLR 1486 at 1487, "Those words 'judicially exercised' are always somewhat difficult to apply, but they mean that the arbitrator must not act capriciously and must, if he is going to exercise his discretion, show a reason connected with the case and one which the court can see is a proper reason." It was true that in the case of *Hood Investment Co Ltd v Marlow Urban District Council* (1963) 15 P & CR 229 Harman LJ said: "The President, being appealed to for reasons, wisely gave none, and I think he was entitled to keep his counsel on that matter." Although he (his Lordship) was himself sitting in that court, he thought that was erroneous in the sense that if the ordinary exercise of discretion was to be departed from, then in the normal way the reasons for that departure should be stated. There was open an appeal to the Court of Appeal. It might be different in the High Court, from which there was no appeal as to costs except with leave of the judge. But in a case such as this, from the Lands Tribunal, it seemed that if the ordinary exercise of discretion as to costs was to be departed from, then the persons so departing should give their reasons so that the parties, if there was an appeal, should know the reasons which influenced them. In the present case no sufficient reason had been vouchsafed in the letter from the tribunal or in the case stated. No reasons were adduced, and it seemed that none existed, whereby the LTE should have been ordered to pay Dr Pepys's costs. Her claim had failed altogether. The right course was that the order in that respect should not stand. The only part of the order which should stand was that from the date of the sealed offer being communicated Dr Pepys should pay the costs; though he (Lord Denning) was glad to have heard from counsel that in the special circumstances of the case the LTE did not propose to enforce any order for costs against Dr Pepys.

ROSKILL LJ delivered a concurring judgment, and SIR JOHN PENNYCUICK agreed with both the judgments delivered.

QUEEN'S BENCH DIVISION

October 30 1974

(Before Mr Justice ACKNER)

GOSLING v SECRETARY OF STATE FOR THE ENVIRONMENT AND ANOTHER

Estates Gazette May 17 1975

(1974) 234 EG 531

Compulsory purchase of clearance area plus added lands—Confusion in both inspector's report and Secretary of State's decision letter as to justification for incorporating added lands in order—Extrinsic evidence not admissible to construe decision letter—Letter cannot be amended under "slip rule"—Objector entitled to have order quashed in so far as it affected his property

This was a motion by Mr Eric George Gosling, a solicitor practising at 248 Walworth Road, London SE17, for an order quashing the London Borough of Southwark (Goodwin Buildings) No 1 Compulsory Purchase Order 1971, made by the second respondents, the London Borough of Southwark, in 1971 and confirmed by the first respondent, the Secretary of State for the Environment, on August 16 1972. The plaintiff sought to quash the order only in so far as it affected Nos 133 and 135 Southwark Bridge Road, London SE1.

Mr G Dobry QC and Mr A D Dinkin (instructed by Lewis, Barnes, Wheeler & Co) appeared for the applicant, and Mr H K Woolf (instructed by the Treasury Solicitor) represented the first respondent. The second respondents took no part in the appeal.

Giving judgment, ACKNER J said: Mr Gosling is a solicitor of the Supreme Court and practises under the name of Lewis, Barnes, Wheeler & Co and Gosling & Co at 248 Walworth Road. He is the owner of two freehold properties, 133 and

A 135 Southwark Bridge Road, SE1, which with other parcels of land are the subject of a compulsory purchase order made by the Southwark Corporation on August 3 1971 and confirmed by the Secretary of State for the Environment on August 16 1972. Mr Gosling wishes to modernise 133 and 135 Southwark Bridge Road for use as offices, and subject to acceptable terms, he is prepared to include 131 and 137 in a modernisation scheme. He moves pursuant to the fourth schedule to the Housing Act 1957 to quash the order. His notice of motion was amended at the outset of this hearing and reads in its amended form as follows: "The grounds of the application are: (1) The compulsory purchase order was B not within the powers of the Housing Act 1957 in that the first respondent"—that is the Secretary of State—"failed to take relevant considerations into account, or alternatively based his decision on legally irrelevant considerations. (2) There was no evidence before the first respondent on which he could decide as he did. (3) The first respondent's decision is not within the powers of the said Act in that he applied the wrong legal test and dealt with the wrong questions, that is to say whether there would be achieved a cleared area of a convenient shape." Mr Gosling naturally relies upon the case of *Ashbridge Investments Ltd v Minister of Housing and Local Government* [1965] 1 WLR 1320, and in particular on C the judgment of Lord Denning MR, where he stated:

> The court can interfere with the Minister's decision if he has acted on no evidence; or if he has come to a conclusion to which, on the evidence, he could not reasonably come; or if he has given a wrong interpretation to the words of the statute; or if he has taken into consideration matters which he ought not to have taken into account, or vice versa; or has otherwise gone wrong in law. It is identical with the position when the court has power to interfere with the decision of a lower tribunal which has erred in point of law.

In relation to the facts, the position is as follows. The order related to the clearance of an area of about 0.3 of an D acre, including half the adjoining roads, containing a five-storey tenement block known as Goodwin Buildings comprising 55 dwellings. Included also in the order were "added lands" having a gross area of about 0.2 of an acre, comprising a three-storey block of four properties, that is 131 to 137 Southwark Bridge Road, containing two shops and living accommodation over, a vacant dwelling together with a single-storey warehouse and an office, and half the adjoining public streets. The land was required as a public open space. The material subsection on which all attention has been focused in the course of this case is the well-known section 43 (2) of the Housing Act 1957:

> E Where the local authority determine to purchase any land comprised in the area declared by them to be a clearance area, they may purchase also any land which is surrounded by the clearance area and the acquisition of which is reasonably necessary for the purpose of securing a cleared area of convenient shape and dimensions, and any adjoining land the acquisition of which is reasonably necessary for the satisfactory development or use of the cleared area.

Thus the added land can fall into one of two categories. It can be land which is surrounded by the clearance area, in which case the acquisition must be reasonably necessary for the purpose of securing a cleared area of convenient F shape and dimension, or secondly it can be adjoining land, in which case its acquisition must be reasonably necessary for the satisfactory development or use of the cleared area. It is incontestable that the added land in this case was not surrounded by the clearance area, as is clearly shown by two maps, maps B and K. The land is a triangular-shaped island site, and the added land forms a corner of the triangle. The added land is therefore adjoining land in the second of the two categories, the second limb of the subsection, and the particular test which is appropriate is to be found in the words "the acquisition of which is reasonably necessary for the satisfactory development or use of the cleared area."

I now turn to the relevant parts of the decision letter of G the Minister, which is dated August 16 1972 and is signed on his behalf by Mr K Stott, senior executive officer. The material paragraphs are paragraphs 3 and 4. Paragraph 3 reads as follows:

> The only objection to the order was on the grounds that the acquisition of some of the added lands was neither essential nor reasonably necessary for the satisfactory development or use of the cleared area, and that their exclusion would not prevent the use of the cleared area either for open space purposes or for some other form of development. Four other representations were made at the inquiry supporting the council's plan to provide the public open space. H

That paragraph is in no way attacked, and there is no inaccuracy or ambiguity which is to be found in it. It is the next paragraph, paragraph 4, on which most attention has been focused. That reads as follows:

> The inspector found that all the property in the clearance area was correctly represented as unfit and concluded that the most satisfactory method of dealing with the conditions in the clearance area was the demolition of all the buildings therein.

So far, no criticism.

> He also concluded that the acquisition of the added lands was reasonably necessary for the purpose of securing a cleared area of convenient shape and dimensions, or for the satisfactory J development or use of the cleared area. The inspector accordingly recommended that the order be confirmed without modification.

If the inspector found, as the Secretary of State stated, that the acquisition of the added land was reasonably necessary for the purpose of securing a cleared area of convenient shape and dimensions, and that this was in itself the justification, he would have been applying the wrong limb of the subsection, because, as I have previously pointed out, that limb only applies if the added land is surrounded by the clearance area, and, as I have previously stated, this land was not. On the other hand, as I ventured to suggest in the unreported case of K *Lockett v Secretary of State for the Environment,* decided in May 1973, the terms of section 43 (2) are very wide. Accordingly the shape and dimensions of the cleared land could well in a given case be so inimical to its proposed development that the acquisition of some adjoining land could properly be said to be reasonably necessary for the satisfactory development of the cleared area; that is the second limb of the subsection. An example was provided in the course of argument where the planning authority sought compulsorily to acquire land for use as a football pitch and the cleared land was just insufficient to provide the area required for such an activity. L The adjoining land was therefore necessary in order to add to the dimensions and/or improve its shape, so that its acquisition could properly be said to be reasonably necessary for the satisfactory development or use of the cleared site.

On the face of the Secretary of State's letter, the inspector has posed to himself an irrelevant question, namely whether the added lands are reasonably necessary for the purposes of securing a cleared area of convenient shape and dimensions. The question upon which he should have concentrated was whether the acquisition of the added land, being adjoining land, was reasonably necessary for the satisfactory development or use of the cleared area. M The mere *ipse dixit* that it made the shape of the cleared land more convenient would not of itself provide the reasonable necessity which had to be established (see *Coleen Properties Ltd v Minister of Housing and Local Government* [1971] 1 WLR 433). The Secretary of State's letter therefore discloses this problem. Paragraph 5 reads as follows: "The arguments put forward by the objector and the council have been considered. The inspector's findings, conclusions and recommendation have been accepted, and the Secretary of State has therefore decided to confirm the order without modification. The endorsed order and map are being returned under separate cover for deposit

A in the council's offices." Thus the Secretary of State shared —if the inspector was in fact muddled—the same muddled appreciation of the operation of the subsection, and thus erred in point of law. The fact that the conclusion was expressed in the alternative, the one limb relevant and the other irrelevant, does not to my mind save the position. It seems to me to be a proper reading of paragraph 4 that the inspector concluded (which conclusions were adopted in paragraph 5 by the Secretary of State) that the mere fact that the added lands provided a convenient shape and dimensions was in itself the answer to the test posed by both limbs of the subsection.

B Mr Woolf, to surmount this difficult position, has first invited my attention to the inspector's report in order to establish that the inspector's conclusions have not been properly recorded in the Secretary of State's letter, and that the inspector in fact properly discharged his functions under the Act. He first drew attention to paragraph 4 of the inspector's report dated April 11 1972, which reads as follows:

> The council has declared that the dwellings within the clearance area are unfit for human habitation and that the best method of dealing with the conditions is the demolition of the buildings within the clearance area. It has also declared that the acquisition of the C added lands is reasonably necessary for the satisfactory development or use of the cleared area.

There, says Mr Woolf, and in my judgment rightly, is a clear indication that at least the council knew that it was the second limb of the subsection which was relevant. Counsel then points to the next paragraph, which reads as follows:

> One objection was received in respect of one property, 133 Southwark Bridge Road, reference number 3, and one late objection was lodged by the same objector in respect of the adjoining property, 135 Southwark Bridge Road, reference number 4. The objections were generally on the grounds that it was not reasonably necessary to acquire numbers 131 to 137 Southwark Bridge D Road to enable the satisfactory development of the cleared area to be carried out.

There, says Mr Woolf, and again rightly, in my judgment, is a clear indication that the objectors knew that it was the second limb of the subsection which was appropriate. Counsel then refers to paragraph 20, which deals with the general case of the council in support of the compulsory purchase order, and reads as follows:

> The acquisition of the adjoining properties, numbers 131 to 137 Southwark Bridge Road and 1 and 2 Quilp Street, is considered reasonably necessary for the satisfactory use or development of the cleared areas.

E Here, says Mr Woolf, is further confirmation that the council had got it right. Then the next paragraph, if more confirmation is required, provides that support:

> There is a serious deficiency of open space in this part of the borough, and to alleviate it it is intended to add the order lands and other land in the vicinity to the adjoining area already defined for public open space purposes in the initial development plan. The order lands form an island site between two busy roads, and it is considered essential to acquire all the properties within the order, as it would not be practical to develop or lay out the area as part of the open space without including the added lands.

F Then the report deals with the objections. There were not objections emanating from the owners of all the added lands. In paragraph 24 and subsequent paragraphs the case for the council in answer to the specific objections is set out. It reads as follows:

> No layout of the proposed park has been prepared, as that would be carried out by a consultant landscape architect, but it would be difficult to design a satisfactory layout without including the sites of numbers 131 to 137 Southwark Bridge Road. It would be difficult to develop the site of Goodwin Buildings for open space purposes if numbers 131 to 137 Southwark Bridge Road were to be retained, as there would be problems in providing rear access to those premises without retaining Quilp Street.

This proposition, namely the difficulty as there described, G does not feature in the many facts found by the inspector. Paragraph 25 makes a further point in these terms:

> Although the frontage of the local park to the major roads would not be significantly reduced by the retention of numbers 131 to 137 Southwark Bridge Road, the view of the backs of the premises from the park would not be very presentable.

This criticism, another justification for the inclusion of the adjoining land, is again not reflected by the inspector in his findings. Then comes paragraph 26, which deals for the first time basically with shape, and there the council states that:

> The exclusion of numbers 131 to 137 Southwark Bridge Road H would result in an open space of an awkward shape even after the rest of the open space land had been acquired, and to have those properties intruding into it would be undesirable.

The council are not using the words in the first limb of the subsection, which are "convenient shape." They are not saying the shape would be inconvenient or the dimensions would be inconvenient. They are saying it would be awkward, and it is quite clear that they are directing their minds to the second limb of the subsection. Paragraph 28 of the report sets out the inspector's opinion on Mr Gosling's objections. I think I ought to set out the whole paragraph, although the first sentence is perhaps a preface: J

> These added lands, together with reference number 6, adjoin the southwestern side of the clearance area, all of which together form a triangular island site bounded on two sides by principal traffic routes and on the other side by Quilp Street. While the fabric of the buildings would appear to be capable of repair and the interiors restored, it seems to me that the acquisition of reference numbers 2 to 5 is reasonably necessary under the terms of section 43 of the Housing Act 1957. It would not be possible in my view to carry out the development of the cleared area without including the added lands, as they enable a cleared area of a convenient shape to be achieved.

What is particularly to be observed about this paragraph K is that before giving, so to speak, his justification, the inspector refers in terms to the relevant section, and he then goes on to give his justification for saying why the acquisition is reasonably necessary having regard to that section. Mr Woolf accepts that the inspector pitches the matter higher than it was pitched by the council by saying that it would not be "possible" to carry out the development, when of course the council were conceding, as one would have thought was obvious, that it could be carried out but that difficulties would be encountered. Having pitched it higher, one then finds that he provides this as the answer to why it is not possible to carry out the development—the cleared area without includ- L ing the added lands would not enable a cleared area of convenient shape; and that is an unexpected and surprising answer, because it does not in fact explain the alleged impossibility. It asserts merely inconvenience, and it uses as the basis for the justification the very words to be found in the first limb, "convenient shape." It seems to my mind clear that the inspector, in not adopting the language of the council, which was "awkward shape," and in not referring to the various difficulties which they raised, was satisfied that all he had to do was to find that the cleared area without the added land did not provide a convenient shape to justify his saying that it was reasonably necessary to acquire the added M lands. In other words, he was applying his mind to the first limb and answering the question by saying that it was reasonably necessary to acquire the added lands in order to secure a cleared area of convenient shape. Mr Woolf invited my attention to paragraph 70, which is the conclusion of the inspector in regard to the compulsory purchase order in its generality, and is not designed to be a decision related to the specific objections of Mr Gosling, as was paragraph 38. Paragraph 70 reads:

> I consider that the acquisition of the added lands is reasonably necessary under the terms of section 43(2) of the Housing Act 1957

in order to give a cleared area of reasonable shape for the satisfactory use or development of the cleared area.

I do not consider I can use that paragraph in order to construe paragraph 38, which is clear in itself. I accept, of course, that I should not approach paragraph 38 as if it were a statutory instrument or some conveyancing document that has to be seen through the expert eyes of conveyancing counsel. However, paragraph 70 is, as Mr Woolf accepts, expressed in language which involves a different emphasis, and it is not dealing with the specific point, namely the objections of Mr Gosling in relation to the adjoining land which is the subject-matter of this application.

If I am wrong in the view which I have just expressed, then I must proceed to the next stage, which presupposes that the inspector did properly carry out his functions and did in fact reach the conclusion that the acquisition of the added land was reasonably necessary applying the second limb test, namely that it was reasonably necessary for the satisfactory development of the cleared site. I turn then to consider the Secretary of State's letter, because it is his letter which is being attacked, and it is his letter which clearly records the muddled and misapprehended position attributed to the inspector. Mr Woolf submitted that he was entitled to provide evidence that paragraph 4 of the Secretary of State's letter did not properly state what it had been intended to say. I must make it perfectly clear that Mr Woolf did not seek to suggest that there was any intrinsic material within the letter to suggest that it did not mean what it stated. For this submission he was proceeding on the basis that the letter itself was clear, but that there was extrinsic material to establish that the Secretary of State had by mistake wrongly expressed himself. Mr Dobry, on behalf of Mr Gosling, contended that I had no warrant to allow any extrinsic evidence, that the Minister's decision was to be found in his letter, and that although I was of course entitled (and indeed obliged) to construe his letter, I was not entitled to go outside it, least of all to admit evidence from his department to say that he did not mean to say what he did in fact say. No authority directly on the point has been found, but Mr Woolf in his most careful and helpful submissions took me to the *Annual Practice,* order 20, rule 11, which deals with the amendment of judgments and orders and provides for clerical mistakes in judgments or orders to be corrected by the court. Counsel also referred to order 20, rule 8, which reads as follows:

> For the purposes of determining the real question in controversy between the parties to any proceedings, or of correcting any defect or error in any proceedings, the court may at any stage of the proceedings and either of its own motion or on the application of any party to the proceedings order any document in the proceedings to be amended on such terms as to costs or otherwise as may be just and in such manner (if any) as it may direct.

It seems clear from the note to that order and rule to be found at page 340 of the current *Annual Practice* that the words "document in the proceedings" have as one would expect, a very restricted meaning. It does not and cannot refer to every document which features in the proceedings. It would not, for example, entitle the court, unless permitted under some other rule of the court or some other rule of law, to amend what purported to be a contractual document or to change the date of a notice to quit on which, by some typing error, the wrong day of the week had been inserted. In my judgment the rules referred to do not touch this particular problem and provide me with no authority to amend the Minister's letter. I did however admit *de bene esse* an affidavit by Mr Stott, the writer of the letter, in case on this point too I should be wrong, so as to enable the matter to be appropriately rectified in the Court of Appeal should this case go further. Mr Stott is a senior executive officer in the Department of the Environment, and it was he who, as I have already stated, signed the letter. The letter was not prepared in fact by him, but by someone junior to him. The procedure apparently is that a standard form of letter is used incorporating certain standard paragraphs, and one of the standard paragraphs here was paragraph 4, which reads in these terms:

> The inspector found that [all] the properties in the clearance area were correctly represented as unfit houses [except . . .]. He concluded that the most satisfactory method of dealing with the conditions in the clearance area [as modified] was the demolition of all buildings therein. The inspector also concluded, having regard to the multiplicity of ownership, that the council were justified in acquiring the unfit houses to secure satisfactory clearance for redevelopment, and that the acquisition of the added lands including reference numbers . . . was reasonably necessary for the purpose of securing a cleared area of convenient shape and dimensions or for the satisfactory development or use of the cleared area. The inspector accordingly recommended that the order be confirmed subject to the modification(s) that . . .

Significantly, there is no asterisk or side note in this paragraph which says "delete as appropriate one of the two alternatives," namely "necessary for the purpose of securing a cleared area of convenient shape and dimensions" and "for the satisfactory development or use of the cleared area." Mr Stott said that the reason for the inclusion of the words "for the purpose of securing a cleared area of convenient shape and dimensions," which is the first limb of the subsection, the irrelevant limb, was that there was "an administrative error," a phrase which I find somewhat lacking in precision. I do not really know whether that means that the administrative error consisted in failing to draw attention in the precedent itself to the fact that one of the alternatives should come out, or that it was the person who prepared the ultimate letter and/or Mr Stott who themselves overlooked striking out the alternative, knowing that one of the alternatives should come out. In my judgment it matters not, because, as I have said, I have no power to amend.

The next point taken by Mr Woolf was that the letter was ambiguous, because paragraph 4 expressed the inspector's conclusion, which was adopted by the Secretary of State in paragraph 5 in the alternative. It was thus not clear which set of circumstances had been taken into account, and counsel submitted that as the second limb was at least referred to, I ought to treat the decision as being correct on the basis that the inspector plus the Minister, or at least the Minister, had the right test in mind. I have already expressed my view earlier in this judgment that it seems to me to be quite clear that the "convenient shape" consequence of the purchase of the added land was treated as being the qualifying matter both in limb 1 and limb 2. Finally, Mr Woolf drew my attention to the words of the fourth schedule to the Housing Act 1957, which are basically the words used in section 245 of the Town and Country Planning Act 1971. The words upon which he specifically placed reliance were these, that if any person is aggrieved by an order on the ground that it is not within the powers of the Act, then he may make an application to the High Court, and where any such application is duly made the court—and I now quote—"may (i) by interim order suspend the operation of the order, or the approval of the plan," etc, or "(ii) if satisfied upon the hearing of the application that the order, or the approval of the plan, is not within the powers of this Act," etc, "quash the order." Counsel said that those are words which on their face show that it is permissive and not mandatory for me to quash the order. I have had the advantage of reading—in fact I was responsible for asking to see—the judgment of Kerr J given on March 7 1974 in the case of *Miller and Others v Weymouth and Melcombe Regis Corporation* [(1974) 27 P & CR 468], and in a characteristically full and helpful judgment the learned judge set out in detail his reasons for reaching the conclusion, with which I respectfully agree, that the word "may" in section 245 (4) of the 1971 Act is permissive and provides the court with a discretion. In the course of giving his judgment, he quoted the well-known

A passage from the judgment of Greene MR in the leading case of *R v Stafford Justices* [1940] 2 KB 33 at 43, which reads:

> In my opinion, the order for the issue of the writ of certiorari is, except in cases where it goes as of course, strictly in all cases a matter of discretion. It is perfectly true to say that if no special circumstances exist, and if all that appears is a clear excess of jurisdiction, then a person aggrieved by that is entitled *ex debito justitiae* to his order.

Mr Woolf says that if he is wrong in his submission that there is jurisdiction in this court to hear extrinsic evidence and on the basis of that evidence to amend the letter of the
B Secretary of State (a submission which perhaps I should add in parentheses is hardly in accord with the concession of his predecessor Mr Slynn in this very case that there was nothing in the nature of the slip rule which would enable the Secretary of State to correct what was really no more than a clerical error), then I should admit the very evidence which I have excluded for the purpose of considering whether I should exercise my discretion and not quash the order. There must be a limit to the forensic gymnastics which are to be imposed upon Her Majesty's judges, and this seems to me to go beyond the limit. If I do, as I have rightly or wrongly so decided in this case, exclude extrinsic evidence on
C the basis that I cannot look at that material in order to amend the Minister's letter, that seems to me to be the end of the matter; that material is not admissible, and the Minister's decision stands as recorded in that letter. I then have to consider in relation to the exercise of my discretion whether there is disclosed, following the *Stafford Justices* case, an excess of jurisdiction, in which case in the absence of special circumstances the applicant, the person aggrieved, would be entitled *ex debito justitiae* to the order being quashed. This is a case where there has been an excess of jurisdiction or a want of jurisdiction. Are there any special circumstances to prevent the usual consequences? To say that special circum-
D stances exist by reason of the very material which I have held to be inadmissible would be the very negation of my decision that extrinsic evidence cannot be looked at to establish the Secretary of State's mistake.

I think that, although perhaps with undue length, I have covered the various points which have been raised, and the consequence which flows from my judgment is that the order in so far as it relates to Mr Gosling's property, the two houses 133 and 135, should be quashed; but I will entertain submission and argument if that is not, as it appears to me to be, what should follow.

E An order was made in the terms contemplated by his Lordship. Mr Gosling was awarded costs against the Secretary of State.

COURT OF APPEAL

March 6 1975

(Before Lord DENNING MR, Lord Justice ORR and Lord Justice SCARMAN)

HUNTER v MANCHESTER CITY COUNCIL

Estates Gazette June 21 1975

F **(1975) 234 EG 903**

Compensation—Unfit property—Owner-occupier payment—Owner who occupied part of house and let remainder entitled to full compensation on the concessionary basis

This was an appeal by Mr Eric Melbourne Hunter, of Lindum Avenue, Old Trafford, Manchester, from a decision of the Lands Tribunal (Mr R C Walmsley FRICS) on July 23 1973 holding that he was entitled to be paid only £464 as compensation on the compulsory acquisition by the respondents, Manchester Corporation, of his dwelling-house in Stockton Street, Moss Side, Manchester.

G Mr H S Singer (instructed by Adam Burn & Metson, agents for Bieber, Travers & Libman, of Manchester) appeared for the appellant, and Mr N R B Macleod (instructed by L Boardman, of Manchester) represented the respondents.

Giving judgment, Lord Denning said: In 1954 Mr Hunter came here from Jamaica. In 1957 he bought a house in a slum area of Manchester, 98 Stockton Street, Moss Side, Manchester. It was a terraced property with two rooms plus bathroom upstairs. Twelve years later, in 1969, the Manchester Corporation made a survey of the area. They took the view that it should be a clearance area, because the houses were unfit for human habitation, and the most satisfactory method of dealing
H with them was by pulling them all down. Mr Hunter heard of the proposal, so he made arrangements to buy another house. He bought 4 Lindum Avenue, Old Trafford, Manchester. His wife and seven children moved into this property on July 10 1969, but he himself did not go with them at that stage, because he was advised by an estate agent that he had better remain in part occupation of 98 Stockton Street. So he let the downstairs of 98 Stockton Street to a Mrs Grenyion, and he himself continued to occupy one of the upstairs rooms, "where he slept, and sometimes his wife visited him there." That continued until July 1971. Meanwhile, on December 3 1969, the corporation had declared the area to be a clearance area. It took some time to get
J the proceedings through, but on May 10 1971 a compulsory purchase order was made, and on March 19 1972 this was confirmed by the Secretary of State. No 98 Stockton Street was included as a house to be pulled down. A month or two later, notice to treat and notice of entry were served, and the acquisition became effective in June of 1972. The question that arises is the amount of compensation to which Mr Hunter is entitled for the compulsory acquisition of his house.

If he had never been in occupation of the house himself, but had always let it out to others, he would only get the site value. That is clear from section 59 (2) of the Housing Act 1957. The
K site value of this house was only £130. But he had been an owner-occupier for many years, and there are provisions in the Housing Acts in favour of owner-occupiers. They are contained in the fifth schedule to the Housing Act 1969. I will not read the details, because it is quite plain that Mr Hunter was a person who qualified for a higher payment. He satisfied paragraph 5 (1) (e) of the schedule, because he had been in occupation throughout the qualifying period, which in this case was April 23 1968 to December 3 1969, when the house was declared to be in a clearance area. He satisfied paragraph 1 (1) (b), because during that period the house was wholly or partly occupied as a private dwelling by Mr Hunter, who was a person entitled to an interest
L in the house. It was wholly or partly occupied by him throughout that period. He occupied it wholly from April 23 1968 until July 1969, and partly from July 1969 to December 3 1969. So Mr Hunter was a qualified person who had an interest in respect of which he was entitled to payment. The amount of the payment is given by paragraph 2. It is an amount equal to the full compulsory purchase value less the site value. The full compulsory purchase value was £1,800, less the site value, which would be £130. So the full payment under paragraph 2 would be £1,670. Now we come to the crucial point in the case. Paragraph 3 (1) says:

M
> The amount which would otherwise be payable under paragraph 1 of this schedule shall be reduced by such part, if any, of that amount as may reasonably be attributed to any part of the house occupied for any purposes other than those of a private dwelling at the date of the making of the compulsory purchase order. . . .

The question is whether this means that the £1,670 is to be reduced. It is to be reduced if "any part of the house [is] occupied for any purposes other than those of a private dwelling at the date of the making of the compulsory purchase order." That date is May 1971. At that date no 98 was occupied in part by Mr Hunter in the upstairs room in which he slept, while the other part, the downstairs room, was occupied by Mrs Grenyion. Both those purposes are those of a private dwelling. In *G E Stevens (High Wycombe) Ltd v High Wycombe Corporation* [1961] 3 WLR 228 Diplock J said at p 233:

> I think occupation of a "private dwelling" means occupation as the dwelling-house of a private person, or occupation by a person in his private capacity as opposed to being occupied for business purposes.

It seems to me that when a man uses part of his house for his own dwelling purposes and lets the other part to a subtenant as a dwelling, the whole house is used as "a private dwelling". So there is no ground for reduction. If part of the house had been used for business purposes, of course, there would be grounds for a reduction. What is argued for the Manchester Corporation is that we should look at the overriding intent. The overriding intent, it is said, was that an owner-occupier should only be compensated in respect of his own personal occupation of the house. He should not be compensated for parts which he lets off to tenants. That is an attractive suggestion, but I am afraid that it does not fit in with the words. Nor is there any such overriding intent. We were referred to Ministry of Housing and Local Government Circular 68/69, but the intent is well put in a comment in the *Encyclopedia of Housing* vol 1 p 2641:

> The object of this [provision] is to mitigate hardship to those persons who have been driven by the extreme shortage of housing to buy unsound and substandard dwellings to live in themselves.

These words apply to the present case. Mr Hunter in 1957 bought this old house in Moss Side in Manchester, no doubt at its then value, because he had nowhere else to live. Twelve years later, when it was going to be pulled down, he naturally looked for another house. In order to buy another house he ought to be able to provide some of the money out of the compensation he gets for his interest in the old house which was going to be pulled down. Although he learned early in 1969 that the house was in an area that was proposed to be a clearance area, he had to remain partly in occupation until the clearance area was declared in December 1969 fully to qualify. But once he did so, it is quite right for him to get compensation for his interest in the old premises even though he let part of it meanwhile to help other people in their housing needs. In my opinion, seeing that this house at the date of the compulsory purchase order in May 1971 was still being used as a private dwelling, there is no ground for reduction. I would only add this, that Mr Hunter appeared in person before the Lands Tribunal, and we have been better placed, because he was represented by counsel before us. On the argument before us, I think the Lands Tribunal did not construe the provisions rightly. The compensation should be determined at the full amount of £1,670.

Orr LJ: I agree; and in adding nothing to what my Lord has said, I intend no disrespect for the Lands Tribunal or for the very able argument we have had from counsel on both sides in the case.

Scarman LJ: I also agree. The respondents submit that the effect of the fifth schedule to the Housing Act 1969 is that an owner-occupier of a house to be demolished can receive compensation over and above the site value only to the extent of his own occupation at the date of the compulsory purchase order. The wording of paragraph 3 (1) is, I think, a great difficulty in the way of this submission, which is really based upon a view of the overriding intent of the provisions of the Housing Act, a view which upon analysis is, I believe, unsound. Mr Macleod, who argued the case very attractively for the respondents, drew our attention to section 68 of the Housing Act 1969, under which it is clear that payments are to be made under the fifth schedule in respect of owner-occupied housing in certain circumstances, and he invited us to infer from that provision and others that the policy was that full compensation should be made available only if the whole of the house was owner-occupied by the person entitled to compensation at the date of the compulsory purchase order. I think that that inference is negatived both by the provisions of the Housing Act 1957, the predecessor of the Housing Act 1969, and by the provisions of the fifth schedule to the 1969 Act. It is clear from the qualifying provisions of each Act that compensation is payable in respect of houses that are owner-occupied as to part only. Then one sees in the schedule to the 1969 Act—the schedule, of course, with which we are concerned—that a person who so occupies a house, that is to say, wholly or partly himself, and has an interest in that house, is entitled under paragraph 2 to the full compulsory purchase value, unless his compensation has to be reduced under paragraph 3. Now, the qualifying period for that entitlement ends with what the statement calls " the relevant date ". The relevant date is very significant: it is the date on which the house was declared as being within a clearance area (in this case, December 3 1969). From that moment on it appears to be reasonable that an owner-occupier—almost certainly a family man—should be looking around for accommodation and should have a reasonable expectation that part of the capital needed for his alternative accommodation will come from the compensation for the loss of the house which sooner or later he is going to be compelled by law to vacate.

Paragraph 3, upon which the respondent corporation relies, and which does provide in certain subsequent events for a reduction of the compensation available to an owner-occupier, appears to me to do no more than ensure that some one who has qualified for compensation is not going to be permitted to make a profit out of his situation. If the house remains in use as a private dwelling, of course he is not going to make any great amount of money if he lets it: he will be restricted by the Rent Acts. If he takes lodgers, he may lose all or part of his compensation on the ground that the house, or part of it, is being put to a business use. If he does put the house after the relevant date to a business use, then the effect of the paragraph is that, to the extent of that use, his compensation is to be diminished: but if he goes on allowing the house (and it is in the public interest that he should) to be occupied as a private dwelling until such time as it has to come down, then the paragraph contains nothing to deprive him of compensation upon the expectation of which he may well have made utterly admirable and reasonable family arrangements. I think therefore, upon a proper analysis, the policy of the Act coincides with the actual wording of paragraph 3 (1) of the schedule, and I, like my Lord, would allow the appeal.

The appeal was allowed, and the decision of the Lands Tribunal varied to provide for compensation of £1,670. Leave to appeal to the House of Lords was refused.

ESTATE AGENTS

A **CHANCERY DIVISION**

November 8 1974

(Before Mr Justice GRAHAM)

CALVERT, HUNT & BARDEN v ELTON

Estates Gazette January 25 1975

(1974) 233 EG 391

Estate agents' branch manager subject to three-year, three-mile covenant in restraint of trade—Validity of covenant upheld—Relevance in this regard (a) of manager's record in his previous employment, (b) of his attitude to the rules of B conduct

This was a claim by Calvert, Hunt & Barden, estate agents, of 57 High Street, Chislehurst, Kent, against a former employee, Mr Eric George Elton, of Pickhurst Mead, Hayes, Kent, for an injunction restraining breach by him of a covenant limiting his business activities for a period of three years after his employment ceased, and for damages.

Mr B J F Galpin (instructed by Zeffertt, Heard & Morley Lawson) appeared for the plaintiffs, and Mr I H Maxwell (instructed by H Montlake & Co) represented the defendant.

Giving judgment, GRAHAM J said: This is an action by a firm of estate agents, Calvert, Hunt & Barden, to enforce a C covenant in restraint of trade against the defendant, Mr Elton, who was employed by them from about January 1972 until about July 1973. One branch of the argument before me raises specifically a point on which there appears to be a dearth of authority, namely, whether in considering the validity of such a covenant it is permissible to take into account circumstances which point to the fact that the employee, prior to the date of the covenant, has been guilty of improper professional activities or other conduct making his trustworthiness in some respects questionable. If it is proper to take such matters into account at all, is consideration of them relevant, not only to the question whether the taking of a covenant in that particular business is permissible or not, but also to the question of the reasonableness D of the extent of such covenant? The plaintiffs argue that it is proper to take such matters into account in relation to both questions, and the defendant that it is not, at any rate in respect of the second question.

The facts are that the plaintiffs are estate agents practising in the London Borough of Bromley and have offices with overlapping areas of activity in Chislehurst and Bickley. This case primarily concerns the Bickley office, which the plaintiffs decided to reactivate in the middle of 1972 after the departure of its former manager at some earlier date. They advertised for a new manager, and after interview, the defendant was taken on in that capacity, with no fixed time-limit to his employment, at a salary E of £1,000 per annum and commission at a flat 10 per cent. As a result of learning, shortly after the start of his employment, certain matters which will be mentioned hereafter, the plaintiffs insisted that the defendant should enter into a service agreement containing a restrictive covenant which would become operative if and when he left the plaintiffs. At this point it is useful to note certain clauses in the agreement, which was not in fact entered into until August 25 1972. Clause 9 contains the covenant, and reads as follows:

"9. AFTER the termination of the Employee's employment hereunder howsoever occasioned the Employee shall not for the period of three years after he shall cease to be employed directly F or indirectly undertake carry on or be concerned engaged or interested either alone or in partnership with or as manager agent or servant of any other person firm or corporation or otherwise in the business of auctioneer valuer surveyor estate agent property manager insurance agent or mortgage broker within a radius of three miles of any branch of Calvert, Hunt & Barden at which he has been employed as Manager and shall not during the said period on behalf of himself or any other person firm or corporation canvass or solicit orders from any person firm or corporation who shall at any time during the continuance of his employment hereunder have been a customer of the Employer nor use any professional information which he may have obtained in the course of and incidental to his employment hereunder to the detriment or attempted detriment of the Employer."

Clause 8 gives a power to dismiss the defendant, if guilty of G conduct causing damage or bringing discredit to the branch office. Clause 2 should also be noted, since it demonstrates that the original salary and commission figures were substantially increased over those at which Mr Elton was originally engaged. In particular the commission was on a rising percentage. The evidence showed that on being asked to enter into a service agreement, Mr Elton had at first demurred but then had agreed, provided there was some advantage given when he did so. The increased figures were the result. The evidence also showed that these figures, actually agreed, were generous for the job in question.

The circumstances giving rise to the matters referred to above H arose as follows. The defendant, who was a member of the professional body known as the National Association of Estate Agents, came into the branch at Bickley as manager, and his only assistant was a secretary. He was therefore in effect the Bickley branch so far as the outside world was concerned. He was described by Mr J Hunt [a partner of the firm of Calvert, Hunt & Barden] as a zealous worker, and in fact as "over-zealous in respect of his own interests." This description was based on action which, Mr Hunt said, the defendant took on several occasions shortly after he had started work at Bickley. The Bickley and Chislehurst branches were in fact run together in the sense that all inquiries and instructions which came first into one office were promptly transmitted to the other to ensure J that there was as great a chance as possible of effecting the business in question, since it did not matter to the profits of the partnership which office finally did the deal. The rule as to commission was, however, that whenever the second office effected business which originally was booked by the first, half the commission was attributed to each office. An office only got the full commission if it effected the deal and had also been the first to receive the instructions in question. Mr Hunt was responsible for overseeing the financial operations of the firm, including those of the Bickley office, and found that on more than one occasion Mr Elton had claimed full commission on deals on which he was entitled to only half commission, because the K original instructions had been received by the Chislehurst office. Another matter to which Mr Hunt testified concerned the essential trust which estate agents must place in each other. The necessity for such trust was emphasised not only by Mr Hunt but also in cross-examination by Mr Crowhurst, an independent estate agent called for the defendant. It is in fact one of the rules of the National Association of Estate Agents that an agent who learns that a vendor, not on his books, has a property for sale, and who finds a purchaser for that property, before introducing that purchaser to the vendor must ask whether the latter has instructed an agent. If told that he has, then he must ask that agent whether he is willing to give him a subagency, and if not, must withdraw from the transaction. In one case a vendor, who came from Otford, complained to Mr Hunt that the defendant had approached him, saying that he had a purchaser who had agreed to buy the vendor's property. The vendor was angry because he had never instructed the plaintiffs or Mr Elton, and Mr Hunt, in accordance with the recognised practice, promptly gave instructions to withdraw from the transaction.

Now at the first interview prior to entering the plaintiffs' employment, the defendant had stated that he had been in partnership with Mr J H Cox, another estate agent trading under the name Wheekers at Catford. When asked why he had left, he said that Wheekers was really only big enough to support one estate agent, and that when it was proposed to bring in both Mr Cox's son and Mr Elton's son, it was obvious that the partnership would not be able to support four partners. The plaintiffs had no reason to doubt this statement at the time. In April or May 1972, however, as a result of the behaviour of the defendant to which Mr Hunt objected, the plaintiffs decided that they must make further inquiries about the defendant from Mr Cox. This they did, and the three partners went to Mr Cox's office and were told, *inter alia*, by him that in fact the defendant had appropriated some of what was properly the firm's money, notably two items from insurance concerns and one sum of about £100 obtained from a client as an *ex gratia* payment. Such payment, Mr Cox considered, clearly belonged to the firm and should have been paid into the firm's account. Further, there had earlier been irregularities in respect of claims for expenses for petrol used in the firm's business. The total figure concerned in all these items was of the order of £200. These matters were confirmed by Mr Cox, who was called for the plaintiffs, and he said that after he had discovered them he made it clear to the defendant that the trust between them had ceased and it was impossible for the partnership to continue. He said that he was willing to continue to employ the defendant's son and to repurchase the defendant's share of the partnership, and that he would say nothing to third parties about it if the defendant left. At first the defendant agreed to go, but later refused to do so, and Mr Cox was compelled to issue a writ asking for dissolution of the partnership. The matter was finally settled shortly before the case was heard, and the partnership was dissolved on May 18 1971. Mr Cox further testified that his agreement with the defendant also contained a restrictive covenant, the material restrictions being three miles and five years. In fact, the three miles in question covered both the office of Douglas Carlton & Associates in Mottingham, at which the defendant now works, and the office which they proposed to open at Lee, a short distance to the north of Mottingham. It follows that if in fact the restrictions in both agreements are reasonable, the defendant is in breach of both of them. At a further interview with the defendant he was told the facts and that the plaintiffs would back him if he wished to bring any action against Mr Cox. This he declined to do, in spite of denying that what Mr Cox said was true.

I should at this point make it clear that I find the facts which I have stated above to have been proved to my satisfaction. I watched the demeanour of Mr Elton, and was not impressed with the content of his evidence or the way in which he gave it. He frequently was unable, as he reiterated on a number of occasions, to remember what had happened when it was sought to pin him down on important matters, and in my judgment he was also often evasive and inaccurate. I have no hesitation in preferring the versions of Mr C E Barden, Mr Hunt and Mr Cox as to what happened whenever, which was frequently, there was a conflict between what they said and what the defendant said.

The matters above referred to all took place prior to August 25 1972, the date of the agreement here in question. That date is important, because the reasonableness or otherwise of the covenant must of course be tested as at that date: see *Scorer v Seymour Jones* [1966] 3 All ER 347, per Salmon LJ at 351. Furthermore, it is well to bear in mind that it is the reasonableness or otherwise of *this* covenant in *this* agreement with which the court is concerned, and that the onus to justify this covenant lies on the plaintiffs. Restrictive covenants in agreements between estate agents and their employees as a class of agreements are not in issue and are not relevant except in the most general sense. What, then, are the principles upon which one must proceed? It was argued by Mr Maxwell for the defendant that, while admitting that a covenant was proper here, the test of reasonableness of such a covenant is objective, by which I understood him to mean that the agreement must be looked at without consideration for any purpose of the personality, habits and history of the particular employee. The question, he said, is what is reasonable between any employer and any employee of the relevant category in this business, in this district, at the relevant time; the employee's character and record are entirely irrelevant and must be rejected from consideration, because otherwise too great an uncertainty would be introduced into the law governing these transactions. Counsel then went on to say that the particular circumstances here were closer to those in such a case as *Strange v Mann* [1965] 1 WLR 629, a case about a bookmaker's manager, than to those in, for example, *Fitch v Dewes* [1921] 2 AC 158, a case concerning a solicitor's managing clerk.

In *Strange's* case there was an area restriction of 12 miles and a time restriction of three years. It was held that the business and the way it was conducted, being principally on the telephone, were such that it was unlikely that customers would seek out the servant after he had left his employer, that he had acquired no personal goodwill or confidential information or knowledge, and that in such circumstances an area restriction was not necessary to protect the employers' business but was really directed against competition *per se*, and the covenant was therefore unenforceable. In *Fitch v Dewes* the covenant was unlimited in time and as to area was expressed to cover "a radius of seven miles of the Town Hall of Tamworth." At page 164 Lord Birkenhead points out that one must first of all decide on the proved facts what the employer is entitled to be protected against, because only then can one decide whether the covenant in question exceeds what is required for that purpose. The business of an attorney, he said, "depends upon the existence of goodwill; upon the association and the intimacy which exist between him who carries on that business and the clients of the firm, an intimacy founded upon many complex considerations not easily to be defined, but very easily to be understood." At page 165 he adds that if the employment came to an end the employee "should not be in a position to use the intimacies and the knowledge which he had acquired in the course of his employment in order to create a practice of his own in that same place and by doing so undermine the business and the connection of the respondent." The restriction was held to be reasonable.

In the cases cited to me, it seems that "public interest" had there been looked at in a broad sense of some economic or business advantage to the public at large, since it was held, for example, that it was in the public interest that proper restrictive agreements should be allowed "because otherwise solicitors carrying on business without a partner would be extremely chary of admitting competent young men to their offices and to the confidential knowledge to be derived from frequenting those offices." A similar broad view was taken by Sellers LJ in *Scorer's* case, referred to above, when at page 349 he stated that the employee in the case before him was unqualified "without such restraints as the society appropriate to one or other of those activities of auctioneer or estate agent would impose." He approved the finding of the judge that "it would not be against the public interest if another unskilled man should be prevented from setting up. The less unqualified persons uncontrolled by any professional rules there are the better." Sellers LJ went on, however, to say that "so long as that aspect is not overstrained it might commend itself to many people." For myself, I agree it is certainly important not to go too far in this respect, since public interest must not be used to validate a covenant which otherwise might be considered doubtful. The true question is whether the covenant is reasonable and necessary in the circumstances to protect the particular employer and is not otherwise against public interest. Public interest may thus sometimes

A invalidate a restrictive covenant, but it can never validate it; but if a particular restriction is reasonable and necessary to protect the particular employer's business it seems to me that, provided any possible bargaining inequality between the parties has been allowed for, it is normally in the public interest that such restriction should be valid. Such a restriction is in those circumstances an exception to the general rule that there must not be interference with individual liberty of action in trading: see, for example, Lord Macnaghten's well-known speech in *Nordenfelt v Maxim Nordenfelt Guns & Ammunition Co Ltd* [1894] AC 535, 565.

Mr Galpin, for the plaintiffs, argued that the true question
B was whether in all the relevant circumstances this covenant was necessary to protect the plaintiffs' business and was not against the public interest. The difficulty is, of course, in defining what are the relevant circumstances. The circumstances which counsel submitted were relevant included the activities of the defendant which Mr Hunt had objected to, and the information of the defendant's previous activities and conduct given to the partnership by his previous employer, Mr Cox. They were relevant, he argued, both to the necessity for the covenant and to its limits. On that basis, he contended that the area and time restrictions of three miles and three years were certainly not wider than necessary to protect the trade connection and reputation of the plaintiffs; but even if such circumstances were not relevant, the
C restrictions were still reasonable and the agreement should be supported. So long as they were not restricting competition *per se*, the plaintiffs were entitled to ask for and could expect to get the full extent of the protection considered necessary in such circumstances. In the light of the defendant's previous conduct and activities, this, notionally at any rate, could be argued to result in a wider restriction being reasonable in his case than in the case of an employee otherwise similar but of unblemished reputation.

In answer to my request for any authority supporting the proposition that an employee's reputation and position were relevant considerations both to the existence of and to the extent
D of the restriction, Mr Galpin, in addition to the *Scorer* case, cited *Haynes v Dorman* [1899] 2 Ch 13. In that case, the father of the defendant was employed by the plaintiff as traveller and manager, and by reason of his employment had much valuable information about the plaintiff's business and customers. The defendant was a young man living with his father, and one of the matters which Rigby LJ appears to have taken into account (see page 27) was the fear on the plaintiff's part that the defendant, who was, at the time when he applied to the plaintiff for a job, in the employment of a rival firm, might in conversation with his father acquire knowledge of facts which might give him, and through him the plaintiff's rivals, an undue
E advantage. This case was in the last century and seems to me to go a long way if the reasonableness of the restriction in issue was really dependent upon such a fear. An examination of the case shows, however, that there were other perfectly good reasons why, in the circumstances of that business and of the nature of the employment of the defendant, it was held to be reasonable to impose the restriction of 25 miles and of unlimited duration, when the son became the employee of the plaintiff.

The matters which are of primary importance in the present case are clearly the nature of the business and the area over which it extends, the confidential information as to the plaintiffs' clients which the employee may be expected to gain, and the
F influence which he may be expected to exert over them. The evidence established that an estate agent's business is very dependent on the personality of its partners and employees. They necessarily, over the years, learn a lot about many of their customers, and about their circumstances and requirements. This is especially true in a dormitory area as opposed to somewhere like the centre of the metropolis. Many of the plaintiffs' sales and purchases were from, as it was called, repeat customers, that is to say from clients who, during their lifetime, remained in the same general area but from time to time, as their circumstances altered, moved from one locality to another and from a smaller to a larger house or vice versa. This was confirmed by the evidence of the defendant himself, as well as by his witnesses, and he agreed that it was a valuable connection. Taken G
by and large, Bickley and Chislehurst are areas where the houses are larger, of better character and more expensive than in Mottingham, in which the defendant's office is situated, or in the nearby area of Lee. As already stated, the Bickley and Chislehurst offices of the plaintiffs were run together, and I think it is quite fair to regard this method of doing business as giving each office an effective sales area larger than it would be if each was in widely-separated and not contiguous areas. This was confirmed by Mr Barden, who said that they had at Bickley a flow of business from Mottingham via the Chislehurst branch. As the map shows, the Mottingham and Chislehurst areas extend nearly to the edge of the three-mile circle from Bickley. H

The evidence as to the areas covered by the various offices with which the witnesses in this case are concerned—for example, Bickley, Chislehurst, Mottingham and New Eltham—was to some extent conflicting. The plaintiffs' witnesses' figures generally tended to be larger than those of the defendant and his witnesses. In these dormitory areas, where there are quite a large number of estate agents, the greatest volume of transactions is generally found in properties situated fairly close to the office. It is equally clear that though there may not have been much of it, the business activities both of the plaintiffs' partnership and of the defendant's witnesses, Mr Crowhurst and Mr A Spencer, extended up to about 10 miles, or on occasion even further, away from their respective offices. These long J
distances, however, I regard as isolated cases and ignore for present purposes. It is impossible to be precise, but it seems, and I so find, that the bulk of the business, including that of the plaintiffs, is done within an area of approximately three miles' radius. The defendant's area was said by the defendant to cover a somewhat smaller radius, perhaps up to 1½ to 2 miles, but Mr Hunt made it clear that the plaintiffs' transactions often covered a three-mile radius and, though not so often, certainly extended up to five miles. Giving the matter the best consideration I can, after listening to the evidence, I consider that a radius of three miles from Bickley, on the facts as to distance of sales from that office, is a perfectly reasonable figure. I think the plaintiffs K
were entitled to impose a restriction of that distance in order to obtain adequate protection against any employee in the defendant's position who would necessarily acquire influence over and personal information about many of the plaintiffs' customers who came and gave instructions to the Bickley office, and, in those cases where he effected such sales or purchases, to the Chislehurst office. The conclusion is based solely on the facts which were proved before me as to the area, as to the nature of the plaintiffs' business at the Bickley office, and as to the defendant's position in that business. If it is permissible to take into account also the character of the defendant and his previous activities as part of the relevant circumstances, then my view that three miles is eminently reasonable is reinforced. L

The true legal position I consider to be as follows. Facts as to previous unprofessional activities and unreliability of an employee in a particular professional position are relevant to, though not determinative of, the question as to whether or not it is reasonable and proper to insist upon a covenant at all, because they indicate what may well take place in a business of that nature. So far as the reasonableness of the extent of the covenant actually taken is concerned, it is, I think, right to say that such facts, if proved, can go no further than justifying the employer in insisting that he is entitled to go to the maximum limits necessary and proper to protect his business interests and reputation. Such limits must, however, still be reasonable, as tested by the case of any professional employee filling that M
particular position, whatever his reputation, if they are not to be held to be too wide and contrary to public interest. So here, evidence was given by Mr Hunt, and confirmed by Mr Crowhurst, that any ex-employee, if he behaves in an unprofessional way in or too near to the area of his erstwhile employer, can only bring discredit on such employer's business. This I can well believe, and it seems to me a general matter which an employer is entitled to take into account in insisting upon a covenant of appropriate area, but I do not think such employer should be entitled, in the case of a particular employee, to impose a covenant wider than would be reasonable in the case of

employees in that category generally. The employer's remedy in the case of a potential employee of bad reputation is to refuse to employ him.

As regards the length of time-limit of three years, there was not really very much dispute, though Mr Maxwell argued that it was too long. The plaintiffs emphasised that the limit of three years was necessary to take care of the repeat customers, and the intention was to specify a period beyond which it was not likely that customers would, if they felt so inclined, follow the defendant into his new employment. Equally, after three years his influence and knowledge of the plaintiffs' customers was not so likely to enable him to follow them. As it is sometimes put, by that time the telephone line will have gone cold. To my mind, and I so hold, there is nothing unreasonable in this limit. If the defendant's character and activities are also relevant to this, then it seems to me the plaintiffs were all the more justified in specifying three years as a minimum, since the defendant had already shown a willingness in his early dealings for the plaintiffs to make use of his knowledge of their customers' requirements in an unprofessional way. I therefore hold that this covenant, the subject of this action, is necessary to protect the plaintiffs' business and reputation, and there is no reason for holding it invalid on the ground that it is against public policy.

An injunction was granted restraining the defendant until July 31 1976 from directly or indirectly undertaking, carrying on, being concerned, engaged or interested, either alone or in partnership with, or as manager, agent or servant of, any other person, firm or corporation, or otherwise, in the business of auctioneer, valuer, surveyor, estate agent, property manager, insurance agent or mortgage broker within a radius of three miles of the plaintiffs' premises at 12 Southborough Road, Bickley. His Lordship told Mr Maxwell that although he had listened very carefully to his (counsel's) submissions in support of a stay, he was not satisfied that this was a case in which the operation of this injunction should be held up for more than seven days, which would enable the defendant to ask the Court of Appeal for further time. The plaintiffs were given an inquiry as to damages, and costs.

QUEEN'S BENCH DIVISION

November 8 1974

(Before Mr Justice FORBES)

HUBBARD AND OTHER v PITT AND OTHERS

Estates Gazette February 8 1975

(1974) 233 EG 595

Picketing on highway of Islington estate agents' offices prima facie unlawful, being of a substantial character and unauthorised by statute—Interlocutory injunction granted against property action group

In these proceedings Mr Ronald Frederick Hubbard, Mr Christopher Eliades and Mr Robert Malcolm Owen, trading as Prebble & Co, estate agents, plaintiffs in an action alleging conspiracy, nuisance and defamation, sought an interlocutory injunction restraining the defendants, Mr James Pitt and eight others, by themselves, their servants or agents, from besetting the plaintiffs' premises at 108-109 Upper Street, Islington; 82 Parkway, Camden; and 543 High Road, Tottenham, London, by picketing on the highway, or otherwise interfering with the plaintiffs' use of the premises.

Mr A T Hoolahan QC and Mr R A Rampton (instructed by B Greenby) appeared for the plaintiffs, and Lord Gifford (instructed by Clinton Davis & Co and Seifert Sedley & Co) represented the defendants.

Giving a reserved judgment, FORBES J said that the case was concerned with the Islington district of London. Like many other areas of the metropolis, this district possessed dwelling-houses which presented an interesting social history in miniature. Many of them were terraced houses built in Victorian days as family homes. Over the years Islington became less fashionable. Many houses had become subdivided into smaller units let at low rents. The tenants of these properties were people with low incomes. More recently, however, Islington had returned to fashion. Terraced houses with many tenants to each house had been converted back to single-family homes. Restored and renovated, they had been sold to people who could afford to buy such properties. Islington had thus become what house agents called an "improving" area. In addition there had been the demolition of older properties and their replacement by modern blocks of flats or offices. No doubt the conversion of down-at-heel tenanted properties to well-maintained owner-occupied houses improved the aesthetics of the urban scene. It also had the effect of drastically reducing the stock of dwelling-houses available for letting at low rents. Still other effects had followed. The high prices which restored and renovated houses fetched when they could be sold with vacant possession represented a powerful incentive to landlords of tenanted property to obtain possession from their tenants. Many of those tenants had some security under various statutory provisions. Some practices designed to obtain possession from tenants were classed as harassment and were illegal. But there were other ways in which landlords could put pressure on tenants which were not illegal. Some people regarded the use of such practices, though legal, as morally unjustifiable.

It was important to see the case against that background, for the defendants were among a group of people concerned at the social problems caused by redevelopment in Islington. It was a loosely-knit collection of individuals who called themselves the Islington Tenants Campaign or Crusade. They felt deeply about the activities of property developers in Islington; they also considered that estate agents who assisted property developers were acting indefensibly. The plaintiff firm was prominent in Islington, so the group decided to picket its offices. On three consecutive days in March 1974 some of the group attended on the public footway in front of the plaintiffs' offices at 108-109 Upper Street, Islington. They held placards and distributed leaflets. Both the placards and the leaflets referred to the plaintiffs in opprobrious terms. Subsequently pickets carrying similar placards and leaflets attended at the same place on every Saturday. All the defendants had at various times been on those picket lines. In those circumstances the plaintiffs had started an action against the defendants. Their statement of claim disclosed three main heads of claim: in conspiracy, nuisance and defamation. They (the plaintiffs) maintained that the picketing was the result of a conspiracy, an agreement to do an unlawful act. The unlawful act complained of was standing on the highway in front of the plaintiffs' premises with placards, and so on. That, said the plaintiffs, was an unlawful user of the highway. The claim in nuisance also involved unlawful user of the highway, but the plaintiffs accepted that they must prove in the ordinary way that they as occupiers of premises adjoining the highway suffered greater damage than the generality of Her Majesty's subjects. The claim in defamation was based on the terms of placards carried by the pickets and of leaflets which they handed out. It was answered in the defences by a plea of justification. So far as the other claims were concerned, the defendants maintained that they had a lawful right to do what they were doing.

On the main issues, the plaintiffs' contentions were (1) that the highway was a piece or strip of land dedicated to the use of the public over which each member of the public had a right to pass or repass and do things reasonably incidental thereto; (2) that any person who entered upon a highway and did or intended to do anything which was not within the above right was a trespasser unless he could show either that the special right was given in the grant of dedication or had been authorised by statute; (3) that a trespasser might not necessarily incur liability, for instance for public nuisance, unless the trespass constituted an unreasonable user of the highway; (4) that there was no right to use the highway to stand, or to stand in line, or to place placards thereon, or to stop pedestrians or to picket, nor was there any right to do any act with a view to compelling a shopkeeper, for instance, to comply with the demands of the person using the highway. Four propositions were put forward on behalf of the defendants: (1) that picketing was not unlawful or actionable unless it took the form of a common law nuisance; (2) that for that purpose, and so far as related to user of a highway, common law nuisance was equivalent or synonomous with unreasonable user of the highway; (3) that the mode of user of the defendants was not unreasonable; and (4) that if it was, it did not cause substantial damage to the plaintiffs. The course of the argument was represented by the following propositions: (1) that to amount to common law nuisance on the highway, any conduct had to be unreasonable; (2) that picketing was a reasonable user of the highway, because there was a special kind of democratic right to picket which was analogous to the right of free speech; and (3) that picketing only became unreasonable, and therefore a nuisance, if it was accompanied by violence or intimidation or amounted to physical obstruction of the highway.

He (his Lordship) was of opinion that one might define the right of the public to use a highway as a right to use it reasonably for passage and repassage and for any other purpose reasonably incidental thereto. If an incidental use was unreasonable in extent, it went beyond the purpose for which the highway was dedicated and became a trespass and therefore unlawful. What conduct in relation to a highway constituted a public nuisance? The defendants said that they were behaving reasonably in the sense that they were not resorting to violence or intimidation and did not station themselves so as completely to obstruct the footpath or the entrance to the plaintiffs' office. But one was not considering what was the behaviour to be expected of a reasonable picket, but whether or not the behaviour of those pickets amounted to an unreasonable user of the highway. Leaving aside for the moment the proposition that there was a democratic right to picket, the defendants based their claim that they had acted lawfully on the assertion that they conducted themselves without violence or intimidation and in a manner which did not obstruct the highway. But it was only necessary to look at the photographs to see that the defendants by their mere presence might well deter clients and potential clients of the plaintiffs from entering the plaintiffs' offices. It was not sufficient to say that the public could easily get by the obstruction, which was one of the claims made by the defendants. The law seemed to be clear: the public had a right to go on every part of the highway, and any act which made it "less commodious" (see *Hawkins's Pleas of the Crown,* 8th ed, book 1, chap 32, s 10) was a public nuisance unless it could be said to be so fleeting and so inappreciable as to fall within the *de minimis* rule. The defendants relied on the judgment of Moulton LJ in *Ward Lock & Co Ltd v Operative Printers Assistants* (1906) 22 TLR 327, in which it was held that there was no cause of action against a union which stationed pickets outside a printing works, there being no evidence that any annoyance or molestation took place. But it was quite impossible to say that that case, which depended for its decision on the facts, was authority for the wide proposition for which the defendants contended.

So far as the general law of picketing was concerned, section 2 of the Trades Disputes Act of 1906 provided: "It shall be lawful for one or more persons, acting on their own behalf or on behalf of a trade union or of an individual employer or firm in contemplation or furtherance of a trade dispute, to attend at or near a house or place where a person resides or works or carries on business or happens to be, if they so attend merely for the purpose of peacefully obtaining or communicating information, or of peacefully persuading any person to work or abstain from working." The important words were "in contemplation or furtherance of a trade dispute," for if peaceful picketing was a lawful exercise when performed by any citizen, it would be unnecessary to pass an enactment which legalised such an exercise only when performed in contemplation or furtherance of a trade dispute. The provisions of the 1906 Act were in their turn repealed by the Industrial Relations Act 1971, and section 134 of that Act provided in somewhat more extended terms an exemption for peaceful picketing from proceedings under an earlier statute and from civil liability for tort; but again only when the picket was conducted "in contemplation or furtherance of an industrial dispute." The statutory position was now governed by section 15 of the Trades Union and Labour Relations Act of 1974, which broadly re-enacted the provisions of section 2 of the Act of 1906. He (Forbes J) had reached the conclusion, in the light of the legislation referred to, that the defendants' conduct, which was not in furtherance of any trade dispute, was not authorised by statute, amounted to an unreasonable use of the highway, and was therefore unlawful unless it could be justified on some other principle.

In these circumstances, the defendants contended that it was the inalienable right of anyone in a democratic society who felt deeply enough about a subject, or disapproved strongly enough of someone's behaviour, to picket any place, whether public or private, in order to express those feelings and to draw public attention towards a matter of substantial public interest and concern. In support of that proposition they relied firstly on an analogy with the principles of freedom of speech, and secondly, on a comparison of their actions with those who, on occasions, picketed the Houses of Parliament or 10 Downing Street or the Soviet Embassy, or those who, by way of demonstration, marched through the streets before holding a public meeting outside the premises of some organisation or power which had attracted particular approbation or disfavour. But it was necessary to remember that there was no such thing in law as unfettered freedom of speech. One's right to say what one liked was circumscribed by, for instance, the laws relating to sedition, contempt of court, obscenity and defamation. Similarly, one's right to be on the highway was subject to the law relating to highways. Whereas the passage of years might make a difference in the utterances which were regarded as obscene, defamatory, contemptuous or even seditious, no such alteration was possible in the law of highways: nothing was better settled than that it was impossible to obtain a prescriptive right to obstruct a highway. The picketing of public buildings, embassies, and so on, and demonstrations, were common and permitted: ergo, said the defendants, they must be lawful. In examining this part of the case it was necessary to make a distinction between the march or procession on the one hand, and the assembly on the highway in front of some premises on the other. It seemed to him (his Lordship) that while an orderly procession could be regarded as a use of the highway for the purpose of passage, and therefore, so long as it was reasonable in extent, a use which fell within the purposes of dedication, the stationing of pickets on the public highway was not a legal exercise of the right of passage and, if it rendered the highway less commodious, amounted to a public nuisance. Further, it was no use adopting a colourable pretence at passage by (for instance) having pickets move around in a circle.

The true reason why demonstrations involving stationary assemblies on the highway were permitted was not that they were not illegal; it was that they were matters in respect of which the remedies available were simply not put into operation. It might well be that the occupiers of 10 Downing Street and the Soviet Embassy regarded it as politically inexpedient to take action. It was quite wrong to argue that a disinclination by those operating in the political field to set proceedings in motion to restrain political conduct of a particular kind meant that such conduct became clothed with a legality which it would not possess in a non-political context. The law appeared to be clear: at common law the use of the highway for picketing was illegal, as it was a use not referable to the purposes for which the highway was dedicated. It was therefore at least a trespass. It might also be an unreasonable user of the highway and therefore a common law nuisance. That would always be a question of fact, and what was or was not a reasonable user of the highway would be determined by reference to the purposes for which the highway was dedicated. As picketing was a use of the highway wholly unconnected with the purposes of dedication, and was, in fact, designed to interfere with the rights of an adjoining owner to have unimpeded access from the highway, it was likely to be found to be an unreasonable user unless it was so fleeting and so insubstantial that it would be ignored under the *de minimis* rule. By statute, picketing on the highway was nevertheless legal so long as it was in contemplation or furtherance of a trade dispute and satisfied the other provisions of section 15 of the Trades Union and Labour Relations Act 1974. Put shortly, therefore, the use of the highway for picketing was illegal unless (1) it was in contemplation or furtherance of a trade dispute in the circumstances set out in the statute, or (2) it was found as a fact it was insubstantial. The defendants were clearly not acting in contemplation or furtherance of a trade dispute, and looking at the affidavits and the photographs, he (his Lordship)

A was unable to say that the picketing could be ignored as being *de minimis*. The photographs showed a fairly formidable array outside the plaintiffs' office, and he had no doubt that the impact was intended to be, and was, considerable. *Prima facie*, therefore, the activities of the defendants in picketing the plaintiffs' office were themselves unlawful as an unreasonable use of the highway, and further, the defendants' agreement to engage in those activities amounted to an unlawful conspiracy.

There remained two final matters. Had the plaintiffs sufficiently demonstrated that they had suffered damage, so as to be entitled to an injunction, and should such an injunction be granted as interlocutory relief? The picket lines shown on the photographs would inevitably have had some substantial effect on the resort
B by customers to the plaintiffs' office, and a regular picket for three hours every Saturday could not properly be called fleeting or evanescent. The matter was really put beyond doubt, moreover, by the terms of a leaflet used by the pickets. Then should an interlocutory injunction be granted? C On the material before the court, and even on the basis of accepting nearly everything in the defendants' affidavits where there was conflict with the plaintiffs, it seemed to him (his Lordship) that the plaintiffs would still be entitled to an injunction if this were the trial of the action. The basis on which the court would grant interlocutory injunctions was well settled. The principle was usually referred to as that of the balance of convenience. He (Forbes J) would emphasise that he had not been concerned with approval or disapproval of either the defendants' attitudes to housing questions in Islington or those of the plaintiffs. The sole issue had been whether or not the use of the highway for picketing which was not in contemplation or furtherance of a trade dispute was a lawful operation. D

The plaintiffs were granted an injunction in the terms sought until trial of the action or further order. Costs were reserved to the trial judge.

FEES AND COMMISSIONS

A **HOUSE OF LORDS**

November 27 1974

(Before Lord REID, Viscount DILHORNE, Lord WILBERFORCE, Lord DIPLOCK and Lord SIMON OF GLAISDALE)

SAUNDERS *v* SOPER

Estates Gazette January 25 1975

(1974) 233 EG 387

Accommodation Agencies Act 1953—Supplying address may not be charged for, but finding accommodation may—Agent's appeal allowed and conviction quashed—McInnes v Clarke explained as a case in which a deposit was charged before B **addresses were supplied—Fact that deposit was returnable if accommodation was not found held immaterial**

This was an appeal by Mrs Frances Maynard Soper, of Brighton Avenue, Southend-on-Sea, Essex, principal of Wilson's Accommodation Bureau, Southend, against a decision of the Queen's Bench Divisional Court on May 1 1974 upholding her conviction by Southend justices on charges of accepting or demanding sums of money in consideration of supplying one Christine Nichols with addresses of houses to let, contrary to section 1 (1) (*b*) of the Accommodation Agencies Act 1953. The Divisional Court's decision was reported at 231 EG 239.

Mr R H Bernstein QC and Mr S J Burnton (instructed by C F T Fisher & Lang, of Southend-on-Sea) appeared for the appellant, and Mr B J Higgs QC and Mr K Dow (instructed by Sharpe, Pritchard & Co) represented the respondent, a police inspector.

LORD REID: For the reasons given by my noble and learned friend, Viscount Dilhorne, I would allow the appeal.

VISCOUNT DILHORNE: The appellant, it was not disputed, carries on and has carried on for some 6½ years a highly respectable agency under the name of Wilson's Accommodation Bureau. On January 25 1973 a Miss Christine Nichols came to see her, wanting accommodation for herself and two friends. The appellant at that time did not know of any likely to suit, but agreed to try and find some, and asked Miss Nichols to call D on her again. Miss Nichols did so on a number of occasions. The appellant, or someone on her behalf, inspected premises at 27 Dawlish Drive, Southend-on-Sea and thought they might suit Miss Nichols, and so Miss Nichols was given that address. On February 13 1973 Miss Nichols signed an agreement in writing which contained the following paragraph:

> In consideration of the services of Wilson's Accommodation Bureau in finding or introducing accommodation acceptable to me/us, I/we agree to pay to Wilson's Accommodation Bureau a fee of the equivalent of one week's rent (minimum £5.25) for the accommodation accepted by me/us. No fee is payable to Wilson's Accommodation Bureau unless and until I or we become the tenant(s) of E the accommodation found or introduced by Wilson's Accommodation Bureau to me/us.

Miss Nichols, having inspected the premises, entered into an agreement to rent them for £7 a week. She thereupon under the agreement into which she had entered became liable to pay to the appellant £7, and she paid that sum on February 20 1973. Later Miss Nichols again came to see the appellant and again told her that she wanted accommodation for three persons. Again the appellant had no suitable accommodation on her books, and Miss Nichols was asked to call again. She did so on a number of occasions, and on the last occasion was given an address, 34 Retreat Road, Westcliff-on-Sea. On March F 16 1973 she signed an agreement similar in all respects to that which she had previously made with the appellant. She became tenant of those premises at a weekly rent of £16 a week, and so became liable under the agreement to pay the appellant £16. This she did not do and has not done, although payment was demanded. On July 12 1973 the respondent, a police inspector, preferred two informations against the appellant, the first alleging that she had on February 20 1973 accepted from Miss Nichols £7 in consideration of supplying her with addresses of houses to let, contrary to section 1 (1) (*b*) of the Accommodation Agencies Act 1953, and the second alleging that she had demanded £16 in consideration of supplying Miss Nichols with addresses of houses to let, contrary to that subsection of that Act. G The appellant was convicted of both offences, and her appeal to the Divisional Court was dismissed. She now appeals to this House with the leave of that court. That she accepted £7 from Miss Nichols and demanded payment of £16 by her is not disputed. The question to be decided is whether, by doing so, she committed offences against this subsection.

The Act is entitled: "An Act to prohibit the taking of certain commissions in dealings with persons seeking houses or flats to let and the unauthorised advertisement for letting of houses and flats." It is to be noted that the object of the Act is not stated to be the prohibition of the taking of all commissions in dealings with such persons, but only of the taking of H certain commissions. The Act is a short one, containing only two sections, the first of which, so far as material, reads as follows:

> 1 (1) Subject to the provisions of this section, any person who during the continuance in force of this Act,—
>
> (*a*) demands or accepts payment of any sum of money in consideration of registering, or undertaking to register, the name or requirements of any person seeking the tenancy of a house;
>
> (*b*) demands or accepts payment of any sum of money in consideration of supplying, or undertaking to supply, to any person addresses or other particulars of houses to let; or
>
> (*c*) issues any advertisement, list or other document describing any house as being to let without the authority of the owner of the house or his agent, J
>
> shall be guilty of an offence [punishable by a fine not exceeding £100, or imprisonment not exceeding three months, or both].
>
> (2) A person shall not be guilty of an offence under this section by reason of his demanding or accepting payment from the owner of a house of any remuneration payable to him as agent for the said owner.

The first matter to consider is what was the mischief at which this Act was aimed. One can only deduce that from its terms. Section 1 (1) (*a*) was clearly directed to preventing persons desirous of accommodation being charged merely for the regis- K tration of their names or their requirements. Section 1 (1) (*b*) was, in my view, equally clearly directed to preventing charges being made just for supplying addresses. It is obvious what section 1 (1) (*c*) was intended to prevent. Commission of any of the three offences created by this section does not in any way depend on the character of the agency. Whether the agency be reputable or bogus, the person who runs it and charges for supplying addresses to persons wanting accommodation commits an offence against section 1 (1) (*b*). The object of section 1 (2) is not at first sight easy to discern, but if it were not there, a person

A who at the request of the owner, say, of a block of new flats supplied the addresses of those flats to would-be occupants would be unable to obtain from the owner or his agent any remuneration for his services.

It was not in this case contended that the agreements signed by Miss Nichols did not correctly and completely state the terms on which the bureau would become entitled to payment from her. The respondent, however, contended that if, in the course of carrying out the contract made, the address of premises to let was supplied, then the demand or acceptance of payment, even though no payment was due unless and until acceptable accommodation had been found and rented by the client, was, *inter alia,* a demand or acceptance of payment for B the supplying of the address. If this be right, it follows that an agent employed by a prospective tenant to find for him, it may be, a large house or one whose requirements are of an unusual character, cannot lawfully demand payment for his services from his client when he finds premises which suit his client and of which his client becomes the tenant. Mr Higgs, in the course of his argument for the respondent, was unable to suggest that in such circumstances there was any mischief in the agent claiming commission which Parliament would want to prevent. In my opinion, the mischief to which section 1 (1) (*b*) was directed was simply to prevent charges being made for supplying addresses and did not extend beyond that. C So, in my opinion, the question for decision in this appeal is simply, did the appellant accept or demand payment for supplying addresses? If instead of supplying one address on each occasion she had supplied 100 or a 1,000 addresses, under the agreements which she signed Miss Nichols would not have been liable to make any payment. She was not asked to pay, nor did she pay, for being given addresses. Under the agreements she became liable to pay not for the giving of addresses but for the finding of accommodation acceptable to her and accommodation of which she became the tenant. On the facts of this case, and in view of the terms of the agreements entered into, consideration of the terms of the D section leads me to the conclusion that this appeal should be allowed; and as I read Lord Widgery's judgment in this case, that would appear to have been his view were it not that he felt himself bound by some previous decisions to which I must now refer.

The first of these was *McInnes v Clarke* [1955] 1 WLR 102. In that case the applicant for premises, when he had stated his requirements and what rent he was willing to pay, was told that the fee payable by him if premises were found for him would be a sum equal to one week's rent, and that a deposit of approximately half the weekly rent would be asked for. If the client was willing to agree to pay the deposit, he was then asked to sign an agreement. After he had obtained accommodation, he E had to pay the balance of the fee. Unsatisfied clients who demanded the deposit back received it. Sometimes before and sometimes after the agreement was signed, a list of addresses would be given to the client. Following this procedure, the appellant, who carried on the business of an estate agent, was convicted on two summonses of offences against section 1 (1) (*b*) of the Accommodation Agencies Act. Apart from supplying a list of addresses, he had done nothing about finding accommodation which was suitable. The Divisional Court dismissed his appeal. Lord Goddard CJ in the course of his judgment said that the agreements in that case showed that what the appellant did was to say to a person, "If you give me a deposit of approxi- F mately half the week's rent, I will send you some names and addresses, particulars of houses, and if you take one of them you have to pay me the rest of the sum, which is to be equivalent to a week's rent. If I cannot satisfy you I will give you your deposit back." Lord Goddard thought the case for the appellant unarguable, because it seemed to him that what the appellant had done was to accept payment in consideration of supplying addresses. Cassels J pointed out that in one of the agreements it was said that the client would get a daily list of addresses for 18 consecutive working days and that that was what the deposit was paid for. Streatfeild J agreed, and said that unless the deposits were paid, no list of addresses would be obtained. Lord Widgery, in the course of his judgment in the present case, said in relation to that case that he did not find it particularly G difficult to accept the proposition that since the payment of a deposit was a condition of obtaining particulars of houses, the payment of the deposit was a consideration for those particulars. He thought that it might be possible to distinguish that case from the present one. I agree. In my opinion, the decision in that case was clearly right and is clearly distinguishable from this case. In that case there was an agreement to pay a deposit before addresses were supplied, and the payment of the deposit did not cease to be a payment because if the client was dissatisfied it was returnable. The judgments in that case make it clear that the decision was founded on the payment of the deposit. In the present case there was no demand for or acceptance of payment just for supplying addresses. H

The next case was *Crouch & Lees v Hàridas* [1972] 1 QB 158. The plaintiffs, a firm of estate agents, had sued for commission for supplying the defendant with particulars of an unfurnished flat which the defendant had rented. The particulars of claim alleged that the defendant had agreed to pay commission for supplying him with particulars of a flat in the event of his taking a lease of it. Davies LJ said that it seemed to him perfectly clear that the plaintiffs did demand a payment for supplying particulars. That, he said, was the whole contract as pleaded by them. That overlooks the fact that it was pleaded by them that the commission was only payable on a lease of the flat being taken. He went on to say that the case was not J distinguishable from *McInnes v Clarke* by reason of the payment of a deposit in that case. He thought that a distinction without a difference. "There was," he said, "no advance payment in the present case, but only an agreement to pay when the defendant was satisfied. That, I think, makes no difference at all, because in each case, unless the defendant obtained a flat and was satisfied with it, there would in the event be no payment at all." With the greatest respect, I do not think that it was accurate to say that the fact that there was no advance payment made no difference at all or that in each case there would have been no payment if the client was not satisfied. In *McInnes v Clarke* there was a payment, which if the client was dissatisfied he K might recover. In *Crouch & Lees v Haridas* no payment of any sum was made or became due unless and until the client became a tenant of the premises of which the address had been supplied to him. In that case the other Lords Justices agreed with Davies LJ. I think that the decision was wrong, for it does not appear to me from the facts set out in the report of the case that any payment was either accepted or demanded just for supplying addresses.

The third and final case to which I must refer is *Arthur Lawrence and Arthur Lawrence Agencies Ltd v Sinclair-Taylor,** where on October 24 1973 the Divisional Court dismissed appeals against convictions for offences against section 1 (1) (*b*) of the Act. There the respondent was looking for accommodation, and L was told that the appellants had on their register a flat to let at £12 a week and if he was given its address he would be required to give the company £18, one and a half times the weekly rent. The respondent was given the address, went to the premises and liked them, and then returned to the appellants' office and paid them £18. It seems to have been the case that if the respondent was not satisfied with the accommodation the £18 would be returned to him. The respondent had signed a letter stating that he agreed to deposit with the appellants £18 in respect of fees for services rendered. This case is on all fours with *McInnes v Clarke* and has the same distinguishing feature as that case has from the present one. It was, I think, rightly M decided.

Each summons in this case charged the appellant with having accepted or demanded money on the dates stated in consideration for supplying addresses when on each of the two dates only one address had been supplied. This makes no difference, and I agree with Lord Widgery when he said in the last-mentioned case that there was nothing in the context of the Act to prevent the operation of the provision in the Interpretation Act 1889 by which the plural is to include the singular. Agreements between accommodation agencies and their clients who seek accommodation must fall into one of the following categories: agreements in writing, oral agreements, and agreements which are partly oral and partly in writing. Where the agreement is wholly in writing and is genuine, then one only has, as here, to construe the agreement to determine what was the bargain made. No matter into which category the agreement may fall, if it provides for the payment of money either before or after the supplying of an address but the payment is not due only when suitable accommodation is found and rented, then, even though the payment is called a deposit and returnable if the client is not satisfied and no tenancy results, it is not difficult to conclude that the payment was for supplying the address and that if such payment is demanded or accepted, an offence is committed. As in my view it was not proved in this case that any payment was demanded or accepted for the supplying of an address, and as it was proved that the payments were accepted or demanded not for that but for finding Miss Nichols suitable accommodation and were due only on her entering into a tenancy, in my opinion she was wrongly convicted, and this appeal should be allowed with costs in this House and the Divisional Court.

LORD WILBERFORCE: For the reasons given by my noble and learned friend, Viscount Dilhorne, I agree that this appeal should be allowed.

LORD DIPLOCK: I have had the advantage of reading the speech of my noble and learned friend, Viscount Dilhorne. I agree with it and would therefore allow the appeal.

LORD SIMON OF GLAISDALE: I have had the advantage of reading in draft the speech prepared by my noble and learned friend, Viscount Dilhorne. I agree with it, and I would therefore allow the appeal.

QUEEN'S BENCH DIVISION

July 12 1974

(Before Mr Justice NIELD)

HEADWAY CONSTRUCTION CO LTD v DOWNHAM AND ANOTHER

Estates Gazette February 15 1975

(1974) 233 EG 675

Land scout employed by developers on commission basis purchases through shell company site in which developers known to be interested—Secret profit on sale to be disgorged—Commissions on separate transactions lost, developers being entitled to avoid the commission contract for fraud

This was a claim by Headway Construction Co Ltd, property developers, of Welwyn Garden City, Hertfordshire, against Mr Geoffrey Downham and his company, Bardcrest Properties Ltd, for damages for breach of contract, damages and other relief for misrepresentation, fraud and deceit, and recovery of a secret profit made upon the sale of land at Stansted, Essex. The defendants counterclaimed commission pursuant to an agreement of July 14 1971.

Mr E A Machin QC and Mr G W Cheyne (instructed by Adam Burn & Metson, agents for Crane & Staples, of Welwyn Garden City) appeared for the plaintiffs, and Mr T S G Baker (instructed by Doyle, Devonshire, Box & Co) represented the defendants.

Giving judgment, NIELD J said: The plaintiff company carried on the business of property developers at Welwyn; the managing director was a Mr Cottrell and another director a Mr Drysdale. The first defendant, Mr Downham, on April 7 1970 took employment with the plaintiffs as a construction manager; he lived then at 33 King's Road, Berkhamsted. The terms of the service agreement appear in a letter of April 3 1970. This agreement was replaced by a commission agreement, the terms of which appear in a letter dated July 14 1971, and under it Downham (and in parenthesis I say the parties will, I know, understand if I omit designation for the sake of brevity) agreed to undertake the task of acquiring land for the plaintiffs on commission and on a part-time basis. The second defendants are a company which was purchased by Downham on June 4 1971. The company has no assets, and in effect Downham and the company are one and the same entity. In the action now before the court, the plaintiffs claim damages, saying that Downham was in breach of his contract of service; that he made a secret profit of £15,500 in relation to the sale of certain land at Stansted, and that he was guilty of misrepresentation, fraud and deceit. In the defence, all these allegations were denied, but at the end of Mr Machin's opening for the plaintiffs, Mr Baker for the defendants told the court that they (the defendants) no longer denied making a secret profit, and said it was admitted that the plaintiffs were entitled to £15,500 less expenses, reducing the figure to £14,369. The defendants, he said, denied any fraud and asserted that the course taken in regard to the purchase of the Stansted land was taken upon the advice of counsel. It was conceded that for all practical purposes the second defendants were in truth the first defendants. There is then a counterclaim in which the defendants claim to be entitled to commission under the second agreement and seek a declaration to that effect and a specified sum of commission of £13,475 in respect of one transaction relating to land at Roundswood Park, Harpenden.

I begin by considering the service agreement of 1970. On March 6 of that year Downham applied for the post of construction manager with the plaintiffs. The terms of the agreement offered to him are set out in their letter of April 3 1970. These terms included a salary of £3,350 and the right to terminate by three months' notice but did not specify Downham's duties. In fact they were generally to supervise building contracts. The terms, it is to be noticed, did not include the finding of sites for the plaintiffs by Downham. Soon after Downham's appointment, the plaintiffs decided upon a change in their operations so as to undertake the private development of land, which would involve the finding of suitable sites, either with already-existing planning permission or without. For this the plaintiffs needed a manager, and Cottrell took the view that Downham was well qualified for the post. At the end of May or early in June 1970 the parties came to fresh arrangements which, as I find, constituted a variation of the first agreement. It was agreed that Downham should search for sites which the plaintiffs might develop privately as building sites. It was further agreed that Downham should report any such sites coming to his notice, that the suitability of the sites should be discussed between Downham and Cottrell, and that it would be for Cottrell to give authority for a bid to be made.

I turn now to the land which is the subject of the claim for recovery of a secret profit. In June 1970 the plaintiffs were engaged upon the development of land in Pennington Lane, Stansted, a project for which Downham had full responsibility. Adjoining the site was an area of land in which the plaintiffs were interested, since, if it could be acquired, it would make an obvious extension to their existing site. The adjoining land was 1.2 acres in extent, and belonged to the Trustees of the Mary Macarthur Holiday Homes for Working Women. It is convenient to call the adjoining land "the trust land." In July 1970 there was a meeting between Cottrell, Drysdale and Downham at which it was pointed out by Cottrell that the acquisition of the trust land would be highly advantageous, and that steps should be taken to discover who the owners of the trust land were. Downham was asked to undertake this task. During the next 18 months, Cottrell asked Downham repeatedly if he had discovered the identity of the owners of the trust land, and Downham said that he had not. Cottrell had asked Downham to make inquiries from the appropriate local authorities. Downham, it is said, did write to the Essex County Council, but not until November 16 1970, and no reply to those inquiries has

A been found. It is said that on December 11 1970 Downham wrote again to the Essex County Council, but no reply to that letter had been found. Towards the end of 1970, Cottrell began to be dissatisfied with Downham's work as construction manager, and particularly with the state of things at Stansted. About May 19 1971, Cottrell and Downham, and probably Drysdale, went together to inspect the trust land. Downham was again asked if he had discovered the owners, and he said no. Cottrell soon after concluded that he must relieve Downham of his post, and on May 28 1971 he said so to Downham, who at once produced from his pocket a letter of resignation. This reads, "Dear Mr Cottrell, With
B some regret I hereby tender my notice of resigning my post with Headway Construction Company Limited, the resignation to become effective at the end of August 1971." That period was extended to the end of September 1971. Cottrell was not certain of the date, but I am satisfied that in June 1971 Cottrell once again asked Downham if he knew the owners of the trust land, and was told he did not.

Although dissatisfied with Downham's work as a construction manager, Cottrell was of the view that Downham had a flair for discovering building sites, and by June or July 1971 the question of a new agreement arose. Cottrell wrote on July 14 1971 a letter setting out the terms of the commission
C agreement, and in part that letter reads:

> "You to spend three days per week in the search for building land. This land to be acquired on an option to purchase, subject to planning permission, including, if necessary, an appeal following the rejection of planning permission. You undertake not to seek or accept full-time employment during the course of this agreement. We undertake to pay you a retaining fee of £2,000 per annum, payable quarterly. This fee to include all personal and travelling expenses. Charges associated with the land will of course be borne by Headway. The acquisition of land will be paid for on a commission basis, and will be payable in each and every case at the completion of sale. The following is the
> D agreed commission scale which will be used as the minimum basis of remuneration, but which may be varied by agreement between us if the circumstances of any one acquisition are particularly favourable to Headway. The commission will apply in two parts: (a) the number of houses contained in the planning permission of each scheme, at a fee of £50 each where under £10,000 of sale value, £75 each where over £10,000 of sale value; (b) the difference in value between the amount paid by Headway and the open market value of the land, to be assessed and agreed, and a fee of 10 per cent of this difference in value to be paid on completion of sale, up to a maximum of £5,000."

Downham's acceptance of these terms is dated July 17
E 1971. It was agreed that the service agreement, as varied, should remain operative until September 30 1971 and that the commission agreement should be operative as from October 1 1971. Mr Baker has submitted that the commission agreement contemplates merely options and not sales, but as it seems to me, both parties have throughout regarded it as dealing with sales or purchases, and Downham himself bases his counterclaim on such a premise.

I now look at the position of the second defendants, Bardcrest. This company was incorporated on May 21 1971. It was a ready-made company, and was acquired by Downham on June 4 1971, that is to say, a week after Downham had
F given notice to the plaintiffs, and, as I find, shortly before Downham once again told Cottrell that he had not discovered the owners of the trust land. The registered office of this company, Bardcrest, was 33 Kings Road, Berkhamsted. The company's file shows a director to be a Mr Howson, said to be a cousin of Downham, and Downham's wife as secretary. On May 15 1972 Mr Howson resigned and Downham was appointed as a director, and the registered office of the company was changed from 33 Kings Road, Berkhamsted to 30 Baker Street, London W1, an accommodation address. Downham told the court that at the end of June 1971 in a public house he heard some discussions about the ownership of the trust land, and after talking to the landlord and others he G
traced the owners through the Charity Commissioners. In June 1971 Bardcrest got in touch with Kershaw & Co, solicitors for the owners, who on July 1 1971 sent Bardcrest a plan of the trust land. On July 13 1971 Bardcrest wrote to the owners. The address given is Suite 5, 29-30 Warwick Street, London W1, but there is a note beneath that: "During July, please reply to 33 Kings Road, Berkhamsted." That letter is signed by B. Howson, and beneath his signature there appear what purport to be the names of the directors: B Howson, J A Greenway and R C Edwards. J A Greenway and R C Edwards, as is admitted by the defendant Downham, were entirely fictitious. So Bardcrest wrote to the owners of the H
trust land, saying "We have pleasure in submitting our formal offer for the purchase of the above land. We are prepared to offer £20,500 subject to contract, and trust that you will be able to accept our offer at an early date." On November 8 1971 Kershaw & Co, on behalf of the owners, wrote to Bardcrest, saying among other things: "Since your offer was made, it is felt that an offer of £24,000 would be an offer which the trustees should accept, subject to the approval of the Charity Commissioners. Perhaps you will kindly let us know by return whether your company would be interested." Then on November 12 1971 the solicitors for Bardcrest wrote to Kershaw & Co saying: "We confirm that J
our client is most certainly interested in pursuing this matter at the price of £24,000, and await hearing from you with a draft contract as soon as possible." On December 7 Kershaw & Co wrote to the solicitors, Masons, acting for Bardcrest, saying: "We are pleased to be able to say that we have heard from the Charity Commissioners that they are prepared to authorise the proposed sale, subject to the result of the publication in the locality, and by advertisement, of preliminary notices. The commissioners require that the notice should be placed on a suitable public notice-board at Stansted, and also advertised in a local paper and the *Financial Times* or the ESTATES GAZETTE." K

In the meantime, namely in November of 1971, Cottrell was told that the owners of the trust land had been traced, and Downham mentioned a price to the plaintiffs of £42,000. Downham was instructed to pursue the offer. On January 1 1972 the *Financial Times* carried an advertisement headed, "Charity Commission; property for sale. (Higher offers invited) building land," and the text included these passages: "Approximately 1.2 acres, Pennington Lane, Stansted, Essex. . . . The Trustees of the Mary Macarthur Holiday Home for Working Women propose to sell the above-mentioned property for £24,000. . . . Higher offers are invited." L
This advertisement came to the notice of Cottrell, who called Downham in and asked him if it was the trust land which was referred to. Downham became ill at ease, and said he would look into it. A week later Downham told Cottrell it was not the land in question, and that the advertisement had been inserted without authority. That was untrue. On February 2 1972 Kershaw & Co wrote to Bardcrest saying that the Charity Commissioners were being asked to approve the sale to Bardcrest at £42,000. Downham told Cottrell that the price had gone up to £44,000, and then to £48,000, and that there was a counteroffer of £56,000. Thereupon the plaintiffs agreed to pay for the trust land the sum of £57,500. M
On February 3 1972 the plaintiffs' solicitors sent a contract to Bardcrest's solicitors agreeing that the plaintiffs would pay £57,500 for the trust land.

In my opinion the only sensible inference to be drawn is that Downham acquired the ready-made Bardcrest company for the express purpose of perpetrating this fraud upon the plaintiffs. Mention was made of two other ventures dealt with through Bardcrest, but these it was said came to nought. Downham deliberately hid from the trust his connection with the plaintiffs, and he deliberately hid from the plaintiffs his interest in Bardcrest. This is admitted to be a case of secret profit. In my judgment it is moreover a case of fraud. A well-recognised definition of the phrase " to defraud " is " to deprive by deceit," and that is exactly what Downham did. When cross-examined by Mr Machin, Downham gave evidence to this effect: " It was my duty to find the owner of the land, and having done so to tell Headway who it was. I broke this duty quite deliberately for my personal gain. In November I told Cottrell it was Bardcrest, but I did not say that was me. When I told Cottrell about the *Financial Times,* what I told him about the *Financial Times* advertisement was untrue. If I told the truth, the deal would be off. It was a lie to secure profit." Part of the defendants' case is that Downham consulted counsel as to the legality of Bardcrest selling the trust land to the plaintiffs. The opinion is not in writing, but a solicitor's note has been mentioned, though not put in. I accept that Downham, with a solicitor, did consult counsel, and that counsel may have given some advice which Downham thought helpful to him. I am also satisfied that counsel did not have the full true facts before him. Again, Downham's evidence was to this effect: " I did not tell counsel it was my duty to tell the plaintiffs if I found out the identity of the owners. I did not say the two names on the Bardcrest notepaper were fictitious. I did not say that if it got out that the company and I were the same the whole deal would be off." I have not heard counsel or solicitor in regard to this part of the history called as a witness.

Before I consider the counterclaim, two points arise on the claim, (1) whether expenses of the sale should be deducted from the £15,500 profit, and (2) whether the plaintiffs are entitled to something more than the figure of £15,500, or the figure reduced by reason of deducting expenses, upon the ground that Cottrell, had he known of the owners and personally conducted the negotiations with proper expedition, would have obtained the trust land for a sum less than the £42,000 which Bardcrest paid for it. These questions must be considered in the light of my findings, which are: (1) Downham was in breach, and this is admitted, of the implied terms in the service agreement that he would serve the plaintiffs with good faith and fidelity: in the case of *Sanders v Parry* [1967] 2 All ER 803, which was cited, Havers J reviewed the authorities; (2) Downham was in breach of the express terms in the service agreement, as varied, that he would search for the owners of the trust land and make known their identity to the plaintiffs; and (3) Downham was guilty of fraud, deceit, and the making of a secret profit.

As to the first issue, namely whether expenses can be deducted by Downham, it is a significant feature of the history here that—and it is quite unusual—there were two separate conveyances, one of the land from the trust to Bardcrest, and a second from Bardcrest to the plaintiffs. This, it is said, would involve a double stamp duty. It seems to me on the material I have that had the plaintiffs negotiated directly with the trust, expenses would not have been incurred, so that on this aspect the damages would be £15,500. On the second issue, namely whether Cottrell could have bought the trust land at less than £42,000, I have this evidence. Mr Cottrell was experienced in land valuation; although he held no professional qualifications he had great experience, and according to him, had he heard in June or July 1971 who the owners were of the trust land, he would have instructed Downham to negotiate a purchase as soon as possible, and there would have been an advantage in that the plaintiff company had the resources to pay, as he put it, " on the nail." At the time land values were rising sharply, and he said: " I am sure we could have got the land for less than £42,000." Two main reasons were advanced, one of which I have mentioned, namely that the resources of the company were sufficient to enable a cash payment to be made; and secondly, that land values were increasing and with proper expedition he (Cottrell) thought he could have secured this property perhaps for as little as £30,000. I have noted that this particular point is not specifically pleaded, but Mr Machin said he had informed Mr Baker of it before the case started. Mr Baker contends that to find a figure here would be mere conjecture. I cannot agree, and I note the case of *Chaplin v Hicks* [1911] 2 KB 786. The proper approach to this problem, I think, is to reach (if the evidence warrants it, of course) a fair and reasonable figure for the loss of the chance to obtain the trust land at less than £42,000. On the evidence, I value that lost chance at £3,000. It follows that on the claim, the plaintiffs are entitled to judgment for the sum of £18,500.

I turn, then, to consider the counterclaim. Downham claims to be entitled to commission in relation to four lots of property: (1) Roundswood Park, Harpenden; (2) Burnham Green, Tewin; (3) Manor Farm, Hurst Cottage and Eastleigh at Bishop's Stortford, and (4) Bishop's Stortford Golf Club. Included in the counterclaim at the outset was a claim for a commission on the purchase of the trust land. Not surprisingly, it is now withdrawn. On the four lots of property which I have mentioned, the commission claimed is quantified only in the case of Roundswood Park. As to the other three, the claim is for a declaration that commission is payable. I consider first the Roundswood Park claim. This site was found in July 1971; there were four owners involved, and on March 10 1972 planning permission was obtained. The plaintiffs agree that Downham dealt with the transaction and carried it through, so that apart from any question of avoiding the contract, Downham would be entitled to commission. That would be based on the term appearing in the letter of July 14 1971, and Downham's entitlement, the plaintiffs say, would be £5,000 and £75 per house—12 houses, £900. That figure, £5,900, was the sum originally claimed, but the defendants, by amendment, now claim £13,475. The difference arises in this way. The first defendant, Downham, contends that the commission agreement should be interpreted as meaning that he was to have £5,000 in respect of each transaction, and not in respect of the site as a whole. Cottrell in evidence said that he had never heard it suggested that the agreement was a parcels agreement until three weeks before the case started, and contends that what the plaintiffs are concerned about is the site and not each acquisition separately. It is pointed out that planning permission is given for the site. The defendants' solicitors' letter of May 8 1972 supports the plaintiffs' contention as to this, and I find that the appropriate figure is £5,900. However, the next question, which is now to be considered, is whether the commission contract is voidable. Mr Baker submits, on the authority of *Nitedals Taendstikfabrik v Bruster* [1906] 2 Ch 671, that where transactions are severable, the fact that some are dishonest does not taint others which are not. With this principle, the court is in agreement. The plaintiffs indeed do not contend that the trust land fraud taints the other transactions: they say that the commission agreement itself is to be set aside for fraud.

Before the court can uphold this contention, it must be shown that there was misrepresentation by Downham before July 17 1971, the date when he agreed to the commission agreement. It must also be shown that such misrepresentation induced, and was intended to induce, the plaintiffs to enter into the contract. Downham maintained, as I have already pointed out, that he first discovered who the owners of the trust land were at the end of June 1971. I do not accept this evidence. As I have said, it seems to me clear that Bardcrest was acquired for the purpose of carrying out this fraud in relation to the trust land, and it follows that by then Downham knew who the owners of the trust land were. Mr Machin points out that by July 13 1971 Bardcrest notepaper had been printed, and he further asked the court to observe that no one was called by the defendants, apart from Downham, to deal with the purpose for which Bardcrest was acquired. I am satisfied that about three weeks before July 1 1971 Downham

A found out who the owners of the trust land were, and that between then and July 17 he falsely told Cottrell that he had not. I further find that this specific misrepresentation was dishonest and was made in order to induce the plaintiffs to enter into the second agreement, that is to say, the commission agreement. It is true that for a long time Downham was in ignorance as to the ownership of the trust land, but when he discovered who the owners were it was his duty to tell the plaintiffs, and if a specific falsehood were not proved (and I find it was proved), silence on the part of Downham would, in my view, amount to misrepresentation inducing the making of the commission agreement. I refer
B to *With v O'Flanagan* [1936] Ch 575, and *Cavendish-Bentinck v Fenn* (1887) 12 App Cas 652. The findings which I have reached on this part of the case are in large measure admitted by Downham, who in the course of cross-examination said: "It was beneficial to have the commission agreement. I would have made £30,000 over three years, and in July 1971 I anticipated securing commission of that order. If Cottrell had known or had any idea about Bardcrest there would be no question of any commission. By early 1971 I had conceived the scheme to buy the land through Bardcrest, and a commission contract came soon after. It was vital that Cottrell should not be told. One reason I did not tell Cottrell
C was so that the commission contract could go on." It is the evidence of Cottrell, and the opinion of Downham in evidence, that the commission agreement would not have been entered into by the plaintiffs had they known that Downham had discovered the owners of the trust land and was in negotiation with them for his own gain. Mr Baker submitted that Downham should succeed in his claim on the counterclaim on grounds of equity, but I cannot but feel that Downham, by his dishonesty, had rendered himself disentitled to invoke a remedy in equity.

Turning, then, to the claim based on the Burnham Green,
D Tewin, land, this site was found by Downham, but in July 1971, that is, during the currency of the service agreement. It involved six acquisitions. Of the three properties Manor Farm, Hurst Cottage and Eastleigh, all at Bishop's Stortford, the first two were found not by Downham but by Strutt & Parker, estate agents, of Chelmsford. All three were found during the currency of the service agreement. Mr Baker points out that although that is so, they were brought within the purview of the commission agreement, as was the Burnham Green site at Tewin, by the plaintiffs' letter of March 13. In regard to these transactions, the plaintiffs assert that not only are they voidable in the circumstances before
E referred to, but that for their inclusion in the commission agreement there was no consideration moving from the defendants. As to voidability, again, when cross-examined, Downham said that if Cottrell had had any idea about Barncrest, he would not have thrown Tewin or Harpenden or Hurst Cottage into the commission agreement. If Cottrell had known about Bardcrest, he would not have written the letter throwing in the Bishop's Stortford Golf Club. It seems to be the situation that Downham did nothing as a result of the plaintiffs' inclusion of these transactions within the commission agreement which he was not already bound to do under the service agreement. In other words, there was no
F *quid pro quo*. Then as to the golf club transaction, Downham was responsible for the idea that part of the golf course might be acquired for building purposes, and this was during the currency of the commission agreement. He played no part, however, in the obtaining of planning permission, but in the correspondence it is correct, as Mr Baker points out, that on March 13 1972 Downham suggested to Cottrell that he should have, in respect of the golf club, a fixed fee of £1,000 and a house fee of £200 a house over £10,000 in value, and £250 over £20,000 in value. To that the plaintiffs replied rejecting the suggestion, and saying: "I also agree that Bishop's Stortford Golf Club is a special case, but I am reluctant to change the format of our agreement too much. G
I confirm that the golf course will be treated as one acquisition within the meaning of our agreement, but I suggest the following modifications: The maximum payment of £5,000 referred to in our agreement be raised to £7,500, £2,000 or £3,000 to be payable immediately," and that was accepted by Downham, and indeed the fee, the special fee of £1,000, was paid by the plaintiffs to Downham, and that is recorded. In this instance, therefore, as in all these other instances, the plaintiffs rely upon Downham's misrepresentation as rendering the commission agreement voidable. The court accepts that contention, and the counterclaim fails.

Judgment was entered for the plaintiffs against the first H
defendant for £18,500, with costs of both claim and counterclaim. Interest was awarded at $8\frac{1}{2}$ per cent from the date of the conveyance of the Stansted land, March 27 1972. Counsel being in no position to offer undertakings with regard to disposal of assets, the defendants were awarded a stay limited to the excess of the amount of the judgment over £5,000, with liberty to apply should they prefer to offer undertakings in respect of the whole.

QUEEN'S BENCH DIVISION J

November 28 1974

(Before Mr Justice CUSACK)

GROSS FINE & KRIEGER CHALFEN v GAYNOR

Estates Gazette March 15 1975
(1974) 233 EG 1015

Did auctioneers secure instructions to include in sale a property sold by another agent by private treaty a few days later?—Yes—Claim to commission, etc, on basis of auction sole agency upheld

This was a claim by Gross Fine & Krieger Chalfen, estate K
agents and surveyors, of Princes Street, Hanover Square, London W1, against Mrs Emily Gaynor, of Green Lanes, St Albans, Hertfordshire, for damages for breach of a contract for sale of "Longwood," Holders Hill Road, Hendon, London NW4, by auction.

Mr R F Nelson (instructed by Slater, Fruitman & Co) appeared for the plaintiffs, and Mr L A Marshall (instructed by Rose & Birn) represented the defendant.

Giving judgment, CUSACK J said: This is a claim for commission made by the plaintiffs, who are a firm of estate
agents. They claim that they are entitled to commission on L
the sale of a property known as "Longwood," Holders Hill Road, London NW4, which was the property of the defendant, Mrs Gaynor. The basis upon which they claim commission is that they say they were appointed not only as agents for the sale of that house, but as sole agents for the sale of it, and that it was in fact sold through the agency of another estate agent, Mr M J Smith-Carington FRICS. The plaintiffs claim that their instructions were to sell the house at an auction which was to take place at the London Auction Mart on June 22 1972. In fact, before that date it was sold privately
on the defendant's behalf by Mr Smith-Carington. The M
defence to the claim is a denial that the plaintiffs were ever appointed agents by the defendant, either sole agents or at all.

The matter arose in the early part of 1972 when the defendant, who is a widow, was anxious to sell the house in which she was then living. It was a matter of some urgency, because she needed the money that would be gained by a sale in order to provide an income for herself. She discussed the matter with her daughter, Miss Gloria Gaynor, and thereafter it is not in dispute that Miss Gaynor took a prominent part in the matters which followed. It has not been pleaded, nor has it been submitted, that she was without authority to act on

A behalf of her mother. It is accepted that she was acting on behalf of the defendant, and it may be that her intervention in the transactions concerning the disposal of this house has been somewhat unfortunate in the result. A decision was taken, after discussion between mother and daughter, that the house should be sold. Some time early in April 1972 Miss Gaynor spoke to Mrs Smith-Carington, the wife of the estate agent to whom I referred earlier. The Smith-Caringtons were friends of the Gaynor family, and as a result of what Miss Gaynor said to Mrs Smith-Carington, Mr Smith-Carington on Tuesday April 18 came to visit the house—I expect he had visited before, but he came on business on this occasion—and B he saw both mother and daughter. According to Miss Gaynor, discussion took place as to how the disposal of the house should be dealt with. Her mother, the defendant, was then adamant that she did not wish it to be disposed of by auction. Miss Gaynor herself took the same view, as she thought it would take too long and place too much of a strain upon her mother. I think Mr Smith-Carington confirms that that was the situation at that stage. A few days later, on April 24, Mr Smith-Carington visited the house again. He took some measurements and considered whether it should be sold as a residence for redevelopment. A little later he made inquiries about the position concerning planning permission, and C ultimately he came to the conclusion that a better price would be obtained if it were sold for redevelopment.

In the meantime, Miss Gaynor had read an article in a newspaper about the success that was being achieved in selling houses, particularly houses with land, at auction. The newspaper article mentioned the name of the plaintiff firm, Gross Fine & Krieger Chalfen, and also referred to sales taking place at the London Auction Mart. As a result of the favourable impression she derived from that article, Miss Gaynor telephoned the London Auction Mart, found out further particulars of the address and so forth of the plaintiffs, D and on April 25 or thereabouts telephoned the plaintiff firm and spoke to Mr Clifford Krieger FSVA, who is a partner in that firm. Mr Krieger's account of the matter given in this court was to this effect. Miss Gaynor wished to explore the possibility of a sale as quickly as possible, and he told her that there was an auction due to take place on June 22. He explained to her that normally about eight weeks was required to make the necessary arrangements to deal with the publicity and print the various documents that were required, and he told her that it was inherent in instructions to sell by auction that the auctioneers should be the sole selling agents. According to him, he explained about the charges for expenses, which E he said he hoped would be limited to £300, and said it was vital, if the house was to be put in the sale of June 22, that instructions should be given at once. According to him, Miss Gaynor said she would like him to proceed with the matter, and asked him to make an appointment with her mother to view the house. Miss Gaynor, as I understand it, was not herself living at the house at the time, she was living elsewhere. Mr Krieger said he also required the name of the solicitors who would be acting, so that that could be included in the published particulars.

Miss Gaynor's account of that telephone conversation is somewhat different, because she said that she had no intention F at that stage, or indeed at any stage, according to her, of giving firm instructions for the house to be auctioned. She telephoned Mr Krieger, asked if he could help, and according to her she said that her mother was thinking of selling the property at Hendon, and asked whether he could go along to see her mother and view the property. Then there was some discussion as to what the nature of the property was, and she described it as a big house divided into flats, with some ground; "a smallish garden back and front" is what she believes she said. He asked her what her mother's telephone number was, and that was given. According to her that is really all that took place. Mr Krieger certainly did make an appointment to see Mrs Gaynor, the defendant, and did go G to see the property. But Mr Krieger, in the course of his evidence, told me that he believed that at that stage, on April 25, a contract had been concluded between him as estate agent and Miss Gaynor on behalf of her mother, to the effect that his firm should auction the property. I do not think that is so. It is not so pleaded, because the contract pleaded is a contract alleged to have been entered into on May 2, a date to which I shall refer later. Certainly I think that Miss Gaynor was going into far more detail in her conversation with Mr Krieger than she now remembers, and that she was given the information which he told me he had given to her over the telephone. The reason why I have come to the conclusion H that a contract was not concluded is that after Mr Krieger had visited Mrs Gaynor on April 27, a letter was written from his firm, dated April 28. I say dated April 28; the date was in fact typed as February 28, but everybody agrees that that was a mistake, and it was April 28. By the time that letter was written he had made his visit to Mrs Gaynor and he says that he had had a look at the house, but he did not have any specific conversation with Mrs Gaynor, or, as I understand it, enter into any specific arrangement with her. Mrs Gaynor's account of the matter was that he came, apparently to her surprise, although according to Miss Gaynor she had been informed of the pending visit. He came, he J talked about figures which Mrs Gaynor did not fully follow or understand, and in the end, according to her evidence, the effect of what she said was, "Go home and write to me."

Now, I can perhaps refer, before I go further, a little more to Mrs Gaynor. Mrs Gaynor gave evidence. She is an elderly lady. It is I think conceded, without any reflection upon her, that as with many other elderly people and some younger people, her memory is not very good. She is obviously not accustomed to legal proceedings, and certainly not to business transactions. In fact, although she quite rightly gave evidence as being the defendant in this case, her evidence was of very K little assistance to the court, because of her want of recollection and her evident confusion about certain matters. Therefore the case must really be determined upon evidence other than that coming from Mrs Gaynor herself. The letter to which I have already referred, that of April 28, was written in such terms that it would be quite inconsistent to find that an agreement had by that date been concluded, because it is a letter inviting instructions. It deals with the auction, refers to the auction which was to take place on June 22, refers to the limitation of expenses to a sum of £300, and refers to the payment of fees at the rate of 2 per cent on the sale price. All those matters had in my view, and I so find, been mentioned L in the telephone conversation with Miss Gaynor. Then it goes on to say: "Inherent in such instructions would be the appointment of this firm as sole agents with sole selling rights on your behalf for a date up to and including eight weeks after the auction." A little later on in the letter occurs this paragraph: "I would be most grateful if you would give this matter your immediate attention, and let me hear from you if possible in writing by return of post, so that I can then instruct the printers, photographers and advertisement programmes to be commenced without delay." The letter was addressed to Miss Gaynor, and a copy was sent to her mother. There was no reply to that letter, and the business on behalf M of the plaintiffs passed from the hands of Mr Krieger to another member of the firm who was assisting him, a Mr J A Salter ARICS. On May 1 Mr Salter himself visited the house and took some measurements, as to which a mistake occurred, to which I shall refer later. On May 2, no answer having been received to the letter of April 28, Mr Salter telephoned Miss Gaynor and inquired as to why there had been no reply to the letter.

In the background throughout these matters was the sense of urgency felt by the plaintiffs if the house were to be included in the sale of June 22, for if there were delay, it would miss what was the vital date, a date which they regarded as vital in the interests of their own client. Mr Salter told me that when he telephoned on May 2, what occurred was this. Nobody can be expected to pledge themselves to exact words, but he probably, he thinks, started by inquiring if Miss Gaynor had received the letter of April 28, and pointing out that there had been no answer to it. He then says he went on to request what he calls confirmation that his firm had been appointed sole agents, and explained exactly what that implied, including the significance of being sole agent in the sense that the Gaynors would be responsible to them whoever sold the house. He says that Miss Gaynor agreed. He could not remember her exact words, but according to him she accepted that the plaintiffs would be sole agents and confirmed their instructions to proceed with the auction sale. She did not raise any question about not being sole agents, and his only other point of inquiry was to ask the name of her solicitor in order that the necessary particulars of sale could be prepared. Miss Gaynor gives a differing account of the matter, because she says that certainly a telephone conversation took place, and there were, according to her, a number of telephone conversations, one of them initiated by her. She is not, I find, entirely clear, nor, I think, does she claim to be entirely clear, as to what was said at any particular conversation. She certainly had this inquiry about whether the letter of April 28 had been received, and she was told about the auction, and was asked "can we go ahead?" and according to her, her answer was that she must ask her mother. She telephoned her mother, her mother still did not want to auction but wanted to sell privately, and her mother said "leave it," and so nothing more was done. That is one account which she gave of the conversation, but she also gave an account of subsequent conversations, and at some stage, she agrees, she did give the name of the family solicitors, but she says only when she was told she would be chargeable for expenses and she answered that that matter must be referred to the solicitors. I should perhaps amplify that a little, because she said it was in the last telephone conversation of any that she had that she told the plaintiffs that she had sent their letter to her mother's solicitors. Mr Yaxley was the legal executive there acting, and she said she told them to contact him, as her mother did not want her to have anything more to do with the matter.

In deciding between the conflicting versions of that vital telephone conversation on May 2, it is necessary to take into consideration what matters may be found for or against the conflicting versions in the outside evidence. On May 2, Mr Salter wrote a letter to Mr Yaxley, to whom I have referred, at the solicitors' firm, Rose & Birn, of Enfield, and he said: "We write to inform you that we have been instructed by our mutual client Miss Gaynor to submit the above property for sale by public auction on June 22. Accordingly I would be obliged if you would kindly forward to me at your earliest convenience the general and special conditions of sale which are to be included in the auction brochure." That letter is wholly inconsistent with Mr Salter having been told on that very day that he was not to proceed with the auction and having been referred to the solicitors solely for the point of dealing with any expenses that might have been incurred. Next, Mr Salter put the question of the advertisements and the brochure in hand, again inconsistent with any belief on his part that he had not been instructed at all. Then later on, as one goes through the correspondence, there came a reply to his letter to Rose & Birn (the letter of May 2); that reply was dated May 8, by which time the private sale had been entered upon. Mr Yaxley, who had taken instructions from the Gaynors, wrote: "Thank you for your letter. We are now instructed that our client has sold by private treaty, and will therefore not be putting this property to auction." It is to be noted, though this is no criticism of Mr Yaxley, that that letter did not contain any statement, "We have never instructed you at all, you have not been our agents at any time," which is in fact the defence in this case. A further letter of May 10, again from Rose & Birn, does say that the plaintiffs had not been appointed sole agents, but it ends up by saying: "Doubtless you will let us have a note of the expenses incurred by you to date. Our clients are not prepared to agree to this property being put to auction, and you should incur no further expenses accordingly." Again, that letter is inconsistent with an assertion that the plaintiffs were never employed at all. If they were employed as agents, I am entirely satisfied that they were employed only on the terms that they were "sole agents." Not only was that their practice in relation to auctions, but the other evidence, including the evidence of Mr Smith-Carington, confirms that that is the practice of agents selling property at auction, as it is impossible to have more than one agent acting.

Now on Thursday May 4, going back a little in date, Mr Smith-Carington, who had at the beginning of the month, either on the 1st or 2nd, received an offer of £54,000 for the property, received a firm offer of £61,000. He went to see Miss Gaynor about it (it happened that Mr Yaxley was there at the time); she spoke to her mother, and it was in those circumstances that the offer of £61,000 was accepted through the agency of Mr Smith-Carington, who in due course drew his commission upon the sale. On Friday May 5, Miss Gaynor telephoned Mr Salter, or he telephoned her—there is a conflict there—I think it more likely that he telephoned her, but they spoke upon the telephone again, and she disclosed that there had been this offer from a private source, and he advised her "not to accept it in her own interest." According to him, what occurred during the course of that conversation was that he told her acceptance would be ill-advised, and she should wait for the advertisement which the plaintiffs proposed to insert on the 13th of the month, and if possible wait for the auction to take place, as he believed a better figure than £61,000 would be achieved. He believed that because at that time he himself had made inquiries about getting planning permission and had satisfied himself that the property was ripe for development. He also says that at that time there was a discussion about sharing the commission. Again the date of that discussion about sharing commission was not entirely clear. It may not have been on May 8, it may have been on May 5. But in any event, Miss Gaynor says that at no time was there any discussion about sharing commission. It was suggested by Mr Smith-Carington in his evidence that the plaintiffs were exerting considerable pressure. Although he qualified the suggestion and I think eventually withdrew it, there was in what he said implicit a suggestion that the plaintiffs were acting improperly in getting into touch with Miss Gaynor as often as they did and reminding her of the date of the proposed auction. I detected in Mr Smith-Carington's evidence a degree of prejudice which he later did not entirely remove. It is right that I should say I do not think the plaintiffs were exerting any improper pressure at all.

Now, the £61,000 was accepted by private treaty as I have indicated. It is against that background that I have to decide where the truth lies as to the terms of the telephone conversation on May 2 and to decide what its effect was. The conclusion I come to is that the evidence of Mr Salter should be accepted, not only because I regarded him as a truthful and accurate witness, obviously embarrassed by the situation in which he found himself as a person junior to Mr Krieger, bearing responsibility in court for this matter; but also by reason of the fact that the correspondence, although not always in the clearest terms, because if correspondence were invariably clear there would be very little litigation, does nevertheless seem to bear out what he says about the transaction that took place. I think probably the real truth of this matter is that neither mother nor daughter, certainly not Mrs Gaynor, nor probably Miss Gaynor, was at all experienced in business

A procedures, and that Miss Gaynor in her endeavours to help, putting it as kindly as I can, really made a hopeless muddle of things out of which she has now endeavoured, unsuccessfully I am afraid, to extricate herself and her mother. I find that she did give instructions on May 2, that the plaintiffs should be sole agents, and that she agreed to their terms as set out in the letter of April 28. It follows from what I have said that the plaintiffs must succeed in this action, and that raises the question of damages, because what is sought is commission on the eventual sale of the property, the private sale for £61,000. The claim amounts to £1,416.65 for commission and expenses. The commission the plaintiffs are
B certainly entitled to, but the the items of expenses which are set out in the further and better particulars require further consideration, because charges are made for advertising space, production costs, and then the charges of making corrections when the auction sale went off. I am not satisfied that the plaintiffs incurred the costs for advertising space and production costs which they claimed, because five other houses were at their instance at auction at the same time, and it seems to me on the evidence which I have, which is rather scanty on this particular matter, that the costs of advertising and production were in the end distributed among the other vendors and did not fall upon the plaintiffs themselves. The sum of
C money involved is comparatively small. The plaintiffs are however entitled to recover, in my view, as special damages the costs which they set out of correcting the block for printing and of the brochure, amounting to £57.48.

When I said just now that the plaintiffs were entitled to their commission, I had not overlooked the submission made to me by Mr Marshall that they really should not have the
full commission claimed, but should have some other figure D
to be assessed as best the court could. That submission was made to me on the basis that they made an error in the measurements of "Longwood" which they published in the preliminary auction particulars. The error arose because the property was described as being a site of 0.435-acre, whereas in fact it was 0.29-acre. Furthermore, it was described as having a frontage of 190 ft, whereas the frontage was in fact 90 ft. The error as to the acreage no doubt crept in as a result of the erroneous measurement of the frontage by adding 100 ft which were not in fact there. Mr Marshall submits to me that that error having been made and circulated,
although it could have been corrected at the auction sale, E
would have made the sale of the property the more difficult, and that it would not have fetched £61,000, as it did in the sale by private treaty. I take the view that although this error is most regrettable, I ought not to assume that it would not have been corrected in time, because the plaintiffs had written (as I have indicated earlier) to the solicitors asking for the various general and special conditions. They would undoubtedly before the sale have inquired further into the position regarding the house. Indeed I do not think I ought to assume that in those circumstances this error would have gone uncorrected, or indeed, that if it had not been discovered
until the actual time of the sale it would have made a differ- F
ence to the price then obtained. It was a buoyant market at that time, and later in the same year the property was sold for a figure of over £70,000. So it is in those circumstances that, as I have already indicated, I think the plaintiffs are entitled to recover their full commission as pleaded, and the special damage limited to £57.48. There must be judgment accordingly, with costs.

HOUSING ACTS

A **QUEEN'S BENCH DIVISIONAL COURT**

November 4 1974

(Before Lord WIDGERY CJ, Mr Justice BRIDGE and Mr Justice SHAW)

WOLKIND v ALI

Estates Gazette February 22 1975

(1974) 233 EG 759

Notice under section 90 of the Housing Act 1957 regulating overcrowding of lodging-houses does not serve to regulate use of the same premises by members of a single family—
B **Conviction of person with care and control quashed by majority**

This was an appeal by Mr Jack Wolkind, clerk of the London Borough of Tower Hamlets, from a decision of the Inner London Crown Court, given on October 19 1973, allowing an appeal by Mr Nimar Ali against his conviction at Thames Magistrates' Court on July 24 1973 on two informations alleging contravention of a notice served under section 90 of the Housing Act 1957.

Mr J A Blair-Gould (instructed by Edward Fail, Bradshaw & Waterson) appeared for the appellant, and Mrs A M
C Williams (instructed by V J Gersten & Co) represented the respondent.

Giving the first judgment, BRIDGE J said that back in 1967 the respondent was in occupation and had care and control of the ground and basement floors of a house at 15 New Road, London E1, and was using two rooms on the ground floor and two in the basement for the accommodation of lodgers. On June 9 1967 the local authority, in effect the present appellants, served notice upon the respondent under section 90 of the Housing Act 1957, prohibiting the use of the basement rooms as sleeping accommodation and restricting the use of the ground-floor rooms as
D sleeping accommodation to not more than two persons in each room. Presumably the restrictions imposed by the notice were observed by the respondent during the six years or more that the premises continued to be used by him as a lodging-house. In 1973, however, that state of affairs came to an end; the respondent ceased to use his premises for lodgers, and was joined by his large family from Bangladesh. At the time the informations were laid against the respondent, his four sons were sleeping in one ground-floor room and he, his wife and his daughter were sleeping in the other ground-floor room. In the magistrates' court the respondent was convicted of an offence against section 90 (4)
E of the Act of 1957, on the basis that he was clearly within the ambit of the language used in the subsection, but his appeal to the Crown Court succeeded, the court reaching the conclusion that, by inference, section 90 did not apply to F
members of one family.

He (his Lordship) thought that the Crown Court was right in its opinion that when one compared the statutory code for controlling overcrowding in dwelling-houses generally with the very different code specially provided for controlling overcrowding in lodging houses under section 90, it was evident that the latter was never intended to apply to any premises except for so long as they were let in lodgings or occupied by members of more than one family. Under section 77, what determined the point at which the number of persons sleeping in a house passed the limit and the house became overcrowded
was a precise mathematical criterion laid down by the Act, G
whereas under section 90 the criteria to be applied in the case of a lodging-house were entirely at the discretion of the local authority. If the argument now presented for the appellant were right, it would lead to a remarkable and anomalous situation, because a notice served under section 90 at a time when premises were used as a lodging-house would continue to apply so as to restrict the number of persons who might be permitted to occupy the premises after they had ceased to be a lodging-house. The result would be that Mr Smith living in one half of a semi-detached house in a suburban road could legitimately house his whole family in H
it without committing any offence under section 78, whereas Mr Brown next door in an identical house and with an identically-sized family would be committing a continuing offence under section 90 by reason only of the historical accident that at some remote date in the past he had used his premises for a short period to let out to lodgers and during that period had been the recipient of a section 90 notice which the local authority had subsequently refused to withdraw. That would be an absurd situation. The appeal should be dismissed.

SHAW J agreed with the judgment of Bridge J.

Dissenting, LORD WIDGERY said that, in his opinion, once a J
valid notice was served under section 90 one had no need to look at the language of other sections. Section 90 was in perfectly plain and unambiguous language. He (the Lord Chief Justice) appreciated the anomaly which would exist if his view were accepted by the court, because by virtue of the section 90 notice having been served the present respondent would be subject to the more stringent restrictions on overcrowding than to the restrictions which would have applied under earlier sections of the Act to family dwelling-houses. What he (his Lordship) would like to see was an amendment to the Act requiring a local authority to withdraw a section 90 notice once it was established that the circumstances no K
longer applied.

The appeal was dismissed with costs.

LANDLORD AND TENANT
GENERAL

A COURT OF APPEAL (CRIMINAL DIVISION)

November 9 1973

(Before Lord Justice ROSKILL, Lord Justice JAMES and Mr Justice TALBOT)

R v SMITH

Estates Gazette January 11 1975

(1973) 233 EG 147

Damaging fixtures not an offence under section 1 (1), Criminal Damage Act 1971, if accused honestly believed fixtures were his own property—Reasonableness of belief is irrelevant provided accused acted honestly

B This was an appeal by David Raymond Smith, of Northey, Laindon, Essex, against his conviction at Woodford Crown Court on June 28 1973 on a charge of causing criminal damage to fixtures in property of which he was tenant, contrary to section 1 (1) of the Criminal Damage Act 1971.

Mr J R T Rylance (instructed by E Edwards, Son & Noice) appeared for the appellant and Mr L Gerber (instructed by the Solicitor to the Metropolitan Police) represented the Crown.

Giving the judgment of the court, JAMES LJ said that it was the first time section 1 (1) had been considered by the Court of Appeal. The appellant, who was 28 years old, was a student. As tenant of a flat and conservatory in Freemasons Road, North Woolwich, London E16, he had installed electric wiring for
C stereo equipment, put up roofing material and asbestos wall panels and laid floor-boards, all with the landlord's permission. These fixtures became part of the landlord's property. When the appellant left, he removed the wiring, causing £130 worth of damage to the roofing, wall panels and floor-boards, in the honest belief that he was damaging his own property. This was the appellant's only defence to the charge, but the defence was withdrawn from the jury by the trial judge (Deputy Judge A H Tibber), who directed the jury to convict. The appellant was conditionally discharged for 12 months and ordered to pay £40 compensation. The *actus reum* in section 1 (1) was "destroys or damages any property belonging to another." By applying the
D ordinary principles of *mens rea*, one found in the words "property belonging to another" elements such as intention, recklessness or absence of lawful excuse which were necessary to the offence. No crime was committed within the ambit of the subsection where a person who destroyed or damaged property belonging to another did so in the honest, though mistaken, belief that the property was his own. Provided that the belief was honestly held, it was irrelevant to consider whether or not the belief was justifiable. The conviction accordingly could not stand.

The appeal was allowed and the conviction was quashed.

E

COURT OF APPEAL

June 12 1974

(Before Lord Justice STAMP, Lord Justice ROSKILL and Lord Justice SCARMAN)

HESLOP AND ANOTHER v BURNS AND ANOTHER

Estates Gazette January 18 1975

(1974) 233 EG 263

House occupied by family for 20 years without any formal arrangement with friendly owner—Family held licensees only, not tenants at will—Executors' claim to possession not barred by statute F

This was an appeal by Mr John Heslop and Midland Bank Trust Co Ltd, as executors of the late Edward Timms, from a judgment of Judge Wingate-Saul at Lambeth County Court on August 24 1973 dismissing their claim against Mr and Mrs Robert Burns for possession of 116 Warner Road, Camberwell, London SE5.

Mr P N Garland QC and Mr G M Shillingford (instructed by Cameron, Kemm, Nordon & Co) appeared for the appellants, and Mr N P Wall (instructed by V J Lissack) represented the respondents.

Giving judgment, STAMP LJ said that the judge had refused G
the executors possession on the ground that their action was barred by the Limitation Act 1939. It was common ground that subject to the Act, the executors owned the property, and that the defendants had occupied it since 1954. The defendants' case was that from the outset they were tenants at will of Mr Timms; that no rent was ever paid; and that by virtue of section 9 of the 1939 Act, the tenancy was accordingly deemed to have determined on the expiration of one year from the commencement thereof. Thereupon, they said, a right of action accrued to Mr Timms which became barred in 1967, after 12 years, under section 4 (3) of the Act. The judge accepted that submission. The executors claimed that there never was a tenancy at will and that the defendants were let into occupation as mere H
licensees, having no legal interest in the land. The defendants conceded that if that were the true position the executors were not statute-barred.

Was there a tenancy at will or a licence? Mr Timms owned properties in Camberwell. In 1951, the defendants had recently come to London and were living in an attic. Mr Burns was a bus conductor, and his wife, who was expecting a child, was looking for light work. Mr Timms engaged her as a cleaner in his office. He seemed to have become fond of her. He bought a cottage in the neighbourhood for her and her husband to live in, and became godfather to their daughter. A year later he moved them into a house in Love Walk, and then in 1954 into the house J
in Warner Road, where they had lived ever since. He used to visit the family regularly, sometimes twice a day. He was extremely generous to them, paying the rates of the houses and not requiring any rent. There was no question of either of them doing any work for him in return for accommodation, so that no service tenancy came into issue. The judge found as a fact that Mr Timms intended to provide a home for the family. In evidence, Mr Burns said that the deceased said about each of the three dwelling-houses when the defendants were in them that "if anything happened to him, the premises would be left to us." Asked, towards the end of her cross-examination, "If the property was regarded as being yours, did it occur to you that K
Mr Timms was being more than usually generous by paying all the rates, etc?" Mrs Burns answered, "No, he had our company; he could come into the house any time he wanted." Se said in evidence that in 1954 the deceased told her he was buying 116 Warner Road "for her," that he told a Salvation Army welfare officer that he had done so, and that at one time he had said, "Don't worry about the house; it's yours." She was shocked when, following the death of the deceased, she found that the property had not been left to her; and he (his Lordship) was bound to say that he had the greatest possible sympathy for

A her in that regard. On the facts before the court, however, he thought it abundantly clear that the parties did not enter into any arrangement, far less any arrangement intended to create a legal relationship, as to the terms on which the defendants should occupy this property. There was no contract, no arrangement, no statement by Mr Timms. The defendants, as he (Stamp LJ) saw it, were allowed to move into the property and occupy it simply as a result of the bounty of the deceased and without any arrangement as to the terms on which they should do so. There was no evidence of any discussion whatsoever taking place as to the terms of the occupation. It was by the effect of the deceased's bounty, or doubtless because of his
B feelings of affection for Mrs Burns, that the home was provided.

The fact that the deceased had already said, in relation to each of the first two houses, that he would leave them to the defendants was no evidence that it was his intention that the defendants should be tenants at will in the meantime rather than licensees. Counsel for the defendants had submitted that since the deceased intended to provide the defendants with a home, he must have intended to give them an interest in the property. But a tenancy at will was no more apt to achieve that purpose than a revocable personal licence to occupy the property; and if one asked the question what interest he intended them to have, it could only, consistently with counsel's submission, be an interest during the rest of the life of the deceased of such a
C nature as would exclude him from any right to turn them out. No such interest was created, and he (his Lordship) found it impossible to infer an intention on the part of the deceased to create such a situation. In his judgment, the proper inference was that the defendants at the outset entered into occupation of the premises as licensees and not as tenants at will; not with a right to exclude the deceased from possession. In the course of the debate the court had been referred to a number of passages in judgments where the effect of exclusive possession in the determination of the question of tenancy or licence had been discussed. As Somervell LJ had indicated in *Cobb v Lane* [1952] 1 All ER 1199, the expression "exclusive possession" in relation
D to an occupier of property might be used in more than one sense. It might mean that the occupant occupied the premises alone or together with his family only, and did not share them with any other person. Such a situation was not inconsistent with the occupation being enjoyed under a mere licence. Or the expression might be used to mean that the occupant had a right to exclude the owner from the premises. Such a situation was clearly more difficult to reconcile with the existence of a mere licence to occupy, but that point did not arise here. The deceased could, as Mrs Burns remarked, "come into the house any time he wanted," and in a conversation of which evidence had been given she had said that he regarded the house as his "second home." The house was not to be the defendants' castle, but a
E house in which he allowed them to live.

The county court judge had attached importance to the fact that documents had been found among the deceased's papers indicating that for income tax purposes he had created a mythical tenancy of 116 Warner Road naming Mr Burns as tenant and recording fictitious payments of rent by him. But since the defendants denied the existence of a tenancy at a rent, and indeed it was common ground that they never paid or agreed to pay rent, he (Stamp LJ) did not think that one was led to the conclusion that as a result of these entries and the payment of tax, the defendants were intended to be tenants at will. Whatever the deceased's motives in making up his books
F in the way described, the facts disclosed could not support the inference that the defendants occupied the premises otherwise than as licensees. The appeal should be allowed.

Roskill and Scarman LJJ concurred, and an order was made accordingly. Execution was directed to issue only with the leave of the county court judge.

QUEEN'S BENCH DIVISION G

March 19 1974

(Before Mr Justice PHILLIPS)

NATIONAL WESTMINSTER BANK LTD v BETCHWORTH INVESTMENTS LTD

Estates Gazette April 5 1975

(1974) 234 EG 43

Break clause in lease—Landlords send notice to tenant company's former registered office by mistake—Was this nevertheless service at the company's "last known address"? —No, because the last known address for the purposes of the clause was that of the company's managing agents, with whom the landlords had dealt for some years. H

By this summons National Westminster Bank Ltd, of Lothbury, London EC, sought a declaration that a notice posted by them in accordance with a break clause in a lease of property in High Street, Godalming, Surrey, was a valid notice for the purposes of the clause. The defendants, Betchworth Investments Ltd, of Dover Street, London W, counterclaimed a declaration that they continued to be the tenants of the premises in question for the remainder of the term granted by the lease.

Mr P Curtis-Bennett (instructed by Rubinstein, Nash & J Co) appeared for the plaintiffs, and Mr D A Wood (instructed by Roche, Son & Neale) represented the defendants.

Giving judgment, Phillips J said: By a lease dated October 5 1966, Westminster Bank Ltd demised to Betchworth Investments Ltd certain property at 77 High Street, Godalming, Surrey, to hold from March 25 1966 for a term of 14 years determinable, nevertheless, as thereinafter provided; and by clause 4, it was provided that:

> If either of them the lessor or tenant shall be desirous of determining the present lease on the twenty-ninth day of September one thousand nine hundred and seventy three and of such its desire shall give six months' previous written notice to the other K of them . . . this present lease shall cease.

By subclause (3) of that clause, it was provided:

> Any notice under this lease shall be in writing and any notice to the tenant shall be deemed to be sufficiently served if left addressed or sent by post to it on the demised premises or sent to it by post at its registered office or last known address in Great Britain or Ireland and any notice to the lessor shall be deemed to be sufficiently served if sent to it by registered post to or left at its registered office.

The plaintiffs in this action, National Westminster Bank Ltd, are the successors in title of the lessors, and the defen- L dants, Betchworth Investments Ltd, are the tenants. By a letter dated March 23 1973, National Westminster Bank Ltd (whom I will call "the bank") gave the defendants (whom I will call "Betchworth") a written notice which was intended to be a notice given in pursuance of clause 4 (3) of the lease, and which, if valid, would have had the effect of bringing the lease to an end on September 29 1973. It was, in fact, posted on March 23. It is common ground that the notice, if received in due time, would have been sufficient to determine the lease on September 29. This case really concerns (and concerns only) the method of service of that letter. There is no factual dispute as to what happened. It was sent M addressed to 157 Victoria Street, London SW1, which, at the date of the lease, was the registered office of Betchworth but which, at the date the letter was sent, was no longer the registered office; this had in fact been moved to 1 Dover Street, London W1, on December 17 1970, some 2½ years previously, but that was a fact not known to the bank. Not only was that address no longer the registered office of Betchworth, but in fact the building had been demolished, and in due course, the letter was returned to the senders, who were the bank's solicitors, and, after amendment, sent on by them to the correct address of the registered office, 1

A Dover Street. But it was not received at that address until April 4 1973, which, it is agreed, is a date too late for the purpose of determining the lease. So the short question in this case is whether the notice which otherwise would have been adequate to achieve its purpose was sufficiently served by being sent to the old registered office, or rather, the address of the old registered office of Betchworth.

That matter turns, in the end, partly on the construction of clause 4 (3) of the lease, and partly on applying the facts. Putting it shortly, what is said by Mr Curtis-Bennett, on behalf of the bank, is that although the address to which it
B was sent was no longer the registered office of Betchworth, it was in fact the last known address in Great Britain or Ireland of Betchworth, and they (the bank) did not know of any other address. That is the submission, put shortly, of the bank. What is said on behalf of Betchworth is this—again putting it shortly and not elaborating the matter—that since May 28 1968 (as appears in a letter which is exhibit 2 to the affidavit of Mr Cotter [an official of the bank]) a firm of chartered surveyors, Mellersh & Harding, of 43 St James's Place, St James's Street, London SW1, had acted in all matters in relation to the lease, except matters of a purely legal character, which were dealt with by the solicitors; and
C that that situation had not changed from 1968, and, at all material times thereafter. Their address of course was known to the bank; and what Betchworth says is that that was an address within clause 4 (3), and therefore it was the last known address. Therefore, whereas service addressed to that address would have been perfectly satisfactory, service addressed to the address of the old registered office is not, because the old registered office is no longer the registered office; and the old registered office, if it could be any place at all, cannot be the last known address, because the last known address is the address which I have just referred to in St James's Street. It was not contended that service at the
D address of the old registered office could in no circumstances be effective, on the grounds that it was demolished.

And so the question is a short, but by no means an easy one. The rival submissions as elaborated, not in detail, in outline—and on each side they have been most helpful—are these. Mr Curtis-Bennett, for the bank, says, well, first of all, the last known address means the last address known to the bank; there is no question of constructive notice, and in that Mr Wood agrees. Then there are three possibilities for proper service: (1) at the demised premises, (2) at the registered office, or (3) at the *tenants'* last known address.
E Counsel stresses the words in subclause (3), "sent by post *to it* on the demised premises or sent *to it* by post at its registered office or last known address in Great Britain." And he submits, therefore, that whatever one may think of the important part which may have been played by the agents of Betchworth, their address cannot, within the terms of the lease as properly construed be an address of the tenant. What Mr Wood says is that it is important to note, as a background to one's approach to this problem, that the general proposition about documents and service is that whether one is talking about orders of court, or documents in an action, or whether one is talking about private documents, in general (and this is true I think) for good service the court requires that there should be actual service. And even in cases, for example, under the Interpretation Act, although there may be presumptions, they are, or nearly always are, rebuttable. And counsel goes on to say that the three alternative methods of service adopted in clause 4 (3) are not chosen at random, but that they are all chosen as being likely to result in actual service of the document being achieved. He (Mr Wood) has cited several authorities as being not directly in point but, as he submits, helping in the background approach to this; I will not take time to cite them. He says they are not directly in point, but they are perhaps indirectly helpful.

In the end, I think Mr Wood has two submissions, and G
the first one really is this: that he would say that the address of the registered office cannot be an "address" within clause 4 (3) unless it is something more than just the registered office and is at least an address at which some of the business of the lease is transacted. If one accepts that test, then this registered office cannot constitute itself an "address," because nothing was ever done at the registered office in relation to the affairs of the lease. I do not myself think that that submission can be right, because I think that the registered office, which necessarily has attached to it an identifiable place, must be an "address." And I do not think there are any peculiarities or specialities about the word "address" H
in clause 4 (3) which would exclude from its ambit the registered office. But Mr Wood's second submission I find much more attractive; and what he says is this, that the word "address" in the subclause is, anyhow, wide enough to include the place where the business of the lease is transacted, and that bearing in mind the general proposition that I have rather briefly summarised about the general need for actual notice, and the various cases which he has cited, the clause should not be construed in a restrictive sense in favour of the landlord, who, after all, is the grantor. That approach I think is a correct one, and on the question of construction (although I feel the full force that Mr Curtis-Bennett puts J
on the words "it" and "its") I am satisfied, first, that there is, to some degree at all events, contrast between "registered office" on the one hand and "last-known address" on the other, although I do not go so far as to accept the proposition that a registered office cannot be an address unless the affairs of the lease are transacted there; and secondly, that the word "address" as used in the subclause is a very wide word indeed. No authority directly in point has been cited, but it must, I think, be that "address" here is wider, for example, than "place of abode"; I think it is wider than "place of business" and wider than "registered office"; looking at it K
either from the point of view of the landlord who will have to serve documents, or of the tenant who will have to receive them, it must be wide enough, in my judgment, to include any address of the tenant which the landlord and tenant have used during the course of the lease while transacting the business and affairs incidental to and arising out of the lease, and this must be so even if the actual occupier of the address in question is not the tenant, but is, for example, an agent of the tenant. So the next question is to see, against that construction of the subclause, what are the facts in this case, and to what extent they match up with that interpretation. L

The facts of the matter are these, as appears first of all from the affidavit of Mr Cotter, who referred to the fact that a letter of May 28 1968 was sent to the bank by Mellersh & Harding, the firm of chartered surveyors who, as already stated, acted as Betchworth's managing agents. Mr Cotter goes on to say:

> Since that date, the file confirms, there has been no direct contact between Westminster or National Westminster Bank Limited and the defendants, all communications being with their managing agents, from whom rent is demanded and received, or solicitors.

Then there is the affidavit of Mr J C Hawkes, who is M
managing director of Betchworth, a chartered surveyor and senior partner in Mellersh & Harding. Mr Hawkes says in paragraph 4 of his affidavit, dated February 18 1974:

> Since that date [that is, since 1968], as Mr Cotter says, all dealings between the plaintiffs and defendants, as between landlord and tenant, have been conducted through the medium of my firm.

It is right to interpose there, although it does not appear in the affidavit, but I was told, and it is accepted, that that is slightly inaccurate, in as much as a certain amount of strictly legal business was conducted between the respective solicitors.

Then the affidavit continues:

> The defendant company has never had a separate place of business; all its business affairs are transacted from my firm's offices.

Of course, that last fact would not, or would not necessarily, have been known to the bank. Now regrettably, none of the correspondence which ensued between 1968 and 1973 is exhibited, but it follows, I think, from those affidavits and from what I have been told, that I ought to accept that in that intervening period of time a correspondence and communication was established between the bank and the managing agents, as a result of which all affairs incidental to the payment of rent, any other problems that arose and so on, except those of a purely legal character which were dealt with by the solicitors, were dealt with between the bank on the one hand and the managing agents on the other. And therefore, really, for a period of years, before the despatch of the letter upon which the bank relied, the bank had been used to dealing with the managing agents in most matters in relation to this lease. That leads me to this conclusion, having regard to the construction that I have preferred of this subclause, that on the material date, which was March 23 1973, the address of the managing agents was an address of the tenant. It must follow, from the other facts to which I have referred, that not only was it an address of the tenant but it was the last known address of the tenant (last known, that is to say, to the bank), because it came to their knowledge, of course, at a time long after the address of the old registered office came to their notice.

Mr Curtis-Bennett has made the telling point, that, plainly before the letter of May 28 1968 was written the only address known to the bank was the address of the registered office, and it cannot be said that as soon as they got the letter of May 28 1968 43 St James's Place, St James's Street, became an address of the tenant known to them. I fully see the force of that observation, but it seems to me, although it is always difficult to pinpoint time, that when one has a long course of dealing between parties it is quite possible to say that at some stage, what was not known at the outset has become known by the end, and I do not really find it necessary, for the purposes of this case, to pinpoint a particular point of time as being the time at which the address became known to the bank. Suffice it to say that in my judgment by the date the notice was sent it was an address, and the last known address to them. I should mention this one final point that counsel relied on, which was a point that he made on the case of *Stylo Shoes Ltd v Prices Tailors Ltd* [1960] Ch 396, a case cited by Mr Wood. Mr Curtis-Bennett has seized on the fact that in section 23 of the Landlord and Tenant Act 1927, the concluding words in section 23 (1) are:

> . . . and in the case of a notice to a landlord, the person on whom it is to be served shall include any agent of the landlord duly authorised in that behalf.

Counsel relies upon that as being statutory recognition of the fact that ordinarily service of such a notice, as that section referred to, on an agent would or might not be service on the landlord. I can see the force of that, but my judgment really rests, in this case, on the construction that I have put on subclause (3) of clause 4. I will not repeat that, but ultimately it is based on the view that "address" there—as I have already indicated, this favours the landlord as well as the tenant—requires a reasonably liberal construction, because it is a word with a broad meaning, and in my judgment, as I have already indicated, it is wide enough to include the address of an agent who has been accustomed to carry on the business of the lease on behalf of the tenant. The consequence of the view that I have taken is that there has not, in this case, been good service of the document upon which the bank can rely in support of their claim that this lease has been determined. For those reasons, in my judgment the claim fails.

A declaration was made on the counterclaim in terms of the negative of the claim, viz that the notice posted on March 23 1973 was not a valid notice for the purposes of the lease in question. The defendants were awarded costs.

COURT OF APPEAL

October 30 1974

(Before Lord Justice CAIRNS, Lord Justice LAWTON and Mr Justice BRIGHTMAN)

COLIN SMITH MUSIC LTD v RIDGE

Estates Gazette April 19 1975
(1974) 234 EG 195

Deserted mistress cannot assert man's statutory tenancy of the rent-controlled premises she occupies as his licensee—On his surrender of tenancy and permanent abandonment of the premises she must give possession to the landlord—Mistress who has borne children is not entitled to be protected as a wife

This was an appeal by Colin Smith Music Ltd from a decision of Judge Forrest at Weston-super-Mare County Court on February 20 1974 rejecting their claim for possession of premises known as First and Second Floor Flat, the Bank House, High Street, Worle, Weston-super-Mare, occupied by the respondent, Mrs Frances Paula Ridge.

Mr P Talbot (instructed by Reed & Reed, agents for Hall, Ward & Fox, of Weston-super-Mare) appeared for the appellants, and Mr C Gosland (instructed by Stiddard & Chew, of Weston-super-Mare) represented the respondent.

Giving judgment, CAIRNS LJ said that in March 1973 a Mr and Mrs Joyner bought the Bank House, the upper part of which was a tenanted flat. By her defence the defendant had asserted that she was joint weekly tenant of this part of the premises with a Mr Venn, with whom she was living. The rent book in Mr Joyner's possession, however, showed Mr Venn as the sole tenant. The defendant, a divorced woman, lived with Venn from 1968 until July 28 1973, when according to her own evidence he left her with no intention of returning. She had two children already, and they lived with her and Venn; and during her association with Venn she had two more children by him. On August 9 1973 the Joyners got Venn to sign a deed of surrender of the protected tenancy, and immediately afterwards they conveyed the premises to the plaintiffs, a company of which Mr Joyner was a director. The county court judge had held that Venn's protected tenancy could be terminated only (1) by judgment, or (2) by surrender combined with actual delivery of possession; and that by leaving the defendant in the flat with her children Venn had failed to deliver possession to the plaintiffs.

It was clear from a number of Court of Appeal decisions that mere physical absence of a tenant from rent-protected premises did not deprive him of his statutory rights. In many of the cases, for example *Brown v Draper* [1944] KB 309, it was a wife who was left in possession. The position of a mistress left in possession seemed first to have been considered in *Brown v Brash* [1948] 2 KB 247, where the woman had in fact left the premises at the time when the landlord sought possession. The principles enunciated in that case were subsequently applied in *Thompson v Ward* [1953] 2 QB 153. There a tenant installed his mistress in premises of which he held a contractual tenancy. Some time later the contractual tenancy was made statutory, and later still the tenant left, leaving the mistress behind. Some time after that he brought an action for possession against her, claiming as statutory tenant. He obtained judgment, but this was set aside on appeal on the basis that though a statutory tenant could in appropriate circumstances be held to retain possession through the medium of an occupying licensee, the person left in the premises must have the function of preserving his rights there against his ultimate home-coming. It appeared, therefore, that a licensee left in possession by a protected tenant when he himself left, intending to remain permanently away from the premises, was no longer entitled to the protection of the Rent Acts. Cases such as *Middleton v Baldock* [1950] 1 KB 657 and *Wabe v Taylor* [1952] 2 QB 735 depended wholly on the relationship of husband and wife. Counsel had sought to say that in these days possession by a mistress who had borne children to a protected tenant was analogous to possession by a wife; but in his (Cairns LJ's) view this was not so. Counsel for the defendant had drawn an interesting comparison with the position of a mistress who, with her children, was residing in the house with the protected tenant up to the time of his death. It had been held in *Hawes v Evenden* [1953] 1 WLR 1169 that she was a member of the protected tenant's family for the purpose of succeeding to the protection of the Act after his death. That of course depended on the construction to be placed on the word " family " in the Act, and he (his Lordship) did not find it of real assistance in this case. In his opinion the defendant could succeed here only if it were laid down as a general principle that any person who had a licence from a protected tenant would continue to have the protection notwithstanding surrender of the tenancy by the protected tenant and his permanent abandonment of the premises. That could not have been the intention of the legislature, and the appeal should be allowed.

LAWTON LJ said that the defendant could look to Mr Venn for money payments in respect of the children he had had by her, but was not entitled to call on him to provide them with a home. There was nothing in the authorities which justified the court in holding that a mistress in her capacity as lodger, or indeed in any other capacity, was in any position analogous to that of a wife. It was clear on principle that the Act did not protect mere licensees, and the appeal should be allowed.

BRIGHTMAN J agreed with both of the judgments delivered. An order was made for possession within 28 days.

COURT OF APPEAL

July 3 1974

(Before Lord Justice BUCKLEY, Lord Justice ORR and Lord Justice ORMROD)

WALKER AND ANOTHER v OGILVY

Estates Gazette April 26 1975

(1974) 234 EG 279

Yearly tenancy of flat in Sleights Hall, Yorkshire—Use of flat by tenant insufficient to amount to use " as a residence " and attract statutory protection—Tenant's intention to reside in flat on retirement either not established, or incapable of colouring occupation only on occasional visits to premises

This was an appeal by Dr John Airlie Ogilvy, of " Cullerne," Birstall, Batley, Yorkshire, from a judgment of Deputy Judge Wrightson at Whitby County Court on November 22 1973 awarding Mr David Filmer Walker and his wife, Mrs Eve Ann Walker, of Sleights Hall, Sleights, Yorkshire, possession of a flat forming part of that property and known as 2 Sleights Hall.

Mr J Pickles (instructed by Lee & Priestley, of Bradford) appeared for the appellant, and Mr F Radcliffe (instructed by Wedlake Bell & Co, agents for Buchanan & White, of Whitby) represented the respondents.

Giving the first judgment, ORR LJ said: This is an appeal by a tenant against a decision of His Honour Deputy Judge Wrightson on November 22 1973 in the Whitby County Court, by which he granted to the landlords an order for possession, which was suspended for two months, of a flat known as 2 Sleights Hall, Sleights, in Yorkshire. That is a house which is situated some miles from the coast. The premises in question were a flat in the house, and that flat was let to the appellant in 1956. He is a doctor, now 60 years of age, practising in partnership with three other doctors at Birstall, Batley, where he lives in a three-bedroom house, " Cullerne," belonging to him, two rooms of which are used as a surgery and a waiting-room in the practice. Those two rooms have been let on a yearly tenancy, determinable by six months' notice, to himself and the other doctors. In 1960 the defendant obtained a written tenancy agreement in respect of the premises in question for a yearly tenancy, determinable by six months' notice. In 1972, the owners of the house, a Mr and Mrs Spaven, sold the house to the plaintiffs in the present action, a Mr and Mrs Walker, who live in it with their three children, and they, by a notice to quit given in September 1972 and expiring in March of the following year, gave what is accepted to have been an effective notice to quit to the defendant. The question which arose at the hearing, and which arises in this appeal, is whether, in the words of section 3 (1) (a) of the Rent Act 1968, the defendant thereafter continued to occupy the dwelling-house—that is the flat—" as his residence," because it is a condition of a statutory tenancy coming into being that he should have done so. That provision is one which had not previously appeared in the Rent Acts, but it embodies a number of decisions in relation to the earlier legislation, the effect of those decisions being preserved and applied to section 3 (1) (a) by subsection (2) of section 3. I find it unnecessary to read that subsection, but that was its effect.

The defendant gave evidence at the hearing as to, first of all, his intention at the time that he acquired the tenancy. He said that he took it with the intention of eventually retiring there, and that he proposes to retire in September of this year. I shall refer presently to what the learned judge had to say about that matter. Evidence was given by the defendant that he employed a part-time gardener and a part-time housekeeper to clean the premises, and as to the extent to which he had stayed there. It is a journey of an hour and 40 minutes from Batley where he has his practice. He also said that in 1960 he had not used this flat at all owing to illness; that up to 1967, apart from that, he had gone on three out of four week-ends in the summer and once a fortnight in the winter; but he said that from April 1967 his visits had become less frequent. He had in fact acquired nine years ago a flat in Spain in which it has been his practice to spend his summer holidays. He further gave evidence that in the last two years he had gone to the premises even less. He said that during his tenancy the longest stay he has made in the premises was 10 days, apart from one period of a month. He gave evidence that it was a place to which he would go for short holidays until he could retire, and that was why he went on paying the rent. His wife gave evidence that he had obtained the premises with a view to retirement. She said they used to go twice a month in the summer and at least once a month in the winter, but that that had been rather diminished in the last few years. For the plaintiffs, evidence was given by Mr Walker that he had not seen the defendant or his car at the premises between May and August 1973; that he had come for a short stay at the end of August for a few days; that staff of his had come in early August, and again between August and September. Mr Walker gave evidence that the user was of the order of once a month in the 18 months preceding the hearing. Mrs Spaven gave evidence that the defendant had earlier used the premises for summer holidays for himself and his family, but that his visits grew less fre-

A quent as the years went by; and she thought that about six years ago he had not come at all for a year. That may have been a mistaken reference to the year in which the defendant admits that he did not go. A Mrs Dunn estimated that the average of the visits paid by the defendant would be once a month at the most.

That being briefly a summary of the evidence given, the learned judge came to the conclusion that this was a case within the line of cases represented by *Beck v Scholz* [1953] 1 QB 570, that is to say, of a house that was used for occasional visits. We have been referred to a number of decided cases on this subject. I cannot derive much assistance in the present
B context from the case of *Brown v Brash* [1948] 2 KB 247. That was a case in which a man who had undoubtedly been occupying premises as his home was sentenced to a term of two years' imprisonment, and the question which arose was whether during that period he remained in occupation of those premises. That decision does not, in my judgment, assist at all on the question with which we are concerned in this case, namely, whether the flat in question, which the defendant only visited from time to time, was occupied by him " as his residence." I derive much greater help from a passage in the judgment of Lord Evershed MR in *Beck v*
C *Scholz,* where he said this at p 575:

> But the question posed, and to be answered by ordinary commonsense standards, is whether the particular premises are in the personal occupation of the tenant as the tenant's "home," or, if the tenant has more than one home, as one of his homes. Occupation merely as a convenience for such occasional visits as I have suggested would not, I think, according to the commonsense of the matter, be occupation as a "home." [Counsel] has drawn our attention to the analogous cases where a man who is about to retire seeks the protection of the Rent Acts for his home or for the place which he has been maintaining as his home, and to which he intends to go when his retirement takes effect.

It is to be noted that the cases referred to at the end of
D that passage are cases of a man who has a home, or some place which he has been maintaining as his home, and intends to retire there. The learned judge said nothing in the present case about the evidence of the defendant as to his intending to retire, until the very end of his judgment, when certain questions were put by counsel. The judge then said that the flat "was taken at that time"—that is 1956—"as a place where he could go to when convenient and for holidays in the summer and to take the children etc. When he took the flat he was only in his middle 40's. It may have occurred to him as a place to retire to as the years went on." And in
E answer to a further question he said: "I do *not* find as a fact that the defendant took the lease with the intention of retiring there." Criticism has been made as to that last answer, having regard to something which the judge had said earlier in his judgment. He had said that all the witnesses had told the truth and no witness had told a lie. Mr Pickles argues that since evidence had been given by the defendant that he nad intended when he took his tenancy to retire to this flat, that must have been something which the judge accepted. But I am satisfied that in saying that the witnesses told the truth what the judge meant was that they intended to tell the truth as they understood it: because quite plainly
F in this case all the witnesses were not giving the complete truth; there were differences in the evidence by different witnesses. That is clearly, in my judgment, the meaning of that passage. I should add that the learned judge in his judgment said that as to the frequency of the visits he preferred the evidence of Mr and Mrs Walker, who referred to a rather smaller number of visits than the defendant.

The judge having come to that conclusion, Mr Pickles does not really seek to say that the conclusion would be wrong but for the evidence of the defendant and his wife about his retirement intentions. He claims that in this case there was evidence before the learned judge, which he should

have accepted, that the defendant took these premises G intending, although he was only then in his 40's, to retire to them at some time, and that that fact would give a different character to his occupation in the meantime. As to the first of those matters, I find it impossible to approach this case on the footing that the evidence as to that intention must be accepted. It was perfectly possible that the defendant thought that that had been his intention when he went there. But in the matter of projecting back a recollection of an event, a witness may be wrong. In my judgment, that was something which the learned judge had in mind in declining to come to a conclusion about that matter. This was a context in which, in my judgment, the onus of proof was on the H defendant. But in any event I find myself unable to accept the argument that if the defendant had intended this as a retirement home that made some change in the character of his occupation. The question here is whether this flat ever became a residence of the defendant. The judge found that it never did. I am quite prepared to accept that if he had found an intention to retire there it was a factor to which he would have had some regard in considering the whole of the facts, but I do not think it is a matter which would go any further than that. In the result, I find it impossible to say that the learned judge in this case erred in his conclusion. The matter, as it seems to me, was essentially one of fact J and degree. There is nothing to indicate that the judge failed to take into account any relevant matter, or took into account something which he ought to have disregarded. For these reasons, it seems to me impossible to say that the conclusion he came to on this matter was wrong. I therefore would dismiss this appeal.

ORMROD LJ: I agree, and I would only add a few words. The learned judge, in my judgment, asked himself the right question as indicated in *Beck v Scholz.* He came to the conclusion that this flat was not personally occupied by the defendant as a home. In so finding, it seems to me on the K evidence put forward by the defendant and by the other witnesses that the judge was plainly right. It would be absurd to say that a man who lives one hour and 40 minutes away from a flat, as in this case, is using it as a home when he goes to it only occasionally, as described by the witnesses, and sometimes does not go for months at a time. When one has in mind the purpose of Parliament in passing the Rent Act, it was never intended, in my judgment, to protect people in occupation of what are in effect holiday houses. I refer to a brief passage in the judgment of Scrutton LJ in *Skinner v Geary* [1931] 2 KB 546, where he said:

> "For the reasons I have given the Act does not in my opinion L apply to protect a tenant who is not in occupation of a house in the sense that the house is his home and to which, although he may be absent for a time, he intends to return. If it were to be held otherwise odd consequences would follow."

He then deals with other matters, and goes on to say:

> "One object of the Acts was to provide as many houses as possible at a moderate rent. A man who does not live in a house and never intends to do so, is, if I may use the expression, withdrawing from circulation that house which was intended for occupation by other people. To treat a man in the position of the appellant as a person entitled to be protected, is completely to misunderstand and misapply the policy of the Acts." M

Those words appear to me precisely to apply to the defendant in this case. I agree that the appeal should be dismissed.

BUCKLEY LJ: I agree with both the judgments which have been delivered.

COURT OF APPEAL

February 7 1975

(Before Lord Justice CAIRNS, Lord Justice LAWTON and Mr Justice MACKENNA)

CITY OF WESTMINSTER ASSURANCE CO LTD v AINIS AND OTHERS

Estates Gazette May 10 1975

(1975) 234 EG 455

Tenants obtain relief against forfeiture upon numerous conditions, some to be performed only in the future—Landlords held to have the right to proceed in the meantime against squatters for possession—Uncertainty as to tenants' status during interim period reflected in dissenting judgment of MacKenna J

This was an appeal by City of Westminster Assurance Co Ltd from a decision of Cusack J in chambers on January 21 1975 refusing to make an order for possession of 17 Upper Grosvenor Street and 19 Culross Street, London W1, against the respondents, Mr Michael Ainis and 12 other named people and other persons unknown.

Mr M Horowitz (instructed by Titmuss, Sainer & Webb) appeared for the appellants, and Mr C R Chope (instructed by P Kingshill) represented the respondents.

Giving judgment, CAIRNS LJ said that the respondents were squatters, but none of the types of issue which commonly arose under order 113 of the Rules of the Supreme Court arose in this case at all. The sole question was whether the appellants had a right to bring these proceedings, or whether the party who could have brought them was a different company. The freeholders of the premises whose possession was disputed were the Grosvenor Estate, and the appellants were their lessees. In 1969 the appellants sublet the premises for a term of 66 years, and in November 1973 that sublease was assigned to a company called International Colloids Ltd. The latter were not satisfactory tenants. They failed to pay their rent; failed to pay for insurance, as the sublease imposed a covenant upon them to do; and failed to keep the premises in repair in compliance with another covenant in the sublease. Accordingly, on August 7 1974 the appellants issued a writ for possession against them for nonpayment of rent and for breaches of covenant. There was no appearance to the writ, and the appellants got judgment for possession against International Colloids in default of appearance.

International Colloids, as they were entitled to do, then applied for relief from forfeiture. Their application came before Master Elton on December 18 1974, when he made an order of some complexity. It was ordered that if International Colloids complied with certain conditions they were to be relieved from forfeiture. The first set of conditions was for payment of four different sums forthwith. Then there were conditions as to reinstatement of a garage on or before March 1 1975; putting the premises in repair before September 1 1975; payment of various items of costs, 14 days after taxation or agreement, the final item of costs being costs in connection with the occupation of the premises by squatters, it being common ground that at that time there were squatters on the premises; and payment of certain surveyors' fees. The order went on to provide that if the conditions were complied with, the first and second defendants (the second defendant was a guarantor of payment of rent and compliance with the covenants in the sublease) should be relieved from forfeiture, and the lease reinstated and all further proceedings stayed, "but if default is made the plaintiffs' judgment herein as against the first defendant dated November 11 1974, and as against the second defendant dated November 15 1974, do stand, and the plaintiffs be at liberty to proceed to execution on the said judgments." It was further ordered that "on default of the defendants complying with any of the above terms at the time specified the defendants' application for relief be dismissed with costs." The defendants did comply with the terms as to payment forthwith of a number of substantial sums, but so far as appeared the time for compliance with any of the other conditions had not arisen. Certainly the time for compliance with the condition that was to be performed by March 1 had not yet arrived, and the condition to be complied with by September 1 would not arise for another seven months. After the making of the order for relief, the appellants issued their originating summons against the present respondents, the squatters, on December 24 1974. What was contended before the judge when those proceedings came before him was that because of the order for relief against forfeiture the party entitled to possession of the premises was International Colloids and not the appellants. Cusack J accepted that submission and declined to make an order for possession.

The position as to how a lease stood when an order for relief against forfeiture had been made but was conditional on a term to be complied with in the future was not clearly dealt with in any of the authorities. Certain things were plain. The issue of a writ for possession on the ground of forfeiture did not terminate the lease: the lease was not terminated until judgment was given for possession. When such judgment had been given, however, the termination related back to the date of the writ. But what about the position when relief was granted? It was stated in *Woodfall on Landlord and Tenant* 27th edn vol 1 para 1959 that "when relief is granted the effect is for all purposes and between all parties to do away with the forfeiture just as if there had never been any." The authority for that proposition was *Dendy v Evans* [1910] 1 KB 263. That was, however, a case where the order for relief took effect immediately upon the order being made, there being no future conditions to be complied with. The only case to which the court had been referred in which he (his Lordship) had found some guidance on the matter was that of *Talbot v Blindell and Others* [1908] 2 KB 114. The headnote read:

> "Where an order granting relief against the forfeiture of a lease was expressed to be made upon the defendants performing certain conditions, and the defendants having performed part of the conditions declined to perform the remainder and desired to waive the relief: *Held* that there was no power to compel the defendants to perform the conditions, and that the order for relief must therefore be treated as abandoned."

In the judgment of Walton J appeared the sentence: "The result is, I think, that I must agree with the construction put on the order by the defendants." In his (Cairns LJ's) view, that was a proper construction. If defendants who had obtained conditional relief declined to perform the conditions, the order for relief fell to the ground. In the meanwhile, until the time came for the performance of any future conditions, the defendants, if they remained on the premises, were there not as tenants under the lease but as tenants at will or on sufferance. That being so, his view would be that, subject to any particular terms of the order covering the interval between the making of the order and compliance with the conditions, it was the plaintiffs who were entitled to possession, and if there was somebody in possession other than the plaintiffs they were entitled to take proceedings to have them ejected. He (his Lordship) confessed that his opinion on the matter had at one stage fluctuated, when the court's attention was drawn to the words in the master's order, "that if default be made the plaintiffs' judgment herein do stand and that the plaintiffs be at liberty to proceed to execution on the said judgment." It was quite possible to construe that as meaning that until default took place the plaintiffs were not at liberty to proceed to execution and accordingly not entitled to possession. But in the end it was the last phrase of that proposition that he did not accept. He did not regard that provision of the master's order as being any more than a fetter upon the actual execution of the judg-

A ment in accordance with normal processes of execution. He was satisfied that, certainly if the premises were vacant at any time between the making of the order giving relief and compliance with the conditions, the plaintiffs would be entitled to go in and occupy the property themselves. If there was in occupation some person or some body of persons who had not been given leave by anybody to go in there, then he thought that the plaintiffs had such a right to possession as entitled them to bring an action for possession and to succeed upon it. For these reasons he would allow the appeal and make the order for possession claimed.

B Agreeing, Lawton LJ said that he inferred from the facts of the case that at about the time the plaintiffs got an order for ejectment against their tenants on November 11 1974 the premises were in fact empty, because if they had not been empty the squatters, who numbered 15 and more, could not have gone into occupation. So long as the order for ejectment stood, the plaintiffs had the right to possession. The best the tenants could hope for was some kind of conditional order. Before the passing of the Common Law Procedure Act 1852, the only relief they could have got was to apply to the Court of Chancery for a stay of execution. Such an order would have taken the form of an injunction
C merely restraining the plaintiffs from levying execution against the tenant, in other words from putting the sheriff in or seeking by processes of law to enforce their order. That, before 1852, would have been the only limitation which the law imposed on them, and he (his Lordship) could see nothing in the statutory provisions since 1852 relating to forfeiture which really altered the position. The effect of Master Elton's order had been to impose a fetter on the plaintiffs' right to possession against the tenant. But it was a fetter imposed solely in favour of the tenant. It was not a fetter imposed in favour of the world at large. The respondents in the present appeal, the squatters, were third parties, and
D as against them he (Lawton LJ) could see no reason at all why the appellants should not be entitled to enforce the right to possession which was given them by the order for ejectment. But for the words in Master Elton's order, namely "that the plaintiffs be at liberty to proceed to execution on the said judgment," he would have thought it clear beyond argument that what the master was intending to do was to restore the plaintiffs to their rights as lessees on the happening of certain events in the future, and that until those events had happened the defendants were not to have the right of a lessee. He could see nothing wrong in that kind of order. He agreed that the appeal should be allowed.

E Dissenting, MacKenna J said that the question was whether, at the date when proceedings were commenced by the appellants against the squatters, that company had a right to immediate possession of the premises. The answer to that question depended on the effect of Master Elton's order. The effect of the last words of the order was that the appellants should not be permitted to proceed to execution on the judgments of November 11 and November 15 until default had been made in the performance of the conditions. He (his Lordship) did not doubt that the master had the power to make the second part of his order, and he thought that its effect was to give International Colloids the right to
F immediate possession. If that were so, the appellants were not entitled to bring proceedings against the squatters on December 24. It would be curious if International Colloids were not entitled under the master's order to have a right to the immediate possession of the premises. If they performed the conditions of the order, the forfeited lease would be reinstated as from the beginning, and that would carry with it the appellants' right to be paid rent in respect of the whole period during which the conditions were being performed. The provisions of the master's order about payment of the appellants' cost in taking proceedings against the squatters were no doubt inserted upon the assumption that proceedings would be taken by the appellants, and upon the further G
assumption that they would be the right plaintiffs, but he (MacKenna J) thought that this further assumption was a mistake. In his opinion Cusack J's order refusing leave to the appellants was right, and the appeal should be dismissed.

The appeal was allowed, and the court made an order for possession forthwith. The respondents were directed to pay the costs.

COURT OF APPEAL

January 29 1975 H

(Before Lord Justice CAIRNS, Lord Justice LAWTON and Mr Justice MACKENNA)

GOFOR INVESTMENTS LTD v ROBERTS AND ANOTHER

Estates Gazette May 24 1975

(1975) 234 EG 607

Tenant's prolonged absence in Malta insufficient to deprive her of protected tenancy of flat in Westminster—Discussion by MacKenna J of date by reference to which tenant's intentions are to be assessed in cases of the kind—Beginning of period of absence, or date of hearing? J

This was an appeal by Gofor Investments Ltd, of Sydenham Road, Croydon, from a judgment of Judge Ruttle at Westminster County Court dated April 25 1974 dismissing their claim against the respondents, Mr Michael Guy Roberts and Mrs Christine Mary Cecile Boyd, for possession of a flat, 2 St Augustine's Mansions, Bloomburg Street, London SW1.

Mr R Hone (instructed by Peter Mallack & Co) appeared for the appellants; Mr C Harris (instructed by Lynch, Hall & Hornby, of Harrow) for the first respondent; and Mr R Pryor (instructed by R C Bartlett & Co) for the second respondent.

Giving judgment without calling on counsel for either K
respondent, Cairns LJ said: This is an appeal by the plaintiffs from a judgment of Judge Ruttle dismissing their claim for possession of a flat against two defendants. The premises are at 2 St Augustine's Mansions, SW1. One of the defendants was Mrs Boyd, who was at one time the contractual tenant of the flat and who succeeded in her defence on the ground that at the time of the action she was the statutory tenant of it. The other defendant, Mr Roberts, who at one stage claimed to have been the tenant but whose defence succeeded on the basis that he was the licensee of Mrs Boyd, now relies only on that second ground. The plaintiffs contend L
that after the contractual tenancy ended Mrs Boyd abandoned possession, and therefore no statutory tenancy on which either she or Mr Roberts could rely ever arose.

Mrs Boyd became the tenant under a lease from a Mr and Miss Newman, who were the plaintiffs' predecessors in title. It was a lease for seven years, expiring on September 29 1972. Mrs Boyd and her husband and family of six children lived in the flat for a number of years up to 1970, but the parents and four of the children then went off to live first in Morocco and then after a short time in Malta, leaving behind two of their sons called Nicholas and Toby. They stayed in the flat for the time being. The lease contained a covenant M
against assigning or subletting or taking lodgers without the consent of the landlords, but in January 1970 Mrs Boyd conceived a plan of assigning the lease to trustees—there were tax reasons for that idea—and then, with the leave, if she could get it, of the Newmans, allowing one or more members of the Boyd family to live there and to take lodgers. On April 8 1970 the Newmans signed a document which on the face of it appears to be a consent to such assignment and occupation. In fact, no assignment to trustees ever took place. But Nicholas and Toby continued to live there after Mr and Mrs Boyd and the rest of the family had left; and

then at about the end of 1970 or the beginning of 1971 Toby left and Mr Roberts came as a lodger. It is perfectly clear from certain of the correspondence that the Newmans were aware of these events, and even if the document that was signed in April 1970 was not in fact a licence to assign and take lodgers, and so forth, Mr Roberts clearly became a licensee of Mrs Boyd with the assent of her then landlords, the Newmans.

In 1971 Nicholas Boyd joined the Army, and thereafter he no longer lived regularly at the flat, but he did come there when on leave. He left the Army in 1972 and spent a few nights at the flat after that. Mr Boyd was in England in the spring of 1972 and he stayed there for some six nights. Mrs Boyd was in Malta throughout this time. Mr Roberts continued to reside at the flat up to the time of the proceedings in January 1973. There was considerable discussion in letters and at an interview which took place in, I think, April 1972 between Mr Boyd, on behalf of his wife, and Mr Newman and Mr Roberts as to what arrangements should be made when the lease came to an end in September of that year. There were various plans for an extension of the lease to Mrs Boyd, or, more correctly, one would say a new lease to Mrs Boyd, or for a lease to Mr Roberts with a view to his assigning it to Mr or Mrs Boyd if and when they came back to England. But these discussions came to nothing. All this time some of Mrs Boyd's furniture remained at the flat. At the time when they left in 1970 there was quite a good deal of it there. Between then and April 1974 a lot of it had disappeared. Whether Mr Roberts had disposed of it, possibly because it was worn out, or whether some other people had taken it away is not clear, but what is clear is that Mrs Boyd never intended any of the furniture to be removed and that some of it did remain there all the time. She said in evidence that there were still two chests of drawers and a carton of pots and pans and kitchen things left, admittedly in a very bad condition, but it has not entirely gone.

In further and better particulars of her defence Mrs Boyd alleged that when she and her husband left for Morocco, and afterwards for Malta, they did not expect to return unless obliged to do so, but expected that their children would come back and live in the flat when they grew up. However, by amendment, made it is true at a late stage, at the start of the hearing before Judge Ruttle, that part of her particulars was deleted, and she substituted the allegation that she had intended from the time when she left in 1970 to return when the children had completed their education. That was after a period of eight or 10 years. Mrs Boyd gave evidence, generally in accordance with her amended particulars, though she went further and said that she now thought it was probable that she would return within two years from that time. That would be by 1976, a period of some six years from when they first left, and some four years or rather less from the time when the lease came to an end. The judge accepted Mrs Boyd as a truthful witness throughout.

At about the time when the lease expired, the end of September 1972, the Newmans sold the freehold reversion to the plaintiffs. The issue before the judge was whether Mrs Boyd had ceased to be resident in the flat or whether it always continued to be her residence, having regard to the fact that she always intended to return to it, and that for at least part of the time the sons and Mr Roberts might be regarded as being there in the capacity of what has been described in some of the cases as caretakers. More especially she relied upon her furniture being left at the flat; that, of course, combined with her intention to return. The relevance of these matters arises from the provisions of two subsections of the Rent Act 1968 and certain cases which have been decided by the Court of Appeal. Section 3 of the Act provides in subsection (1) as follows:

"Subject to sections 4 and 5 below "—and they are not relevant —" (a) after the termination of a protected tenancy of a dwelling-house the person who, immediately before that termination, was the protected tenant of the dwelling-house shall, if and so long as he occupies the dwelling-house as his residence, be the statutory tenant of it"

Then by subsection (2):

"In paragraph (a) of subsection (1) above and in Schedule I to this Act, the phrase 'if and so long as he occupies the dwelling-house as his residence' shall be construed as requiring the fulfilment of the same, and only the same, qualifications (whether as to residence or otherwise) as had to be fulfilled before the commencement of this Act to entitle a tenant, within the meaning of the Increase of Rent and Mortgage Interest (Restrictions) Act 1920 to retain possession, by virtue of that Act and not by virtue of a tenancy, of a dwelling-house to which that Act applied."

So the effect of that subsection was to say that in considering whether a person did retain a dwelling-house as his residence, one had to look at the authorities prior to 1968 to see what the position was. It was laid down in a number of cases from the 1920's onwards that the fact that a person was not physically in residence at any particular time did not mean that the house had ceased to be his residence. I take as a convenient summing-up for the purpose of this case one sentence from the judgment of Ormerod LJ in the case of *Tickner v Hearn* [1960] 1 WLR 1406 at page 1410: "In such cases "—and that is cases of the kind I have been referring to—" there must be a real hope of return coupled with the practical possibility of its fulfilment within a reasonable time." That was further elaborated by a judgment of the Court of Appeal in *Brown v Brash* [1948] 2 KB 247, a judgment of the court delivered by Asquith LJ. The passage I am going to cite is on pages 254 and 255, because it seems to me that there is much in this statement of the law which is applicable to the present case. Asquith LJ said:

We are of the opinion that a "non-occupying" tenant *prima facie* forfeits his status as a statutory tenant. But what is meant by "non-occupying"? The term clearly cannot cover every tenant who, for however short a time, or however necessary a purpose, or with whatever intention as regards returning, absents himself from the demised premises. To retain possession or occupation for the purpose of retaining protection the tenant cannot be compelled to spend 24 hours in all weathers under his own roof for 365 days in the year. Clearly, for instance, the tenant of a London house who spends his weekends in the country or his long vacation in Scotland does not necessarily cease to be in occupation. Nevertheless, absence may be sufficiently prolonged or unintermittent to compel the inference, *prima facie*, of a cesser of possession or occupation. The question is one of fact and of degree. Assume an absence sufficiently prolonged to have this effect. The legal result seems to us to be as follows: (1) The onus is then on the tenant to repel the presumption that his possession has ceased. (2) In order to repel it he must at all events establish a de facto intention on his part to return after his absence. (3) But we are of opinion that neither in principle nor on the authorities can this be enough. To suppose that he can absent himself for five or 10 years or more and retain possession and his protected status simply by proving an inward intention to return after so protracted an absence would be to frustrate the spirit and policy of the Acts, as affirmed in *Keeves v Dean* [1924] 1 KB 685 and *Skinner v Geary* [1931] 2 KB 546. (4) Notwithstanding an absence so protracted the authorities suggest that its effect may be averted if he couples and clothes his inward intention with some formal, outward, and visible sign of it; that is, installs in the premises some caretaker or representative, be it a relative or not, with the status of a licensee and with the function of preserving the premises for his own ultimate home-coming. There will then, at all events, be someone to profit by the housing accommodation involved, which will not stand empty. It may be that the same result can be secured by leaving on the premises, as a deliberate symbol of continued occupation, furniture; though we are not clear that this was necessary to the decision in *Brown v Draper* [1944] KB 309. Apart from authority, in principle, possession in fact (for it is with possession in fact and not with possession in law that we are here concerned) requires not merely an "*animus possidendi*" but a "*corpus possessionis*," namely, some visible state of affairs in which the *animus possidendi* finds expression. (5) If the caretaker (to use that term for short) leaves or the furniture is removed from the premises, otherwise than quite temporarily, we are of the opinion that the protection, artificially prolonged by their presence, ceases, whether the tenant wills or desires such removal or not.

I refer also quite briefly to two other cases which Mr Hone has relied upon. One is *Beck v Scholz* [1953] 1 QB 570. That was a decision of this court in which what had happened was that the tenant had left furniture and a caretaker and desired to retain possession because he or she occasionally found it convenient to resort to the premises; and this court held that in those circumstances there was not sufficient to constitute a continuance of residence there. It does not seem to me that that is really helpful to us in this case, because the whole point of it was that the presence of the caretaker and the furniture was not with a view to ultimate intention to return there to use the place as the home of the tenant, but was for convenience at a time when it was not being occupied by him except occasionally. The other case was *Dixon v Tommis* [1952] 1 All ER 725. This, it appears to me, is a case which assists the defendants rather than the plaintiffs here. I read from the headnote:

> The statutory tenant of a dwelling-house within the Rent Restrictions Acts went to live in other premises, leaving his furniture in the house, and his son and the son's family took up residence in the house. On a claim by the landlord for possession on the ground that the tenant had ceased to reside in the house, the tenant gave evidence that he contemplated retiring from his profession in about three years' time and he would then be unable to maintain the other premises and intended to return to the house, of which meanwhile his son was merely a caretaker on his behalf. The county court judge having refused an order for possession: *Held*, the question whether the intention of the tenant to return to the house, although after so long a period as three years, constituted sufficient *animus revertendi* was one of degree for the county court judge, and it was impossible to say that, in answering that question in the affirmative, he had erred in law.

Unless it be said that in this case the fact that it might be difficult to establish if there was anybody there as a caretaker distinguishes this case, or unless it can be said that a period of eight or 10 years is so different from a period of three years that different considerations must apply to it, *Dixon v Tommis* seems to me to support the defendants' case here.

The judge founded his decision in favour of the defendants first on the evidence of Mrs Boyd that she always intended to return, and that intention was demonstrated by a letter on which the judge placed considerable reliance, in my view rightly so, a letter written on March 25 1972 to Mr or Miss Newman. It is sufficient for me to cite just one or two sentences from a long letter. One sentence was this: "The real snag is our having a home in England for the older ones to return to when they start work and for us to return to when the youngest, nearly 10 and very bright, has finished school." Then later: "I hope we could find a nice family to keep our home nice for us." In my view, the judge was fully entitled, on Mrs Boyd's oral evidence, supported by that letter, to reach the conclusion that at all material times she did have a definite intention of returning to this country and returning to this flat within something like eight to ten years of the time when she left. I think the material time would be from the expiration of the lease at the end of September 1972 onwards. Probably if the intention were departed from at any time between then and the hearing, then it would mean her defence would fail. Then in addition to the intention to return, which was the *animus possidendi*, for the *corpus possessionis* the judge relied to some extent on the fact that Nicholas and Toby, and indeed Mr Boyd, had been there from time to time, but mainly upon the presence of furniture. It is true that in *Brown v Brash* Asquith LJ put it as a matter of some doubt as to whether furniture alone would constitute the *corpus possessionis*, but for my part I see no reason why it should not do so.

At the end of the day this really resolved itself into a question as to whether the period of eight or ten years was so prolonged a time that it could not be said that Mrs Boyd was continuing to be in residence throughout that period, and reliance was placed on that part of the judgment of the court in *Brown v Brash* at p 254 where Asquith LJ said: "To suppose that he can absent himself for five or ten years or more and retain possession and his protected status simply by proving an inward intention to return after so protracted an absence would be to frustrate the spirit and policy of the Acts." It is important to observe that the learned Lord Justice went on in his next proposition to say: "Notwithstanding an absence so protracted the authorities suggest that its effect may be averted if he couples and clothes his inward intention with some formal, outward, and visible sign of it." So the court in that case was certainly not treating an absence of five or ten years as being necessarily so prolonged as to mean that it was an unreasonable time within the meaning of Ormerod LJ in *Tickner v Hearn*. What it comes to, I think, is that it is, as was said in *Dixon v Tommis*, a question of fact for the judge in all the circumstances of the particular case. With this matter of reasonableness the judge dealt clearly and fully in a passage which I cite from one of the two notes of his judgment, and the one which I gather counsel on both sides preferred for the purpose of this particular part of it. He said:

> In 1970 the theory was that the absence would be for not more than 10 years. It was suggested that that was not a reasonable time. I must judge reasonableness in the context of all the circumstances. This is a family where the husband is on a pension and who were intending to be away for about 10 years. That is not unreasonable in that context. It might be different if they were merely going for pleasure or something of that sort. I can see nothing unreasonable in saying that this flat does not cease to be their home for the time that their children are being educated in Malta. Now I am told that Toby and Hamish Boyd may go and live there. It may be that the Boyds themselves will return sooner than previously contemplated. I accept there are financial limitations in Malta more general than a few years ago. However, the flat has remained their second home.

In my opinion, the learned judge was applying the right tests and there was evidence from which he could reach the conclusion that he did, and there is no ground for saying that he was wrong in the judgment that he gave in favour of Mrs Boyd. It is conceded by the plaintiffs that if Mrs Boyd succeeds, then Mr Roberts succeeds as her licensee. Accordingly, I would dismiss the appeal.

LAWTON LJ: I agree. There were times during the course of Mr Hone's submision on behalf of the landlords when I had to remind myself that I was not trying this case. Had I been doing so I might have come to a different conclusion to that to which the trial judge came. The question for this court has been whether the judgment of the trial judge should be upheld. That question has to be resolved in accordance with well-known principles. The first problem has been whether the learned judge misdirected himself in law. The next question has been whether there was any evidence to support his findings.

One of the difficulties in this case has arisen from the fact that over the years since 1920, when the status of the so-called statutory tenant had first to be considered, there have been many cases decided as to when a person loses the status of a statutory tenant. Sometimes the cases have been reported, either in the ordinary law reports or in journals of professional bodies which take an interest in cases affecting property. From the law reports such cases have got into textbooks. In the textbook which is so frequently used nowadays by counsel, namely *Megarry's Rent Acts*, there are a number of pages which set out illustrations of particular types of cases in which it has been held that the tenant has lost the status of a statutory tenant. There is a danger, which this case manifests, of particular examples of decisions on fact being made into a corpus of law. The right approach is not to examine what judges in other cases have decided on different facts but to ask what was the issue before the judge. Section 3 (1) of the Rent Act 1968 makes clear what it was, to wit that a man shall have the benefit of a statutory tenancy "if and so long as he occupies the dwelling-house as his residence." In my opinion that raises a simple issue of fact which does not require subtle distinctions to be drawn by lawyers. The administration of this Act of Parliament, which affects so many hundreds of thousands of people, is not made any easier for county court judges if subtle distinctions are made.

The matter, unfortunately, is somewhat complicated by section 3 (2), which sets out a statutory test for construing what is meant by the words "if and so long as he occupies the dwelling-house as his residence." That statutory instruction is as follows: "The phrase . . . shall be construed as requiring the fulfilment of the same, and only the same, qualifications (whether as to residence or otherwise) as had to be fulfilled before the commencement of this Act to entitle a tenant, within the meaning of the Increase of Rent and Mortgage Interest (Restrictions) Act 1920, to retain possession, by virtue of that Act and not by virtue of a tenancy, of a dwelling-house to which that Act applied." As my brother Cairns has said, subsection (2) of section 3 does require the courts to look to see what the law was before the Rent Act 1968 came into operation. The definitive judgment on this matter is to be found in the judgment of this court in *Brown v Brash,* to which Cairns LJ has already referred and from which I need not quote, save to remind myself that Asquith LJ, in delivering the judgment of the court, pointed out that the question is one of fact and of degree.

A county court judge, when he comes to consider the issue as to the status of a statutory tenant, has to ask himself certain questions. It was conceded by Mr Hone in argument that in this case the learned trial judge did ask himself the right questions. He did ask whether Mrs Boyd had a real intention of returning, and he answered that question in the affirmative. He also asked himself whether there was any *corpus possessionis,* and he answered affirmatively in favour of Mrs Boyd. So there can be no question in this case of his having misdirected himself as to the law. This court next has to inquire whether the evidence was sufficient to support his findings. Another way of putting this issue is to ask whether his findings on the evidence were so unreasonable as to be perverse. The learned judge gave most careful attention to the evidence in this case and he came to the conclusion, surprising perhaps to some, that there was a real intention to return and a sufficient *corpus possessionis.* I might not have come to the same conclusion had I heard the evidence. Perhaps other judges would not have come to the same conclusion; but I find it impossible to say that there was not evidence upon which those findings could have been made, and accordingly I can see no reason for interfering with the judgment. Mr Hone called our attention to a number of other cases on different facts where other judges have come to conclusions which are not in harmony with the conclusion in this case. That kind of analysis of other cases decided on fact is not helpful: it is a commonplace of forensic experience that every case has got to be decided on its own facts, and, as was pointed out in the course of argument, in all the reported cases which have been decided against the tenant it is easy to point to an additional fact which might have made all the difference. I would dismiss the appeal.

MACKENNA J: We are told by two of the judgments in *Tickner v Hearn* [1960] 1 WLR 1406, that of Ormerod LJ at page 1411 and that of Upjohn LJ at page 1416, that it is a relevant consideration whether a tenant who is residing elsewhere intends to return to the demised premises "within a reasonable time." A question arises on this view of the law which I can explain by an example. At the time when the contractual tenancy comes to an end the tenant may be living elsewhere, but may intend to return at some future date, say in eight years' time. The question of his right to occupy as a statutory tenant may not be raised at that particular time. When it is raised, say a year later, the tenant may still be living elsewhere, but now his intention may be to return in another two years' time. In the case I have supposed, is the question for the court whether a period of eight years is a reasonable time, or one or two years? With respect to Cairns LJ, in my judgment two years is the relevant period, and this view is, I believe, consistent with the judgments in *Tickner's* case. The tenant's intention, and other similar matters must, I think, be considered at the date when his right to continue as a statutory tenant is challenged.

If this is so, it is easier to reach a conclusion in Mrs Boyd's favour in the present case. Her intention in 1972, when her statutory tenancy began, was to return when the education of all her children was completed, which was likely to be 1980, which is still very far ahead. Her intention in 1974, when the case was tried, was to return in two years, which is not very far away, and might, without straining the law too much in the tenant's favour, be considered a reasonable time even when added to the two years which had already expired. If she comes back to St Augustine's Mansions in 1976, she will have been absent for four years from the date in 1972 when she became a statutory tenant, which is a shorter period than that of the tenant's absence in *Tickner's* case. In that case the successful defendant had become a statutory tenant of the premises in question in January 1954. She had not lived there since June 1954. She had become a patient in a mental hospital in August 1954 and was still living in the hospital in 1959 when her right was challenged in the county court. When the case was heard, it was uncertain when, if ever, she would return, yet her claim to be the statutory tenant succeeded. If a decision could be given in Mrs Hearn's favour consistently with the Act, I do not see why Mrs Boyd's claim to the protection of the Act should fail. I would only add that in my judgment a reasonable time means one which is reasonable having regard to the policy of the Act, which is to provide people in this country with homes where they can live. I would dismiss the appeal.

Both respondents were awarded costs.

QUEEN'S BENCH DIVISION

July 11 1974

(Before Mr Justice MAY)

RICHARDS v DE FREITAS

Estates Gazette May 24 1975

(1974) 234 EG 613

Forfeiture—Landlord's right to refuse rent tendered on tenant's behalf—Entitled to raise question on whose account money was tendered, that of the tenant or of a company licensee—Further issue whether landlord could reject a second tender made after issue of writ claiming forfeiture but before service—Yes, despite principle that service, not mere issue, is what affects the forfeiture—Numerous interesting points canvassed in squabble (essentially) over liability for costs

This was an appeal by Mr Derek Graham Richards, of Forest View, Chingford, London E4, from a decision of Master Lubbock in the High Court on April 23 1974 giving Mr Urban Ray de Freitas, of Rolls Park Avenue, Chingford, E4, leave to defend an action for forfeiture of a lease of factory premises in Manor Park, London E12.

Mr J B W McDonnell (instructed by Bernard Solley & Co) appeared for the plaintiff, and Mr J S Colyer (instructed by Durrant-Piesse) represented the defendant.

Giving judgment, MAY J said: This is an appeal from a decision of Master Lubbock of April 23 1974 ordering that the defendant should have leave to defend the action. I heard the argument in this appeal in chambers on June 26, and it has now been adjourned for judgment in open court at the request of the parties. The facts can be shortly stated. The plaintiff is the freeholder of small factory premises at 3-7 Chesterford Road, London E12, of which, at least until the events to which I shall shortly refer, the defendant was the tenant under a lease for 21 years from May 1969 at an annual rent of £2,000, payable quarterly in advance. The lease contained a repairing covenant and the usual proviso for re-entry entitling the plaintiff to forfeit the lease and to re-enter upon the demised premises if the defendant should fail to pay the rent within 21 days of its becoming due, whether formally demanded or not. In the past the plaintiff and the defendant had carried on business together in a company, Ross's Frozen Foods Ltd. When the plaintiff and the defendant separated the latter took this company over, and it has in fact been it which has at all material times occupied the relevant factory premises, although, as I have said, the actual leasehold term has remained vested in the defendant. There is no dispute that the plaintiff was well aware that it was the company that was occupying the premises, and indeed, until the end of 1973 the quarterly rent of £520 was always paid by a company cheque.

On November 22 1973 Barclays Bank Ltd appointed one Wilson to be the receiver and manager of the company under powers contained in a debenture held by them. Mr Wilson sent the plaintiff the usual form of letter telling him of his appointment on November 26 1973. Solicitors for the plaintiff replied on December 12 to the effect that the company was in no way indebted to the plaintiff, and reaffirmed, as was the fact, that their client was the owner of the relevant property, that his tenant was the defendant, and that they understood the company had occupied those premises as the defendant's licensee. Some time in December 1973 the defendant forwarded the plaintiff's rent demand for that quarter to Mr Wilson, the receiver, and as this was for rent of the premises occupied and operated by the company the receiver's cheque for the quarter's rent was sent to the plaintiff on about December 31 1973 with the receiver's compliments slip, upon which it was briefly stated that the cheque was for the rent due for the Christmas quarter. No point has been taken at any time that this tender of the rent was by cheque and not in cash. On January 3 1974 the plaintiff's solicitors wrote to the receiver referring to this tender of the rent and asking him whether he was paying it on behalf of the defendant or on behalf of the company. The letter went on to say, first, that if the rent had been paid on behalf of the defendant it would be accepted, but that if it had been paid on behalf of the company it would not be accepted; and secondly, that they (the solicitors) were that day serving a schedule of dilapidations upon the defendant, a copy of which they enclosed. Unfortunately, in circumstances which are not material, the receiver did not reply to this letter. Consequently, on January 28 1974 the plaintiff's solicitors wrote again to the receiver, telling him that as they had not had any such reply they were returning his cheque, and that they were now instructed to issue proceedings against the defendant for arrears of rent and forfeiture.

On February 5 1974 the writ in this action was issued claiming forfeiture of the lease on the ground of nonpayment of rent for the Christmas quarter. Before that writ was served, the receiver wrote to the plaintiff's solicitors on February 6 explaining the reasons for his delay. He went on to say that he understood that the company had been in the practice of paying the quarterly rent on behalf of the defendant, as was indeed the fact, and that it was because of this that he, as receiver, had taken steps to see that the rent had in fact been paid. In further reply to the plaintiff's solicitors' letter of January 3 the receiver accepted in his letter that the defendant was the lessee and that the company occupied the premises under licence. He consequently enclosed a fresh cheque for the full quarter's rent. In so far as the schedule of dilapidations was concerned, he said that he had authorised and instructed the defendant to carry out the necessary work. As I have indicated, however, by this time proceedings had been started, and on February 12 the plaintiff's solicitors replied to the receiver in these terms: "We thank you for your letter dated the 6th instant with enclosure"—which was, of course, the second cheque—"which we return herewith. We issued proceedings prior to receipt of your letter, and the lease is therefore forfeited unless Mr de Freitas wishes to apply to the court for relief. Such application will be opposed. Please accept this letter as formal notice that your licence to occupy the premises is determined." The writ in this action was ultimately served upon the defendant on February 15 1974.

This brief recital of the facts cannot but suggest that it is perhaps unfortunate that the court has had to be troubled with this matter and that the parties have not been able satisfactorily to compose their difficulties. As will be apparent, there is very little between them. For the plaintiff it is contended that, having received the receiver's formal letter of November 26 1973 and then his cheque for the rent without any explanation, the plaintiff was entitled at the least to be concerned whether by the acceptance of that payment he might compromise his position vis-à-vis the defendant on the one hand and the company on the other. Consequently, it is submitted, the plaintiff was in such circumstances entitled to reject that tender of the rent then due. Counsel then argues that his client was also entitled to reject the second tender of the rent enclosed in the receiver's letter of February 6 1974, first, because by that time the 21 days for payment of the Christmas rent referred to in the proviso for re-entry in the lease had expired; and secondly, because by that later date the schedule of dilapidations had been served upon the defendant, accompanied by the usual notice under section 146 of the Law of Property Act 1925. In the result, contends the plaintiff, the defendant's lease was forfeited by the service of the writ upon him on February 15 1974; his only right now is to seek relief from that forfeiture; and this should only be granted upon payment by the receiver of all the arrears of rent and mesne profits due to date, upon the execution of the repairs listed in the schedule of dilapidations, and upon payment by the defendant of the plaintiff's costs of these proceedings.

For his part, the defendant contends that upon a true analysis of the situation as at January 3 1974 the acceptance by the plaintiff of the receiver's cheque for the rent could in no way have prejudiced the former's position; in the circumstances, that tender of the rent was a good tender, and accordingly, the plaintiff was not thereafter entitled to forfeit or seek to forfeit the lease. If this should be wrong, then the defendant by his counsel secondly submits that upon the authorities the second offer of the rent on February 6 1974, being between the issue of the writ the day before and its service nine days later, was also a good tender which the plaintiff was not entitled to reject. Accordingly, for this reason, also, the service of the writ did not forfeit the lease. If either of these two contentions should be correct, then the proper order would be to dismiss the plaintiff's claim with costs. If, however, I were to come to the conclusion that both these contentions were wrong, then nevertheless the state of the law applicable to the undisputed facts which I have outlined is by no means easy or certain; the defendant's contentions, it is submitted, are at the least well arguable in law; and consequently the defendant should in any event have leave

A to defend so that the whole matter may be more fully argued than was possible before me in chambers. In truth, as I see it, the real issue now between the plaintiff and the defendant is substantially which of the two of them should be liable for the costs of these proceedings. As I have indicated, I think that it is unfortunate they have reached the point which they have and that substantial costs have been incurred. I can see no good reason why this matter could not have been satisfactorily compromised between the parties at an early stage. Nevertheless, as will appear, even though the practical issue may be so limited, the various contentions of the two sides do raise interesting and indeed difficult **B** points of landlord and tenant law, which of course it is my duty to consider and decide, notwithstanding the view that I may take of the conduct of one or other or both of the parties.

The first question which therefore arises is whether the plaintiff was entitled to refuse the receiver's cheque for the rent due at Christmas 1973; or at the least, whether he was entitled to return it after his solicitors had not received a reply to their letter of inquiry of January 3 1974 within the 21 days referred to in the proviso for re-entry after December 25 1973. Mr McDonnell, on the plaintiff's behalf, submits that a creditor, particularly a lessor, cannot be expected to **C** accept tender of a debt due to him proferred by any and every third party. In the first place, he submits that a creditor is not bound to accept a tender which would not in law operate to discharge the relevant debt. In this connection he refers me to a passage from the judgment of Parke B in *Simpson v Eggington* (1855) 10 Ex 845 at 847, which is in these terms:

> The general rule as to payment and satisfaction by a third person, not himself liable as a co-contractor or otherwise, has been fully considered in the cases of *Jones v Broadhurst, Belshaw v Bush* and *James v Isaacs;* and the result appears to be, that it is not sufficient to discharge a debtor unless it is made by the third **D** person, as agent, *for and on account* of the debtor and with his prior authority or subsequent ratification.

The italics are those of the report. This dictum of Parke B was followed by Humphreys J in *Smith v Cox* [1940] 2 KB 558, which was a landlord and tenant case and concerned the payment of rent to a landlord by a third party. On this first question Mr McDonnell secondly submits that a creditor is not bound to accept a tender from a third party when such acceptance would operate to the former's detriment to any greater extent than would payment by the actual debtor in person. Indeed, he goes further and contends that common- **E** sense requires that a creditor is not bound to accept tender from a third party when he reasonably fears that such acceptance might prejudice his position.

I have been referred to no authorities on the question when, if at all, a creditor is entitled to reject a tender by someone other than his debtor except for the two cases which I have mentioned and which can perhaps be said to assist only somewhat obliquely. In my judgment, however, a creditor is not bound to accept a tender from a third party either when on the facts that he then knows, or ought to know, the acceptance will not discharge the debt due to him, or alternatively, again on such facts, where acceptance would **F** prejudice his position vis-à-vis his debtor to a greater extent than would tender by that debtor in person. In such circumstances I do not think that a creditor must be deemed to know the complete factual situation if he does not do so: he is entitled to make his decision on the facts as he knows or ought to know them. Secondly, if the proferred tender will not discharge the relevant debt, it cannot in my view properly be described as a tender. Finally, I see no reason in justice or commonsense why a creditor should be required unnecessarily to act to his detriment in such a case. In the present case, therefore, I have first to ask myself whether on the facts which were or ought to have been known to the **G** plaintiff or his advisers the acceptance of the receiver's cheque on some date in January 1974 prior to the expiry of 21 days after Christmas 1973 by the plaintiff would have discharged the defendant's liability to the latter at that time for the quarter's rent. Following the dictum of Parke B in *Simpson v Eggington.* I do not think that on the facts as they in truth were there can be any doubt that acceptance of the receiver's cheque would have discharged the defendant's liability in respect of the defendant's quarter's rent. Paragraph 15 of the defendant's affidavit, sworn in these proceedings on June 21 1974, which has not been disputed, is enough to make this clear:

> **H** (15) In December 1973 when I received the rent demand for the quarter's rent due on December 25 I passed it on to Mr Wilson's staff, who were working at [other premises], and I requested them to pay it from the company's account. Mr Wilson as receiver of the company had my authority, as did the company, to pay all rent due on the relevant premises on my behalf.

But although this may have been the actual situation, I do not think that the plaintiff and his advisers were, nor ought they at that stage to have been, fully aware of it. The plaintiff's solicitors' letter to the receiver of January 3 1974 asked, "Would you please let us know whether you are paying this money on behalf of [the defendant] or on behalf of the company," which was a very pertinent question in the light **J** of the dictum of Parke B, and I see no reason to doubt the genuineness of that inquiry. In the particular circumstances which obtained as between the defendant, the company, the receiver and the plaintiff at this time, I do not think that the plaintiff or his advisers either were or ought to have been aware that the receiver was paying the rent either as the defendant's agent for or on his account or with his prior authority (at this stage no question of subsequent ratification can have arisen). On this ground, therefore, I am of the opinion that the plaintiff was entitled to reject the receiver's tender at least until his solicitors had had a reply to their letter of January 3 1974 within 21 days after December 25 **K** 1973.

In these circumstances, it is perhaps unnecessary for me to express a view on the second question, namely, whether acceptance of the receiver's tender on the facts as they were, or ought to have been, known to the plaintiff and his advisers would have prejudiced his position as against his tenant, the defendant. Nevertheless, in case I may be wrong on the view I have expressed on the first question, I will say that I do not think that in the postulated circumstances the plaintiff would have been so prejudiced. This could only have been so if his acceptance of the receiver's cheque either might **L** have created a new tenancy in some person other than the defendant, or might have waived some default by the defendant under his lease which would otherwise have entitled the plaintiff to forfeit the term. In so far as the creation of any new tenancy was concerned, the relevant issue would have been *quo animo* the rent was received. There can be no suggestion in the present case, in my opinion, that if the plaintiff had accepted the receiver's cheque he would have done so with any intent of creating any new tenancy in anyone. As to the possible waiver of any forfeiture, there could be none if the alleged ground of forfeiture was the failure to pay rent within the relevant 21 **M** days. In so far as forfeiture on the ground that the defendant was in breach of covenant not to assign his term without the plaintiff's consent was concerned, he (the plaintiff) would only have waived the right to forfeit if he had known of the material breach of covenant. Here the plaintiff did not know whether there had been any such assignment; this was the very thing that his solicitors were trying to find out. Finally, although a schedule of dilapidations accompanied by a notice under section 146 of the Law of Property Act 1925 had been served, it was invalid in that it failed to comply with the requirements of the Leasehold Property (Repairs)

Act 1938 as extended by the Landlord and Tenant Act 1954. Consequently, whether or not the defendant was in breach of the repairing covenants in his lease, the plaintiff was not then, nor, so far as I am aware, is he now, in a position to forfeit the lease on that ground. Consequently, no acceptance by him of any rent in January 1974 could have prejudiced his position on this score. In the result, I take the view that if I am wrong and the plaintiff was not entitled to reject the receiver's cheque on the basis of the first contention put forward on his behalf, then also would he not have been entitled to reject it on the basis of his second contention.

Be that as it may, the next question with which I have to deal in this case is whether the plaintiff was entitled to reject the second tender by the receiver of the rent which had been due at Christmas 1973 and which the latter offered between the issue of the writ herein on February 5 1974 and its service on the defendant on February 15 1974. It is clear from the decision of the Court of Appeal in *Canas Property Co Ltd v KL Television Services Ltd* [1970] 2 QB 433 that, contrary to the view previously held by some, it is the service of the writ upon the defendant tenant in proceedings for possession on the grounds of a breach by the latter of a covenant in his lease that is the notional re-entry which completes the forfeiture, and that it is not merely the issue of the writ. It was consequently submitted on behalf of the defendant that there exists in these cases between issue and service of the writ what may be described as a *locus poenitentiae* during which it is always open to the tenant, at least where the ground of forfeiture is nonpayment of the rent, to tender the rent due and thus put an immediate end to the effectiveness of proceedings which have been begun. I do not agree. In my judgment the decision in the *Canas* case did not alter the basic principles applicable to forfeiture. It merely corrected the then held view of the point in time at which the forfeiture took place. Once a lessor becomes aware of a breach by his lessee of any of the covenants in the lease, then he has an election. He may either elect to allow the term granted by the lease to continue, merely suing for damages for the breach of covenant; or alternatively he may elect to forfeit the lease, either by actual re-entry or by a notional re-entry effected by the service of the writ in ejectment proceedings. He is not bound to choose the former in preference to the latter. In these circumstances, the issue of the writ is a mere incident in the procedure; it does not signal the start of any special period in so far as the relationship between lessor and lessee is concerned. If at any time after a lessor becomes aware of a breach of covenant on the part of his lessee he accepts rent from the latter, then it is clear and indeed irrebuttable evidence that he has elected not to forfeit but to continue the lease. When he becomes aware of the lessee's breach of covenant, he is permitted by the law a reasonable time in which to make up his mind in which direction he proposes to elect. That that reasonable period of time may encompass the period between issue and service of the writ is, in my opinion, neither here nor there. If, having issued the writ to the knowledge of the lessee, the lessor delays for an unreasonable period before he serves it, then this may well be evidence that he has changed his mind and proposes not to forfeit the lease but to allow it to continue. This is, however, a matter of fact to be considered in each case.

Applying these principles to the facts in the present action, by February 5 1974 the defendant lessee was in breach at least of the covenant to pay the rent. He may at the same time have been in breach of the repairing covenant. In consequence the plaintiff lessor decided to forfeit the lease, and the writ was issued. Ten days later it was served and the notional re-entry completed. Having made his election, there was in my judgment no obligation whatever upon the lessor in that period of ten days to accept any second tender of the rent from the receiver. He was entitled to say to the defendant lessee that he had so conducted himself as to incur a liability to forfeiture which he, the lessor, proposed to enforce. To accept rent thereafter would, as I have indicated, have negatived his clear election. Once the right of forfeiture has accrued, no lessor is bound to give up that right or, to put it in another way, no lessee is entitled to relief from that forfeiture save on the special conditions laid down in the case of *Gill v Lewis* [1956] 2 QB 1. I should add that, anticipating the *locus poenitentiae* argument to which I have referred, counsel for the plaintiff sought to meet it by the inference which he suggested could be drawn from a very short dictum of Lord Denning MR in the *Canas Property Co* case at page 441 of the Law Reports. The relevant passage reads:

> *Ogan v Raite & Holt* in 1903 was decided by Grantham J. It is only reported in 67 JP Journal 100. It says that 'by the issue of the writ the plaintiff exercised his option to determine the lease.' That was wrong. But the decision is right. The landlord did not accept the rent. So the lease was forfeited when the writ was served. The defendants had to seek relief.

In *Ogan v Raite* the defendant tenants failed to pay one quarter's rent due on a particular day. After the expiry of the 21 days reserved for providing for re-entry, in that case also the plaintiff issued a writ claiming possession, and on the following day the defendants tendered the rent by cheque. Counsel in the present case submitted that when in the *Canas* case the learned Master of the Rolls said of *Ogan v Raite,* "The landlord did not accept the rent," he must in the context be taken as meaning or to have inferred that in that case the landlord *was entitled* not to accept the rent, and the only basis on which he could have been so entitled was that it was tendered after the expiry of the 21 days reserved by the proviso for re-entry. I cannot accept this argument. The report of the judgment in *Ogan v Raite* is very short, but it is to be noted that it does record that after its tender the defendants' cheque was returned marked "cancelled." This can only have been because it was stopped, for some reason or other, by the defendants. Consequently in that case the relevant rent had not been paid by the defendants, and thus could not have been accepted by the landlord plaintiff by the time the action came before Grantham J for trial. In these circumstances, I am quite satisfied that that was all that Lord Denning meant by the seven-word sentence relied on by the plaintiff in our case, that "*As a matter of fact* the landlord did not accept the rent," and that no other inference such as was suggested can be drawn from it. Nevertheless, for the reasons which I have already given, I do not think that there was any obligation upon the plaintiff lessor in this present case to accept the second tender by the receiver for the rent due at Christmas 1973 in the period between February 5 and 15 1974. It further follows from all that I have said that in my opinion the lessor in this case validly forfeited the relevant lease when he served the writ in these proceedings on the defendant lessee on February 15 1974.

Although I have reached this conclusion, should the defendant nevertheless still retain his leave to defend granted by the learned master on the ground that is put forward, namely that the law is by no means clear, that his case is well arguable and that in consequence he should be given the further opportunity to do so, no doubt more fully and at greater length? I have no doubt that he should not. The facts in the present case are simple and not in dispute. If my view of the relevant law is wrong it can be corrected. Mindful of the dictum of Russell LJ in *Bigg v Boyd Gibbins Ltd* [1971] 1 WLR 913 at 915, it would in my judgment now be a complete waste of further costs were I to allow the master's order to stand. Subject to argument on the precise terms of my order, I therefore propose to allow this appeal and to give judgment for the plaintiff for possession of the relevant premises, and also whatever is now the correct figure for arrears of rent and mesne profits in this present action. The

A **defendant, however, may have relief from such forfeiture if he pays both those arrears of rent and mesne profits and, if not already paid, the further sum of £320 arrears of rent** for which judgment was obtained for the plaintiff against him in the Bow County Court on February 12 1974, but to which it has not hitherto in this judgment been necessary to refer.

Relief from forfeiture was granted on terms that arrears of rent and mesne profits to date, including the sum of £320 referred to, be paid within three weeks. No order was made as to the costs of the proceedings before the master, and the
B plaintiff was awarded only half the costs of the appeal to his Lordship.

COURT OF APPEAL

January 23 1975

(Before Lord Justice CAIRNS, Lord Justice LAWTON and Mr Justice MACKENNA)

NATIONAL WESTMINSTER BANK LTD v BETCHWORTH INVESTMENTS LTD

C Estates Gazette May 31 1975
(1975) 234 EG 675

Notice exercising right to break lease—Whether sent to tenant's "last known address"—Yes—Last known address of tenant company its registered office, not the address of an associated firm of managing agents with whom the landlord had normally dealt

This was an appeal by National Westminster Bank Ltd from a judgment of Phillips J dated March 19 1974 declaring in favour of the respondents, Betchworth Investments Ltd that a notice of March 23 1973 purporting to exercise a
D right to determine a lease of premises at 77 High Street Godalming, Surrey, was not validly served by the appellants in that it was incorrectly addressed. The judgment below was reported at (1975) 234 EG 43.

Mr C F Dehn QC and Mr P Curtis-Bennett (instructed by Rubinstein, Nash & Co) appeared for the appellants, and Mr D A Wood (instructed by Roche, Son & Neale) represented the respondents.

Giving judgment, CAIRNS LJ said: This is an appeal from a decision of Phillips J on cross-claims for declarations. The learned judge dismissed the plaintiffs' claim and made a
E declaration on the defendants' counterclaim. The question at issue was whether a notice served by the plaintiffs as landlords on the defendants as tenants to break a lease was validly served or not. The plaintiffs claimed that it was; the defendants claimed that it was not. The judge decided that it was not, and the plaintiffs now appeal to this court. The result depends on whether the address to which the notice was sent was the last known address of the defendants within the meaning of the lease. The lease was a 14-year lease granted on October 5 1966 by the plaintiffs' predecessors to the defendants—the plaintiffs' predecessors, because the plaintiffs are National Westminster Bank Ltd and the plain-
F tiffs' predecessors were the Westminster Bank Ltd. While the lease was a lease for 14 years from March 25 1966, it was subject to provisions for earlier determination. Those provisions were contained in clause 4 (1) of the lease in these terms: "If either of them the lessor or the tenant shall be desirous of determining the present lease on the twenty-ninth day of September one thousand nine hundred and seventy-three and of such its desire shall give six months' previous written notice to the other of them"—then I need not read the next few lines—" then and in such case on the said twenty-ninth day of September one thousand nine hundred and seventy-three this present lease shall cease," and I need not read the concluding words of the subclause. Then G
in subclause (3) of clause 4 was this provision:

> Any notice under this lease shall be in writing and any notice to the tenant shall be deemed to be sufficiently served if left addressed or sent by post to it on the demised premises or sent to it by post at its registered office or last known address in Great Britain or Ireland (whether Eire or Northern Ireland) and any notice to the lessor shall be deemed to be sufficiently served if sent to it by registered post to or left at its registered office.

It was recited at the beginning of the lease that the defendants, Betchworth Investments Ltd, had their registered office at 157 Victoria Street, London SW1, and that was the H
only address of the defendants given in the lease. The evidence in the case was by way of affidavits, and there is no substantial dispute on the facts. According to the affidavit of Mr Cotter, who is the assistant regional premises manager of the plaintiffs, there was correspondence between the plaintiffs and the defendants at that address during the next 18 months or so. I understand that statement of Mr Cotter's as meaning that during that time letters were sent by the plaintiffs to the defendants addressed to 157 Victoria Street, and replies were received by the plaintiffs from that address. On May 24 1968 the plaintiffs wrote to the defendants about paying the cost of certain works, and the letter was sent to J
157 Victoria Street. The reply, however, came from a firm of chartered surveyors, Mellersh & Harding, from their address at 43 St James's Place, SW1. It is therefore apparent that the letter had been sent on from the registered office in Victoria Street to the chartered surveyors, who were the managing agents for the defendants. At no time was any express indication given by the defendants to the plaintiffs that they had changed their address from 157 Victoria Street. However, after that time, May of 1968, there was no direct communication between the plaintiffs and the defendants. All the communications that took place were between the plaintiffs and either those managing agents or the defendants' K
solicitors, mainly the former. Accordingly, the plaintiffs' records continued to show 157 Victoria Street as the defendants' address. In March 1973 the plaintiffs decided to exercise their option to break the lease, and instructions were given to their solicitors to serve the appropriate notice by post at that address, and the letter was prepared and posted on March 23 1973, addressed to the defendants at 157 Victoria Street. It is admitted that the notice was sent in good time to reach that address by March 29, and that if it was correctly addressed the notice was validly served. Although, by a respondents' notice in this court, it was contended that actual delivery was needed, that argument was L
not pursued by Mr Wood on behalf of the defendants here.

Now in fact by March of 1973, 157 Victoria Street was no longer the defendants' registered office, and indeed was no longer in existence; the building had been demolished; and the envelope containing the notice was returned by the Post Office to the plaintiffs. The plaintiffs then discovered that by that time the defendants' registered office had been changed to 1 Dover Street, and had in fact been so changed in December 1970. The plaintiffs sent the notice then to that office, but of course it did not arrive there till some days later, in fact April 4, which was too late to be good service M
giving the required six months' notice. The defendants' case, which was accepted by the trial judge, was that by March of 1973 the defendants' last known address within the meaning of the clause was 43 St James's Place. The plaintiffs concede that 157 Victoria Street was not in any sense the defendants' address in March of 1973, but they say it had been at one time their address, and it was known to the plaintiffs to be their address, and it had never ceased to be their last known address. It is said that it would only cease to be their last known address if the plaintiffs got to know that some other address had become the defendants' address, and, say the plaintiffs, 43 St James's Place never became

A the defendants' address at all, and if it did, it never became known as their address to the plaintiffs. Reliance is placed, and in my view very properly placed, upon section 107 of the Companies Act 1948, which provides:

> A company shall, as from the day on which it begins to carry on business, or as from the fourteenth day after the date of its incorporation, whichever is the earlier, have a registered office to which all communications and notices may be addressed.

There may, of course, be more than one address of a company, and the relevant clause of this lease recognises that. There may be another place where the company carries on business. But that does not deprive the registered office of B having the character of a place to which communications and notices may be addressed. Now, did 43 St James's Place ever become the address of the defendants? Mr Hawkes, the managing director of the defendant company, says that at no material time did the defendants have a separate place of business. The whole of their business was conducted through the firm of chartered surveyors as their managing agents, that being a firm of which Mr Hawkes himself was the senior partner. The judge accepted that the registered office of a company could be an address within the meaning of this clause, and consequently could be, in certain circumstances, C the last known address within the meaning of that clause; but he considered that the expression would also cover, in the context of this lease, any address of the tenant which the landlord and the tenant used in the course of the lease while transacting business arising out of the lease, and that because from 1968 onwards all such business, except certain legal business conducted between solicitors, had been dealt with between the plaintiffs and the managing agents, the address of the managing agents became at some stage (he did not decide exactly when) an address of the defendants, and since the plaintiffs learnt of it after they learnt of the registered office, it was therefore the last known address D within the meaning of the clause.

With all respect to the judge, I cannot accept either of these conclusions. There is no suggestion that any letter was ever addressed to the defendants at St James's Place or that any letter signed as their letter was sent from there. There is no evidence that their name was displayed there; and it may be observed that it is provided by section 108 (1) (a) of the Companies Act that "every company shall paint or affix, and keep painted or affixed, its name on the outside of every office or place in which its business is carried on, in a conspicuous position," and so forth. I am quite unable to accept that when business is conducted by manag- E ing agents on behalf of a company the address of the managing agents becomes the address of the company, unless indeed the company were to inform a correspondent "letters may be sent to us addressed care of so-and-so," in which circumstances I think it might become an address within the meaning of this clause. Even if it could be said that St James's Place had become an address of the defendants and to have become known to the plaintiffs as such, I do not see how it would displace 157 Victoria Street as another known address. Plainly a person or a company can have more than one last known address. I am quite unable to accept the judge's reasoning that the last address which F becomes known to the landlord is the last known address.

Mr Wood, who has presented a vigorous argument to the court on behalf of the defendants, has said that the places chosen by the parties for the purpose of this clause as the places to which notices may be sent are chosen as places at which the tenant could be expected to get those notices. With that I entirely agree; but that does not mean that every place at or through which a tenant could be expected to get a notice thereby becomes an address. It has to be within the description of one of these three places, the demised premises, the registered office or the last known place of address; and, of course, if people are honest and businesslike there will rarely be any failure of actual communication. Nobody G for a moment suggests in this case anything other than honesty on the part of the defendants, and as to being businesslike I can only say that it would have been more businesslike on their part if they had informed the plaintiffs of the change of their registered office, and indeed, if they desired to say so, that they were carrying on business thenceforth from 43 St James's Place, though if they were carrying on business from St James's Place then they ought to have their name up at the premises. There is no evidence one way or the other as to whether they had. Mr Wood suggests that the defendants' address is simply a place where they can be reached. I would not accept that. For instance, taking an H individual, as distinct from a company, it may be known to a person dealing with him that he could always be reached in a particular public house, that he was at that public house every evening and that a letter addressed there would always reach him. But that would not become his address for the purpose of a clause of this kind. I am quite unable to accept that the address of agents can become the address of the tenants simply by being the place to which and from which correspondence is carried on by those agents as agents for the tenants. Reference has been made to dictionary definitions of the word "address," and all those definitions are roughly the same, the Oxford English Dictionary having as J its relevant definition "The name of the place to which a person's letters are directed." I think that means letters which are directed to that person by name at that particular address, not letters to an agent, even if he be a managing agent acting on his behalf.

Mr Wood suggests that a registered office is not in itself an address. I think it must be. I refer again to section 107 of the Companies Act. The very nature of a registered office is the place to which communications and notices may be addressed. That is the definition of it. And by saying, "our registered office is so-and-so," as the defendants did K in this lease, they are saying, "that is our address to which communications may be sent." Then further it is suggested that if a registered office always is an address it was unnecessary surplusage for both "registered address" and "last known address" to be mentioned in the clause. I think they are both mentioned for perfectly good reasons. It is convenient that the landlord should be able to serve the tenants at their registered office because the registered office is something which is always ascertainable. They could always discover that from the Companies Register. On the other hand, they may not know the company's registered office. They may not want to take the trouble of going to the Com- L panies Register to discover it, and therefore it is convenient for them to be able to serve the notice at the place of business which, in nine cases out of ten, will be the actual place where day-to-day business is being carried on by the company, and in the tenth case will be the last place where the landlords had communication with them. It seems to me perfectly sensible that both alternatives should be provided. I, for my part, am entirely satisfied that the last known address of the defendants within the meaning of this clause was the Victoria Street address, and accordingly the notice was validly served. I would therefore allow the appeal, set aside the declaration made on the counterclaim, and make instead the M declaration that is prayed for in the plaintiffs' statement of claim.

LAWTON LJ: I agree. In my judgment, this appeal turns solely upon the construction of the words "last known address" in clause 4 (3) of the lease. Those words are words in ordinary usage in English. In no way are they words of art. They have no special meaning in law. They must be construed in the way ordinary people accustomed to the accurate use of the English language would construe them. In my judgment, no useful purpose is served by a detailed semantic analysis of the phrase. What would the ordinary

A person regard as the last known address of this company? Such a person might know, and must be deemed for this purpose to know, the provisions of section 107 (1) of the Companies Act 1948. Once such a person had been told what the law requires of a company in respect of a registered office, such a person would undoubtedly say, " the registered office is the address of the company "; and he might go on to say, " and any address of a company must be a place with which the company has a close connection." The offices of the managing agents would not, in ordinary usage, be described, in my judgment, as the company's address. It would be the address through which the company could be reached, B but not sufficiently close to be the company's address. In those circumstances, it seems to me that the lack of proximity of the managing agents' offices to the company is sufficient to resolve this case. I too would allow this appeal, and make an order as my brother Cairns LJ has stated.

MACKENNA J: I do not think that the defendant company can succeed unless either (1) Victoria Street never was an " address " of the company within the meaning of clause 4, or (2) St James's Place became one of its addresses. It may be that something more would have to be shown for the company to succeed, but unless one or other of these two proposi- C tions is established the company must fail. I hold (1) that at the date when the lease was granted Victoria Street was an " address " of the company within the meaning of clause 4, and (2) that St James's Place has not been shown ever to have become one of its addresses. The first of these conclusions depends on the view that the address of the company's registered office is an " address " within the meaning of clause 4, even though no business of the company was ever transacted at that office. The second depends on the view that St James's Place has not been shown to have been more than the address of the company's agents, and that something else would have to be shown to make it the company's own D address. It follows from these findings that Victoria Street was an address of the company known to the plaintiffs, and the only address of which they ever knew, and therefore that it was that company's " last known address," so that the notice sent to it was properly served. I would allow the appeal.

The appellants were awarded the costs of the appeal and of the claim and counterclaim below.

CHANCERY DIVISION

E July 29 1974

(Before PENNYCUICK V-C)

WILLIAMS AND ANOTHER v. WELLINGBOROUGH BOROUGH COUNCIL

Estates Gazette June 7 1975
(1974) 234 EG 743

Council house sold to sitting tenants with aid of mortgage from council themselves—Common-form 5-year right of pre-emption in transfer—On default by mortgagors, council nevertheless may not claim to exercise power of sale by F executing transfer back to themselves at original price—Purported transfer ineffective

This was an action by Mr Dennis Williams and his wife, Mrs Rita Williams, for a declaration that a transfer executed by the defendants, Wellingborough Borough Council, purporting as mortgagees to revest the freehold interest in 56 Cowper Road, Wellingborough, in themselves at the price originally paid by the plaintiffs, was void and of no effect, and for an injunction restraining the defendants from taking steps to register themselves as freehold proprietors.

Mr P St J Langan (instructed by Maxwell, Batley & Co, agents for Smith, Chamberlain & Co, of Wellingborough) appeared for the plaintiffs, and Miss L E Appleby (instructed G by Sharpe, Pritchard & Co, agents for the council solicitor) represented the respondents.

Giving judgment, PENNYCUICK V-C said that in 1972 the plaintiffs, who were tenants of the council, sought to buy the property they occupied. On November 20 1972 the council executed a transfer of the property in common form for £5,300. The transfer contained a clause that the purchasers would not, within five years, sell or otherwise dispose of the property without first offering it to the council. On the same date the plaintiffs executed a legal charge in common form on the property in favour of the council as mortgagees, the purchase price being made repayable by monthly instalments H of £38.83, which had since risen to £47.46. The plaintiffs executed a further charge in favour of J S Hodge Securities for £1,000 on a second mortgage. Subsequently Mrs Williams became ill, and the plaintiffs got into arrears with their repayments; by May 15 1974 the sum of £341.54 was owing. The council served notice requiring payment of that amount. Mr and Mrs Williams failed to comply with the notice, whereupon the council, in purported exercise of the power of sale contained in the mortgage, executed a transfer of the property to themselves for £5,300, the original purchase price. The council then offered the plaintiffs a tenancy of the house, which they refused. Shortly afterwards the plaintiffs issued J the writ in the action, by which they sought a declaration that the transfer back to the council was void and an injunction to stop the council taking further steps to register themselves as freehold owners of the property.

The relevant provisions were contained in section 104 of the Housing Act 1957, subsection (1) of which enacted:

> Where a local authority have acquired or appropriated any land for the purposes of this Part of this Act, then, without prejudice to any of their other powers under this Act, the local authority may, with the consent of the Minister, sell or lease any houses on the land or erected by them on the land, subject to such covenants and conditions as they may think fit to impose in K regard to the maintenance or use of the houses, and upon any such sale or on the grant of any such lease they may, if they think fit, agree to the price or any premium being paid by instalments or to a payment of part thereof being secured by a mortgage of the premises.

Subsection (3):

> On the sale of a house in accordance with this section . . . a local authority may in any case, and shall if so required by the Minister, impose conditions . . . (c) precluding the purchaser . . . from selling or letting the house during [a period not exceeding five years from the completion of the sale] unless he has . . . offered to resell or sell the house to them and the authority have refused L the offer or have failed to accept it within one month after it is made, and prescribing or providing for the determination of the price to be paid in the event of the acceptance of such an offer.

The Minister of Housing and Local Government gave certain directions pursuant to section 104 contained in a circular No 54/70, dated June 30 1970, including the following:

> Appendix I. Disposal of houses provided under Part V of the Housing Act 1957. The Minister of Housing and Local Government, in exercise of his powers under section 104 of the Housing Act 1957 and of all other powers enabling him in that behalf, hereby gives to all local authorities . . . the following general consents, to operate as from July 1 1970: (1) To sell, or to lease M for a term of 99 years or more, any house provided under or appropriated for the purposes of Part V of the Housing Act 1957 (a) to a sitting tenant . . . (2) To sell, or to lease for a term of 99 years or more, such a house and to such a person as are mentioned in subparagraph (1) at less than the current market price, consideration or rent subject to the following conditions: . . . (b) on such a sale of a house the local authority shall under section 104 (3) of the Housing Act 1957 impose conditions to provide that (i) during a period of five years from the date of the completion of the sale the house by virtue of a restrictive covenant inserted in the conveyance for the purpose shall not be resold at a price in excess of the amount at which it was

A sold by the authority plus such increase for improvements (if any) as may be agreed . . . or determined by the Minister . . . ; (ii) the local authority reserve to themselves in the conveyance a right of pre-emption precluding the sale or the grant of a lease of the house during that period of five years unless the owner has first offered to resell the house to the local authority and the authority have refused the offer or have failed to accept it within one month after it is made.

Section 101 of the Law of Property Act 1925, so far as relevant, provided:

(1) A mortgagee, where the mortgage is made by deed, shall, by virtue of this Act, have the following powers, to the like B extent as if they had been in terms conferred by the mortgage deed, but not further (namely): (i) A power, when the mortgage money has become due, to sell . . . the mortgaged property . . . by public auction or by private contract. . . .

Clause 6 of the transfer to the plaintiffs said:

Provided further that in case the borrower shall . . . (2) be in default for one month in payment of any interest or other moneys herein covenanted to be paid by the borrower . . . then and in any such case the whole of the moneys payable or to become payable hereunder shall be deemed to be forthwith due and owing and the council may at any time thereafter without any previous notice to or concurrence on the part of the borrower. . . . (b) exercise the power of sale . . . conferred upon mortgagees by the C Law of Property Act 1925 and so that the power of sale may be exercised whether the council are in possession or not and without the restrictions imposed by section 103 of the said Act and so that for the purposes of such Act the whole of the moneys so becoming due as aforesaid shall be deemed to be moneys hereby secured.

The question now for decision was one of law, namely, was the very unusual transaction the council had purported to carry out an effective exercise by them of their power of sale? It was not in question that the power of sale had arisen by reason of default on the part of the plaintiffs in the payment of the mortgage instalments. Counsel for the plain- D tiffs referred to the well-known principle by which it was not competent for a mortgagee execising a power of sale to sell to himself. He (counsel) submitted that there was in fact no sale in the present case, the transaction being a nullity. Counsel for the local authority argued, on the other hand, that in the very special circumstances and having regard to the policy which apparently underlay the Housing Act 1957, this should be treated as a valid transaction.

The general principle that a mortgagee could not sell to himself was established beyond dispute. It would be sufficient to make two citations from cases of high authority and two citations from textbooks. In *Farrar v Farrars Ltd* (1888) 40 E Ch D 395 at p 409 in the Court of Appeal Lindley LJ said: "It is perfectly well settled that a mortgagee with a power of sale cannot sell to himself either alone or with others, nor to a trustee for himself . . . nor to anyone employed by him to conduct the sale. . . . A sale by a person to himself is no sale at all, and a power of sale does not authorise the donee of the power to take the property subject to it at a price fixed by himself, even although such price be the full value of the property." In *Henderson v Astwood* [1894] AC 150 in the Privy Council Lord MacNaghten said at p 158: "The so-called sale was of course inoperative. A man cannot sell F to himself, either in his own person or in the person of another." *Halsbury's Laws of England* 3rd ed (1959) vol 27 p 307 para 575 stated, "A mortgagee cannot sell to himself, either alone or with others, nor to a trustee for himself." Finally, there was the statement in *Snell's Principles of Equity* 27th ed (1973) p 398: "The sale must be a true sale. A purported sale by a mortgagee to himself, whether directly G or through an agent, is no sale at all, and may be disregarded."

It would be difficult to find a set of more uncompromising statements of principle. Counsel for the local authority referred to a series of cases on the analogous but by no means identical position where a trustee sold directly or indirectly to himself, and in particular she relied on *Holder v Holder* [1968] Ch 353, a case in the Court of Appeal, where it was held that in certain special circumstances the rule of equity that a trustee was debarred from purchasing trust property did not apply. The circumstances concerned were that the individuals involved had taken virtually no part in the administration of an estate of which an executor, after H certain small acts of administration, had purported to renounce probate. He (his Lordship) could not accept that the decisions cited justified him in disregarding in the present case the authorities directly concerned with sales by a mortgagee to himself. He (the Vice-Chancellor) thought that those authorities governed the facts before him, which in one particular respect afforded a striking instance of the application of the relevant principle. This was not even the case which sometimes arose of a mortgagee, directly or indirectly, bidding at an auction for purchase of the mortgaged property. There was nothing remotely resembling a sale here. The council simply executed a transfer of the property back to J their own name, and all else apart, it was impossible to describe that as a sale "by public auction or by private contract" within the meaning of section 101 of the Law of Property Act 1925.

Counsel had gone on to submit that if the plaintiffs were precluded from making a profit on the resale of the property within five years of their purchase of it, it would be anomalous that they should indirectly be entitled to make a profit on the sale of that property by the council as mortgagee, the council having been put into the position in which they were more or less obliged to sell by the plaintiffs' own default in K payment of instalments. Why, counsel asked, should the mortgagors be entitled to the surplus in the latter case, where they would not be entitled to make a profit in the former case? It was possible that the court's decision was not in accordance with what might be called the general policy of the Housing Act 1957. There was, however, nothing in the Act, and nothing in the transfer or legal charge, which would justify him (his Lordship) in reaching any different conclusion. Having regard to the apparent policy of the Act, he abstained from expressing any concluded view as to what would be the position if the transfer and legal charge had contained an express provision authorising the council to L do that which it had done. There was no such express provision. Counsel had submitted, finally, that the plaintiffs were now precluded by estoppel from claiming any surplus which there might be from a sale of this property by the council. But the only elements relied on as setting up such an estoppel were the terms of the transfer and charge themselves, and it was impossible to derive from those terms anything which went beyond what was there provided.

His Lordship made a declaration, as sought by the writ, that the transfer dated June 3 1974 was void and of no effect. It would not, he said, be useful to go into the question whether the transfer should strictly be regarded as voidable or void, because it M was quite clear that the matter having come before the court, the court would be obliged to avoid this transfer if it had any validity at all, and make an injunction in the terms sought by the writ. Nothing prejudiced the right of the council to exercise any powers of a mortgagee which they could otherwise exercise.

LANDLORD AND TENANT
BUSINESS TENANCIES

A **COURT OF APPEAL**

November 8 1974

(Before Lord Justice CAIRNS, Lord Justice LAWTON and Mr Justice WALTON)

BAILEY ORGANISATION v UNITED KINGDOM TEMPERANCE & GENERAL PROVIDENT INSTITUTION

Estates Gazette February 1 1975

(1974) 233 EG 507

Interim business rent—Court's refusal to exercise discretion and determine a rent relating back to a date before the " freeze "—Fact that landlord could obtain an advantage in this way, if B **true, not a ground for refusing to fix a rent—Landlords' appeal allowed**

This was an appeal by the United Kingdom Temperance & General Provident Institution from a decision of Judge Lyall Wilkes at Newcastle upon Tyne County Court refusing to fix an interim rent under section 24A of the Landlord and Tenant Act 1954 for premises in Grey Street, Newcastle upon Tyne, occupied by the respondents, the Bailey Organisation.

Mr C B Priday (instructed by Roche, Son & Neale) appeared for the appellants, and Mr A G Steinfeld (instructed by Grunhut, Makepeace & Duffy, of South Shields) represented the respondents.

C Giving the judgment of the court, CAIRNS LJ said that Part II of the Landlord and Tenant Act 1954 placed restrictions on the ending of business tenancies. Where, apart from the Act, such a tenancy could be brought to an end by notice given by the landlord, the landlord in order to terminate the tenancy had to give notice in a special form, requiring the tenant to state whether he would be willing to give up possession and informing the tenant whether the landlord would oppose the granting of a new tenancy. Then the tenant might apply to the county court for a new tenancy, and if the landlord did not oppose that application the duration, rent and other terms of the new tenancy, if not agreed by the parties, would be determined by the court. If notice to terminate had been given and an appli- D cation for a new tenancy had been made, then the old tenancy was continued until three months after the application had been disposed of. It was found that some tenants took advantage of section 64 by spinning out the proceedings on an application for a new tenancy and so continuing to enjoy possession of the premises at a rent far below what was likely to be fixed for a new tenancy. Accordingly by the Law of Property Act 1969, section 3, there was inserted in the 1954 Act a new section 24A which provided that once a landlord had given notice to terminate he might apply to the court to determine a rent which it would be reasonable for the tenant to pay while the tenancy continued by virtue of section 24, and the court might determine E a rent accordingly. This was what had come to be known as an interim rent. Section 24A (2) provided (so far as was applicable to the present case) that a rent determined in proceedings under the section should be deemed to be the rent payable under the tenancy from the date on which proceedings were commenced.

In 1971 the Bailey Organisation held the premises 10 Grey Street, Newcastle upon Tyne, as tenants of the United Kingdom Temperance & General Provident Institution upon a 20-year lease from March 25 1962 at a rent of £650 a year for the first five years and thereafter £850. The lease however contained a break clause under which the landlords could terminate it at the end of the 10th year, ie on March 25 1972. On September F 1 1971 the landlords gave written notice to terminate the tenancy on the following Lady Day, the notice being in the appropriate form required by section 25 of the Act of 1954. By letter of September 10 1971 the tenants' solicitors notified the landlords that the tenants would not be willing to give up possession and wished to have a new tenancy. On November 4 1971 the tenants applied by originating application in the county court for a new lease, proposing a term of 50 years at an initial rent of £1,000 a year. The landlords' answer, dated November 19 1971, did not oppose the grant of a new tenancy, but objected to the term and rent proposed. For the next 11 months, no further steps were taken in the court proceedings. On October G 11 1972 the landlords gave notice that they intended to apply to the county court for an interim rent to be fixed under section 24A. This application did not come on for hearing until January 29 1974. Ever since November 1971, or earlier, negotiations about the terms of the new lease had been going on, and by January 1974, subject to agreement on terms as to insurance, etc, the parties were agreed on a rent of £2,000 a year. Further, the tenants were willing to have the interim rent fixed at that figure and the landlords were willing to accept it. So on January 23 1974 the tenants' solicitors wrote to the registrar saying that they consented to the interim rent being fixed at £2,000 a year from October 11 1972, being the date of the landlords' notice of application under section 24A, and asking to be H excused from attendance at the hearing on the 29th. As the matter then stood, everything seemed set fair for the fixing of an interim rent.

On January 24 1974 the registrar of the county court, or somebody on his behalf, telephoned the landlords' solicitors and asked the reason for the application. The landlords' solicitors complied with the court's request by writing a long letter which made it clear that the reason why the landlords wished at that stage to have the interim rent determined was because they had been advised that if this were done the rent of £2,000 a year would be deemed to be the rent payable from October 11 1972, and that date was before the rent freeze which came into force J on December 1 1972 and froze each business rent at the rate payable on November 5 1972. The tenants and their advisers were quite unaware of that letter, and Mr Steinfeld, for the tenants, made it clear that they would have opposed the application if they had realised why it was being pursued; it was for this reason that the matter had been fully contested in the present proceedings. The county court judge refused the application to fix an interim rent. He began by saying that the jurisdiction was discretionary. He observed that the mischief at which section 24A was directed was dilatoriness on the part of the tenant; here, he said, it was the landlord who was dragging his feet. It would not seem that there was any evidence to support K that finding, and Mr Steinfeld had not sought to uphold the decision on that ground. But then the judge went on to say that he declined to exercise his discretion in favour of fixing the interim rent because he took the view that, if the landlords were right in their contention, such an order would enable them to recover the full £2,000 a year despite the rent freeze, and this was a device to defeat the freeze to which the court ought not to accede. The landlords appealed.

Their Lordships did not find it necessary to decide whether, if the interim rent was fixed at £2,000 a year, it would be payable at that rate or only at £850 a year while the Counter-Inflation (Business Rents) Order 1973 remained in force. Nor was it desirable to reach a decision on that point in a case where the matter had not been fully ventilated, since one side had argued that the standard rate for the purposes of the order would, if the appropriate determination were made, be £2,000 a year, and the other side had said only that that might be the result. In the court's view, if the effect of determining an interim rent was to fix the 1973 standard rate at the same figure, that was no good reason for refusing to determine it. As between the parties, a rent of £2,000 a year was admittedly a fair one. Section 24A made no provision for the interim rent to be fixed by agreement (this contrasted with sections 33 to 35, which provided for the duration, rent and other terms to be agreed by the parties and decided by the court only in default of agreement), and it was to be observed that before the question of the freeze had come up both parties assumed that, despite the large extent of agreement about the new lease which had been achieved, it was the proper course for the interim rent to be determined by the court. Certainly their Lordships saw no reason why the landlords should not ask the court to determine it. They were of opinion that the freeze point was not a matter for the court to take into consideration in exercising its discretion. If on the fair construction of statutory provisions a person could gain an advantage by obtaining an order of the court which, apart from the statutory provisions, was a perfectly proper order, the fact that he might obtain that advantage could not be a proper ground for refusing to make the order. If on the other hand, as the judge thought at the end of his judgment, £850 would remain the standard rate notwithstanding the determination of the interim rent at a higher figure, no ground remained for refusing the determination: the position was just as it seemed to be when the tenants consented to the order being made. Accordingly the court allowed the appeal and determined the interim rent at £2,000 a year.

COURT OF APPEAL

December 6 1974

(Before Lord Justice BUCKLEY, Lord Justice BROWNE and Mr Justice BRIGHTMAN)

MANCHESTER GARAGES LTD v PETROFINA (UK) LTD

Estates Gazette February 1 1975

(1974) 233 EG 509

Business tenancy—Landlord company's intention to occupy premises for purposes of own business may be proved by evidence of regional manager to whom appropriate authority has been delegated by the board of directors

This was an appeal by Manchester Garages Ltd, tenants of premises known as the Clayton Service Station, Ashton New Road, Clayton, Manchester, from an order of Judge Bailey at Manchester County Court dated March 6 1974 rejecting an application for a new lease of the property on the ground that the respondent landlords, Petrofina (UK) Ltd, formerly Petrofina (Great Britain) Ltd, had established their intention to go into occupation for the purposes of their own business.

Mr B C Maddocks (instructed by Lickfolds, Wiley & Powles, agents for Taylor, Hindle & Rhodes, of Manchester) appeared for the appellants, and Mr G C Spafford (instructed by Isadore Goldman & Son, agents for Joseph Norton & Co, of Liverpool) represented the respondents.

Giving judgment, BUCKLEY LJ said that the landlords let the property in question to the tenants in 1960 for 21 years with certain powers to determine. They exercised these powers in such a way that the contractual term came to an end on March 1 1974. They gave a proper notice of termination under the Landlord and Tenant Act 1954, intimating that they would oppose a grant of a new tenancy. They gave as their reason for opposing a grant of a new tenancy the reason which was contained in section 30 (1) (g) of the Act, namely, that they intended to occupy the holding for the purpose of a business to be carried on by themselves. The ground of the appeal was that there was insufficient evidence before the court to justify the finding of the judge that the landlords had the requisite intention.

The memorandum of association of the landlord company clearly empowered it to carry on a retail trade of serving petroleum and other associated products, and there was no question that the intention, if the company had the intention, was one which was within its powers. No resolution of the board of the company was passed relevant to this matter at all, no evidence was given in the court below by any director of the landlord company, and there was no evidence from anybody informing the court of what the intentions of the board as a board were. The only evidence before the court having a bearing on the matter was the evidence of Mr Victor John Moodie, who was north-west regional manager of the landlords' business. The landlords were a company who carried on business through many outlets throughout the country. Mr Moodie said that he was totally responsible for the landlords' affairs in the north-west. He had held that office for 10 years, and his decisions had never been reversed. In cross-examination he said: "I have made decisions like this before and have full authority to do so." The decision which he did make was that the landlord company should itself go into occupation of the service station and should carry on the business there of running the station, retailing petrol and other like products to the public. Judge Bailey in his judgment said that Mr Moodie had full authority and that he was impressed with Mr Moodie's evidence on the issue of the company's intention to do what the company alleged in their opposition to the application. He came to the conclusion that the intention was a genuine intention and that accordingly the landlords had made out their case under section 30 (1) (g).

The contention in support of the appeal had been that since the landlords were an incorporated body, their intention must be found either in the form of a decision of the board, or of an intention in the minds of the members of the board, or, it was said, an intention formed by an authorised agent of the board provided he was what counsel described as "a head-office mind representing the mind or will of the company." Counsel argued that Mr Moodie was a regional manager, that he was not at the nerve centre of the company, and that his decision could not be regarded as establishing the intention of the company, because it could be over-ridden by the board. He (counsel) submitted that the intention of an agent of the company, acting within his authority, was not the intention of the company so long as his decision could be over-ridden. Reference was made to a number of authorities bearing upon the way in which an intention of a corporate body could be established in evidence. Those authorities showed that the question to be decided in such a case as this was whether the landlord at the date of his notice of opposition and at the date of the hearing had a firm and settled intention to carry out whatever particular activity it was said that he did intend to carry out, whether it be occupying premises for the purpose of his own business, demolishing them, or whatever it might be. Particular mention might be made of the decision of the Court of Appeal in *H L Bolton (Engineering) Co Ltd v T J Graham & Sons Ltd* [1957] 1 QB 159, in which, Mr Maddocks submitted, the agents of the company, whose minds were said to constitute the mind of the company, were agents at a directorial level.

He (his Lordship) thought that the agents in that case were acting not as directors in what they did, but as managers. Denning LJ in his judgment appeared to be considering their actions and their states of mind in their managerial capacity. There certainly was evidence before the court in the present case that Mr Moodie had complete responsibility for the conduct of the company's business within his region, and the question of what should be done with the Clayton Service Station was a matter which arose within his region and was within the ambit of his authority. The judge proceeded, quite properly, on that basis; indeed, it did not appear that that view of the matter was at all disputed before him. In his (Buckley LJ's) opinion, if some act which was contemplated was one which was within the authority of an agent of a company whose decision was said to evidence the intention of the company, if it was a matter within his authority to carry out without any further authority of any superior authority in the company,

A such as the managing director or the board of directors, then his intention was the intention of the company for the relevant purpose, at any rate if it was not shown to be inconsistent with some concurrent contrary intention in some superior authority who could gainsay him in respect of the particular act. If the intention of the agent was shown to be a firm and settled intention—which was not, as he (his Lordship) thought, by any means the same as an irrevocable intention or one which could not be overruled—then it was the firm and settled intention of the company. In the present case, the evidence indicated that it was Mr Moodie's firm and settled intention that the company should recover possession of this service station and should
B carry on trade there itself. There was no indication that the decision of Mr Moodie was contrary to any policy or any project of any higher authority in the company so that he would be likely to have been over-ruled in the matter. In these circumstances, in his (Buckley LJ's) judgment, the judge was fully justified in treating that evidence as being evidence of a firm and settled intention on the part of the landlord company to use the premises, when possession of them was recovered, for the purpose of the landlord's own business. For these reasons he would dismiss the appeal.

Agreeing, BROWNE LJ said that it seemed to him undesirable to lay down fixed categories of who could and who could not be an agent whose intentions could be treated as the intentions
C of the company. One of the factors which the county court judge was entitled to take into account in this case was the fact that the landlord company had in fact applied for and got planning permission for what was called "reorganisation" of the existing petrol service station. The evidence of Mr Moodie, coupled with that fact, was ample evidence on which the judge was entitled to find that the landlord company had the necessary intention.

BRIGHTMAN J agreed, and the appeal was dismissed with costs.

D

COURT OF APPEAL

October 31 1974

(Before Lord SALMON, Lord Justice STEPHENSON and Mr Justice MacKENNA)

SUN ALLIANCE & LONDON ASSURANCE CO LTD v HAYMAN

Estates Gazette March 8 1975

(1974) 233 EG 927

Business tenancy—Notice to terminate given in old form
E **referring to period of two months from "receiving" of notice—Majority hold this form of words to be substantially to the same effect as the new form, which refers to the "giving" of the notice—Landlords entitled to possession—Court's remarks about wording of forms prescribed for the purposes of the Act of 1954**

This was an appeal by Miss Freda Hayman, tenant of business premises at 32 The Parade, Leamington Spa, from a judgment of Judge Lind-Smith at Warwick County Court holding that her landlords, Sun Alliance & London Assurance Co Ltd, were entitled to possession of the property.

Mr A Dinkin (instructed by Field & Sons, of Leamington
F Spa) appeared for the appellant, and Mr I McLaren (instructed by Russell Livingstone, Wood & Co, of Malvern) represented the respondents.

Giving the first judgment, STEPHENSON LJ said that the landlords served the tenant with notice under the Landlord and Tenant Act 1954 to terminate her tenancy, and it was not suggested that the notice failed to comply with the requirements of section 25 of the Act otherwise than in one single respect. The document served required the tenant "within two months after receiving this notice" to notify the landlords in writing whether or not she would be willing to give up possession of the premises on the relevant date, March 25 1973. That form of words was in fact out of date. G
The notice was a reprint, filled in and corrected in some particulars immaterial to this appeal, of Form 7 in the appendix to the Landlord and Tenant (Notices) Regulations 1954 (SI 1954 No 1107) made by the Lord Chancellor under section 66 of the Act. Regulation 4 provided: "The forms in the appendix to these regulations, or forms substantially to the like effect, shall be used for the following purposes, that is to say . . . (vii) A notice under the provisions of section 25 of the Act . . . shall be in Form 7." Unfortunately the 1954 regulations were no longer in force at the date of the landlords' notice. They had in point of fact been replaced
twice, first by SI 1957 No 1157 and subsequently by SI 1969 H
No 1771. In the 1969 version, Form 7 referred not to the "receiving" of the section 25 notice, but to the giving of it. The tenant said that the document served upon her was not "substantially to the like effect" as the form prescribed by the current regulations. She pointed out that sections 25 (1) and 66 (1) of the 1954 Act required notice to be given in a form prescribed by the Lord Chancellor, and that that form was required to include "such an explanation of the relevant provisions of this Act as appears to the Lord Chancellor requisite for informing persons [to be served] of their rights. . . ." She submitted that the notice served was invalid,
because it required her to give her counternotice under sec- J
tion 24 (1) (a) of the Act within two months after "receiving" the landlords' notice, and warned her that her application for a new tenancy would not be entertained unless she notified the landlords of her objection to giving up possession within two months after "receiving" their notice and that she must apply for a new tenancy not less than two nor more than four months after "receipt" of the notice; whereas the notice should have stated that those times ran from the "giving" of the landlords' notice.

The 1969 wording of the form of notice brought it into
line with the wording of section 29 (2) and (3) of the Act, K
whereas the 1954 wording departed from it, as was pointed out by the Court of Appeal in *Price v West London Investment Building Society* [1964] 1 WLR 616 and recognised by the Lord Chancellor in certain amendments made by SI 1967 No 1831. But neither those amendments nor the decision in *Price's* case determined the question posed by the tenant in this case. The importance of the answer was great. If the change back to the obsolete wording had produced a form of notice which was substantially to the like effect as the new wording, the landlords' notice was good and the tenant could not apply for a new tenancy because she was out of
time; if the effect of the old wording was substantially L
different from the new, the notice was bad, the landlords must serve a fresh notice and the tenant could apply for a new tenancy. Counsel for the tenant had rightly conceded that on the facts of the present case there was no difference in time between the date when she was given notice and the date when she received it; counsel for the landlords conceded, equally rightly, that if the judge held that he could take that fact into account in answering the question raised by the tenant, he was wrong. For, as it seemed to him (his Lordship), what the court had to do was to construe the few relevant words in the regulations and in the two forms of notice and
decide whether in their ordinary significance the old words M
which were in fact used did mean substantially the same as the new words which should have been used.

Counsel for the tenant submitted that there was a real difference between the giving of a notice and its receipt. If a tenant went away on holiday and found the notice on his return home, he would not receive it until he found it, but the landlord would have given it when it reached the tenant's home. If the notice was in the old form, the tenant on reading it would reasonably think that he had two months from the day he found it to go into action, whereas the true time limit was shorter by the length of his holiday and the time the notice had been at his home. That was at first sight a powerful argument, without considering the complications introduced by supposing that the tenant failed to find the notice on his return because it had gone under the linoleum near his door: *Lord Newborough v Jones* [1974] 3 WLR 52. But he (Stephenson LJ) had come to the conclusion that the argument must be rejected. It was one provided for the tenant by the reluctance of the legislature to refer simply, directly and consistently to the service of notices in the particular provisions of statutes and statutory instruments which prescribed notices. In his judgment, the effect of section 66 (4) of the 1954 Act was that a notice under the provisions of the Act was both given and received when it was served in accordance with section 23 of the 1927 Act, and to anyone who knew the law, the time when it was given and the time when it was received were one and the same, namely the time of service, when the giving of the notice was in law complete.

There was therefore, on a true construction of the old form, no material departure from the statute and no material difference from the new form. It was a distinction without a difference. It described the same act from the landlord's and tenant's points of view. The giving and receiving of the notice were two aspects of the same action and were simultaneous, like " the giving and receiving of a ring " in the Form of Solemnisation of Matrimony in the Book of Common Prayer. The one gave, the other received. The tenant took what the landlord gave when he gave it, and he need not be there to take it from the landlord. The time when this two-sided act was done was the time when it was deemed to be done by the statutory provisions as to service. The tenant could not say that he had not received the notice when the Act said that it had been served on him. The effect which a court of law must give to the different words in order to see whether their effect was substantially alike was their true effect in the context of landlord and tenant. Those doubts which the 1967 amendments sought to remove with " an explanation of the relevant provisions of the Act "—an explanation that he (his Lordship) regarded as inadequate—would never have existed if Parliament had spoken plainly of " service " and " served " in relation to notices instead of " giving " and " given " in section 25 and the regulations. Why not a plain statement that all notices referred to in this Act were given and received when served in accordance with section 23 of the Landlord and Tenant Act 1927, if it was too much trouble to repeat the words of that section instead of resorting to the cryptic cross-reference in section 66 (4)?

It was said that it was hard on the tenant, who was not necessarily conversant with section 23 of the 1927 Act, to read different words in the same sense in the light of that section: the different words would, or might reasonably, have a different effect on him or her. But if it was legitimate to look at the effect of the difference in wording on the mind of a hypothetical tenant who was ignorant of the relevant law, the answer to that contention was that such a tenant would not see any difference between the time of giving notice and the time of receiving it. To him or her a notice would not appear to be given unless and until it was received, for instance when found after returning from a holiday. That, it was true, would not be when it was served according to law. But as counsel for the landlords pointed out, this notice would not be bad simply because it was misleading, but only if it misled in different directions; if it misled the tenant into thinking that she could take action when in law she was out of time, it would be good provided that it would have misled in the same way if it had been in the proper form. That was not an attractive argument, but (he Stephenson LJ) saw no answer to it. Certainly it provided strong support for so amending the Act and regulations, the forms and the notes, as to leave landlords and tenants in no doubt when time began to run for and against them. He would dismiss the appeal.

Agreeing, Lord Salmon said that he would add that the notice was misleading in both its forms, because neither referred on its face, or in the explanatory notes accompanying it, to the fact that it had been served in accordance with the law, and that therefore the tenant could not be heard to say that the notice had not been validly served upon him, nor deny that it had been given to, and therefore received, by him. This was particularly alarming, as section 66 (2) made it evident that the legislature recognised that the notice was likely to be served on persons who might have difficulty in obtaining legal help and still more difficulty in understanding the notice by themselves. This they could only do by ploughing through the Act until they reached the, to them, obscure cross-reference in section 66 (4), and then obtaining access to the Landlord and Tenant Act 1927, section 23, both very unlikely contingencies. If this had not been recognised by the legislature, there would have been no point in section 66 (2) enacting that the form of notice should include an explanation of the relevant provisions of the Act for the purpose of informing tenants of their rights and obligations under those provisions. He (his Lordship) strongly supported Stephenson LJ's recommendation that the Act and regulations, the forms and explanatory notes should be revised as soon as possible in the way he suggested. There might well be difficulty in finding Parliamentary time to amend the Act and to make the consequential amendments in the forms now. The explanatory notes, however, particularly note 2, could be quite easily amended in the immediate future, and should be, so that tenants were clearly informed of their statutory rights and obligations when a landlord's notice under the Act was served upon them.

Dissenting, MacKenna J said that a notice was a notification, a making known, a communication, of some matter by one person to another, and in the ordinary way a notice was not received until the person to whom it was to be communicated became aware of that matter. In other words, he received the notice when the matter was brought to his attention. Where the notice was given personally, the giving and the receiving would as a rule be simultaneous. But where, as here, the law provided that notice might be given in other ways—by leaving a document at the tenant's place of abode, or by sending it to him through the post—the view might reasonably be taken that the giving and the receiving might not be simultaneous, that the notice would be given when the document was left at the tenant's house or delivered by the postman, but that it would not be received until the tenant actually became aware of its existence. To tell a person taking that view that he must act within two months from the giving of the notice was not the same as telling him that he must act within two months from its receipt. For this reason it seemed to him (his Lordship) that the old form which used the word " receiving " was not substantially to the like effect as the new form which used the word " giving." The judge had decided this case against the tenant because on its particular facts the giving and the receiving of the notice were truly simultaneous. He (MacKenna J) thought that that was wrong. If the form used was not substantially to the like effect as the prescribed form, the notice was invalid. It was irrelevant that, on the facts of the particular case, the giving and the receiving were simultaneous, so that it did not matter to the tenant which words were used. The case was to be decided on a comparison of the two forms without regard to those facts. He would allow the appeal.

The appeal was dismissed.

A **CHANCERY DIVISION**

January 16 1975

(Before Mr Justice GOULDING)

WATNEY AND ANOTHER v BOARDLEY

Estates Gazette March 8 1975

(1975) 233 EG 935

Option in lease of business premises to purchase property—Exercise of option not an "instrument of surrender," therefore not caught by contracting-out provisions of Law of Property Act 1969, amending Landlord and Tenant Act B **1954—Tenant who exercised option but failed to complete purchase must give possession**

In this adjourned order 14 summons Mrs Christina Margaret Watney, of Finstock Manor, Finstock, Oxon, and Mr Sanders Watney, of Bisterne Close, North Burley, near Ringwood, Hampshire, suing as executors of Oliver Vernon Watney deceased, claimed possession of business premises known as the Corner Garage, Charlbury, Oxon, from the occupier, Mr Geoffrey Mansel Boardley.

Mr W D C Poulton (instructed by Lee & Pembertons) appeared for the plaintiffs. The defendant did not appear and was not represented.

C Giving judgment, GOULDING J said that he had to decide questions of law which in his view were not without difficulty. The relevant lease was dated July 3 1967 and was made between the plaintiffs as lessors and two predecessors in title of the defendant as lessees. By clause 1 of the lease, premises known as the Corner Garage and the Brice & Price Workshops at Charlbury were demised to the lessees for a term of seven years from April 1 1967 at an annual rent of £600. By clause 5, the lessors covenanted as follows:

> If the lessees or either of them shall at the end of the term desire to purchase the reversion in fee simple of the demised D premises . . . and shall not less than three months prior to the determination of the lease give to the lessors three months' notice in writing of such desire then the lessors shall on the expiry of the notice and on payment of the sum of £15,000 together with all arrears of rent and interest on the said sum . . . convey the premises to the lessees in fee simple free from any encumbrance. It is hereby agreed that such sale and purchase as the result of the exercise of the option shall be in accordance with and upon the latest edition of the National Conditions of Sale current at the time with the exception of National Condition 13 and National Condition 16 (3) and 16 (4).

On October 30 1970 an agreement under seal was executed varying in certain respects the terms of the lease. The E agreement was expressed to be made between the plaintiffs as lessors and Mr John Gordon Blencow, an intermediate owner of the term of years, of the other part. It provided that the purchase price of the reversion on exercise of the lessee's option under the lease should be whichever was the lower of the sum of £15,000 and the market value of the premises with vacant possession. The agreement contained certain stipulations as to the manner of ascertaining the market value, and also provided that the option might be exercised at any time during, or at the termination of, the term. In the events that happened, the provisions of that agreement were not material to the ascertainment of the price. F On April 6 1973 the defendant, who had at about that time become the owner of the term, through his solicitors gave notice to the plaintiffs in accordance with clause 5 of the lease that he desired to purchase the reversion of the premises in fee simple. According to the terms of the lease, completion should have taken place three months after the date of the notice exercising the option, ie on July 6 1973. The defendant, however, was not ready with his money at that date, and the plaintiffs gave him various indulgences in an endeavour to bring the purchase to a satisfactory conclusion. However, their patience did not result in actual completion.

On January 7 1974 the plaintiffs' solicitors gave 28 days' G notice to complete pursuant to the National Conditions of Sale. That notice was not complied with. Some further correspondence followed, and in the end, on May 8 1974, the plaintiffs' solicitors notified the defendant's solicitors that, in view of the defendant's failure to complete, the plaintiffs had elected to treat the contract between them and the defendant as at an end. By that letter possession was demanded forthwith, and in any event not later than May 24 1974. The defendant did not give up possession. He subsequently asserted that following the expiry of the original seven-year term he was entitled to the benefit of the Landlord and Tenant Act 1954, and he caused a notice to be served H demanding a new tenancy under that Act. As already observed, clause 5 of the lease provided that on the expiration of the notice, which occurred on July 6 1973, rent ceased to be demandable. Rent had to be paid up to the expiration of the notice, and thereafter interest at 8 per cent had to be paid on the purchase-money up to the date of completion. He (his Lordship) thought that the effect of such a clause was that at the date when rent was to cease the tenancy also ceased, so that a tenant who had exercised the option was thereafter in possession not as tenant but as purchaser let into possession prior to completion. That analysis of the position derived support from authority on other clauses in J other leases: see *Cockwell v Romford Sanitary Steam Laundry Ltd* [1939] 4 All ER 370 and *Nightingale v Courtney* [1954] 1 QB 399. However, the defendant's contention, appearing from the correspondence, was that the tenancy under the lease of 1967 was continued in existence by the Act of 1954 notwithstanding the expiry of the contractual term of seven years. The relevant section was section 24, but subsection (2) of that section qualified subsection (1) as follows:

> "The last foregoing subsection shall not prevent the coming to an end of a tenancy by notice to quit given by the tenant, by surrender or forfeiture, or by the forfeiture of a superior tenancy." K

In that form, subsection (2) would have been a complete answer to the suggestion made in correspondence on behalf of the defendant, for he (his Lordship) was satisfied that the effect of clause 5 of the lease, followed by the notice exercising the option, was to bring about a surrender of the tenancy on July 6 1973. Difficulty nevertheless remained, because of the impact of amending legislation contained in the Law of Property Act 1969. This legislation was passed, as he (Goulding J) understood it, to make it more difficult to avoid by agreement the protection given to tenants of business premises by the Act of 1954. The Act of 1969 added the following words at the end of section 24 (2) of the L Act of 1954:

> " . . . unless (a) in the case of a notice to quit, the notice was given before the tenant had been in occupation in right of the tenancy for one month; or (b) in the case of an instrument of surrender, the instrument was executed before, or was executed in pursuance of an agreement made before, the tenant had been in occupation in right of the tenancy for one month."

In other words, the present law, so far as relevant, was that a protected tenancy could not be brought to an end by an instrument of surrender if that instrument was executed in pursuance of an agreement made before the tenant had been in occupation in right of the tenancy for at least one M month. A possible answer to the plaintiffs' claim for possession was thus that the tenancy under the lease was brought to an end by an instrument of surrender, namely the notice exercising the option, executed in pursuance of an agreement contained in the original lease and thus made before the first month of occupation under the lease had elapsed.

Three alternative answers had been made by counsel for the plaintiffs to that argument. First, it was said that there was here no instrument of surrender. True it was, counsel submitted, that the effect of giving notice to exercise the option was that the term came to be surrendered, but the notice itself did not effect the surrender, which indeed was something to take place three months later, and the notice on the face of it was wholly different from what any ordinary layman or lawyer would describe as an instrument of surrender. Therefore there was no instrument of surrender, and the qualifications imposed on section 24 (2) by the Act of 1969 had no relevance. The second argument was that the option clause in the lease could not fairly be described as an agreement in pursuance of which the exercise of the option took place. Reliance was placed on the familiar analysis of an option as being an irrevocable offer. It was said that that offer was accepted by notice exercising the option: then, and only then, did any agreement whereunder surrender took place come into existence. The third argument was that the agreement, if any, in pursuance of which the notice was given to exercise the option was constituted not by the lease but by the agreement of 1970 incorporating and republishing the relevant terms of the lease. That varying agreement was of course made long after the expiration of the first month of the tenancy.

He (his Lordship) did not propose to say anything about the second or third points. The second, in particular, he regarded as a matter of considerable difficulty, having regard to the purpose of the amending legislation and the lack of precision in the words "in pursuance of." In his view, the plaintiffs should succeed on their first contention. In the older language of the law of landlord and tenant, a surrender in consequence of the exercise of the option to purchase would properly be described as a surrender by operation of law, having effect as such because the relation of vendor and purchaser paying interest on the purchase-money was inconsistent with the continuation of a tenancy under which rent was payable. If that was not the right way of looking at the matter, it might be that the surrender should be considered as taking place by virtue of clause 5 of the lease operating after the service of a notice exercising the option. On neither analysis would it be right, in his judgment, to describe the notice exercising the option as an instrument of surrender. It was not directed to surrender as such, but to purchase of the reversion. It did not, at the date when it took effect as a notice, bring about a surrender. All it did was to produce such a state of affairs that the tenancy could not continue after the contractual completion date. Accordingly it would be an unjustified and strained interpretation of the term "instrument of surrender" to describe the notice as such an instrument in the present case. The Act of 1969 therefore did not in the circumstances deprive the landlord of the benefit of section 24 (2) of the Act of 1954, whereby the Act did not prevent the coming to an end of a tenancy by surrender. The plaintiffs were entitled to possession of the premises and an inquiry (1) as to the damages they had sustained by reason of the defendant's failure to complete the purchase pursuant to the option and notice, and (2) as to the mesne profits due to them in respect of the defendant's occupation during any period subsequent to the discharge of his agreement to purchase on May 8 1974.

The plaintiffs were awarded the costs of the summons. The costs of the inquiry were reserved.

CHANCERY DIVISION

June 13 1973

(Before Mr Justice GOULDING)

SIMS-HILDITCH v SIMMONS

Estates Gazette March 29 1975

(1973) 233 EG 1173

Business tenancy—Lease for 16 years from Christmas Day 1956—Landlord's notice to terminate "on the 25th day of December 1972" held good—Ground for possession in section 30 (1) (g) held established though landlord sought premises for the purposes of a business to be carried on by him in partnership with his wife—Further point as to compensation certificate under section 37 of Act of 1954—Court obliged to take point as to position of Jersey executor of deceased tenant—Absence of UK grant of representation bars grant of certificate or (semble) other relief under the statute—Illuminating discussion as to action to be taken by the court in such circumstances—Landlord's anxiety for "tenant's" present summons to be dismissed

This was an originating summons by Mrs Ellinor Betty Sims-Hilditch, widow, and purported executrix of John Henry Sims-Hilditch, seeking, in substance, a declaration that notices served on her by the defendant, Mr John Philip Simmons, purporting to terminate business tenancies of land at Crowhurst Park, Crowhurst, Sussex, were invalid and of no effect; alternatively, an order for the grant of a new tenancy.

Mr M Lyndon-Stanford (instructed by Capel Cure, Peachey & Bates) appeared for the plaintiff, and Mr L Hoffmann (instructed by McKenna & Co) represented the defendant.

Giving judgment, GOULDING J said: This is an application by originating summons entitled "In the matter of the Landlord and Tenant Act 1954," and it relates to claims regarding business premises made under Part II of that Act as subsequently amended. The defendant, Mr John Philip Simmons, is the owner of property known as Crowhurst Park, near Battle, in Sussex, and from time to time he granted tenancies of several parts of his property to the late Major Sims-Hilditch, whom I will refer to as "the tenant." The present plaintiff is the widow of the tenant; she claims to be the executrix of his last will, and counsel stated that she has proved that will in the island of Jersey, though not yet in any part of the United Kingdom. To the question of the plaintiff's position as executrix I shall return later.

By her originating summons, as amended, the plaintiff claims several different heads of relief. I can conveniently take them one by one. The first is stated in paragraph 1A of the summons, and it claims a declaration that the tenancy of premises generally known as the stable flat, Crowhurst Park, granted by the defendant to the tenant by an agreement dated December 16 1960 is a business tenancy to which Part II of the Landlord and Tenant Act 1954 applies, with a further declaration that that tenancy is still subsisting. It became clear from the uncontradicted evidence of the defendant that the necessary facts to support that claim do not exist; in particular, the flat in question has not been occupied for a considerable time. Accordingly, the claim under paragraph 1A of the summons was not pursued. The next claim is in paragraph 1B of the originating summons and is for a declaration that three notices purported to be given by the defendant under the provisions of section 25 of the Act of 1954 are invalid and of no effect. Those notices relate to three separate tenancies which are agreed to be business tenancies within Part II of the Act of 1954. The principal tenancy agreement or lease is dated December 23 1958 and comprises the mansion house known as Crowhurst Park and certain adjoining grounds. Then there is a second agreement dated February 6 1959 comprising a piece of land adjoining the parcels demised by the first document and thirdly, there is a brief tenancy agreement relating to a walled garden at Crowhurst Park and some stable accommodation. That third instrument is dated July 2 1961. A separate notice, as I have said, was served in respect of each of those tenancies, but all the notices are *mutatis mutandis* in the same form, and it will be sufficient to refer to that regarding the principal property, the mansion house and grounds. The notice is given on a printed form and drawn in accordance with the relevant statutory regulations. It reads:

A Landlord and Tenant Act 1954. Landlord's notice to terminate business tenancy. To Mrs Ellinor Betty Sims-Hilditch, executor of Major John Henry Sims-Hilditch deceased, of Melbourne House, St John, Jersey, tenant of premises known as Crowhurst Park, Crowhurst, Sussex, together with the pleasure grounds belonging to the said mansion house.

1 I, John Philip Simmons, of Breadsell Farm, Breadsell Lane, St Leonards-on-Sea, Sussex, landlord of the above-mentioned premises, hereby give you notice terminating your tenancy on the 25th day of December 1972.

2 You are required within two months after the giving of this notice to notify me in writing whether or not you will be willing to give up possession of the premises on that date.

B 3 I would oppose an application to the court under Part II of the Act for the grant of a new tenancy on the ground that I intend to occupy the premises for the purposes or partially for the purposes of a business to be carried on by me in them.

4 This notice is given under the provisions of section 25 of the Landlord and Tenant Act 1954. Your attention is called to the notes overleaf.

The notice is dated January 21 1972 and is signed by the defendant's solicitors as his agents for the purpose. I need not read the common-form printed notes on the back. To explain the criticism which is made of that form of notice C I must refer to section 25 of the Act of 1954. Subsection (1) contains provisions which are necessary because of the enactment in the preceding section, section 24, that a tenancy to which Part II of the Act applies shall not come to an end unless terminated in accordance with the provisions of that part of the Act. The subsection takes that up and says, subject to a proviso that is not now material:

> The landlord may terminate a tenancy to which this Part of this Act applies by a notice given to the tenant in the prescribed form specifying the date at which the tenancy is to come to an end (hereinafter referred to as "the date of termination").

Subsection (2):

> D Subject to the provisions of the next following subsection a notice under this section shall not have effect unless it is given not more than twelve nor less than six months before the date of termination specified therein.

Then follow two subsections which are clearly designed to ensure that the statutory notice which terminates a business tenancy shall not abridge the contractual rights of the tenant. Subsection (3) concerns periodic tenancies which could have been brought to an end by notice to quit apart from the Act, and then subsection (4) deals with tenancies for a fixed term in the following words:

> E In the case of any other tenancy a notice under this section shall not specify a date of termination earlier than the date on which apart from this Part of this Act the tenancy would have come to an end by effluxion of time.

Subsection (5):

> A notice under this section shall not have effect unless it requires the tenant, within two months after the giving of the notice, to notify the landlord in writing whether or not, at the date of termination, the tenant will be willing to give up possession of the property comprised in the tenancy.

Subsection (6):

> F A notice under this section shall not have effect unless it states whether the landlord would oppose an application to the court under this Part of this Act for the grant of a new tenancy, and if so, also states on which of the grounds mentioned in section thirty of this Act he would do so.

It is of course those last two subsections, (5) and (6), that require the statements contained in paragraphs 2 and 3 of the form of notice which I have read. The attack which is made on the notice is quite a short one, and it arises under subsection (4) of section 25, although to decide it it is necessary to bear in mind the rest of the provisions of the section. What is said is that the notice offends subsection (4) in that it specifies a date of termination earlier than the date on G which, apart from the Act, the tenancy would have come to an end by effluxion of time. I therefore have to look at the instruments by which the respective tenancies were granted to see what the last-mentioned date would have been. The principal lease of 1958 demised the property comprised therein for the term of 16 years "from the 25th day of December 1956." The second instrument provided that the tenancy was granted for the period of the first lease; the third instrument was also made by reference, namely for the duration of the lease on the mansion house. Accordingly, I am concerned in each case with a term of 16 years from Christmas Day 1956. It is agreed, having regard to the use of the pre- H position "from," that the first day of that term was St Stephen's Day, commonly called Boxing Day, December 26 1956, and accordingly the last day of the term is Christmas Day, December 25 1972. The tenant, or his successor in title, was entitled to remain in possession until the very last moment of Christmas Day 1972. In other words, the moment of termination of the contractual tenancy by effluxion of time would have been that point of time, that instant of midnight, which divided Christmas Day from Boxing Day. Which then is the date on which, in the language of subsection (4), the tenancy would have come to an end?

I have considered that question, with helpful arguments J from counsel, in the light of all the other language of section 25, but in the end it seems to me that the words of subsection (4) are themselves sufficiently clear. Mr Hoffmann, for the defendant, puts this question: "On which day of the calendar does the year come to an end by effluxion of time?" The natural answer, in his submission, is on December 31, not on January 1. I accept that submission. I think the natural usage is that for which Mr Hoffmann contends. Further, there is sufficient authority to show that it is proper so to use language in relation to tenancies. I refer to the case of *Sidebotham v Holland* [1895] 1 QB 378. That was a decision K of the Court of Appeal on a yearly tenancy, but the language and the reasoning used, in my judgment, can fairly be extended to the case of a fixed term. Accordingly, when in paragraph 1 of the notice the defendant gave notice to the plaintiff terminating each tenancy on Christmas Day 1972, he did, in my judgment, refer to the date on which, apart from the Act, the tenancies would have come to an end by effluxion of time, and he did not offend against subsection (4) of section 25. Paragraph 2 of the notice was relied on by Mr Lyndon-Stanford for the plaintiff, because it requires the person receiving the notice to state whether she will be willing to give up possession of the premises on that date. L Mr Lyndon-Stanford says that that points to giving up possession *before* the last moment of Christmas Day and therefore the construction I have put on paragraph 1 must be wrong. I do not think so. It would be an affirmative answer to paragraph 2 to state the occupier's willingness to give up possession at the last moment of Christmas Day, and that would make the paragraph harmonise with the Act and with paragraph 1 of the notice, both interpreted in what I have held to be the ordinary use of language. Accordingly, the claim made by the plaintiff in paragraph 1B of the originating summons must fail. I should mention one other authority that was cited on that point, *Ladyman v Wirral Estates Ltd* M [1968] 2 All ER 197. In my judgment it does not assist either side here, because the particular point which I have had to deal with was in effect conceded by the tenant, as appears at the foot of page 198 of the report

The next claim on the originating summons, and the only remaining one apart from formal claims for further or other relief and costs, is contained in paragraph 1C of that document: "Alternatively,"—that is, on the footing that the section 25 notices were good—"an order for the grant by the defendant to the plaintiff pursuant to Part II of the Act of 1954 of a new tenancy." The defendant resists that, as has already appeared from the notices, on one ground and one ground only: that stated in paragraph (g) of subsection (1) of section 30 of the Act of 1954. The subsection begins: "The grounds on which a landlord may oppose an application under subsection (1) of section 24 of this Act"—that is, an application for a new tenancy—"are such of the following grounds as may be stated in the landlord's notice under section 25 of this Act or as the case may be, under subsection (6) of section 26 thereof, that is to say "—paragraph (g)—" subject as hereinafter provided, that on the termination of the current tenancy the landlord intends to occupy the holding for the purposes, or partly for the purposes, of a business to be carried on by him therein, or as his residence." The words "subject as hereinafter provided" relate to subsection (2) of section 30, which is not material in the present case. Subsection (3), added by the Law of Property Act 1969, enables the landlord, where he has a controlling interest in a company, to treat the business to be carried on by the company as if it were a business to be carried on by him personally. The defendant has given evidence and been cross-examined, and I find without hesitation that he has a genuine and substantial intention for the occupation of the several premises for the purposes of a business. The real objection that is made by the plaintiff to that contention of the defendant is that in the plaintiff's submission the defendant wants the occupation of the premises for a business to be carried on not by himself alone, but in partnership with his wife.

The expressions used by the defendant in cross-examination were, as one might expect in the case of a husband and wife living happily together, not altogether unambiguous as regards partnership. On the other hand there is, in my judgment, a significant piece of evidence in the shape of a printed form of application for caravan pitches which the defendant says he has had printed with a view to using it in the business that he desires to pursue. It is headed: "Site application form to J P and E J Simmons. Please reserve a site for my caravan at Crowhurst Park," etc. Although the point is not altogther an easy one, taking that form put in evidence by the defendant together with his oral evidence, I find that his present intention does relate to a business to be carried on by the defendant in partnership with his wife. However, in my judgment that is not an obstacle to his contention here. The defendant, in the ordinary use of language, will nonetheless occupy the premises and will nonetheless carry on business in the premises for the existence of the partnership. Where two persons carry on business in common as partners each of them occupies the firm's premises and each of them carries on business. Accordingly, the analogy between the present controversy which settles on the words "business to be carried on by him," and the dispute considered by the Court of Appeal in *Clift v Taylor* [1948] 2 KB 394 is not a false one. It was there held on a different statutory enactment relating to the rights of tenants that occupation by a firm of which the landlord was a partner was occupation by the landlord himself within the meaning of the enactment. Similarly, I hold here that the carrying-on of a business by the defendant in partnership with his wife will be a carrying-on of business by the defendant himself within the meaning of section 30 (1) (g). Thus the plaintiff's case fails also under paragraph 1C of the originating summons.

The plaintiff asks, in those circumstances, for a certificate under subsection (4) of section 37 of the Landlord and Tenant Act 1954, the same Act which I have been discussing hitherto. Subsection (1) of section 37 is long, and I will not read it, but it provides that in certain cases a tenant of business premises who fails to obtain a new tenancy under the Act will be entitled on quitting the holding to recover compensation in money from the landlord. Then subsection (4) says: "Where the court is precluded from making an order for the grant of a new tenancy under this Part of this Act in the circumstances mentioned in subsection (1) of this section, the court shall on the application of the tenant certify that fact." Thus the certificate can be given only on the application of the tenant, that is, the person in whom the relevant tenancy is vested. Here I come back to the difficulty I feel owing to the fact that the tenant is dead and the plaintiff, claiming to be his executrix, has not obtained a grant of probate in this country. It is pressed on me that if the defendant does not object, it is really nothing to do with the court whether the regular course has been followed in that respect or not. I find difficulty in accepting that contention. There is no doubt that the claim made by the plaintiff here is as executrix and that the notices served by the defendant on her were addressed to her as executrix. Whatever may have been the position before the Judicature Acts, however, in my judgment the present High Court in which the common law, equitable and ecclesiastical jurisdictions are united cannot disregard an omission to obtain probate if it is brought to the court's attention. Of course, the plaintiff can sue before probate, but in the ordinary course she cannot obtain relief from the court without producing a grant. In my judgment the parties cannot, as it were, contract out of those requirements of the law which are not imposed for their protection alone. I therefore propose to order that upon due proof of the plaintiff's title as executrix of the will of the tenant she be at liberty to apply in chambers for a certificate under subsection (4) of section 37 of the Landlord and Tenant Act 1954, and except for that purpose I stay all further proceedings under the originating summons.

Mr Hoffmann drew his Lordship's attention to the provision in section 64 (1) of the Act of 1954 extending current tenancies to the date on which an application for a new tenancy was finally disposed of in the courts. The defendant, said counsel, would not like to face a situation in which it could be claimed that the present proceedings had not been finally disposed of, so that possession could not be regained in four and a half months' time. If the plaintiff's application for a certificate were made in the present proceedings it could hardly be said that there had been a final disposal, and he (Mr Hoffmann) sought an order dismissing the proceedings and leaving the plaintiff to make an originating application for a certificate once United Kingdom representation had been taken out.

Mr Lyndon-Stanford said that as far as perfecting title was concerned, the plaintiff was (counsel thought) in a position to appoint herself as trustee of the lease, which would cure the technical defect to which his Lordship had drawn attention. He (counsel) respectfully suggested that if his Lordship was minded to take the title point, the application should not have been entertained in the first place; certainly now the condition of application for a certificate should be only that the plaintiff produce or appoint a proper trustee.

Goulding J said that he was not sure about that. He was inclined to dismiss the summons, rejecting the application for a certificate on the ground that he had no admissible evidence that the plaintiff was executrix of the late tenant. That would not hold matters up unless Mr Lyndon-Stanford could and did appeal.

Mr Lyndon-Stanford said that he certainly would appeal, the amount of the compensation in question being more than £4,000. In any event, he (counsel) doubted whether the summons could be dismissed; he thought no order would be the correct course, because there was a summons in the action still outstanding, an application by the defendant for an interim rent.

Mr Hoffmann said that that raised an interesting point. When section 64 was enacted, the only application which could be made under Part II of the Act was an application for a new tenancy, and plainly that was what was referred to. Under the new section 24A inserted in 1969, the landlord might apply to the court for determination of an interim rent, and by section 24A (2), "a rent determined in proceedings under this section shall be deemed to be the rent payable under the tenancy from the date on which the proceedings were commenced." The words "the proceedings" in that subsection referred to the application under section 24A and not to the originating summons applying for a new tenancy, and this was so even though section 24A applications were

A commonly made by summons in the originating summons proceedings, that was to say, in the tenant's proceedings applying for a new lease. So here, the question would be whether the "application" in section 64 included a section 24A application which would have the effect of keeping an originating summons in which such an application had been made alive solely for the purpose of determining an interim rent. He (counsel) proposed to resolve this question in the present proceedings by withdrawing his application for an interim rent. He remained concerned by the prospect of an appeal from the dismissal of the summons. If there was an appeal going into next term . . .

Mr Lyndon-Stanford asked leave to confer briefly with Mr Hoffmann. After a short discussion, he said that he thought there B was a prospect of agreement on some issues, and the question of the certificate could be left to be mentioned in a few days' time when his Lordship had ruled on the substantive points which remained. That left only the question of costs. He (counsel) asked for the costs of the application for an interim rent, and for an order that that application be not renewed. As for the costs of the action generally, he asked for an order that he pay only a proportion of the defendant's costs, on the basis that he was entitled to challenge the defendant's intention to occupy and see whether it was a firm one.

Mr Hoffmann said that the practice on this issue was referred to in the *White Book* at page 897, rubric 62/2/64, and at greater length in volume 2 of *Woodfall*, page 1383. The law as there C stated had however been affected by the amendment of 1969. Under the earlier procedure the tenant had to go to court to contest the landlord's intention to occupy, or he had no right to a section 37 (4) certificate. That situation had now been remedied. As for the more general point in *Woodfall* about the plaintiff having to establish his case to the court's satisfaction, that was true of any action; nevertheless, a plaintiff who was opposed and was successful normally got his costs.

Goulding J said that he would make no order that day, but that the order he was disposed to make would, among other matters, dismiss the originating summons with costs. In a further judgment delivered, after renewed argument, on July 31, his Lordship held (1) that the court could not, despite the defen- D dant's attitude, recognise the plaintiff as "tenant" for the purposes of the Act of 1954 until she had perfected her title by taking out United Kingdom probate, and (2) that the plaintiff could not appoint herself into the position, occupied by the deceased, of trustee of the tenancy for a company, since the title of an executor to appoint a trustee under section 36 of the Trustee Act 1925 could be established only by a proper grant of representation in the United Kingdom. Accordingly, the plaintiff could pursue her claim to a section 37 certificate only upon proof of probate in this country. An order was made by which the plaintiff's summons, including the application for a certificate, was dismissed with costs, without prejudice, however, to the plaintiff's right to apply for a certificate by further originating summons on proof of appropriate representation in the United Kingdom; and the defen- E dant's summons for determination of an interim rent was (with his consent) dismissed with costs. The defendant's consent to this course was given on the basis that the originating summons was dismissed, and Mr Hoffmann observed that he (the defendant) would seek a reconsideration of the position in the event of an appeal.

COURT OF APPEAL

January 23 1975

(Before Lord EDMUND-DAVIES, Lord Justice BUCKLEY and

F Lord Justice ORMROD)

SNOOK v SCHOFIELD AND ANOTHER

Estates Gazette April 19 1975

(1975) 234 EG 197

Landlord and Tenant Act 1954—Section 25 notice in old form—"Substantially like" new form—Case governed by result of recent appeal—Order for possession granted

This was an appeal by Mr Peter George Snook, of South Parade, Southsea, Hampshire, from a judgment of Judge McLennan at Portsmouth County Court on December 5 1973 G dismissing his claim for possession of shop premises at 118 Highland Road, Southsea, occupied by the respondents, Mr Kevin Donald Schofield and his wife, Mrs Linda Susan Schofield.

Mr M Driscoll (instructed by Hubert Way, Malpas & Stanley, of Portsmouth) appeared for the appellant, and Mr N Huskinson (instructed by Barlow, Lyde & Gilbert, agents for Warner, Goodman & Co, of Portsmouth) represented the respondents.

Giving judgment, Lord Edmund-Davies said that the premises were used for business purposes and fell within the H ambit of Part II of the Landlord and Tenant Act 1954. The appellant served notice under section 25 of the Act to terminate the respondents' tenancy. The section provided by subsection (1) that such notice should be "in the prescribed form," and by subsection (5) that it should not have effect unless it required the tenant within two months after the giving of the notice to notify the landlord whether or not, at the date of termination, he would be willing to give up possession of the property comprised in the tenancy. The Landlord and Tenant (Notices) Regulations 1957 prescribed a form of notice (Form 7) the relevant part of which read: "You [the tenant] are required within two months after J receiving this notice to notify me in writing whether or not you will be willing to give up possession of the premises on that date." This was the form that was served by the appellant. The 1957 regulations had however been superseded by regulations made in 1969, and these prescribed a form of notice the relevant part of which read: "You are required within two months after *the giving of* this notice. . . ." The submission made on behalf of the tenants in the present case was that the section 25 notice served was not in a form like the prescribed notice, and the county court judge upheld this argument, holding that the old form was not substantially to the same effect as the form of 1969. K After the appeal had been set down and was about to be heard, precisely the same point was dealt with by the Court of Appeal in *Sun Alliance & London Assurance Co Ltd* v *Hayman*, judgment in which was delivered on October 31 1974 [see 233 EG 927]. Giving judgment on behalf of the majority, Stephenson LJ said that on the true construction of the old form, "there was no material departure from the statute and no material difference from the new form. . . . It described the same act from the landlord's and tenant's point of view. The giving and receiving of the notice are two aspects of the same action and were simultaneous." The effect of that judgment was that the section 25 notice which L was there given and which was similar to the notice given in the present case, was substantially to the same effect as the notice prescribed under the current regulations; indeed, counsel were agreed that that was the result and that the decision in the *Sun Alliance* case governed the outcome of the present appeal. There must be an order accordingly.

Buckley LJ and Ormrod LJ agreed. The appeal was allowed and the landlord granted an order for possession within 28 days. He was awarded the costs of the county court hearing, but did not seek the costs of the appeal.

M

COURT OF APPEAL

January 27 1975

(Before Lord Justice RUSSELL, Lord Justice STAMP and Sir JOHN PENNYCUICK)

LEWIS v MTC (CARS) LTD

Estates Gazette April 19 1975

(1975) 234 EG 199

The "competent landlord" provisions of the Landlord and Tenant Act 1954 mean what they say, so that a business subtenancy can be terminated by the head landlord before

expiry of a contractual mesne interest—No new interest to be inferred from the fact that subtenants continued to pay rent to mesne landlords while disputing the Act's effect

This was an appeal by MTC (Cars) Ltd, of Portland Place, London W1, from a decision of Templeman J in the Chancery Division on June 14 1974 granting possession of 4-8 Ledbury Mews North, Bayswater, London W11, to the respondent, Mrs Mabel Jenny Lewis, of Llanarth, Cardiganshire. The judgment below was reported at (1974) 231 EG 1417.

Mr J S Colyer (instructed by Goodman, Derrick & Co) appeared for the appellants, and Mr M Essayan (instructed by Underwood & Co) represented the respondent.

Giving judgment, RUSSELL LJ said that the respondent plaintiff was the freeholder of business premises of which the appellant defendants held an underlease for a term ending on December 20 1968, three days before their immediate landlords' term expired. After the expiry of their contractual terms the defendants and their immediate landlords continued to pay their respective rents, the latter becoming tenants of the plaintiff from year to year and the defendants, who were in possession, remaining as subtenants under a statutory continuation of their subtenancy by virtue of the Landlord and Tenant Act 1954. On November 13 1972 the plaintiff gave the defendants' immediate landlords notice to quit terminating their yearly contractual tenancy on December 23 1973, and on November 21 1972, acting as "competent landlord" within the meaning of the Act, she gave to the defendants a notice under section 25 determining their tenancy on May 31 1973, a date which was of course prior to that on which their immediate landlords' tenancy was to determine. The defendants now asserted that there was no power to serve a notice under section 25 with a termination date earlier than the expiry of any contractual mesne tenancy there might be.

Subsections (3) and (4) of section 25, which were the statutory provisions governing in the case of all tenancies what termination date could be specified in a notice under the section, appeared to him (his Lordship) to be designed to recognise and prevent the overriding of a tenant's contractual rights. They were not directed to the situation where the tenant's contractual rights had come to an end and only the artificial continuation of the tenancy under the statute existed. In such a case, he (Russell LJ) thought the statute imposed no fetter on the date of termination to be specified in the notice. It only required the date of the notice to be appropriately related in point of time to the selected termination date. It might at first sight appear strange that by a section 25 notice the defendants' subtenancy could be terminated before the end of the mesne landlords' contractual tenancy and by the plaintiff's own direct action without reference to the mesne landlords, thereby, for example, depriving the latter of profit rental while they remained liable for the rent to the plaintiff under the head lease. However, the plaintiff was only in a position to give a section 25 notice to the defendants because of the relatively short interest of the mesne landlords in reversion. By paragraph 3 of the sixth schedule to the Act, moreover, the acts of the plaintiff as the competent landlord were made binding on the mesne landlords, and by paragraph 4 of the same schedule the mesne landlords, so long as they had not consented to those acts, were made entitled to compensation from her for any loss that may be caused to them. Those provisions deprived what appeared to be the plain result of the language of section 25 of any strangeness or oddity that might otherwise have occurred. Accordingly the judge was quite correct in holding that the date specified in the plaintiff's section 25 notice could not in law be criticised.

A second point had been taken by the defendants based on the fact that after May 31 1973 and up to December 23 1973 they (the defendants) had continued to pay sums by way of rent to the mesne landlords. That fact, it was contended, involved the creation of a new contractual tenancy between the mesne landlords and the defendants which was protected by the Act. It was abundantly plain that after service of the section 25 notice the defendants had maintained that it was a thoroughly bad notice, and if it were so they were perfectly entitled to assert that their tenancy continued to be protected by the statute, and, being protected by the statute, naturally they would continue to pay rent at the old rate to the mesne landlords as long as the latter remained the owners of the reversion. He (his Lordship) could see no justification for attributing the receipt and payment of rent after May 31 1973 to the creation of a new subtenancy when the defendants themselves asserted that what they were doing was simply paying the sums which they would be obliged to pay on the footing of the statutory continuation of their tenancy. A passage from the judgment of Denning LJ in *Marcroft Wagons Ltd v Smith* [1951] 2 KB 496 at 506 might be aptly quoted, although it was a rent restriction case:

> "If the acceptance of rent can be explained on some other footing than that a contractual tenancy existed, as, for instance, by reason of an existing or possible statutory right to payment, then a new tenancy should not be inferred."

Although that was in a different context it seemed to be an extremely sound approach which could be applied to the instant case.

STAMP LJ and SIR JOHN PENNYCUICK agreed, and the appeal was dismissed. Leave to appeal to the House of Lords was refused.

COURT OF APPEAL

October 31 1974

(Before Lord Justice MEGAW, Lord Justice STAMP and Lord Justice ROSKILL)

G ORLIK (MEAT PRODUCTS) LTD v HASTINGS & THANET BUILDING SOCIETY

Estates Gazette April 26 1975

(1974) 234 EG 281

New lease of business premises—Tenants who have applied to court in time may vary their proposals as to terms of new lease either before or during the hearing of their application—But new lease cannot include a licence granted by the landlords' predecessors in title to park vehicles on adjoining land not forming part of the tenants' holding

This was an appeal by the Hastings & Thanet Building Society from a decision of Judge McCreery at Newport, Isle of Wight, County Court on November 27 1973 holding that the terms of a new lease of business premises occupied by the respondents, G Orlik (Meat Products) Ltd, should include a right over the appellants' adjoining land not to be found among the terms of the current tenancy.

Mr P Medd QC and Mr B Coles (instructed by Percy Walker & Co, of Hastings) appeared for the appellants, and Mr F M Ferris (instructed by Robinson, Jarvis & Rolf, of Newport) represented the respondents.

Giving the judgment of the court, STAMP LJ said: This is an appeal from the judgment of Judge McCreery in the Newport, Isle of Wight, County Court, delivered on November 27 1973. The appeal concerns an application by G Orlik (Meat Products) Ltd, whom we shall call "the tenants"—the respondents in this appeal—for a new tenancy of business premises under the provisions of Part II of the Landlord and Tenant Act 1954, as amended by the Law of Property Act 1969. The appellants in this court—we shall call them "the landlords"—are the Hastings & Thanet Building Society. They were the respondents in the county court proceedings. They objected to the grant of a new tenancy, doing so on the

A ground set out in section 30 (1) (c) of the Act: that is, they asserted substantial breaches by the tenants of their obligations under the current tenancy. The learned judge on the evidence heard by him rejected that ground, the sole ground, of objection. The landlords do not in this appeal seek to attack the judge's decision in that respect. Accordingly, on this appeal it is accepted that a new tenancy should be granted. The issues in the appeal relate solely to the terms of the new tenancy. The particular matter in dispute is that the tenants claim that the new lease should contain a term permitting them, subject to limitations, to park motor vans used in their business in such a way that they will, as to part of B their width, stand over a part of a yard which belongs to the landlords, adjacent to the land comprised in the lease. The landlords contend that no such term can properly be included in the new lease.

Before reciting the facts which are relevant to the main argument on this appeal, we shall deal with an argument put forward by Mr Medd on behalf of the landlords as his second ground in support of the appeal. It was, he admitted, highly technical. The point is simply this. In all other respects, in contrast with what is all too frequently found in cases under Part II of the 1954 Act, all the time provisions of the Act for the giving of notices and the initiation of C proceedings were adhered to on both sides. But in one respect, according to the landlords, the tenants failed, and the failure, it is contended, was fatal to their success upon—or indeed to their being heard by the county court upon—the only matter now in issue: that is, their claim for the vehicle-parking terms to be included in the new lease. The original application of the tenants to the court under section 24 (1) (a) of the Act, was made within the time limits stated by section 29 (3). It was in the form (Form No 335) prescribed by order 40, rule 8 (1) of the County Court Rules. But as originally submitted, the proposals included in paragraph 3 of the form contained no reference to any term relating to D the parking of vehicles, whether on the landlords' land or otherwise. The proposals put forward were confined to rent. When the application came on for hearing before the judge, the tenants sought to claim rectification of the pre-existing lease so that it should be altered in such a way as to provide for the parking rights. It was pointed out that the court did not have jurisdiction to entertain such a claim. The tenants then, it would seem, abandoned any suggestion of rectification of the lease, but sought instead leave to amend the proposals contained in their application so as to include amongst them a term relating to the parking of their vehicles on, in part, the E landlords' adjacent land. The application to amend was opposed by the landlords, one ground being the ground urged in this court: namely, that because of the time limits in section 29 (3) of the Act, which had by then expired, there was no jurisdiction to give leave to amend.

It does not appear to have been suggested that the landlords were taken by surprise, but, of course, if there were lack of jurisdiction the absence of any prejudice created by the granting of leave would not be relevant. In our judgment, no ground has been shown for holding that there is no jurisdiction to allow an applicant to amend, after the four-month period, the proposals contained in an application to F the court made within that period. Counsel for the landlords is right in saying that the Act is strict, and uses strict language, in its time provisions, but we are unable to see anything in the Act which deprives the court of jurisdiction to grant leave to the applicant to amend the detail of his proposals after the expiration of the four-month limit. The decision of this court in *Williams v Hillcroft Garage Ltd*, reported in *The Times* newspaper as decided on January 27 1971 [see 218 EG 1163], if not a direct authority for that proposition, provides, at the least, most persuasive indication that it is correct. Accordingly, we reject this second ground put forward in support of the appeal.

We now come back to the relevant facts which should be G narrated for the purpose of considering the first ground put forward in support of the landlords' appeal. The facts as found by the county court judge may be summarised as follows. Since 1963 the tenants have carried on the business of wholesale butchers upon premises in Union Road, Ryde, in the Isle of Wight. These premises have been and are held under a title with which this application is not concerned, and we refer to them as "the tenants' main premises." Adjoining the tenants' main premises was a parcel of land consisting of a yard with an entrance on Union Road belonging to Victoria Wine-Tylers Ltd. In June 1964, faced with difficulties in parking vans used by them for the purposes H of their business, the tenants opened negotiations with Tylers with a view to renting the yard for parking their vans. Tylers had a garage and store behind the yard and, subject to a way from the entrance of the yard to this garage and store being kept clear and open at all times, Tylers on July 28 1964 indicated that they were agreeable to the tenants parking their vans on the site. They suggested the payment of £100 per annum for this privilege and expressed their understanding that there was sufficient room for four vans. Negotiations were protracted. Mr Orlik, the tenants' managing director, thought it would also be of advantage to use part of the yard as a cold store, and after, at one time, J resiling from that suggestion he again put it forward. The learned judge summarised the result of the negotiations by saying that he was satisfied that Mr Orlik's original intention in respect of the land was to use it only as a parking place, that that intention was neither replaced nor altered by the subsequent plan for a cold store, and that by the autumn of 1966 Mr Orlik and Mr Edwards (who was Tylers' property manager) had agreed in principle upon a lease which would entitle the tenants to use the land not only for a cold store but also for the parking of three vans. The judge found that it was further agreed that the three vans should be parked in such positions as would allow Tylers to park their own K van at the back of the yard. We would emphasise that what the judge held had been agreed in principle was a lease containing the provisions to which he referred.

The judge went on to say that, agreement having thus been reached, the tenants built their cold rooms early in 1967. The cold rooms were built along the side of the yard adjoining the tenants' main premises, so reducing, so long as they remained there, the area of yard available for parking. There was, however, still room for the parking of three vans in agreed positions, two of them alongside the cold rooms and the third on the site of the wash room, to which we will refer. So L parked, this left room for Tylers to take a vehicle across the yard from its entrance in Union Road to Tylers' garage and store at the rear of the yard. Some time after the completion of the cold rooms—the judge found the exact date uncertain and immaterial—the tenants built alongside them—again, we would add, on Tylers' land and on the side of the yard adjoining the tenants' premises—a wash room. This further reduced the area available for parking and thereafter the tenants could not, without obstructing Tylers' access, park more than two vans: that is to say in the two agreed positions alongside the cold rooms.

The tenants' existing lease was dated December 11 1967. M It contained a demise to the tenants of the premises to which we shall refer more particularly in a moment for a term of seven years from September 29 1966 at a rent of £150 per annum. Because it was common ground in this court that in view of the construction of the cold rooms and wash room its effect was to preclude the parking of any of the tenants' vans in the yard, we need only summarise its provisions. What was demised to the tenants was not the whole of the yard subject to a right in Tylers giving them access to their garage and store room, but what was substantially a rectangular piece of the yard adjoining the tenants' main premises, embracing the area of the cold room and wash room buildings and a very narrow strip of land beyond. Excluded from the demise was the smaller rectangular part of the yard, which we will call the landlords' part of the yard, at the far side of the yard from the tenants' main premises. Included in the demise to the tenants was the right to use the landlords' part of the yard for the purpose of access to and egress from the demised premises. The narrow strip of land beyond the cold-room and wash-room buildings included in the demise was not wider than was necessary to park a motor cycle . It was far too narrow for a van. The lease contained a covenant both by the tenants and Tylers designed to prevent the obstruction of access to their respective premises "by permitting or suffering the standing or parking of vehicles or storage of merchandise" on the landlords' part of the yard. The learned county court judge found that from the time in 1967 when the tenants built their cold room they parked their vans daily in the yard in the agreed positions, and that after the construction of the wash room they continued to park on the two positions alongside the cold room: that is to say, partly upon the landlords' part of the yard. We complete the history of the matter by stating that in 1969 the reversion expectant upon the determination of the lease was assigned to one Darken, and that in September 1970 it was assigned to the landlords.

The tenants' claim to have included in the new lease a term conferring on them the right to park two vans in the agreed positions, ie partly on the landlords' part of the yard, is based on the submission that they already have that right in connection with their tenancy. Accordingly, so the argument runs, a term conferring that right ought to be included in the new lease either under section 32 (3) or under section 35 of the 1954 Act. We will refer to those sections later in this judgment. We will, however, first consider whether the submission that the tenants now have a right of parking on the landlords' part of the yard during the continuance of the current lease is well founded. It was submitted, and the learned county court judge accepted this submission, that an irrevocable licence to park two vans in the two agreed positions for the term of the lease ought to be inferred. Expressing himself as satisfied that the agreement as to parking entitled the tenants to park their vans in the yard, and that that agreement was put into effect early in 1967 and acted upon by both parties to the lease, the judge inferred the grant by Tylers to the tenants of a licence to park three vans in the yard, such parking to be in such a position as not to interfere with access for Tylers' own van. The terms of the licence were, he said, subsequently modified to provide for the parking of two only of the tenants' vans. Having regard to the circumstances in which the licence was granted and to the importance to the tenants' business of the rights which it conferred, the judge was satisfied that the licence was intended to run throughout the term of the tenants' lease and he held it was irrevocable and binding upon Tylers' successors in title.

There are, in our judgment, two fatal objections to this finding. In the first place it is axiomatic that where there is a contract which is afterwards reduced into a deed, then, unless there is something to show that the parties intended to make some term which was not to be included in the deed and which was, to quote the words of Bowen LJ in *Palmer v Johnson* (1884) 13 QBD 351 at 357, "to go on to any extent" after the execution of the deed, the rights of the parties are governed entirely by the deed: see also the judgment of Brett LJ in *Leggott v Barrett* (1880) 15 Ch D 306 at 311. Here the right to park *was* to be included in the deed; and if the lease did not give effect to the common intention of the parties the tenants' remedy was to seek its rectification. Secondly, if there was such a licence as was found by the judge, it was contractual only, arising out of a contract to which the landlords, as assignees of Tylers, were not parties. There was no finding, and no evidence to support a finding, that the landlords even knew of the existence of a licence when they acquired the reversion, and we cannot accept Mr Ferris's submission that it was for the landlords to establish that they were not bound by it. The absence of evidence that the landlords when they acquired the reversion knew anything of the negotiations leading to the execution of the lease is likewise, in our judgment, fatal to a submission that the tenants have a right to have the existing lease rectified and that this is a right to which effect ought to be given in the new lease. No authority was cited to show in what circumstances, if any, a party to a deed may have it rectified when the rights of the other contracting party thereunder have been assigned to a third party; but it would plainly be contrary to principle that there should be rectification on the ground of mutual mistake where the assignee at the time of the assignment was unaware of the right of rectification or of the mistake out of which the right arose.

Alternatively Mr Ferris relied upon a proprietary estoppel. What, as we understand it, is said is that the tenants incurred expenditure in the mistaken belief that they had or would have the right to park in the two agreed positions and that the landlords are estopped from denying the existence of a right to do so. We do not think it necessary to elaborate on the submission, because one of the conditions essential to establish a right by the effect of such an estoppel was simply not shown to exist. In the passages in *Snell on Equity,* 27th ed, pp 565-568, which were alone relied upon in support of the estoppel that was sought to be established, it is pointed out that one of the conditions which is required to be satisfied is that the person against whom the right is claimed should have known that the claimant was incurring the expenditure in this mistaken belief, but nevertheless stood by or participated in the expenditure; and there is, in our judgment, no room for the application of the doctrine where, as here, a deed quite simply does not give effect to the common intention of the parties. The equity relied on is based on unconscionable behaviour or fraud, and there was in this case no scintilla of evidence, nor any finding, that either party to the lease was aware that it did not give effect to their common intention, and nothing to indicate unconscionable behaviour on the part of Tylers. There is nothing to indicate that after the execution of the lease the tenants incurred expenditure or altered their position, or that Tylers stood by knowing they were doing so in the mistaken belief that they had the right of parking which they claim. Moreover, even assuming the existence of such an equitable right, there was no evidence that the landlords purchased with notice of its existence.

We conclude that neither by the effect of a grant contained in the lease nor by the effect of any irrevocable licence or any equity have the tenants as against the landlords any such right of parking as is claimed. If it be right that the tenants have no right of parking such as is claimed, subsection (3) of section 32 of the 1954 Act, which was relied upon on behalf of the tenants and requires that where the current tenancy includes rights enjoyed by the tenant in connection with the holding, those rights shall (in the absence of agreement between landlord and tenant) be included in the new tenancy, affords the tenant no assistance. Section 35 of the Act is in more general terms. It provides as follows:

> The terms of a tenancy granted by order of the court under this Part of this Act (other than terms as to the duration thereof and as to the rent payable thereunder) shall be such as may be agreed between the landlord and the tenant or as, in default of such agreement, may be determined by the court; and in determining those terms the court shall have regard to the terms of the current tenancy and to all relevant circumstances.

The object of Part II of the Act is to give security of tenure to business tenants by (*inter alia*) conferring power on the court to order a new tenancy of the property comprised in "the holding": that is to say (see section 23 (3)), of the property comprised in the existing tenancy excluding any part not used for business purposes; and however widely expressed, section 35 cannot, in our judgment, consistently with the scheme found in Part II, be construed to enable the court to enlarge the holding, for example by ordering the grant of an easement over the landlords' land or by conferring rights over the landlords' land not hitherto enjoyed. Mr Ferris relied upon the judgment of Upjohn J in *Re Albemarle Street* [1959] Ch 531 as authority for the proposition that section 35 gave the court the widest possible discretion as to the terms to be included in the new tenancy. There, the tenants under the terms of their existing tenancy had a right to exhibit advertising signs outside a part of the landlords' building of which the demised premises formed part. The right was, as the judge held, one which had no connection with the demised property and was purely personal and merely a licence. The judge, pointing out that the great width of section 35 was shown by the fact that the court was entitled and bound to have regard, not only to the terms of the current tenancy, but to all relevant circumstances, held that he had jurisdiction to make an order for a new lease containing this term of the current lease. It is not necessary to comment on that decision except to point out that there the tenant had under the current lease the very right which it was sought to have included in the new lease, whereas in the instant case the tenants had no such right. In our judgment, the new tenancy ought not to contain a new term as to parking, and we therefore allow the appeal to the extent of varying the order in the court below accordingly.

The appeal was allowed with costs in the Court of Appeal. The matter was remitted to the county court for reconsideration in the light of the court's judgment.

CHANCERY DIVISION

October 7 1974

(Before PLOWMAN V-C)

HILLMAN v DAVISON: RE HENNESSEY'S AGREEMENT

Estates Gazette June 14 1975
(1974) 234 EG 825

Long lease at a premium—Right to buy purchaser out in certain events held avoided by provisions in section 17 of the Landlord and Tenant Act 1954 against contracting-out—A surrender is effective, but an agreement to surrender is caught by the statute

By this summons Mr Mayer Hillman, of 44 Primrose Hill Road, London NW3, sought a declaration that certain options contained in an agreement for a lease made on April 26 1961 between himself and the defendant, Mrs Brenda Davison, formerly Hennessey, of 56 King Henry's Road, London NW3, were void and of no effect.

Mr W A Blackburne (instructed by Gouldens) appeared for the plaintiff, and Mr P St J Langan (instructed by Bower, Cotton & Bower) represented the defendant.

Giving judgment, PLOWMAN V-C said that the originating summons questioned the validity of certain options in an agreement for a lease dated April 26 1961, made between the defendant as vendor and the plaintiff as purchaser of three rooms on the top floor of 44 Primrose Hill Road, London NW3, for 99 years from March 25 1961 at a premium of £2,500 and a ground rent of £25 a year. That agreement contained the following provisions in clause 7:

It is hereby mutually agreed by the vendor and the purchaser (a) that if the purchaser shall at any time during the term granted by the said lease wish to vacate the said premises he shall forthwith give to the vendor notice of such his desire and the vendor shall have an option (such option to be exercised within three months of the receipt of such notice) to purchase the residue of the term granted by the said lease from the purchaser for the sum of £2,500 such sum to be paid as to £1,000 on the exercise of such option and as to the balance of £1,500 by equal half-yearly instalments over two years; (b) that if the vendor shall at any time during the term granted by the said lease wish to sell the said building with vacant possession she shall forthwith give to the purchaser notice of such her desire and the vendor shall have an option (such option to be exercised within three months of the receipt of such notice by the purchaser) to purchase the residue of the term granted by the said lease from the purchaser for the sum of £2,500: Provided that the exercise by the vendor of such options shall not prejudice her rights against the purchaser under the said lease in respect of any antecedent breach of any of the tenant's covenants therein contained.

The agreement for a lease was followed on July 7 1961 by a lease which did not, however, mention the options in the agreement. The reason why the originating summons had been brought was that the plaintiff wanted to spend £5,000 on alterations to the premises and wanted to know, before he did so, what his legal position was, having regard to the terms of the options. He was not motivated by desire to repudiate the bargain, but had brought his proceedings for the reason indicated, and whatever the result had offered to pay the defendant's costs. It was common ground that his tenancy was protected by Part I of the Landlord and Tenant Act 1954, and it was submitted on his behalf that the options were void as infringing section 17 of the Act, which provided:

The provisions of this Part of this Act shall have effect notwithstanding any agreement to the contrary: Provided that nothing in this Part of this Act shall be construed as preventing the surrender of a tenancy.

Mr Blackburne, for the plaintiff, submitted that while the proviso to section 17 said that "nothing in this Part of this Act shall be construed as preventing the surrender of a tenancy," that meant an actual surrender and did not include an agreement to surrender. He (counsel) submitted that an agreement to surrender was within the substantive part of section 17 and that both the options conferred by the agreement for a lease in the present case were in effect agreements to surrender and so caught by the Act. Reference had been made to analogous provisions in Part II of the Act. Section 24 provided:

(1) A tenancy to which this Part of this Act applies shall not come to an end unless terminated in accordance with the provisions of this Part of this Act; and, subject to the provisions of section 29 of this Act, the tenant under such a tenancy may apply to the court for a new tenancy (a) if the landlord has given notice under the next following section to terminate the tenancy, or (b) if the tenant has made a request for a new tenancy in accordance with section 26 of this Act. (2) The last foregoing subsection shall not prevent the coming to an end of a tenancy by notice to quit given by the tenant, by surrender or forfeiture. . . .

Certain conditions were then specified. Section 38, which was also in Part II of the Act, said:

(1) Any agreement relating to a tenancy to which this Part of this Act applies (whether contained in the instrument creating the tenancy or not) shall be void in so far as it purports to preclude the tenant from making an application or request under this Part of this Act or provides for the termination or the surrender of the tenancy in the event of his making such an application or request or for the imposition of any penalty or disability on the tenant in that event.

The words "purports to preclude the tenant" in that subsection had been construed by the Court of Appeal in *Joseph v Joseph* [1967] Ch 78 at 79 as meaning "has the effect of precluding the tenant." He (his Lordship) agreed with Mr Blackburne that that was an analogous case, and agreed with him that when section 17 said that nothing in Part I of the Act was to be construed as preventing the surrender of the tenancy,

A that meant an actual surrender of the tenancy, and that an agreement to surrender the tenancy *in futuro* was caught by the earlier provision of section 17 saying that " the provisions of this Part of this Act shall have effect notwithstanding any agreement to the contrary." The question which then arose was whether as a matter of construction clause 7 of the agreement for a lease was an agreement by the plaintiff that he would *in futuro,* in certain events, surrender the tenancy. His (Plowman V-C's) answer to that was, clearly it was. If either of the events happened which gave rise to an option in favour of the defendant landlord, the defendant on the exercise of that option was entitled to vacant possession of B the property, and that implied a surrender of the tenancy to enable the defendant to acquire vacant possession. Counsel for the defendant submitted that there was a difference in the wording of sections 38 and 17: that section 38 was framed in terms of effect, whereas section 17 was framed in terms of intent. He also submitted that one ought, if possible, to C construe a statute in such a way as to leave freely negotiated commercial arrangements standing. That was so, but at the same time the provisions of the Act had to be construed in the light of the policy of the legislature in passing the legislation for the protection of tenants. The effect of these submissions would be to enable a landlord to drive a coach and horses through the legislation by providing, in every agreement for a lease which was within Part I of the Act of 1954, provisions such as those which were found in clause 7 of the present agreement. He (his Lordship) was driven to the conclusion that the options concerned were caught by D section 17 of the Act. He was asked to say that " the options contained in clause 7 of an agreement for sale dated April 26 1961 and made between the defendant and the plaintiff were void and of no effect," and he granted a declaration to that effect.

LANDLORD AND TENANT

RENT ACTS

A **QUEEN'S BENCH DIVISIONAL COURT**

November 20 1974

(Before Lord WIDGERY CJ, Mr Justice MAIS and Mr Justice CROOM-JOHNSON)

METROPOLITAN PROPERTY HOLDINGS LTD v FINEGOLD AND OTHERS

Estates Gazette March 1 1975

(1974) 233 EG 843

Fair rents—Scarcity element—Allowance to be made under this head must reflect "broad, overall general scarcity affecting a really substantial area," not merely a local scarcity such as might be produced by the foundation of an American school in St John's Wood, London

This was an appeal by Metropolitan Property Holdings Ltd, of Prince Consort Road, London SW7, against a decision of a committee of the London Rent Assessment Panel dated July 23 1973 fixing fair rents for a large number of flats in a block known as South Lodge, Grove End Road, St John's Wood. The respondents were Mr W Finegold and other tenants living in the property.

Mr R H Bernstein QC and Mr M Singh (instructed by D J Freeman & Co and Grangewood Allen & Co) appeared for the appellants. The respondents did not appear and were not represented. Mr H Woolf (instructed by the Treasury Solicitor) appeared as *amicus curiae.*

Giving judgment, LORD WIDGERY referred briefly to the nature of the appeal and continued: The appellants complain that in reaching its conclusions as to the fair rents of each of these several flats the committee misdirected itself in its application of section 46 of the Rent Act 1968. Before I deal with the merits of the matter, it is convenient to take up a point made by Mr Woolf on behalf of rent assessment committees in general, that there ought to be some recognised practice as to the type of documents put before the court in cases of this kind. It is the fact, and I do not say this in any critical sense, that the paper produced and put before us in this case has greatly exceeded that which was necessary; and it may be necessary in the future to issue a practice direction to put the matter on a formal basis. Meanwhile, I accept and confirm Mr Woolf's submission that one ought in general in these cases to be able to deal with the point of law raised by having before the court the notice of motion, the reasons for the decision of the committee and such features of the evidence put before the committee as are necessary to develop the point of law. If that approach had been made in the present case, it would, I feel sure, have resulted in a considerable saving in time, money and typing.

But to return to the issue, the flats in question, as I have said, are in St John's Wood, and there was recently built in St John's Wood a substantial school, restricted in its entry to the children of American families in London. The fact that that school was built has undoubtedly rendered this part of St John's Wood far more attractive to American families than it might otherwise have been, because of the facility of the education of their children which this school provides. Although there is no specific finding to this effect on the part of the committee, it is a reasonable inference, I think, F from the material which has been put before us that the presence of this school has almost certainly put up the market rental values of flats and houses in the neighbourhood. And it may be, although again there is no specific finding to this effect, that this has produced locally an element of scarcity in the sense that more Americans want to come and live in St John's Wood and have the facility of having their children educated at that school than the accommodation vacant and to let in St John's Wood would permit. I think it only right to approach the problem on the footing that both those assumptions are good, namely, that the school in its own immediate surroundings has produced an increase in the G number of Americans, and secondly, it may well have produced an element of scarcity, or accentuated an element of scarcity which previously existed. The bone of contention between the parties is this, that applying its mind to section 46, the rent assessment committee has reached the conclusion that the presence of the school has produced an element of scarcity of the kind mentioned in section 46 (2), and has thought it right, in fixing the fair rents of these flats, to make a deduction, and one can go no further than that, on account of that scarcity. To see why that contention is put forward and what the answer to it might be, one has to look at the section itself. It is quite short. Section 46 (1) H provides:

> "In determining for the purposes of this Part of this Act what rent is or would be a fair rent under a regulated tenancy of a dwelling-house, regard shall be had, subject to the following provisions of this section, to all the circumstances (other than personal circumstances) and in particular to the age, character and locality of the dwelling-house and to its state of repair."

I would observe on that, straight away, that that seems to be saying in Parliamentary language that one must have regard to the sort of factors which tend to push rents up or down on the market. One must have regard to the age of J the premises, and that may have an effect up or down according to whether the premises are old or modern. One must have regard to their character and their locality. Their locality is important because a house situate in pleasant surroundings, and with the advantage of local amenities, may very well command a higher rent than an identical house in a less attractive setting. Looking for a moment at the American school to which I have referred, if the committee took the view that the presence of that school made the houses in the surrounding area, and in particular these flats, more attractive, and thus likely to command more rent, then so far as subsection (1) of section 46 is concerned the fair K rent ought to reflect that factor. In other words, looking for the moment only at subsection (1), any amenity (as the word has been used frequently in this argument), any advantage which the premises inherently have, either in their construction, their nature, their scale, their situation, their proximity to a school, a zoo or a theatre, whatever it may be, all those factors which would tend in the market to increase the rental, are factors to be taken into account by the committee in fixing the fair rent. To what extent they are taken into account is, of course, the duty of the committee to decide, but that these are matters which are eligible for consideration is beyond doubt. At this point, as I have already said, the presence of the American school would, on the face of it, tend to put up the fair rent because it would be an amenity making the premises more attractive. Then one comes to subsection (2):

"For the purposes of the determination it shall be assumed that the number of persons seeking to become tenants of similar dwelling-houses in the locality on the terms (other than those relating to rent) of the regulated tenancy is not substantially greater than the number of such dwelling-houses in the locality which are available for letting on such terms."

This is the provision which is intended to eliminate what is popularly called "scarcity value" from the fair rent fixed by the committee, and it is a provision which has given rise to a great deal of difficulty in practice. I think that before one begins to consider the difficulties, and before one begins to consider the section in detail, one must have clearly in mind what Parliament's obvious intention was in including this provision in the Act. It seems to me that what Parliament is saying is this: if the house has inherent amenities and advantages, by all means let them be reflected in the rent under subsection (1); but if the market rent would be influenced simply by the fact that in the locality there is a shortage, and in the locality rents are being forced up beyond the market figure, then that element of market rent must not be included when the fair rent is being considered. Parliament, I am sure, is not seeking to deprive the landlord of a proper return on the inherent value and quality of his investment in the house, but Parliament is undoubtedly seeking to deprive a landlord of a wholly-unmeritorious increase in rent which has come about simply because there is a scarcity of houses in the district and thus an excess of demand over supply. Bearing that in mind, one turns to the point in the committee's decision which is under attack in this appeal, and indeed the point upon which Mr Woolf has usefully made submissions on its behalf. The committee has come to the conclusion, and expressed it in more than one way, that the presence of the American school has created a local scarcity of premises. When I say "local," I deliberately do not attempt at this stage to define it further. But what was in the committee's mind undoubtedly was that the attraction of the school has produced a local scarcity of houses, and that there are consequently in the premises now under review all the elements contained in section 46 (2) of a scarcity which ought to be eliminated when the fair rent is assessed. I find the committee's views on this somewhat difficult to follow, I must confess, and indeed to get them fully one has to go back to what was said in an earlier case in which the same issue was raised. In the ordinary way one would not think it right to consider an earlier case when deciding a later, but it is useful here because, as has been shown, some light is thrown upon the working of the committee's mind upon this point. In a letter of January 5 1973, giving an explanation of the committee's views in the earlier case, one finds this passage:

"The committee accepted that there is a substantial amenity value in the St John's Wood area, but it is satisfied that there is a considerable degree of 'scarcity' within section 46 (2) of the Rent Act 1968. In the committee's opinion the matters relied on by Mr Bernstein as constituting evidence of 'scarcity' do not constrain it to limiting the difference between market rents and fair rents merely to the extent reflected in the valuations made by Mr Derisley. The committee considers, in the light of its own general knowledge and experience, and having regard to the evidence and its inspection of the property, that the rents determined by it properly reflect both 'amenity' and 'scarcity'."

I pause there to say, so far so good, but the next paragraph is illuminating:

"The American school is undoubtedly an amenity for those wishing their children to attend it, but the committee considers that the presence of the school tends to limit accommodation in the area available to tenants in general and thus to create an element of 'scarcity'."

It is because I have read that paragraph in the course of the argument that I venture to say that the committee seems to be acting on a supposedly very local scarcity created by the presence of the American school. In its decision in the instant case, one gets a clear touch of the same reasoning in this paragraph:

"The committee does not accept Mr Bernstein's contention that the committee in 1972 was wrong in attributing an element of scarcity to the presence of the American school. The committee accepts that 'locality' for the purposes of section 46 (2) is very substantially wider than the St John's Wood area. It is of the opinion that, within this wider area, the American school must to some extent reduce available accommodation similar to that in South Lodge."

Again emphasis is laid on the effect of the school in its immediate locality. For the reasons I have already given, I do not think that Parliament was concerned with this kind of local scarcity when section 46 was passed. If one is looking for the unearned, unmeritorious increase in rent which might accrue to landlords if section 46 (2) had never been passed, one must, I think, take a very much wider sweep than the sort of area to which the committee seems to have applied its mind in this case. Of course, if you look at half a dozen streets round the American school, you may well find a scarcity. As you go out to a greater radius round the school, then the effect of the school is less and less. But as I emphasised, we are not looking at the effect of the school as such; we are looking for scarcity in the locality which results from an excess of demand over supply. It seems to me, with all deference to the committee, that it has somewhat lost sight of the fact that the sort of scarcity we are concerned with is a broad, overall, general scarcity affecting a really substantial area, and that it wrongly focused its attention on an extremely limited area which would not, I think, qualify as a "locality" for the purposes of section 46 (2).

What should be done? What can we add which may be of some assistance in the future? I think that committees will find their consideration of section 46 somewhat easier if they start with the propositions clearly in mind that amenity advantages which can increase the fair rent under section 46 (1) do not result in a set-off under section 46 (2) merely because the amenity advantages of a particular house or district attract more people than can live there. The test on scarcity is to be taken over the locality as a whole, and that, as I emphasised, is a broad area. What area? We have been referred to *Palmer v Peabody Trust* [1974] 3 All ER 355, where, dealing with the word "locality" in section 46 (1), I said that the exact extent of the locality was something which was primarily for the committee to fix. I would repeat that in regard to the fixing of the locality under section 46 (2), but at the risk of repetition, I do emphasise that when the committees fix their locality for the purpose of deciding whether there is an overall scarcity or not, they must pick a really large area, an area that really gives them a fair appreciation of the trends of scarcity and their consequences. It may be, although I would not for a moment attempt to define the limits of the area precisely, that when operating section 46 (2) committees will be well advised to draw their inspiration from the area with which they are familiar in their work. Of course, different parts of the country require different considerations, but there will be many instances in practice where the most reliable area for the committees to choose, and the one upon which they are likely to achieve the most accurate result, is the area from which their work regularly and normally comes.

However, to return to this case, it seems to me that the committee has erred in, I say at once, the one relatively small respect which I have tried to describe, and I think that

A the matter must be reconsidered in the light of that error. The question arises as to whether this matter should go back to the committee in order that it might reconsider the point for itself, or whether it should go to an entirely new committee. While I hear and respect Mr Bernstein's submissions that from the litigant's point of view it is always somewhat unsatisfactory to go back to the original tribunal, it is the fact that we constantly do that in this court, particularly in regard to magistrates' courts, when some relatively small error has been made, and all that is necessary is for the court to look again at the point upon which they were in error. I think that when one has regard to all the work
B which has been done in this case, the inspections by the committee and all the other matters, the proper thing to do here would be to allow the appeal and send the case back to the committee to review its approach to section 46 (2) in the light of the judgments of this court.

MAIS J: I agree.

CROOM-JOHNSON J: I agree.

QUEEN'S BENCH DIVISIONAL COURT

November 25 1974

C (Before Lord WIDGERY CJ, Mr Justice MAIS and Mr Justice CROOM-JOHNSON)

METROPOLITAN PROPERTY HOLDINGS LTD v LAUFER AND OTHERS

Estates Gazette March 15 1975
(1974) 233 EG 1011

Fair rents—" Perverse " of rent committee to ignore " abundant expert evidence " tendered on behalf of landlords? —No—" Typical simple case " in which landlords said £x, tenants £y, and committee were entitled to fix their own
D **figure, saying merely that in the light of their experience they thought it the right one**

This was an appeal by Metropolitan Property Holdings Ltd, of Prince Consort Road, London SW7, against a decision of a committee of the London Rent Assessment Panel dated September 20 1973, fixing fair rents for premises occupied by the respondents, the respective tenants of 34 flats in Buckingham Mansions, London NW6, 68 flats in Marlborough Mansions, NW6, and 29 flats in Avenue Mansions, NW3.

Mr R H Bernstein QC and Mr M Singh (instructed by D J Freeman & Co) appeared for the appellants; Mr S Walsh (instructed by Seifert, Sedley & Co) for the respondents; and Mr H K Woolf (instructed by the Treasury Solicitor) as *amicus curiae*.

Giving judgment, LORD WIDGERY said that the appellants were represented by counsel before the committee and also had highly-qualified expert witnesses to support their case. One of these witnesses was Mr George W Burrow, a chartered surveyor [partner of Tyler & Co, London EC3]. His approach to the problem of what was a fair rent was first of all to try to assess a fair market rental, having regard to the conditions specified in section 46 (1) of the Rent Act 1968, and then to make a deduction for such element in the rent as was attributable, in his opinion, to scarcity. His conclusions on each limb of the problem were based, in the end, on his professional opinion. He looked at certain actual rentals to assist him or confirm his theories, but each step, in the last analysis, depended on personal professional opinion and experience. In his evidence, he chose as an example the six-roomed flats. In assessing a fair market rental he thought that for the flats improved by the landlords, the figure would be about £1,400 per annum, and for the unimproved flats about £1,200 per annum. A 10 per cent discount for scarcity would, he considered, be appropriate in each instance. He justified this by saying that expert G
valuers had for some years been allowing five to ten per cent for scarcity in lettings of this kind. He did not think there had been any recent increase in the scarcity element in the rent, so that 10 per cent was sufficient and right for the present case. He (his Lordship) was not going to criticise anything in all this, because Mr Burrow was approaching the problem before him by a well-recognised and thoroughly reputable route. The tenants, too, were supported by professional evidence, specifically the opinion of Mr Ronald Wilkie-Long, a chartered surveyor [partner of Watsons, of London W1]. He had had some regard to current market rentals, but he put the main emphasis on recently-registered H
rents. That was also a perfectly reputable and proper way of approaching the problem; which approach one adopted depended on the circumstances of each case. Because Mr Wilkie-Long looked at recently-registered rents in the area, there was no need for him to make an allowance for scarcity, because the registered rents themselves reflected any scarcity element. It was evident that Mr Wilkie-Long thought scarcity was more important and more widespread than Mr Burrow had considered it to be. He spoke of the element of scarcity having increased in recent years, whereas Mr Burrow had taken a different view. Based on these considerations Mr Wilkie-Long produced figures for fair rents substan- J
tially below Mr Burrow's.

The committee's decision contained a detailed assessment of the physical qualities of the flats. Quite clearly they had inspected every single flat and they weighed the advantages and disadvantages of the various factors. In their conclusions, they said that they had decided to confine their considerations to the subject flats. When dealing with the scarcity element, they considered on balance that the evidence showed that the element of scarcity had slightly increased rather than diminished. They decided on fair rents after having regard to all the circumstances and applying their K
own knowledge and experience. The rents they fixed were well below those proposed on behalf of the landlords, who now raised a number of complaints. They said that the committee had failed to give reasons that dealt with the substantial case which they (the appellants) had made; that the only inference from such reasons as were given was that the committee did not properly direct themselves on the issues before them; and that the committee had overthrown highly-qualified expert and other evidence as to both market rents and scarcity without saying why they had done so. In addition to these points, which were raised by the notice of motion, the landlords made further complaints in an affidavit L
by a Mr Singh [of the landlords' solicitors]. Firstly, that there was cogent evidence of market rents in actual lettings, amply supported by their witnesses, so that it was perverse of the committee to decide that the market value of an average six-roomed unimproved flat was other than £1,200 per annum. Secondly, that committees some 4-5 years previously were fixing fair rents about 5-10 per cent below market rents; there was no evidence that scarcity of this kind of accommodation had increased—on the contrary there was a great weight of positive evidence that it had not—so there was no justification to fix the rents at 35-40 per cent below market rents. Thirdly, that a rent which represented a M
steadily decreasing income in real terms could not be described as fair: the rent officer's fair rents in 1972, net of the service element, represented an increase of about 17 per cent over the rent fixed by a committee in 1967 in respect of 43 Marlborough Mansions, while in the same period the retail price index rose by about 38 per cent. Leaving that last complaint aside for the moment, the general tenor of the landlords' complaints was that there was abundant expert evidence supporting their view and that either it was perverse of the committee to reject that view, or in rejecting it they failed to give reasons for so doing.

Paragraph 9 of the sixth schedule to the Act of 1968 directed committees to make inquiry as they thought fit and to consider any information supplied or representation made to them. If it appeared to them that the rent registered or confirmed by the rent officer was a fair rent they were to confirm it; if not, they were to determine a fair rent of their own. In other words they were to pay attention to everything they were told, look at all the material put before them and then come up with what they thought was a fair rent. The obligation on them to give their reasons came under section 12 of the Tribunals and Inquiries Act 1971, which said that it was their duty to furnish a statement, either written or oral, of the reasons for their decision if requested. This provision applied to a wide range of tribunals, and the nature and extent of the reasons to be given had in some measure to be governed by the nature of the problem in hand. It was well established that rent assessment committees were entitled, indeed bound, to have regard to their own experience and knowledge. They acquired a really remarkable knowledge of prevailing rents in their own area simply from the experience of doing their job. It had been firmly established for many years that a tribunal could and should take into account their own experience and knowledge. That did not mean that they could throw over expert evidence; however, if the landlord's expert said the rent should be £X, and the tenant's expert said it should be £Y, and they themselves thought that neither figure was right, not only could they choose a figure in between, but they should do so. In a typical, simple case like that there were no reasons the committee could give except that they were not satisfied with either of the alternatives put forward in evidence and on their own expert knowledge preferred another figure. If the decision was based simply on the committee's own views, having regard to the evidence before them, then there were no reasons which the committee could give except to say that they thought their figure was the right one.

As far as the landlords' specific points were concerned, therefore, it seemed quite clear to him (Lord Widgery) that the first three grounds of complaint could not be sustained. The landlords had also complained in their notice of motion that the committee overthrew highly-qualified expert and other evidence. Perhaps the committee did, but if they took the view that it was necessary to do so to produce the right answer themselves, then they acted fully within their discretion. As to the landlords' allegation that the rent fixed by the committee was unfair because it did not take into account inflation and the rising cost of living as expressed in the retail price index, one could only say that there was no obligation on the committee to tie themselves to the retail price or any other index. Finally, no one could say the committee's decision was perverse. It was supported by the evidence of Mr Wilkie-Long to a very large extent and, lest it be thought otherwise, there was no obligation on the committee to quantify any element of scarcity which they allowed. Indeed, if they approached the problem by reference to other registered rents in the area, the question of quantification of scarcity did not arise. He (the Lord Chief Justice) was quite satisfied that the decision could not be attacked in law and accordingly the appeal should be dismissed.

MAIS and CROOM-JOHNSON JJ agreed, and the appeal was dismissed with costs. Leave to appeal was refused.

LANDLORD AND TENANT

RENT REVIEW

COURT OF APPEAL

October 23 1974

(Before Lord Justice MEGAW, Lord Justice LAWTON and Lord Justice ORMROD)

KENILWORTH INDUSTRIAL SITES LTD v

E C LITTLE & CO LTD

Estates Gazette June 7 1975

(1974) 234 EG 743

Six months' notice of rent review procedure required by clause in lease, "provided that failure to give such notice shall not render void the right of the landlord to require determination of a new rent as aforesaid"—These words mean what they say—Landlord nearly three months late with his notice, but still entitled to operate review procedure

This was an appeal by E C Little & Co Ltd, of Greenway, Southbank Road, Kenilworth, Warwickshire, from a judgment of Megarry J on January 21 1974 holding in favour of the respondents, Kenilworth Industrial Sites Ltd, of 42 Queens Road, Coventry, that a notice dated October 10 1972 was effective to bring into operation rent review provisions contained in a lease of factory premises in Farmer Road, Kenilworth, for 21 years from January 19 1968.

Mr E J Prince (instructed by Martin & Nicholson, agents for Pitt & Derbyshire, of Birmingham) appeared for the appellants, and Mr C P F Rimer (instructed by White & Leonard, agents for J Arch & Co, of Coventry) represented the respondents.

Giving judgment, MEGAW LJ said that clause 1 of the lease in question provided for a rent of £2,980 to be paid for the first five years, from January 19 1968 until January 18 1973. Clause 5 provided:

> "Not more than 12 months nor less than six months before the expiration of the fifth, tenth and fifteenth years of the term the landlord shall serve upon the tenant a notice to agree the rent . . . for the ensuing five years and thereupon the parties hereto shall agree a new rent as aforesaid . . . provided always that any failure to give or receive such notice shall not render void the right of the landlord hereunder to require the agreement or determination as aforesaid of a new rent."

On October 10 1972, which was less than the required six months before the expiration of the fifth year of the term, the landlords served notice on the tenants to agree a rent for the second five years of the term and proposing a rent of £6,176 per annum. Megarry J granted the landlords a declaration that that notice was effective to bring into operation the provisions of clause 5 for determination of a new rent, and the tenants now appealed.

It was clear that in clause 1 of the lease express words fixed the rent for the first five years only. If clause 5 for some reason could not be operated, it would seem that the consequence was that no rent at all would be payable, at least for the next five years and possibly not for the remaining 16 years. But that would be to ignore the proviso to the clause, which said that even though the time limit for the serving of notice had not been observed, that was not to render void the right of the landlords to require the agreement or determination of a new rent for the succeeding five years. Counsel for the tenants had submitted that clause 5 was in the nature of an option and so was affected by the special principles of law which (he said) governed all option agreements. He conceded, however, that if on a true view the clause was not to be regarded as in the nature of an option, he could not complain of Megarry J's decision. He (his Lordship) thought it impossible to treat the clause as an option. It was to his mind clear, as the judge held, that failure by the landlords to comply with the requirements of the opening words of the clause did not prevent them from thereafter, outside the time for service of notice, validly demanding agreement or arbitration of a new rent. The appeal was without substance, and should be dismissed.

LAWTON and ORMROD LJJ agreed, and the appeal was dismissed with costs.

LICENCE

A
COURT OF APPEAL

October 24 1974

(Before Lord DIPLOCK, Lord Justice CAIRNS and Lord Justice BROWNE)

GREATER LONDON COUNCIL v JENKINS

Estates Gazette March 22 1975

(1974) 233 EG 1095

Councils who let licensees into " short life " property may use summary possession procedure once licence has been effectively terminated—GLC nevertheless held on facts to have begun proceedings too soon

B This was an appeal by Mr Alan Jenkins occupier of 11 Elgin Avenue, Paddington, London W9, from a judgment of Judge Leslie at Bloomsbury and Marylebone County Court on September 13 1974 awarding the owners of the property, the Greater London Council, an order for possession.

Lord Gifford (instructed by the North Paddington neighbourhood law centre) appeared for the occupier, and Mr B K Levy (instructed by A G Gardner) represented the council.

Giving judgment, LORD DIPLOCK said that it was common ground that the occcupier went into the house in question
C in the spring of 1973 as licensee of the council as part of an arrangement for using houses which had been acquired for demolition and replacement, known familiarly as " short-life properties," for temporary accommodation for students and homeless persons. The first point taken on his behalf was that the special procedure for recovery of land permitted by order 26 of the County Court Rules was not available in cases where a person remained on as a trespasser after determination of a licence to occupy the land. He (his Lordship) agreed with the construction put upon the corresponding order 113 of the Rules of the Supreme Court by Pennycuick V-C in *Bristol Corporation v Persons Unknown* [1974] 1
D WLR 365 that it was clear beyond a peradventure that the order applied to such cases; he disagreed with the Vice-Chancellor's own suggestion, however, that there was a discretion in the court whether to permit the use of the summary procedure in a case where there had originally been a licence to occupy. The second point, and the real issue in the case, was whether or not the licence had been effectively withdrawn by the date when the council commenced proceedings, ie August 14 1974. The position was that a registered charity known as Students Community Housing (" SCH ") negotiated with the council on behalf of the occupier of 11 Elgin Avenue and numerous other properties. In correspon-
E dence with SCH the council had imposed a " deadline " for occupation of No 11 which was ultimately extended until May 31 1974, the date on which the council now claimed the occupiers' licence ran out. Subsequently there were discussions in which SCH apparently sought to extract from the council an undertaking by which vacant possession of the house was made dependent on the offer of alternative accommodation. No such undertaking was given, but letters were written in July and August which referred to vacant possession being given by September 1 1974. He (Lord
Diplock) thought that a recipient of such a letter would be F
entitled reasonably to conclude that his licence to occupy was extended to that date, and it followed that the appeal succeeded on the technical ground that proceedings were begun about two weeks too soon. It was to be noted that before the county court judge emphasis was laid, no doubt for tactical reasons, on the argument that an agreement had been reached whereby the occupier could remain until alternative accommodation was offered.

CAIRNS LJ delivered a concurring judgment, and BROWNE LJ agreed. The appellant was awarded his costs in both courts.

G

QUEEN'S BENCH DIVISIONAL COURT

February 18 1975

(Before Lord Justice WIDGERY CJ, Mr Justice ASHWORTH and Mr Justice BRIDGE)

BROMSGROVE DISTRICT COUNCIL v CARTHY AND ANOTHER

Estates Gazette June 21 1975

(1975) 234 EG 907

H

Land without site licence used by gypsies without permission from owners—Nature of " reasonable steps " to be taken by owners to evict occupiers—Physical force not required—Legal action may be, and justices should remember the new procedure in order 113, RSC, but everything depends on the facts of each case —Owners who failed to go to law absolved by the magistrates—Council's appeal fails

This was an appeal by way of case stated from a decision by Redditch justices on June 26 1974 dismissing an information preferred by the appellants, Bromsgrove District Council, against the respondents, Elsie and Beatrice Carthy, of Tamworth, Staffordshire, alleging that as joint occupiers of land at Wapping Lane, Beoley, Worcestershire, they permitted it to be used as a caravan site without a licence, contrary to the requirements of the Caravan Sites and Control of Development Act 1960.

Mr R Sears (instructed by Sharpe, Pritchard & Co, agents for the clerk to the council) appeared for the appellants, and Mr B Healy (instructed by Argyle & Sons, of Tamworth) represented the respondents.

Giving judgment, LORD WIDGERY said that the respondents' land was not accessible by public transport and they did not have any transport of their own. They therefore employed a local agent to administer the land on their behalf. There was no caravan site licence for the land, and none had been applied for. A number of caravans belonging to gypsies, between one and eight in number, had been parked on the land since August 1973. The respondents found this out and instructed their agent to remove the caravans. The agent did his best on the spot. He went to the site and tried to persuade the gypsies to leave. The justices found that he tried all reasonable means to get the gypsies to leave, though he drew the line at physical force. The respondents did not take legal advice with a view to evicting the gypsies, but they did nothing positive to indicate that they

A approved of the caravans, and a notice indicating that the land was private was put up. A feature which had been mentioned in the case was that the respondents' agent had tried to get planning permission to build a dwelling on the land, but this had been refused by the local authority. It was said that if the respondents had been granted planning permission they would have taken more steps to get the gypsies off the land. The justices found that although the respondents had not resorted to proceedings for an injunction they had taken all reasonable steps to get the caravans off the land, and they (the justices) dismissed the information. The council appealed.

In deciding a case of this kind, one must have regard to what
B steps the occupier might reasonably have taken. Failure to take steps that were unreasonable did not amount to permitting, and it was the justices who were entitled to decide what was reasonable and what was not. Clearly the landowner was not required to resort to physical force; that was not a step the owner could reasonably be expected to take. However, a failure to take legal action for eviction could obviously amount to a failure to take reasonable steps, and it was argued that an owner who failed to take legal steps to evict gypsies must be held to be permitting their occupation. He (his Lordship) thought that that went too C
far. Everything must depend on the nature of the case. The prospects of success in any legal action had to be considered, and the likely cost to the respondents taken into account. It was no good bringing legal proceedings against a floating population like gypsies only to find that the people against whom the legal proceedings were successful were replaced by others. In the present case, he (his Lordship) had no reason to doubt that the justices applied their minds carefully to the case, and therefore their judgment must stand. He would add that since the introduction of order 113 of the Rules of the Supreme Court, the procedure for obtaining possession from squatters and other floating populations had been made a great deal easier than it D
was; when the question of reasonable steps came up, notice should be taken of the new proceedings when considering what steps were reasonable. In this case the justices did not consider the new order, but he (Lord Widgery) thought it likely that if they had, they would have come to the same decision, and in those circumstances he would dismiss the appeal.

ASHWORTH and BRIDGE JJ agreed, and the appeal was accordingly dismissed.

NEGLIGENCE

QUEEN'S BENCH DIVISION

November 1 1974

(Before Mr Justice WILLIS)

WEEDON AND ANOTHER v HINDWOOD CLARKE & ESPLIN

Estates Gazette April 12 1975

(1974) 234 EG 121

Professional negligence—Did surveyor fail to take advantage in clients' interests of the Baptist Chapel case when handling in 1969 a compensation claim based on 1962-63 notices to treat?—Yes—Measure of damages taken as a 36 per cent increase in the value of the land concerned over the relevant period—£7,440 and costs awarded against Blackheath firm

This was a claim by Mr Charles Stuart Weedon and Mr Stanley William Weeks, suing as executors of Charles John Rice deceased, against Hindwood Clarke & Esplin, surveyors, valuers and estate agents, of National Westminster Bank Chambers, Blackheath Village, London SE3, for damages for professional negligence.

Mr P Curtis-Bennett (instructed by Lewis Rubin, of Bexleyheath) appeared for the plaintiffs, and Mr W A Blackburne (instructed by Herbert Smith & Co) represented the defendants.

Giving judgment, WILLIS J said: This is an action for damages for professional negligence brought by the executors of Charles John Rice deceased against a firm of estate agents and chartered surveyors, Hindwood Clarke & Esplin. Briefly, the allegation of negligence arises from the circumstances in which it is said that, in valuing land for the purposes of compensation for compulsory acquisition in December 1969, the defendants, that is to say Mr N Singleton, senior partner of the defendant firm, failed to exercise the proper standard of care to be expected of a valuer.

The case has unusual features, and I should indicate briefly the background of the matter. In premises in Bexleyheath, at the junction of Mayplace Road West and Erith Road, the late Mr Rice and his son carried on a commercial and industrial undertaking on land which had at all relevant times a number of uses. They dealt in secondhand building materials; they manufactured concrete kerbs and sold them from the site; they either let or licensed—it perhaps matters not—to a firm called Andersons part of the site for the purpose of the sale of secondhand cars; there were two relatively dilapidated cottages on part of the site, and there were a number of buildings used for various purposes, garages, stables and sheds of one sort and another; it was a fairly typical backyard form of industry which had grown over the years and was carried on in a somewhat haphazard way. Indeed, part of the site had been the subject in 1958 of an enforcement notice by the local authority, because the land was zoned for residential purposes and there was alleged to be a nonconforming use. Some steps were taken to comply with the enforcement notice, but clearly nonconforming use was continued thereafter with apparently no protest or action from the local authority. It is not a material fact in this case, but that is part of the background. It was a site which the local authority desired to acquire for housing purposes. It was not made the subject of a single compulsory purchase order. Two compulsory purchase orders were made, one in 1960 in respect of what was described as plots 4 and 5, and the other, which concerned plots 1 and 2 and the frontage land, in 1962. Both those orders were confirmed by the appropriate Minister, and the matter proceeded in the ordinary way with notices to treat being served in respect of the first order in June 1962 and in respect of the second order in April 1963.

Claims were made by Mr Rice. In respect of the first order a claim of some £44,000 was made. At that time, Mr Rice retained the services of a well-known firm of City surveyors, Debenham, Tewson & Chinnocks, and a partner, now a consultant, in that firm, Mr R F Hill, negotiated for him. There is no doubt that the negotiations which took place between Mr Hill and Mr W E Ellis of the district valuer's office were conducted on the footing of the law as it applied in those days. That law was to be found in section 5 of the Land Compensation Act 1961, which, as is well known, reproduced the code of compensation for compulsory acquisition laid down in section 2 of the Acquisition of Land (Assessment of Compensation) Act 1919. Three only of the six rules of that code are applicable to this case: "(2) The value of land shall, subject as hereinafter provided, be taken to be the amount which the land if sold in the open market by a willing seller might be expected to realise"; "(5) Where land is, and but for the compulsory acquisition would continue to be, devoted to a purpose of such a nature that there is no general demand or market for land for that purpose, the compensation may, if the Lands Tribunal is satisfied that reinstatement in some other place is *bona fide* intended, be assessed on the basis of the reasonable cost of equivalent reinstatement"; and (6), the disturbance rule, "The provisions of rule (2) shall not affect the assessment of compensation for disturbance or any other matter not directly based on the value of land." Acting on the basis of rule (2) as set out in that section, Mr Hill and the then district valuer reached a figure of £20,550 as being the value of the land. They deferred the question of any assessment of compensation under rule (6), and there never, in the circumstances, became an agreed figure between Mr Hill and the district valuer for disturbance. It is apparent that Mr Rice was not altogether happy about the figure of £20,550 and thought that Mr Hill should have done better. Mr Hill took the view that he could not do better. Indeed, there is a letter in the correspondence which indicates that Mr Hill thought that if they went to the Lands Tribunal they would not get anything more than he had been able to negotiate, and that in certain circumstances possibly the figure agreed might not be reached before the tribunal. None the less, finality was not achieved. The years went on, and Debenham, Tewson & Chinnocks dropped out. By 1968 the local authority wished to get things moving again. Mr Rice had died in 1964. The district valuer therefore approached his executors, Mr Weeks, an accountant, and Mr Weedon, a solicitor, to get the matter finalised, because the council wanted the land for development. In these circumstances, the defendants were approached to negotiate the compensation, because, as I have indicated, the figure of £20,550 was a provisional figure which had never been accepted.

By 1968, when Mr Singleton undertook his instructions and started to negotiate, a very important thing had happened, that is to say, at long last the challenge which had been threatened over many years to the current practice of paying compensation under rule (5) of the Land Compensation Act had succeeded before the Court of Appeal in the case of *West Midland Baptist (Trust) Association (Incorporated)* v *Birmingham Corporation* [1968] 2 QB 188. It is unnecessary to refer to the details of that case, beyond saying that it was a unanimous decision of the Court of Appeal in relation to the injustice of the practice of assessing compensation under rule (5) whereby, once notice to treat had been served, the cost of reinstating was taken to be the cost as at the date of the notice to treat. That manifest injustice, said the Court of Appeal, was to be put an end to forthwith so far as they were concerned. Additionally, Salmon LJ in particular made a strong attack upon the justice of the practice of taking the date of the notice to treat as the date for assessing compensation under rule (2). That was the legal position, therefore, when the negotiations started, and it would be supposed that a very experienced surveyor like Mr Singleton and the district valuer with whom he was negotiating would both have been aware that the law as to rule (5) had probably been changed and the practice as to rule (2) might possibly be changed. It would also have been known that the *Baptist Chapel* case was on its way to the House of Lords. By the time Mr Singleton had concluded his negotiations, and recommended figures to the executors which they accepted, namely £21,000 for the value of the land and £6,000 for disturbance, the House of Lords had pronounced in relation to rule (2). That was in August 1969, and the figure agreed by Mr Singleton with the district valuer was arrived at in December 1969.

Mr Rice had been unhappy about a similar settlement, and the question was raised with Mr Singleton by Mr Weeks whether he had had in mind the decision in the *Baptist Chapel* case in negotiating the figure he had recommended for acceptance, because it seemed very strange to the trustees that the figure agreed in 1969 by Mr Singleton was only £450 more than the figure which had been agreed by Mr Hill in 1965 on the basis of 1962-1963 values; it would have seemed, so thought the plaintiffs, that with the general rise in land values which was common knowledge, the figure ought to have been a great deal more. Mr Weeks therefore queried whether perhaps Mr Singleton had either not been aware of the *Baptist Chapel* case, or if he had, had not given effect to it, and had indeed accepted from the district valuer 1962-1963 values. Faced with that inquiry from the executors, Mr Singleton made it perfectly clear in correspondence that he had been aware from the start of his negotiations of the *Baptist* case and of the impact it had had upon the practice of valuation for the purposes of compensation, and that he had taken it into account in the course of his negotiations. He told me in evidence that he valued on the basis of what he described as " current," that is to say 1969, values, and that the similarity between his and Mr Hill's figures was due to the fact that the general rise in property values was not reflected in the sort of property owned by Mr Rice.

That is the background of the case. Mr Curtis-Bennett submits that if the valuation which was made by Mr Singleton in 1969 was right, two things follow. Either the valuation of Mr Hill and the district valuer in 1965 on the basis of 1962-1963 values at £20,550 was wrong, or if that was not wrong, then land values have not increased or at least have not increased *quoad* this particular property. Alternatively, counsel says, Mr Singleton must have been applying the wrong criteria, and by that Mr Curtis-Bennett means that despite what Mr Singleton has said, he was indeed applying 1962-1963 values, despite the fact that the *Baptist Chapel* case had been at least before the Court of Appeal. G
What Mr Curtis-Bennett then says is this. Even though Mr Singleton was faced, or might have been faced on the basis of that case, with the district valuer saying, " Well, it is only in the Court of Appeal, and I cannot possibly take account of current values until we know what the House of Lords has said," it would have been the action of a prudent valuer to have made known to his clients that that was the position and to have deferred, as Mr Hill has told me that he deferred in a number of negotiations at that time, coming to a final decision until it was known whether or not the House of Lords approved the majority view in the Court of Appeal. H

In January 1969 Mr Singleton supplied to the executors a written statement setting out the position under compulsory purchase orders as viewed by him. Paragraph 2 reads: " The freehold interest is to be valued as at the dates of the notices to treat, namely, 1962-1963." Paragraph 3, so far as material, reads: " The compensation should represent the market value of the whole land in 1962/63." The statement is a long document, and it is unnecessary to refer to all of it; but in paragraph 9, after setting out the various figures which had been bandied about from time to time as to the value of this land, including the provisional agreement of the district valuer in 1965 of £20,550, and " D.V.'s best offer to date "—that is J
to say, in the current negotiations—the same figure, £20,550, Mr Singleton went on to say: " If the land is sold entirely for residential purposes, the likely density would permit approximately 16 two-bedroomed flats. Because of the awkward shape of the land and the additional cost, a residential developer in 1962-63 would probably have been prepared to pay £1,000 per unit, but no more. This would give a price of £16,000 as compared with the district valuer's original offer of £15,000." Mr Singleton agrees, of course, that that was perfectly obviously a 1962-1963 valuation. He went on to say in paragraph 10: " Although part of the land K
has no commercial use rights, a better valuation can be made by working on existing use rights section by section, and this gives a figure of about £19,000 as compared with the district valuer's best offer to date of £20,550." Plainly, as I think, that refers to the district valuer's offer in the 1962-1963 provisional agreement; despite the fact that Mr Singleton, when cross-examined upon that paragraph, says that he was not contrasting on a 1962-1963 basis a residential valuation on the one hand with a commercial section-by-section valuation on the other, I regret to say that I think he plainly was. He sought to explain the phraseology by saying that, somehow or other, between paragraph 9 and paragraph 10 he had L
moved mentally from a 1962-1963 to a 1969 current-value concept of valuation. With a valuer of Mr Singleton's experience, I find it impossible to accept that he would not, if that had been his process of thought, have made it clear in that document. The statement concludes by saying: " A recent discussion with the district valuer indicated that he considers that he has gone as far as he can at £20,550, although possibly he would round this up a bit in order to reach final agreement." That, Mr Singleton says, is a 1969 current-value figure.

The criticism which is made of Mr Singleton in relation to that document is this. If he were indeed mindful through- M
out these negotiations of the Court of Appeal's decision and of what might well be a much more favourable basis of compensation by reason of being able to take advantage of any increase in land values since the date of the notice to treat, why was there not a word of that possibility in the memorandum? There is no indication at all of a caveat that the executors might well be advised to await a final settlement until the House of Lords had pronounced upon this problem. Everything in that document, as Mr Curtis-Bennett submits and as I accept, points to the fact that the figures were based on 1962-1963 notice to treat valuations. There-

A fore, as it seems to me, if I am not to accept that document at its face value, it is essential for Mr Singleton to be able to convince me that, whereas nobody could be blamed for thinking that he was using 1962-1963 figures throughout that document, he and Mr G Shadlock, the assistant district valuer, were in fact negotiating and did in fact agree the final figure on the basis of 1969 values. Mr Shadlock confirmed that he and Mr Singleton negotiated on the basis of values current in 1969. Mr Singleton was cross-examined about his figure of approximately £21,000 for the 1969 valuation of this land and about the breakdown which was supplied contemporaneously in 1965 by Mr Hill, and, within a few hundred
B pounds or so, the figures are exact. Mr Singleton says that it is a pure coincidence that the notice to treat figures are in fact almost pound for pound the same as his figures for 1969. If indeed he was taking account of the *Baptist* case, and produced the same figures for all practical purposes as Mr Hill and the district valuer did in 1965, then inevitably it must follow that for some reason or another the value of this particular land did not share in the general rise in land values. It is common ground with all the valuers that in South London there was an average increase in land prices generally of about 6 per cent per annum over the period that
C I have to consider, which gives something like a 36 per cent increase between 1962-1963 and 1969. Mr Singleton and the two valuers who supported him on this matter, Mr Shadlock and Mr Jennings, who is an experienced local valuer, say that this land is to be treated exceptionally. Mr Shadlock went so far as to say that although, apart from increased dilapidation of the buildings, the condition of the land and the nature of its user were exactly the same in 1962-1963 as in 1969, as I so find, nevertheless because the business was making a loss virtually no increase in value had taken place. I find this evidence quite impossible to accept. Mr Jennings's evidence was to the effect that it was not this particular piece
D of land, but land of this type, with rather a ramshackle mixed use upon it and with a lot of backland, which had not shared in the general increase.

What then is the position as to the basis upon which Mr Singleton was valuing in 1969? He told me that although he was himself aware of the Court of Appeal's decision and of the fact that it would, if approved, drastically affect the practice under rule (2), his experience of the valuation office at the time when he started upon his investigations was that its valuers were not prepared to work on the new basis. He went on to say that the valuation office was waiting for the appeal to the House of Lords which was known to be pending,
E and that when he prepared the memorandum in January 1969 he was hopeful that the House of Lords decision would confirm the new basis under rule (2); although he was optimistic, however, he got no concession from the district valuer, who still maintained that he must approach the valuation as at the date of the notice to treat, even though claiming that the figures previously agreed by his predecessor had exceeded the 1963 value. Mr Singleton went on to say, when asked why in those circumstances nothing appeared in the memorandum about that matter, that he did not want to complicate the situation for Mr Rice, so he made no mention of the *Baptist* case, although in fact he was working on
F 1968-1969 values. I have to ask myself this: if Mr Singleton was working on 1968-1969 values, and the district valuer, as would appear from the evidence which I have just quoted, was not prepared to budge from the notice to treat value, how was it that their figures were so near, because they were never more than between £1,000 and £2,000 apart? One explanation for this apparently strange position is the view held by all three valuers called for the defence that the district valuer and Mr Hill were wrong in 1965. Mr Singleton advanced a figure of some £1,500 as the excess valuation; Mr Shadlock, though understandably he did not wish to criticise a colleague, particularly as he was not in that office in 1965, thought that Mr Ellis was "a bit on the high side," while Mr Jennings was prepared to say that he was something like £5,000 out in 1965. Equally, he thought that Mr Singleton and the district valuer in 1969 were also about £5,000 too high. I am bound to say that I can get very little assistance, in spite of his local expertise, from the evidence of Mr Jennings.

I accept that there was an overall increase in the value of land between 1962-1963 and 1969. I accept that there is no sensible distinction to be drawn between the categories of land which shared in that increase. *Prima facie,* therefore, I ask myself why I should not apply a 6 per cent increase to this land over the years between 1962-1963 and 1969. Having heard the explanations given in relation to this particular piece of land by Mr Shadlock, or land of this type by Mr Jennings and Mr Singleton, I can find no reason why I should make any modification to a 6 per cent per annum increase in value between the two dates of valuation. If that is so, it seems to me that to reach a figure in 1969 which was virtually the same as that reached in 1962-1963 must *prima facie* be wrong. How does the defence seek, in part at all events, to meet that position? Mr Blackburne, who, if I may say so, has argued this case and presented it with great persuasiveness and ability, invites me to say that I ought to accept the evidence of the three valuers on his side that the district valuer was too high in 1965. I am unable to do so. I propose to accept the provisionally-agreed figure in 1965 as being, as near as one can get, correct. It was agreed at arm's length between the district valuer, Mr Ellis, and a very experienced valuer, Mr Hill, and I can find no reason why I should accept the criticism that is made years later of that valuation by people who were not concerned at the time. I therefore propose to anchor myself to the figure of £20,550 agreed in 1965 as being the fair value of the land for the purposes of the assessment of compensation at 1962-1963 values.

Because the matter has been raised and closely argued, I feel compelled to make a finding, although I regret the necessity of doing so, as to the basis upon which the 1969 figure of compensation was assessed. I am not able to accept Mr Shadlock's evidence that he was valuing for the purposes of the district valuer's office on current values. Mr Singleton has told me that so far as he recalls—although Mr Shadlock does not—there was a directive from the Inland Revenue to district valuers following the Court of Appeal decision that they were to continue to value on the traditional basis of rule (2). If there was, it would accord with commonsense and what could be expected in a fluid situation. I think Mr Shadlock was sticking to his instructions and was throughout the negotiations valuing on a 1962-1963 basis. It is apparent from the correspondence, and in particular the memorandum of January 1969, that it was not until Mr Singleton was questioned about the correctness of the 1969 valuation that he asserted that he was well aware of the *Baptist* case during the negotiations and that it was taken into account both by him and the district valuer to the advantage of the executors. I am bound to say that I am unable to draw any other inference from the memorandum and the relevant correspondence which followed it, and from the evidence which I have heard, than that throughout the negotiations which led to the final figure of some £21,000 for the land out of a total of £27,000, the valuation was on 1962-1963 values and no effect was given to the *Baptist Chapel* case. On that finding, Mr Blackburne has conceded that if Mr Singleton either (1) knew of the *Baptist Chapel* case, and either did not take it into account in his valuation or, as I think may well have been the case, was pushed off it by a district valuer who would not budge, or (2) did not know about the case as he should have, then a finding of negligence must follow. It is always exceedingly distasteful to make a finding of this sort against a professional man, and I do so with the greatest

A regret, but I feel that I have no option on the evidence which is presented to me but to find that there was a failure on Mr Singleton's part to take the steps which a prudent valuer should have taken, either because he was not aware of the *Baptist* case and its implications for valuation at the time, or because, being aware of it, he did not apply it as he should have. In either event, his conduct amounted to a failure in his duty to take normal professional care and exercise his normal professional skill.

What, then, is the measure of damage? Mr Hill has put before me a valuation of £32,000 as being what, in his view, was a proper valuation in 1969. As I indicated during the case, B I do not feel myself equipped to deal with valuations on one side or another, nor even if I were do I feel that the evidence which it has been possible to put before me was in sufficient detail, particularly in relation to comparable properties, to enable me to do so. Therefore, as it seems to me, I am only able to approach the matter of damages on a general basis. Mr Blackburne has urged upon me that I should embark upon a consideration of various aspects of the valuation and prefer his clients' valuations to that of Mr Hill. As I have already indicated, I do not propose to embark upon a detailed valuation because I am unable to do so. C Where I am, I think, on safe ground is in accepting what is common to both sides, namely, the average increase in values over the years amounting to some 36 per cent. I have already indicated that I also regard myself on safe ground in accepting £20,550 agreed between Mr Hill and the district valuer in 1965. A possible view would be to say that I accept Mr Hill entirely, but I do not think that that would be a proper approach in the circumstances of this case. As Mr Curtis-Bennett puts it, it may well be as optimistic a valuation as Mr Jennings's valuation was pessimistic. I propose to base myself on the increase in the value. The figure I have been given by way of calculation—and I think it D is roughly right—is £7,400, which is 36 per cent of the 1965 valuation. This seems to me to be the proper measure of damages as reflecting the amount by which the sum recommended for acceptance in 1969 fell short of what was the market value of the land at that time assessed under rule (2). In the circumstances, therefore, this claim must succeed, and there must be judgment for the plaintiffs for the sum of £7,400.

The plaintiffs were awarded the costs of the action.

RATING

A

COURT OF APPEAL

December 3, 1974

(Before Lord DENNING MR, Lord Justice ROSKILL and Sir JOHN PENNYCUICK)

GILBARD (VO) v AMEY ROADSTONE CORPORATION LTD

Estates Gazette March 22 1975

(1974) 233 EG 1095

Rating—Gravel pit worked out in 18 months—Assessment on royalty basis not limited by reference to dwindling stock B of gravel—Commonsense valuation method applied despite technical arguments based on dates of valuation officer's proposals

This was an appeal by a local valuation officer from a decision of the Lands Tribunal dated April 3 1960 determining the rating assessment for 1970-71 and 1971-72 of a sand and gravel pit at Mytchett, Surrey, occupied by the respondents, Amey Roadstone Corporation Ltd. The decision of the Lands Tribunal was reported at (1973) 227 EG 1479.

Mr A P Fletcher (instructed by the Solicitor of Inland Revenue) appeared for the appellant, and Mr Q J Iwi C (instructed by Copleys, of St. Ives) represented the respondents.

Giving judgment, LORD DENNING said: In 1970 a great motorway, the M3, was being built. Large quantities of sand and gravel were needed, and some came from a pit at Mytchett, a few miles from Aldershot. Extraction started on April 6 1970 and went on at a great rate. In nine months the needs of the motorway were satisfied. More was extracted for other users. By September 30 1971, all the sand and gravel had been extracted and the pit was closed. So extraction lasted for only 18 months in all. The question now is, what was the rateable value of the pit for the 18 D months that it was worked?

There is a well-known formula for ascertaining rateable value. It goes back for centuries, but it is now contained in section 19 of the General Rate Act 1967. The rateable value of a hereditament is to be taken to be the net annual value, and the net annual value is:

> . . . an amount equal to the rent at which it is estimated the hereditament might reasonably be expected to let from year to year if the tenant undertook to pay all usual tenant's rates and taxes and to bear the cost of the repairs and insurance and the other expenses, if any, necessary to maintain the hereditament in a state to command that rent.

E That formula, so simple in appearance, is impossible to apply to some hereditaments. For instance, there are undertakings such as waterworks which could not conceivably be let to a tenant at any rent at all. In such cases it has been found necessary for rating experts and the courts to have recourse to hypotheses of a more or less violent character. Water undertakings are valued on the profits basis: see *Kingston Union Assessment Committee v Metropolitan Water Board* [1926] AC 331 at 338. Likewise with brickfields and gravel pits, which are continually being exhausted by the extraction of clay or gravel. Such hereditaments are not let at a rent from year to year, but are worked on a royalty basis. In order to rate them, the rating experts and the courts assess the net annual value on the current rate of extraction of gravel. The practice is to estimate the amount of gravel which will be extracted during the current year. To do this, the valuer will take the actual output during the previous calendar year (January 1 to December 31) of which the figure is known, and then assume that extraction will continue at the same rate during the current year. The figure will be in cu yds, say 120,000 cu yds. The valuer will ascertain from commercial men what royalty would be payable on such output, so much for every cu yd, say 5p a cu yd. Then he will multiply the cu yds by the royalty, and thus arrive at the annual value; in the example it would be £6,000. So long as the pit is being worked at a steady rate of extraction this method works well. But difficulties arise when the extraction rate is much higher in one year than the next, or when all is extracted in less than a year, say in 10 months. In that case the assessment must be made as nearly as may be on the actual output for the particular year. This is made clear by the leading case of *R v Westbrook* (1847) 10 QB 178 in which Lord Denman CJ said at pages 203-4:

> A case was supposed of a brickfield worked out in less than a year to meet the demand of some enormous contract for a public work; the consequence would be that the land would have a very much increased value for the year, and it would be only reasonable that it should bear an increased rate for that year; in the following year its value might sink almost to nothing, and the rate ought to fall proportionately, even to nothing if, the brick earth being exhausted, the land, like an exhausted coal mine, should become entirely unproductive.

After giving an example, the Chief Justice continued:

> But no injustice would be done if in every year the occupier could be assessed according to the actual value in that year; and it is the duty of the overseers to arrive as nearly at this as they can.

Difficulties also arise when a brickfield or a gravel pit is nearing the end of its life. On April 1 in any year it may have little left to extract. There may be only 20,000 cu yds, which may be finished in a couple of months. To get the net annual value, you must then assume that the rate of extraction over the two months, say 10,000 cu yds a month, would be continued over a whole year. So the annual rate would be 120,000 cu yds. A royalty of 5p a cu yd would give a net annual value of £6,000. But when it comes to payment of the rate, the ratepayer would be charged only for the two months for which he had worked it; see section 18 (2) of the General Rate Act 1967. That is £1,000. It would have been quite wrong to take the 20,000 cu yds as the net annual value. In *R v Waddon* (1875) LR 10 QB 230 at 243, Mellor J said:

> The rate must be made on the net annual value, and therefore it cannot be made upon the net quarterly value: and it appears to me that the net annual value is the same whether the profit is earned wholly in one quarter of a year or spread over the year.

Such being the general principle, I must say a word about the procedure for ascertaining the rateable value. If any person (including a valuation officer) is aggrieved by any value ascribed in the valuation list, he may at any time make a proposal for the alteration of the list. It may be in April, November, March or any month in the rating year. Evidence is then taken as to the net annual value as at the date of the proposal: see *Barratt v Gravesend Assessment Committee* [1941] 2 KB 107. That figure then has effect from the commencement of that rating year, that is, from April 1 in that year: see section 79 (1) of the Act of 1967. There are exceptions in the case of alterations to houses, and so forth, as to which see section 79 (2) (b) and section 68 (4) (b), but none of these concerns a gravel pit.

This brings me to the facts of the present case. The first rating year in question was from April 1 1970 to March 31 1971. On December 21 1970 the valuation officer made a proposal for that year. He estimated that 200,000 cu yds would be extracted during the year. On that large quantity, the gravel company would pay a royalty of 1s 8d per cu yd, which would give a figure of £16,670 rateable value. The gravel company asserted that that was erroneous. They said that the value of the hereditament was to be taken as at the date of the proposal, December 21 1970, and that at that date most of the gravel had already been extracted. The amount left to be extracted was estimated to be only 50,000 cu yds, and the company said that the rateable value should be assessed on the gravel which remained in the pit, and not on the gravel which had already been extracted. They would pay a royalty of 1s 2d per cu yd on 50,000 cu yds, and that gave a figure of £2,920 rateable value. The next rating year was from April 1 1971 to March 31 1972. On May 27 1971 the valuation officer made a proposal. At that time the pit was nearly worked out. There were estimated to be only 30,000 cu yds left at that time. The valuation officer in fact ascertained later in the year that 21,927 cu yds had been extracted in the six months from April 1 1971 to September 30 1971, when the pit was closed. At that rate there would be 43,854 cu yds extracted in a full year. The valuation officer took that figure, applied a royalty of 1s 8d a cu yd, and arrived at a figure of £3,660. (This had to be reduced by a quarter under the Mines and Quarries (Valuation) Order 1971, making £2,745.) The gravel company objected. They said that the figure should be assessed at the 30,000 cu yds estimated to be left on May 27 1971. They put the royalty at 1s 2d a cu yd. They thus assessed at a figure of £1,750. (This had to be reduced by the said quarter, making £1,312.) Thus the difference between the parties is this. The valuation officer contended that the rateable value of the gravel pit should be assessed according to the estimated amount which would be extracted in a full year at the current rate of working, irrespective of whether there was a large quantity left or only a little left which would be soon worked out. If it was worked out in a short time, the gravel company would have to pay only a proportionate part of the rate: see section 18 (2) of the Act of 1967. The gravel company contended, however, that the rateable value should be assessed according to the estimated amount of gravel still remaining to be extracted. If there was only a little left, the rateable value would be very small, but it would have effect for the whole of the current year. The gravel company said that it had been so decided by the Court of Appeal in the *Farnham Flint Gravel & Sand Co Ltd v Farnham Union* [1901] 1 KB 272.

I cannot accept the contention of the gravel company. It leads to absurd results. Take this simple instance. A gravel pit contains an estimated quantity of 120,000 cu yds. It is worked at 10,000 cu yds a month. If a proposal is made by the valuation officer at the beginning of the rating year, ie, on April 2, the rateable value would be a royalty on 120,000 cu yds. Now suppose that a proposal is made by the gravel company many months later, at the end of the following February and towards the end of the rating year. The valuation officer would still say that the rateable value was a royalty on 120,000 cu yds, because that was the annual rate. But the gravel company would say that the rateable value was only a royalty on 10,000 cu yds, being the quantity of unexhausted gravel which remained at the end of February. That rateable value, say the gravel company, would apply for the whole of the current year from the previous April 1 to March 31. So that the gravel company would pay no rates on the 110,000 cu yds which they had extracted during the previous 11 months, but only on the 10,000 cu yds which remained. Such a result is so absurd that the gravel company's contention must be wrong. I am not satisfied that the *Farnham Flint* case does lead to such a result, but if it does, it must be wrong too. As I have often said, and I would repeat it now, in my opinion this court is not absolutely bound by a previous decision of its own. It is not bound when that decision is shown to have been plainly wrong. In the *Farnham Flint* case Channell J dissented. He had an unrivalled knowledge of these branches of the law. He stated the principle thus:

> In rating a property in respect of a beneficial use which is self-destructive, the overseers have to ascertain the amount of use the actual tenant makes of it; then ascertain what a tenant would give for liberty to make that amount of use of it from year to year, independently of the question whether there is or is not enough of the property remaining for the tenant to be able to go on making the use for the whole year.

He took the case of a cemetery which was nearly full but had only room for six months' burials. He said at p 283:

> It would not be valued at half the annual value in the last year because there was only room for six months' burials, any more than a man would be rated at half the annual value of his house because he had only six months' term remaining of his lease.

There is of course a technical difference between the two cases, as Collins LJ pointed out, so that the analogy is not perfect. But the good sense of Channell J's remarks cannot be gainsaid. The cemetery "would not be rated at half the annual value in the last year because there was only room for six months' burials." So a gravel pit is not to be rated at half the annual value in the last year merely because there is only gravel enough for six months' working. It is to be rated at its full annual value, but the rates are only to be charged for the period in which it is occupied for extracting gravel. It is to be noted that the latter principle has been applied by the legislature in regard to coal mines: see the National Coal Board (Valuation) Order 1963 (SI 1963 No 636). I think that the principle laid down by Channell J was right. We should follow him and not the others. It gives a just result. The rateable value is the annual value, but the rates are only paid for the number of days that the gravel is being extracted. It may be possible to distinguish the *Farnham Flint* case on one ground or another, but it is a tedious exercise on which I do not think it necessary to embark. I would allow the appeal and hold that the rateable value of the gravel pit should be the valuation contended for by the valuation officer, that is, £16,670 for 1970-71, and £2,745 for 1971-72.

ROSKILL LJ: These two appeals by a valuation officer by way of case stated by the Lands Tribunal raise a question of principle of some importance regarding the correct method of determining the rateable value of a rapidly wasting hereditament which was described in the special case as "Sand and gravel pit and premises, Mytchett Pit, Mytchett." Mytchett is in Surrey. It is important to note that that description accords with the definition of "hereditament" in section 115 (1) of the General Rate Act 1967. It is the pit and premises in question whose rateable value has to be determined, not the totality of the land where that pit and those premises are situate. The valuation officer had made certain proposals the details of which will be found on pages 4 to 6 of the special case. I do not repeat those details in this judgment. I observe, however, that the first relates to the rating year 1970-71 and the second to the rating year 1971-72. The respondent occupiers objected, and the matter went to the local valuation court. By that time the appellant valuation officer had amended his proposals as set out on page 6 of the special case. The local valuation court determined the matter in favour of the occupiers. But both parties appealed thence to the Lands Tribunal, the occupiers on the ground that the resultant determination was too high, the valuation officer on the ground that it was too low. Ultimately the issue resolved itself into one of law, the Lands Tribunal having to determine which of two possible bases of assessment was correct in point of law. The member of the Lands Tribunal before whom the matter came on appeal from the local valuation court accepted the contention of the occupiers. He accordingly directed that the entries in the valuation list should be made which will be found on page 17 of the special case. These were based on the agreed quantities of gravel remaining for abstraction on the two relevant dates, namely 50,000 cu yds on December 21 1970 for the rating year 1970-71 and 30,000 cu yds on May 27 1971 for the rating year 1971-72. It should be explained that those dates were the dates of the respective proposals for the two rating years in question. The valuation officer now challenges the member's conclusions in this appeal to this court.

This valuation problem arises in a peculiarly acute form on the facts of the present case. A rating year runs from April 1 in one year to March 31 in the next. The extraction of gravel from this pit began on April 6 1970. Owing to the urgent needs of the adjacent motorway, the M3, the extraction rate built up very quickly, and the requisite extraction of gravel for that motorway ended in December 1970. In the course of this working it was discovered that the quantity available for economic extraction was less than half the quantity originally estimated. This had the consequence that before the proposal of December 21 1970 was made, the bulk of the available gravel had already been extracted. The balance was disposed of in accordance with current market demands put forward from time to time, and the pit was finally closed, exhausted, on September 30 1971. The output in the rating year 1970-71 was just under 200,000 cu yds, and in the rating year 1971-72, down to September 30 1971, just under 22,000 cu yds.

Stated summarily, the occupiers' contention was that section 19 (3) of the Act of 1967 provides that in the case of hereditaments not falling within section 19 (2) of that Act, the net annual value shall be "an amount equal to the rent at which it is estimated the hereditament might reasonably be expected to let from year to year if the tenant undertook to pay all usual tenant's rates and taxes and to bear the cost of the repairs and insurance and the other expenses, if any, necessary to maintain the hereditament in a state to command that rent." That value has to be determined by references to what the hypothetical tenant would pay on the date of the relevant proposal for a yearly letting, and what such tenant would then be prepared to pay would depend upon the amount of gravel then remaining to be extracted, which in the instant case was some 50,000 cu yds on December 21 1970, being the proposed date for the year 1970-71, and some 30,000 cu yds on May 27 1971, being the proposed date for the year 1971-72. No tenants as at those two dates would conceivably agree to pay a rental for the extraction of quantities of gravel other than those which were then in fact or were then reasonably believed to be available for extraction. Stated equally summarily, the valuation officer's contention was that the correct way to ascertain what a hypothetical tenant would pay was first to ascertain estimated annual output for the year in question and multiply that figure by the estimated royalty per unit of output, and then, since one was dealing with a pit which was worked only part of the two rating years in question, calculate the output of the pit in each rating year at the same average rate of output as was in fact maintained during that part of the rating year for which the pit was actually worked. It is important to distinguish, it was said, between the net annual value, which had to be calculated on the hypothesis that the pit was let from year to year, and the amount which the occupiers of the pit would actually pay by way of rates, which would be only in respect of the actual period of their occupation.

The problems to which section 19 (3) of the 1967 Act and various of its predecessors, such as section 22 (1) of the Rating and Valuation Act 1925, give rise are notorious, especially in relation to hereditaments whose net annual values have to be ascertained but which in practice are not let from year to year or may in practice never be let at all. Some of these problems are discussed by Lord Pearce in his speech in *Dawkins v Ash Brothers and Heaton* [1969] 2 AC 366 at 381-2, and need not be repeated here. It is doubtless because of these and other problems that section 35 (1) of the 1967 Act empowers the Minister by order to make provision for determining the rateable value of hereditaments to which that section applies, including coal mines occupied by the National Coal Board and other hereditaments consisting of or including mines and quarries. Mr Fletcher drew our attention to the National Coal Board (Valuation) Order 1963 (SI 1963 No 636), printed in *Ryde on Rating,* 12th ed 1968, p 1367. That order, by virtue of section 117 (3) of the 1967 Act, has effect as if made under section 35 of that latter Act. Mr Fletcher said that article 16 of that order prescribed the appropriate machinery for resolving the present problem in the case of National Coal Board coal mines where production started during a rating year. He stated that his contention was that what was there enacted by statutory instrument to be applicable to National Coal Board coal mines applied equally in principle to a gravel pit such as that with which this appeal is concerned, notwithstanding that no comparable order had been or was to be deemed to have been made under the 1967 Act. Pressed to state why, if this were so, a statutory instrument was required to produce the desired result, in the case of National Coal Board coal mines but no such statutory instrument was required to produce this result in the case of gravel pits, Mr Fletcher was, if he will forgive my saying so, unable to produce a satisfactory answer. Nonetheless, as my Lord, the Master of the Rolls, pointed out during the argument, the adoption over the last century or so of the profits basis of assessment, notwithstanding the total absence of statutory authority for so doing, is eloquent of the problems of valuation which have to be faced in applying the statutory formula to cases which it does not readily fit.

It seems to me that two fundamental principles must underlie the approach to the present problem. The first is that the court is to seek as the net annual value that rent which a hypothetical tenant would pay. The second is that the assumed letting is from year to year. To my mind the strength of the case for the valuation officer, and the weakness of the case for the occupiers, can best be demonstrated by the fact that if the occupiers be right, and the net annual value for 1970-71 has to be determined by and only by reference to the amount of gravel remaining on December 21 1970, the occupiers will escape assessment upon their beneficial occupation which they gained down to that date from all the extraction from April 6 1970 onwards. Indeed, as Mr Fletcher pointed out during the argument, the later in the rating year a proposal is made, the greater the value of the beneficial occupation from the beginning of the rating year to the date of the proposal upon which the occupiers will escape all assessment, since by reason of section 79 (1) of the

1967 Act, an alteration resulting from the belated proposal will date back to the beginning of the rating year in question, cases such as the present not falling within section 79 (2) of the 1967 Act (see also in this connection *Barratt v Gravesend Assessment Committee* [1941] 2 KB 107). This was said on behalf of the occupiers to be the result of a statutory anomaly following the change in the law regarding the retrospective effect of an alteration to the valuation list which was made in 1925. Whether or not it be right to call this a statutory anomaly, it seems to me to be so manifestly an absurd result that it is only to be accepted if there is no acceptable alternative available. By contrast, the occupiers contended that even accepting, without admitting, the apparent absurdity of this result in relation to the first year, when one came to consider the second (and in this case the final) year it was just as absurd, if the valuation officer's contention were right, that the occupiers should have the net annual value of that second or final year assessed on the basis of an output in excess of that ever likely or expected to be attained, and that in real life no tenant would ever pay a rent calculated by reference to a quantity different from and greater than that which he would be likely to extract from the pit which was the subject matter of the proposed assessment. But a net annual value has somehow to be determined for the final year, and the statute requires the assumption of a letting from year to year in order to arrive at that net annual value. If, as is the fact, the pit will become exhausted in that final year, then the argument for the occupiers, pushed to its logical conclusion, means that the requirements of the statute can never be complied with for that final year, because the occupiers ought never to be assessed for that year by reference to more than they would actually extract in that year, and that quantity they would *ex hypothesi* extract in a period less than a calendar year.

Some means of avoiding this apparent inability to conform with the language of the statute must be found, and to my mind the valuation officer's contentions provide a basis for complying with the requirements of the statute in this respect which will not cause hardship to the occupiers, because they will only pay rates for such part of the final year for the period in that year in which they remain in beneficial occupation of the pit. I can therefore see nothing in principle which requires me to reject the valuation officer's contentions, and much in them which avoids the anomalies of which I have spoken whilst also complying with the basic statutory requirements. I therefore turn to such authority as there is in order to see whether that authority either supports these contentions or compels their rejection. The two principal authorities are *R v Whaddon* (1875) LR 10 QB 230 and *Farnham Flint Gravel & Sand Co Ltd v Farnham Union* [1894-1904] R & K 217 (Divisional Court) and [1901] 1 KB 272 (Court of Appeal). It must be said at once that with the utmost respect to all the judges respectively involved, neither of these decisions is very satisfactory or very easy to understand. Both are critically discussed at length in *Ryde* at pages 235-9. The latter decision binds the court for what it actually decides, and with all deference to the contrary view of the Master of the Rolls, it is not open to this court to say that it was wrongly decided or to prefer the dissenting judgment of Channell J in the Divisional Court. The problem is to ascertain what that case really did decide. But before turning to consider that problem I will first consider *R v Whaddon.* It is important to observe that in *R v Whaddon* the assessment was an assessment upon 10 acres of coprolite-producing land. The tenant was not always in occupation of the same 10 acres of land, but he had the right to occupy 10 of the acres at any given time. The question was whether he should be rated upon the whole 10 acres or only upon the 3½ acres which he was actually working in the relevant rating period, which in that case was a quarterly period and not a period of 12 months. A majority of the Court of Queen's Bench, Cockburn CJ dissenting with reluctance, held that the tenant was properly assessed upon the whole 10 acres. Though *Ryde* at page 239 suggests that *R v Whaddon* may have to be treated as overruled by the *Farnham* case, this court in the *Farnham* case never said that *R v Whaddon* was wrongly decided. It merely distinguished it. It is enough for present purposes to say that I can find nothing in *R v Whaddon* which assists in the determination of the present appeal.

The *Farnham* case is more difficult, not least for the reasons first, that there was a division of opinion in the Divisional Court between Bucknill J and Channell J; secondly, that the report in *Ryde and Konstam's Rating Cases* to which we have referred does not contain the passage in the judgment of Channell J strongly criticised by Collins LJ in this court; and thirdly, that there was, as appeared from the judgments in this court, considerable confusion as to what Blackburn J either had said or had meant to say or was to be taken as having said in an earlier case much relied upon by Channell J in the Divisional Court, something which was held in this court either not to have been said by Blackburn J or if said by him to be wrong. To my mind the only safe approach to the decision in the *Farnham* case is to look at the facts by reference to which the decision of this court was given. The learned Master of the Rolls, Sir Archibald Levin Smith, said that he proposed to adhere to the facts stated in the special case and not to go outside them. I therefore turn to see what the facts were. At the relevant date, which at that time was the date when the rate was made, the occupiers were in occupation of 3½ acres of land. In 2½ of those acres the gravel had already been exhausted, but the occupiers still used those 2½ acres for storing gravel which they had already dug. They were still gaining gravel from the remaining single acre. A dispute arose as to the basis upon which they were to be assessed. The occupiers contended that the assessment should be based either on the value of their occupation of the single acre, ascertained from what they paid for it together with the cost of making good, plus the value of the 2½ acres, ascertained upon the basis of their value when let for storage purposes, or on the value of the gravel in the single acre plus the value of the 2½ acres let as above. The rating authority contended first, that the value was to be ascertained by taking the output of gravel from the land in their occupation during the previous year (which it is to be observed in passing was not the same 3½ acres), and assessed upon the basis of the total royalty the occupiers would have paid to the owners of the land if instead of buying the gravel (and here again it is to be observed in passing that they had agreed to buy the gravel from the owners of the land and not just to extract it), they had agreed to extract the gravel in return for payment of a royalty on each yard of gravel dug—a method adopted in other cases in the neighbourhood. The second contention of the rating authority is not, if I have understood the report correctly, immediately relevant.

It should be observed, therefore, that the rating authority in their main contention were arguing in favour of a net annual value arrived at by reference to the output of gravel in a different year from different land from that which the occupiers occupied during the year of assessment. Furthermore, the assessment in that case was upon the land, and not as here, as already pointed out, on the pit and premises. Quarter sessions rejected the first contention of the rating authority as wrong in law, but accepted that if they were wrong in their conclusion, the assessment appealed from was correct. They also rejected the rating authority's second contention. As appears from page 275 of the report, quarter sessions held that the annual value of the land in the appellants' occupation at the date of the making of the rate ought to be taken as the amount of rent or royalty at which the same could then be reasonably expected to be let to a

A tenant for one year, regard being had to the value of the exhausted 2½ acres for storage purposes. On this basis they found as a fact what the rent was at which that land could be expected to be so let, and they fixed the net annual value accordingly. In other words, on the facts of the *Farnham* case, quarter sessions were able to reach a conclusion what the net annual value was for the land in question without resort to what had happened in relation to different land in the previous year. It was that conclusion of quarter sessions on those facts which this court ultimately upheld as correct in point of law. As already pointed out, there was no need to resort to the formula contended for by the rating authority
B when there was evidence which enabled the requirements of the statute to be more simply complied with in respect of the 3½ acres in question in relation to the year in question.

Now I do not read the judgment either of Sir Archibald Levin Smith MR or of Collins LJ in the *Farnham* case as refusing to allow the adoption under any circumstances of the formula for which the rating authority contended in that case. Indeed, one passage in the judgment of Collins LJ suggests otherwise. At page 282 the learned Lord Justice said: "The main point of difference between my brother Channell and the court of quarter sessions is whether the
C annual value is to be ascertained by reference to what was paid in the year preceding the date when the rate was made or by reference to what may reasonably be expected to be obtained in the following year. But for the judgment of the learned judge and the quotation from the judgment of Blackburn J I should have thought the matter clear, because the only cases in which recourse can be had to actual profits or earnings in the past are those in which ordinary methods fail. In such cases antecedent profits can be used as a stepping stone to arrive at a conclusion as to the future." In other words, in a case where what Collins LJ called "ordinary methods" fail, resort may be had to the previous year as a
D stepping-stone for the future. In the present case, the facts of which are very different from those of the *Farnham* case, the basis of assessment contended for by the occupiers, as already pointed out both by my Lord and earlier in this judgment, produces an absurd result. It did not do so in the *Farnham* case. I can see no reason why in law in the present case that result should not be avoided by adopting the formula supported by evidence before the Lands Tribunal and advanced on behalf of the valuation officer in the present case. Though, therefore, I reach my conclusion by a different route from my Lord, the Master of the Rolls, because I respectfully decline to say that the *Farnham* case
E was wrongly decided—not only does it bind this court for what it decides, but in my judgment it was correctly decided on its facts—I agree with the result which he has reached. It follows that I agree with the conclusion in his judgment that the appeal should be allowed and the rateable value for the pit should be that contended for by the valuation officer, that is to say, £16,670 for the rating year 1970-71 and £2,745 for rating year 1971-72.

SIR JOHN PENNYCUICK: I too would allow this appeal. I agree with the judgment of Roskill LJ, and like him, I have reached my conclusion by rather a different road from the Master of the Rolls. I will read at the outset section 19 (3)
F of the General Rate Act 1967:

> The net annual value of any other hereditament shall be an amount equal to the rent at which it is estimated the hereditament might reasonably be expected to let from year to year if the tenant undertook to pay all usual tenant's rates and taxes and to bear the cost of the repairs and insurance and the other expenses, if any, necessary to maintain the hereditament in a state to command that rent.

It is not in dispute that where a proposal is made at any given date during a rating year, the net annual value under section 19 (3) must be determined by estimating the rent at

G which the hereditament might be reasonably be expected to let on a yearly basis at the proposed date, and that the annual value so determined must be treated as having effect for the current rating year. For example, if the proposed date is February 1 1975, rent is estimated on a yearly basis for a period commencing on that date and the rent so estimated is applied to the rating year, that is from April 1 1974 to March 31 1975. This process results in an obvious anomaly where the lettable value is falling steeply, but section 19 (3) is mandatory in its terms, and I see no escape from its application where the circumstances admit of the estimation of a yearly rent on the proposed date. There may be a variety
H of methods by which the yearly rent of special types of hereditaments can properly be estimated in accordance with section 19 (3), but I do not think it is legitimate to redress an anomaly by substituting a different method of ascertaining rateable value for any given year in place of the statutory method. The Court of Appeal so decided in relation to a quarry in the case of *Farnham Flint Gravel & Sand Co Ltd v Farnham Union* [1901] 1 KB 272. I think that was a correct decision on the facts of the case, and it seems to me to be binding on this court. I will refer to this case again. With great deference to the Master of the Rolls, I wish to dissociate myself from his observations in connection with *Farnham*.
J The position is however different where for some reason it is impossible to estimate the rent of a hereditament on a yearly basis at the relevant proposal date. In those circumstances the machinery under section 19 (3) will not work, and it is necessary to devise other machinery in order to determine fairly the rateable value of the hereditament for the relevant year.

Turning to the present case, the hereditament is a gravel pit described as "Sand and gravel pit and premises, Mytchett Pit, Mytchett." As I understand the statement of facts, the respondent company acquired about 60 acres of land at
K Mytchett in 1969, but part only of this site constitutes the pit. The site also comprises residential development, which is presumably separately rated, and I think some waste land. At any rate, I proceed upon the footing, as I understand has been accepted throughout, that the relevant hereditament consists of the pit and nothing else. That means the parcel of land constituting the pit. The pit was in operation, either literally or for all practical purposes, throughout the rating year April 1 1970 to March 31 1971. I put it that way because work on the construction of weighbridge and offices began in March 1970 and extraction of gravel began on April 6 1970. On December 21 1970 the valuation officer
L made a proposal for alteration of the valuation list by including this hereditament. The effect of this proposal was that under section 19 (3) the value of the hereditament for the rating year April 1 1970 to March 31 1971 fell to be determined by reference to the rent at which the pit might be expected to let on a yearly basis commencing on December 25 1970. But here arises the difficulty. On December 25 1970 the pit was well on the way to being worked-out, its estimated life being less than a year. In the event, the pit was closed on September 30 1971. This being the position, the company contends that the rateable value for the year April 1 1970 to March 31 1971 should be determined, in
M accordance with section 19 (3), by estimating the rent which a tenant would pay for a year's possession of the parcel of land constituting the pit as it stood on December 25 1970. The valuation officer contends that a determination in literal compliance with section 19 (3) is inappropriate and that the rateable value should be determined on the basis of extraction rate and hypothetical royalty over the year from April 1 1970 to March 31 1971. The company's basis produced a figure of £2,920, that of the valuation officer a figure of £16,667. The precise method by which these figures were reached is not in issue. The company's basis thus provides virtual immunity from rates in respect of the profitable

A year's working, while the valuation officer's basis fairly reflects the year's working.

If in the circumstances prevailing on December 25 1970 it were right to treat the pit as a hereditament which would continue to exist for a whole year, I should see no answer to the company's contention, however anomalous the result. The mandatory direction in section 19 (3) would be applicable, and moreover the case would be covered by the decision in the *Farnham* case. But in fact on December 25 1970 the pit had an estimated life of much less than a year. As soon as the gravel should be worked-out, the parcel of land would of course remain in existence as a disused pit, but the B rateable hereditament consisting of the pit would cease to exist. It seems to me to follow that section 19 (3) cannot be applied literally, for there would be no hereditament capable of forming the subject-matter of a yearly tenancy. It then becomes necessary to devise other machinery for determining the rateable value of the pit for the year April 1 1970 to March 31 1971; and I see no reason why one should not adopt that advanced by the valuation officer, which produces a fair result, and is, as I understand it, in accordance with established practice. No other alternative was canvassed. The *Farnham* case is distinguishable on the ground that the C rate was levied upon an area of 3½ acres, only 1 acre of which was currently in use as a quarry. But the remainder was in use for storage purposes. The entire area was treated as the relevant hereditament, no suggestion apparently being made that this hereditament could not continue to exist throughout the year commencing on the proposal date. I D suspect that a gravel pit is normally treated as a separate hereditament from adjoining land in the same ownership, so that this question does not normally arise.

It only remains to consider the position for the rating year April 1 1971 to March 31 1972. In this year the proposal date was May 27 1971. At April 1 1971 the estimated quantity of gravel remaining unworked was 30,000 cu yds. At the proposal date the pit was almost worked-out, and it will be remembered that it was closed on September 30 1971. The company contended that the rateable value for the year April 1 1971 to March 31 1972 should be determined by E reference to the gravel actually unworked. The valuation officer contended that the value should be determined by applying the extraction and royalty basis over a twelve-month period and subdividing his resulting figure so as to correspond with the number of days in which the gravel was actually worked. The company's basis produced a figure of £1,750, that of the valuation officer a figure of £3,660, less in each case a reduction of a quarter under the Mines and Quarries (Valuation) Order 1971. The valuation officer's formula corresponds to the general scheme laid down by section 18 of the General Rate Act 1967 where a hereditament is occupied for part only of a year (cf *Ryde on Rating*, F 12th ed, p 750), and I think one should treat it as appropriate in the present case.

The appeal was allowed with costs above and below. Leave to appeal to the House of Lords was refused.

REAL PROPERTY AND CONVEYANCING

A **CHANCERY DIVISION**

June 21 1974

(Before Mr Justice WALTON)

NIELSON-JONES v FEDDEN AND OTHERS

Estates Gazette January 4 1975

(1974) 233 EG 49

Memorandum signed by husband and wife empowering husband to sell jointly-owned matrimonial home held to amount neither to an assignment to him of her beneficial interest nor to a severance of the beneficial joint tenancy—Remarks of importance concerning law and practice of severance generally

B This was a claim by Mrs Angela Nielson-Jones, formerly Todd, against Mr Edgar Warwick Fedden, co-trustee with her of a fund representing the proceeds of sale of the Old Rectory, Kirby Wiske, Thirsk, Yorkshire, and the three executors of her late husband, Roger Alexander Todd, for a declaration that she was absolutely entitled to the fund in question.

Mr R Scott (instructed by Williams & James, agents for Barber, Robinson & Co, of Harrogate) appeared for the plaintiff; Mr C Aldous (instructed by Kingsley, Napley & Co) for the first defendant; and Mr M Essayan (instructed by Slaughter & May) for the second, third and fourth defendants.

Giving a reserved judgment, WALTON J said that in 1971 Mr C and Mrs Todd were separated and the marriage was heading for dissolution. But it was a civilised situation, and there was no question of parsimony on the part of Mr Todd being responsible for the breakdown; he had given his wife £20,000 in securities and placed the matrimonial home, the Old Rectory, Kirby Wiske, in joint names, beneficially as well as legally, although he had provided all of the purchase price. Mr Todd was then still living in the house and it was obvious that he would have to find a smaller home. He discussed the question of sale with his wife, and the result was a memorandum, signed by both, dated August 31 1971:

> "[The husband] to use his entire discretion and freewill to either D sell the Old Rectory, Kirby Wiske, and employ the proceeds realised to his new home, if it so is decided to sell, in order to provide a home of substance for his children when visiting to stay and himself to live in and, perhaps, to work from if it be not the Old Rectory, Kirby Wiske. Not applicable to domestic contents."

The first two questions to be considered, in the light of the family background, were (i) was that memorandum an assignment by Mrs Todd to her husband of her beneficial interest as a beneficial joint tenant, (ii) alternatively, did the memorandum amount to a severance of her interest as a beneficial joint tenant? Whatever else was doubtful, it was clear that the memorandum was intended to give Mr Todd *carte blanche* as to E selling the property, and he duly exercised the authority so given him by entering into an agreement for sale of December 15 1971 at a price of £20,000. Formally, Mrs Todd was a party to the agreement, but there was no doubt that it was negotiated entirely by or on behalf of Mr Todd alone.

Whatever the parties might have intended, and whatever the true construction of the memorandum, read in the light of all admissible surrounding evidence, might be, its precise meaning and effect were far from clear to the parties' legal advisers, and so, from about mid-September 1971 onwards, the parties' legal advisers were in negotiation about the financial affairs of the parties. So far as Mrs Todd was concerned, she had made F it quite clear that she wanted her financial position entirely separated from that of Mr Todd, and it seemed plain that he reciprocated that approach to the overall position. Looked at in that light, both of them had made it perfectly plain that they were looking forward to an ending of the joint tenancy in the proceeds of sale of the property. However, there were without prejudice negotiations being carried on between them all the time, which negotiations it was admitted never resulted in any actual agreement. During the course of those negotiations, Mr Todd required some additional finance for removal expenses from the Old Rectory, and by agreement between himself and Mrs Todd, they each took £200 out of the deposit paid by the G purchaser of the Old Rectory. It had further been stated that out of this deposit there were also paid two small bills, one of Mr Todd's and one of Mrs Todd's, both for work at the Old Rectory, by agreement of the parties. The third question arose out of the negotiations and the attitudes taken up by the parties in the course of those negotiations, coupled with the payments out of the deposit. Did they, in any shape or form, have the effect of severing the beneficial joint tenancy, if the memorandum did not have any effect beyond giving Mr Todd authority to sell on behalf of his wife and himself?

To complete the history, he (his Lordship) must add that Mr Todd died suddenly on March 21 1972. Mrs Todd appointed H the first defendant, Mr Fedden, as a co-trustee with her in order to enable the sale into which she and Mr Todd had entered to be completed, and completed in due course it was. The present dispute, accordingly, related to the purchase-money, or rather, having regard to the payments out, to the balance of the purchase-money. Probate of Mr Todd's will was on October 19 1972 granted to the second, third and fourth defendants. Under that will the residue of Mr Todd's estate went on discretionary trusts for the benefit of his four children, now aged between fourteen and eight. Mrs Todd remarried on March 24 1973 and became Mrs Nielson-Jones, but it would be convenient to continue to refer to her as Mrs. Todd.

On the first question, there was no dispute as to the law. It J depended simply upon the intention of the parties, or, rather, since this would be an act of pure bounty on her part, on the intention of Mrs Todd. "The language is immaterial if the meaning is plain" (*per* Lord Macnaghten in *Wm Brandt's Sons & Co v Dunlop Rubber Co Ltd* [1905] AC 454 at 462). The vital words alleged to be dispositive out of the document were: "[The husband] to use his entire discretion . . . to . . . sell the Old Rectory . . . and employ the proceeds realised to his new home . . . in order to provide a home for . . . himself to live." With considerable hesitation, he (his Lordship) had come to the conclusion that these words were not sufficiently clear to constitute an assignment. Doubtless, if one assumed that the K parties had determined to separate out their financial affairs, one could step by step produce the answers that yes, the intention was that Mr Todd could have used as much as he required of the proceeds of sale, and yes, there would be no point in taking the conveyance of the new property into the joint names since all hope of the marriage continuing had come to an end. But this was, to his (Walton J's) thinking, beside the point. It might have been the obvious thing to do, but it never was done; and in his opinion, all that the memorandum did, so far as this first point went, was to agree that the

A requisite amount of the proceeds of sale should be made available to Mr Todd for the purpose of purchasing a new house, but without any agreement as to the ownership of the new house.

The questions of severance involved in the other two questions must, to some extent, be dealt with together. Persons could be joint tenants only if the three unities of possession, interest, and title were present. It was customary, following *Blackstone*, to add the unity of time to the commencement of title, but there were exceptions to this requirement in the case of conveyances to uses or in the case of wills (see *per* Lord Hatherly, then Page-Wood V-C, in *Kenworthy v Ward* (1853) B 11 Hare 196). In order to effect a severance before the 1925 legislation came into force, it was necessary to destroy one of the three unities. Section 36 (2) of the Law of Property Act 1925 was a rather elliptic piece of legislation, but since it was not in dispute that the methods of severance of a joint tenancy in personal estate before 1926 were precisely the same as the methods of severance of a joint tenancy in real estate (see *Williams on Personal Property*, 18th ed (1926) p 524), the final effect of this subsection was merely to add another method, the method of notice in writing or other acts effectual in equity, to the ways in which the severance of a joint tenancy in real estate might be effected. Why this highly convenient method of severance was not also extended to personal estate, he C (his Lordship) was at a loss to understand. This disparity led to a highly ingenious argument on the part of Mr. Essayan to the effect that it would be so highly inconvenient if this really were the law—because presumably once the real property had been sold in exercise of the trust for sale it became converted into personal property—that one must read section 36 (2) as really declaratory of the law, so that the same methods were available in all cases. From there, counsel went on to submit that severance in general was much easier than had previously been thought. While he (Walton J) was unable to accept counsel's analysis, the point raised was a very valid one: severance by notice was possible only where a legal estate, not D being settled land, was vested in joint tenants beneficially. Once there had been a sale, that condition no longer applied, so that severance after sale apparently could not be effected by notice. This was obviously very inconvenient.

If, as had already been held, the memorandum was not an assignment of Mrs Todd's interest in the proceeds of sale of the house to her husband, could it nevertheless be read as a severance of their joint beneficial interests? Was it an agreement to the effect that each of them thereafter was to be solely entitled to his and her respective one-half share in such proceeds? With the best will in the world, he (his Lordship) found himself wholly unable to give the memorandum such E a construction. It was in fact rather easier to read the memorandum as an assignment than as a severance, so that as he could not give it the first construction, almost *a fortiori* he could not give it the second. It appeared to him that the document dealt solely with the use by Mr Todd of the whole of the proceeds of sale, and that with regard to ownership, use was wholly ambiguous: hence, it could not be implied from the fact that Mr Todd was to have the use of the whole of the money either that title thereto was assigned to him or that he was entitled to have his own half absolutely, and Mrs Todd her own half share absolutely.

There remained the third point: was it possible, either from F the correspondence alone, more particularly from the determination therein manifested by both parties that their respective financial affairs were going to be kept separate, whether or not coupled with the fact that they were both negotiating as to what precise share of the proceeds of sale each should take, and whether or not coupled with the actual distributions out of the deposit paid by the purchaser of the Old Rectory, to say, in accordance with the established authorities, that there was a sufficient course of conduct by Mr and Mrs Todd as to lead to the implication either (i) of an agreement to sever, and hence a severance, or (ii) of a declaration by either of them of an intention to sever, which declaration, submitted Mr Essayan, was also an established method of severance? As to (i), he (Walton J) thought that when parties were negotiating G to reach an agreement, and never did reach any final agreement, it was quite impossible to say that they had reached any agreement at all, certainly not an agreement to sever, merely because they had, in the course of those negotiations, reached an interim agreement for the distribution of comparatively small sums of money. As to (ii), he assumed in favour of Mr Essayan that the correspondence did indeed disclose an unequivocal declaration by Mr Todd to the effect that he wished to sever the joint tenancy so as to make himself master of a one-half share of the net proceeds of sale of the property. The question then was, could such a declaration—a unilateral declaration—ever be effective to sever a beneficial joint tenancy? H It appeared to him (his Lordship) that, in principle, there was no conceivable ground for saying that it could. So far as he could see, such a mere unilateral declaration did not in any way shatter any one of the essential unities. Moreover, if it did, it would appear that a wholly unconscionable amount of time and trouble had been wasted by conveyancers of old in framing elaborate assignments for the purpose of effecting a severance, when all that was required was a simple declaration.

The question was not untouched by authority. In *Moyse v Gyles* (1700) Prec Ch 124 one joint tenant made a deed of gift of his moiety to his wife, with the expressed intention of severing the joint tenancy, in order to make provision for her. J As the deed was made in favour of the wife, it was void at law; and the deed being made without consideration, it was held that equity could not relieve. Again in *Partriche v Powlet* (1740) 2 Atk 54 it was quite clear that Lord Hardwicke LC regarded severance as only effected (if one left out of account the acquisition of a mergeable interest) by either (i) an agreement to that effect, or (ii) actual alienation. Down to *Re Wilks: Child v Bulmer* [1891] 3 Ch 59 it was fair to say that the whole current of authority was against severance by means of such declaration as was now envisaged. With great respect, he (Walton J) felt that in *Hawkesley v May* [1956] 1 QB 304 Havers J had, in deciding as he did that a letter written by the plaintiff's sister K was a sufficient act on her part to constitute a severance of a joint tenancy, entirely misapprehended the judgment of Lord Hatherly (then Page Wood V-C) in *Williams v Hensman* (1861) 1 John & H 546. The first method of severance of which Lord Hatherly was talking was actual alienation, or something equivalent thereto. Havers J appeared to have been very poorly assisted by counsel on this point, for they did not cite any of the relevant authorities to him, so far as the report showed. It was also very difficult to understand the reasoning on which his conclusion was based, in that having regard to the decision in *Walmsley v Foxhall* (1870) 40 LJ Ch 28 which he himself quoted, it appeared that as soon as any income became payable L to joint tenants of corpus, there was an automatic severance as regarded such income. Hence, how a mere direction to trustees to pay such already automatically-severed income into a separate bank account could conceivably relate to a severance of corpus, he (his Lordship) failed to understand. Of course, this was, so far as the judge was concerned, a merely minor matter, because he immediately proceeded to put the severance on an entirely irreproachable footing, namely the payment-out to the sister of her share of the funds in September 1942, which obviously smashed the unity of possession.

In *Re Draper's Conveyance* [1969] 1 Ch 486, Plowman J unfortunately applied the dictum of Havers J to a case between M husband and wife. In those proceedings the wife issued a summons under section 17 of the Married Women's Property Act 1882, asking for a determination of all questions between herself and her husband in respect of the interest of each of them in the property, which was of course the former matrimonial home, and an order that the property be sold and the proceeds distributed in accordance with their respective interests. Plowman J held that the summons, coupled with the wife's affidavit, operated in equity as a severance. Once again, none of the relevant cases were cited to the judge; *Re Wilks*, in particular, should have been drawn to his attention. Since he (Walton J) had already concluded that Havers J's dictum

A was wholly unwarranted, he regretted he could place no greater reliance upon it when blandly repeated in *Re Draper's Conveyance*. In the result, he held that Mrs Todd, now Mrs Nielson-Jones, had become, by survivorship, beneficially entitled to the whole of the net proceeds of sale of the former matrimonial home, the Old Rectory, Kirby Wiske, save to the extent to which such proceeds had already been distributed. He answered question 1 posed by the originating summons accordingly.

CHANCERY DIVISION

B December 10 1973

(Before Mr Justice WALTON)

SAMUELS (W A)'s TRUSTEE v SAMUELS

Estates Gazette January 11 1975

(1973) 233 EG 149

Wife claims to have improved husband's house (a) before his desertion (b) between desertion and bankruptcy, and (c) thereafter—Claim under (a) based on wife's work in husband's business—Too remote—Claim under (b) based on payments made (in effect) out of maintenance received from husband,
C **valid to the extent of £227—No claim under (c)—Section 37, Matrimonial Proceedings and Property Act 1970 applies only as between husband and wife, not as between one spouse and the other's assignee in bankruptcy**

By this motion the trustee of the property of Mr Wesley Altamont Samuels, a bankrupt, sought a declaration that the freehold premises known as 60 The Chase, Clapham, London, were the property of the bankrupt now vested in the trustee pursuant to section 38 of the Bankruptcy Act 1914; alternatively, a declaration as to the respective interests of the trustee and the respondent, Mrs Rosanna Samuels, in the premises.

Mr E C Evans-Lombe (instructed by Booth & Blackwell)
D appeared for the trustee, and Mr R N Thomas (instructed by Kanter, Jules & Co) represented the respondent.

Giving judgment, WALTON J said that the bankrupt married the respondent on October 6 1966. At that time he was running a small wholesale wine-distributing business in which Mrs Samuels worked hard for no direct remuneration. Between them, however, they took some £3,000 per annum out of the business for their living and other expenses. The business was ultimately wholly unsuccessful, and Mr Samuels was made bankrupt upon a creditor's petition, the act of bankruptcy having been committed on June 13 1972. The receiving order was made on September 28 1972, and the actual adjudication of Mr Samuels
E as a bankrupt was made on the 30th of that month. This was a case where hard evidence was in a good many respects lacking, and that was particularly so in relation to the conduct of the business, in particular as to the precise point of time at which it became insolvent. The total deficiency was £4,740.20, and this was probably building up over a considerable period of time; ultimately he (his Lordship) was of the opinion that at all material times the business, after making provision for the £3,000 per annum drawn out by Mr and Mrs Samuels, was insolvent.

The present proceedings concerned a property, 60 The Chase, which was acquired by Mr Samuels, together with a Mr Newell,
F on March 25 1970 for £7,000. The position was that the vendor of the house had received an intimation from the local authority that it was essential that he should carry out certain structural works which (as ultimately effected) cost in all something of the order of £2,000. He was unwilling to do this. Mr Newell was a tenant of the property, and he joined with Mr Samuels in purchasing the property from the vendor, the purchase price being left by the vendor on a 100 per cent mortgage of the property. Mr and Mrs Samuels already had a matrimonial home together in Kilburn, but Mr Samuels intended that 60 The Chase should be a, if not the, matrimonial home as well. At some stage, which it was not possible to pinpoint with any certainty, after the property was purchased, the necessary structural work, which was in relation to a wall, was duly carried out. It involved G a certain amount of professional builder's work, stated to be to the extent of £1,200, although no bill for this (or any other) sum was produced. Of this amount, it was stated that only £700 had actually been paid, and this sum was paid by cheque drawn on Mr Samuels' account with the National Westminster Bank. Besides this professional work, materials costing in all some £700 were used. The most expensive items were some girders costing £300, but these were obtained by exchanging an old truck belonging to the business for them. The remaining £400 came out of the business. On March 16 1971 Mr Newell transferred his interest in the property to Mr Samuels for no consideration apart from an indemnity against his liability under the mortgage H in favour of the vendor. The reason for this transaction was that Mr Samuels required to borrow some additional money from the bank and was proposing to give a second charge on the property to secure it. Mr Newell was unwilling to burden himself with any further liability and so this transaction was carried out. Mr Newell remained in the premises thereafter in the character of a tenant.

In October 1971 Mr Samuels deserted Mrs Samuels. She naturally stopped working in the business, and inquired from him upon what she was expected to live. He told her that she could have the rents from 60 The Chase, amounting at that time to about £7 a week. In the circumstances, having regard to Mr Samuels' *prima facie* duty to maintain his wife, he (his J Lordship) thought that this should be regarded as maintenance provided by him for her. Mrs Samuels, who was an energetic and resourceful woman, proceeded to carry out repairs and improvements, redecorations and refurnishings of the property, using probably some of the rents so abandoned to her, borrowings, and her own earnings. As a result of her efforts she had undoubtedly increased the value of the premises, and by the time the trustee in bankruptcy started to collect the rents in October 1972 she had doubtless improved the rental value of the premises. In proceedings commenced earlier in 1973 in the Family Division of the High Court she claimed, under section 17 of the Married Women's Property Act 1882 and section 4 K of the Matrimonial Proceedings and Property Act 1970, a share in the premises. Those proceedings were stayed with a view to the decision of Mrs Samuels' claim in the Chancery jurisdiction, and the present proceedings were launched by a notice of motion of October 19 1973. Mr Thomas, for Mrs Samuels, had wisely concentrated solely upon, and placed his client's case solely on the basis of, section 37 of the 1970 Act, which provided:

> "It is hereby declared that where a husband or wife contributes in money or money's worth to the improvement of real or personal property in which or in the proceeds of sale of which either or both of them has or have a beneficial interest, the husband or wife so contributing shall, if the contribution is of a substantial nature and L subject to any agreement between them to the contrary express or implied, be treated as having then acquired by virtue of his or her contribution a share or an enlarged share, as the case may be, in that beneficial interest of such an extent as may have been then agreed or, in default of such agreement, as may seem in all the circumstances just to any court before which the question of the existence or extent of the beneficial interest of the husband or wife arises (whether in proceedings between them or in any other proceedings)."

The question therefore arose as to what contribution to the improvement—not repair—of the property was made by Mrs Samuels. That question neatly divided into three: (a) in relation to the initial structural repairs; (b) in relation to the period M between October 1971 and June 13 1972; and (c) in relation to the period after June 13 1972. As regarded the initial structural alterations, part thereof was paid for by Mr Samuels' cheques drawn on his own bank account, which account was, as intended, later secured by a second mortgage on the property. Mrs Samuels maintained in the witness box that this overdraft facility was created, or intended to be created, for the benefit of the business, to enable it to import more wine from Italy. Therefore, argued Mr Thomas, since the only basis on which the overdraft was obtained was the business and the income it was producing, in which he submitted Mrs Samuels had an interest, the payment of this part of the cost amounted to a substantial indirect contribution by Mrs Samuels. And this was, he said, *a fortiori* with the remainder of the payments which were made, in cash or truck, from the business.

He (his Lordship) regretted that he was wholly unable to accept this chain of reasoning. Accepting for the moment that as the result of the decision in *Nixon v Nixon* [1969] 1 WLR 1676 if a wife worked full time in her husband's business for no wages she might acquire some interest in it, all that could mean was that, as between herself and her husband, if the business at the end of the day came to be sold or otherwise realised, or if surplus assets were withdrawn therefrom, she would be entitled to some proportion thereof. It could not possibly mean that if, far from being solvent, the business was insolvent, nevertheless she had an interest in moneys which should never have been withdrawn from it. Not, at any rate, unless she was claiming such an interest in the business as would render her also personally liable to its creditors. The situation would in substance be no different if Mr Samuels had withdrawn £1,000 from an insolvent business and put the money into, say, premium bonds. On his bankruptcy, nobody could possibly deny that that money ought to go to his trustee in bankruptcy to pay his creditors, and he (Walton J) doubted if anybody would even think of claiming some fraction thereof for Mrs Samuels. The fact that the money went into the house and not into premium bonds could make no difference. In addition, so far as the moneys which came out of the National Westminster Bank were concerned, although of course Mr Samuels used this account for his business, he was personally liable for the money. It was exactly and precisely the same as if he had borrowed the money from a bank with whom he did not keep the business account. The fact that the facility might have been arranged in connection with the affairs of the business, since it was not in any way charged upon the business (quite apart from the fact that there was really nothing there to charge), was *nihil ad rem*. He (his Lordship) therefore rejected *in toto* any suggestion that Mrs Samuels made any contribution to the initial structural repairs. She plainly did not.

As regarded the second head of claim, her claim was quantified in a bundle of receipts and other documents and certain cheques. Mr Evans-Lombe, for the trustee in bankruptcy, submitted that Mrs Samuels might well have made all such payments out of the rents she received and that therefore the expenditure by her came out of the property itself and not from her. He (his Lordship) did not accept this submission. He thought that the rents she received were received by her by way of maintenance, and that it was impossible to ignore these payments by her on this ground. The court was left, so far as this item by itself was concerned, with expenditure of some £227 by Mrs Samuels towards the improvement—as distinct from the decorative repair—of a house worth over £9,000. Was this a sum "of a substantial nature" within the meaning of section 37? With some hesitation, he (Walton J) thought it was, and he was fortified in this conclusion by the fact that Mr Evans-Lombe certainly did not proffer any argument to the contrary. It was, however, clearly extremely marginal.

There remained the third and last category, namely those items relating to a date after June 13 1972. The crucial question which arose was, clearly, had that date any significance anyway? Mr Evans-Lombe put the matter thus. Although undoubtedly the provisions of section 37 must be given effect to by any court before whom the matter of its application was raised (note the concluding words "whether in proceedings between them or in any other proceedings"), it related to an accretion or decretion of interests as between husband and wife, and was no longer applicable where the contest was not one as between husband and wife. Having regard to the provisions of the Bankruptcy Act 1914, as from June 13 1972 the husband's beneficial interest in the property, whatever it then was, vested in the trustee in bankruptcy, and as from that date section 37 became inapplicable. It was inapplicable, because the words "in . . . which either or both of them has or have a beneficial interest" were no longer satisfied. Of course Mrs Samuels had some beneficial interest in the property, but *vis à vis* the trustee in bankruptcy she had it as Mrs Samuels, not as the wife of anybody who had an interest in the property. Put more shortly and elegantly, that section applied only as between husband and wife, and did not apply between a wife and a third party, even if that third party were an assignee of the husband's. Mr Thomas, for Mrs Samuels, argued, on the other hand, simply that there was nothing in the section which became inapplicable on the assignment of Mr Samuels' interest to the trustee; Mrs Samuels, the wife, then had a beneficial interest in the proceeds of sale of the property as the result of her contributions under the second head of claim already considered, so that the extent of her final share in such proceeds was to be determined by her post-June-13-1972 contributions as well as her pre-June-13-1972 contributions.

Of these two arguments, he (his Lordship) preferred that of Mr Evans-Lombe. In his view section 37 was applicable as between husband and wife only, and conferred no rights to an increased share of the property as between a wife and her husband's trustee in bankruptcy by reason of the matters therein referred to. The section did not in his view effect any general alteration in the law. Assume, for example, that A and B (not husband and wife) were joint tenants at law of a property which they held upon trust for sale with the net proceeds belonging to the two of them as tenants in common. If B improved the property, he did not obtain a greater interest in the proceeds of sale, but on the usual equitable principles he would be recompensed out of the proceeds of sale for any improvements effected by him to the extent to which they had enhanced the sale price. He (his Lordship) could not think that section 37 was intended to produce the result that if the improvements had been effected by B's wife, and not by B himself, any different result would follow as between A and B. But, of course, as between B and B's wife, section 37 would apply.

On these conclusions and materials the next stage arose: what share of the property was it just to attribute to Mrs Samuels? It would, of course, at any rate on the peculiar facts of the present case, be much easier to answer that question if what was notionally distributable between the parties was the equity of redemption in the premises after discharge of the two mortgages charged on the property. But the language of the section prohibited that. It was the beneficial interest in a share of the proceeds of sale of the property which had to be fixed. And although of course, having regard to the width of the language of the section, this would by no means be a rigid determination, he (his Lordship) would think that in any simple case the best guide would still be furnished by comparing the respective financial contributions of both parties, not, as in the usual case, to the purchase price, but to what might best be described as the cost of the finished product. Before he came to work out that sum—as a guide, of course, and not as in any way a compulsory line of approach, having regard to the wide wording of section 37—he would first notice how narrow were the limits within which he was compelled under section 37 to proceed if his decision was to have any meaningful operation at all. He was told—there was no strict evidence about it—that the market price of the property was now thought to be about £15,000. If, therefore, the trustee in bankruptcy was to receive anything at all from the sale of the house, having regard to the fact that the total of the charges thereon now amounted (including interest) to some £12,000, he must receive over 4/5ths. This at once demonstrated that Mr Thomas's suggestion that Mrs Samuels should receive one-third was pitched far too high. Even if one divided the balance remaining between them equally—and it must be borne in mind that his contribution throughout was far, far greater than hers—she would only receive 1/10th. There was certainly no obvious reason why she should receive more than this.

Approaching the matter via the strict amount of contributions to the finished product, Mr Samuels contributed the original £7,000 plus £1,900 for the structural repairs; Mrs Samuels contributed £227 to the finished product price of £9,127. Very roughly, this would give Mrs Samuels a 1/30th share. On the whole, he (Walton J) thought justice would be done by giving her a 1/20th share. This would mean, in monetary terms, that she would get back something like £750, which was a three-fold return on her investment, and the trustee in bankruptcy would get about £2,250, which was again not far off three times the

A amount which had been taken out of the business for the purposes of the property and would not be discharged (via payment-off of the mortgages) in full. To the extent to which Mrs Samuels' expenditure after June 13 1972 had added to the sale price of the property, the trustee must recompense her therefor. And strictly, if the parties could not agree upon a figure, there should be an inquiry as to that extent, and this must rank as a charge on the net proceeds of sale of the property after the discharge of the two mortgages and before the residue was split in the proportions indicated.

Mr Evans-Lombe asked his Lordship to grant an order in solemn form directing a sale, the conduct of the sale to be given to the trustee and the respondent to do all such things as were necessary B to procure such sale with vacant possession. As to the time within which possession should be delivered, counsel continued, the trustee had always been prepared to give a reasonable period of time, so that Mrs Samuels might be given an opportunity to purchase the bankrupt's interest. In his submission the trustee in bankruptcy stepped into the bankrupt's interest as trustee of the property, though this area of the law was not at all clear. The wife would be entitled in this case, the court having found that she had a share, to insist upon the appointment of a fresh trustee to represent, in a sense, her interests.

His Lordship said that the bankrupt would have to be joined as a respondent, and he could not make an order now. He understood Mrs Samuels was anxious to acquire the property if she possibly C could, and the practical course would be to stand the motion over. By the time the case came back the trustee and Mrs Samuels might have agreed as to what was to happen.

The motion was accordingly adjourned. Mr Evans-Lombe said that there might be no need to trouble the court further. His Lordship made no order as to costs.

CHANCERY DIVISION

December 21 1973

(Before Mr Justice PLOWMAN)

SMITH v MORRISON AND ANOTHER;

D SMITH v CHIEF LAND REGISTRAR AND OTHERS

Estates Gazette January 11 1975

(1973) 233 EG 153

Competing purchasers—First purchaser's application to lodge cautions held up by difficulties of identification, second purchaser obtains a clear official search and lodges transfer for registration within priority period—Second purchaser entitled to do this despite his knowledge of first purchaser's claim to have a binding contract—"Honest doubt" as to validity of claim held within the ambit of "good faith"—Further points by first purchaser on practice at Land Registry unsuccessful

E These were claims by Mr William Henry Smith, (1) against Mr George Edward Morrison, of Longland Farm, Chertsey Lane, Staines, for specific performance of a contract dated September 20 1972 for sale of the farm together with a bungalow, 194a Chertsey Lane; and (2) against the chief land registrar, for a declaration that cautions lodged by the plaintiff in respect of the titles comprised in Longland Farm had priority over an application by Frederick Coyle & Co Ltd, property developers, to be registered as proprietors of the land in question, and for an order restraining the registrar from registering any disposition of the land otherwise than pursuant to the declaration. Coyles F were added as defendants to the first and second actions, and Mr Morrison as a defendant to the second action.

Mr M Albery QC and Mr L H Hoffmann (instructed by Lucien Fior & Co) appeared for Mr Smith; Mr D Nicholls QC (instructed by Rooks & Co) for Coyles; and Mr P Howell (instructed by the Treasury Solicitor) for the registrar. Mr Morrison took no part in the proceedings.

Giving judgment, PLOWMAN J said that in 1972, the defendant, Mr Morrison, was the owner of Longland Farm. It was about five acres, and he ran a pig farm there with about 1,000 pigs. He also owned a nearby bungalow, 194a Chertsey Lane. The titles of these properties (with one exception, which could be ignored) were registered under the Land Registration Act 1925.

G On September 20 1972 Mr Morrison contracted to sell the farm and bungalow to the plaintiff, Mr Smith. The price of Longland Farm was £34,000, and the bungalow was sold for £7,000, making £41,000 in all. Two months later, on November 23 1972, Mr Morrison contracted to sell the same properties to the second defendants, Coyles, for £42,000. The proceedings raised the question which of these contracts was entitled to priority.

The relevant facts were as follows. Coyles were a private company incorporated in 1927, whose business was that of builders and contractors. They were also property developers, and the director in charge of their development activities was a Mr Andrews. In 1971 Coyles were interested in acquiring or obtaining H options over a number of different sites in the vicinity of Longland Farm with a view to developing the whole as a building estate if they could get planning permission. As part of the scheme, the acquisition of Longland Farm was necessary, or at least desirable, for the purpose of providing access to a part of the area known as the Wedge. On July 17 1972, Mr Andrews wrote to Mr Morrison confirming that he (Mr Andrews) had offered Mr Morrison £33,000 for Longland Farm and £7,000 for the bungalow, making £40,000 in all, subject to contract, and that Mr Morrison had accepted that offer. But very shortly afterwards the chances of obtaining planning permission were poor, and on July 27 Mr Andrews wrote to Mr Morrison withdrawing his offer. In September, however, the prospects of planning perJ mission improved, and early in that month Mr Andrews reopened negotiations with Mr Morrison. But on September 20, Mr Smith, who had heard that the farm was for sale, went to see Mr Morrison and entered into an oral contract with him to purchase the farm and bungalow for £41,000. This was to be apportioned as £34,000 for the farm and £7,000 for the bungalow. Mr Morrison had entered an appearance to the present action, but he delivered no defence and was not represented; the position on the pleadings, therefore, was that Mr Morrison admitted the contract on which Mr Smith sued.

Immediately after entering into the oral agreement, Mr Smith, who was a farmer, went to see a friend in the property business, K obtained from him two prints of a form of contract and took these back to Mr Morrison. They were filled in partly by Mr Morrison and partly by Mr Smith and handed to Mr Smith in exchange for his cheque for £4,100 as a 10 per cent deposit. The first, relating to Longland Farm, was headed with the vendor's address:

"George Edward Morrison, Longland Farm, Chertsey Lane, Staines, Middlesex. I (vendor) hereby acknowledge receipt of the sum of (£3,400) being the deposit and part payment of the sum of (£34,000) for which sum I have agreed to sell my freehold property known as Longland Farm, Chertsey Lane, Staines, Middlesex, app. 5 to 5½ acres to you with vacant possession at completion, with completion to take place (1/4/73) weeks from now"

L —and the figures "1/4/73" were repeated. Then:

"Signed G. E. Morrison; date 20/9/72"

—and then the purchaser's name and address were set out:

"W. H. Smith, Halebourne Farm, Bagshot Road, Chobham, Surrey."

The other document, relating to 194a Chertsey Lane, was in similar form except that the amount of the deposit acknowledged was £700 as a deposit in part payment of the sum of £7,000 and the date for completion was not filled in. The date for completion in the memorandum relating to Longland Farm, M "1/4/73," was suggested by Mr Morrison and agreed by Mr Smith, because Mr Morrison wanted to find another farm to which to move his pigs, and needed time. A month or so later, Mr Andrews was again in touch with Mr Morrison, to whom he wrote on October 26 1972. He said that he understood that Mr Morrison had had several offers from developers, but assumed he would not sign any contract until he had found his new farm. He referred to changes in the possibility of obtaining planning permission and suggested that Mr Morrison should contact him before signing any contract, as he (Mr Andrews) might well be able to improve on the offers. Shortly afterwards they met, and Mr Morrison said that he had found a suitable farm for his pigs and was now in a position to sell. A price of

A £42,000 was negotiated, subject to contract. No mention was made of Mr Morrison's agreement with Mr Smith.

Two or three weeks later, there was a telephone conversation between Mr Smith and Mr Andrews. There was a conflict of evidence as to who telephoned whom and what was said, but it was common ground that Mr Smith told Mr Andrews that he had purchased the property. Mr Andrews said in evidence that he did not believe Mr Smith. However, Mr Andrews went to see Mr Morrison the next day and told him about the telephone call. Mr Morrison assured him that he had not agreed to any deal with Mr Smith. Mr Andrews then got in touch with Coyles's solicitors. On November 16, Mr Morrison wrote to
B Mr Smith a letter which was never received. It said:

"As you were unable to contact me yesterday after my call to see you, I am sorry to have to tell you that the proposed purchase of the above property is now cancelled. This decision was brought about by the vendors of the property that I intended to buy. Your deposits will be refunded in due course."

On November 23 1972 Mr Morrison and Coyles signed a formal contract for the sale and purchase of the farm and bungalow for the sum of £42,000, of which £6,250 was apportioned to the bungalow. The date fixed for completion was December 11 1972, that also being the date on which Mr Morrison was due to complete the purchase of the new farm.
C Exchange of contracts took place at the offices at Staines of Dale & Newbery, Mr Morrison's solicitors, in the presence of Mr Hall, a legal executive of the firm, Mr Morrison, Mr Cooke, a partner in Rooks & Co, Coyles's solicitors, and Mr Andrews, representing Coyles. At that meeting, Mr Morrison stated that he had not entered into any other contract, but had received a cheque as a deposit for which he had given a receipt and the deposit was to be returned.

Mr Cooke then went on holiday, and in his absence the Coyles purchase was dealt with by Miss Foinette, an assistant solicitor, under the supervision of Mr Pritchard, who was not only a partner in Rooks & Co but also a director of Coyles and chairman of its parent company. The two firms of solicitors then
D set about the ordinary conveyancing process appropriate to the sale of registered land. Meanwhile, on November 27, Dale & Newbery wrote to Mr Smith returning his deposit "as the matter is not proceeding." On December 1, Mr Smith's solicitors, Lucien Fior, returned the cheque for the deposit, asserting that "your client has signed a binding memorandum of sale to sell this property to our client," and calling for the abstract of title. On December 4 Dale & Newbery replied asking for a copy of the supposed memorandum, and on December 5 they applied to the chief land registrar for official searches. On December 6 Lucien Fior wrote enclosing a photocopy of the memorandum
E and observing that this had been "lodged at the Land Registry." On December 11 Mr Smith issued his writ claiming specific performance by Mr Morrison. By December 13 Rooks & Co had obtained clear searches in respect of all the titles except that of the bungalow (on which Mr Smith's contract had been noted), and on that day, knowing that Mr Smith had threatened proceedings but not knowing that he had issued a writ, they completed the sale of the farm, but not of the bungalow, to Coyles. On December 18 they lodged the transfer to Coyles at the Land Registry for registration. As a result of the present proceedings, however, Mr Morrison remained on the register as the registered proprietor of all the properties. He went out of possession on
F January 31 1973, and on the same day Coyles took possession and were still in possession.

Meanwhile what had happened at the Land Registry was that Lucien Fior, on behalf of Mr Smith, had lodged notice of his contract with Mr Morrison in terms which related only to title SY 246971, which was confined to the bungalow, and did not specify any other title numbers. As far as the bungalow was concerned, it was common ground that Mr Smith thus obtained a priority which he still retained. But difficulties then arose in relation to Longland Farm, which was comprised in six different titles. There was some confusion as to which part was comprised in which title, and as already noted, the title to one small piece of the land was not registered at all. Requisitions followed, and the matter was still under investigation at the registry when G Rooks & Co lodged their application for official searches on December 5. Not till December 12, the day before completion of the sale to Coyles, did Lucien Fior lodge an application for cautions against dealings with the land comprised in the various title numbers of Longland Farm. On January 1 1973 the official search priority period expired, and the registry, taking the view that Lucien Fior's applications for cautions had come too late to affect the priority conferred by Coyles's applications for official searches, cancelled Mr Smith's applications for cautions. On January 4, Mr Smith issued his originating summons against the chief land registrar. His case in his two actions was that Coyles were not purchasers in good faith; that their application H for registration as proprietors was not in order, and that consequently, on both or either of those grounds, they obtained no priority under the Land Registration (Official Searches) Rules 1969. Further or in the alternative, it was alleged that Coyles wrongfully induced Mr Morrison to break his contract with Mr Smith by transferring Longland Farm to Coyles instead of to Mr Smith, and an injunction was claimed restraining Coyles from taking any steps to procure its registration as proprietor and requiring it to withdraw its application for registration.

With regard to the question of good faith, counsel for Mr Smith submitted that a purchaser in good faith within the meaning of the rules must be an honest purchaser, and that a pur- J chaser who knew that his vendor had already sold to a third party, or who showed reckless indifference in the face of a *prima facie* valid claim by a third party to be a prior purchaser, was not honest. He submitted (and it was common ground) that the relevant date was the date of completion, and he argued that at completion Coyles acted with reckless indifference to Mr Smith's claim, being aware of its existence. He (his Lordship) approached the question of the meaning of "good faith" in the light of Lord Denning's general remarks in *Central Estates (Belgravia) Ltd v Woolgar* [1972] 1 QB 48 at 55, to the effect that the phrase was often used in statutes but rarely defined, and that the judges had to do the best they could to interpret K the legislature's will in regard to it. Having considered all the authorities cited by counsel on both sides, he (his Lordship) had come to the conclusion that a purchaser who acted honestly acted "in good faith" within the meaning of rule 2 of the 1969 rules, and on the facts he was satisfied that Mr Pritchard, whose decision it was to complete, was not dishonest. At the time he took his decision he had no ulterior motive but an honest doubt as to the validity of Mr Smith's claim; there was an issue whether the memoranda signed by Mr Morrison were complete when he signed them, and Mr Morrison was emphatic that he had never entered into any binding contract with Mr Smith. In all the circumstances, Mr Pritchard was entitled to take advan- L tage of Coyles's rights under the rules of 1969. The next question was whether in so doing he had wrongfully induced a breach of Mr Morrison's contract with Mr Smith. It was clear from the authorities that inducement involved some direct and deliberate interference with the execution of a contract. In his (Plowman J's) judgment, Mr Pritchard's state of mind at the time of, and in relation to, the transfer to Coyles was not such as to render Coyles liable to a claim for inducing breach of contract. As already stated, Mr Pritchard's state of mind was one of honest doubt, and that was not enough to bring him within the principles of liability established by the relevant decisions.

Mr Smith pleaded next that for the following reasons Coyles's M application for registration as proprietor was not "in order" within the meaning of rule 5 of the rules of 1969:

"(a) The transfer did not comply with rule 121 (1) of the Land Registration Rules 1925 in that it was not in Form 35.

"(b) The application did not comply with rule 259 of the Land Registration Rules 1925 in that it included neither a copy of the second defendant's memorandum and articles of association nor a certificate from the second defendant's solicitors that it had power to hold land.

"(c) The name of the transferee on the transfer was not the correct name of the second defendant."

As regarded point (a), in fact Form 19, which was the ordinary common form of transfer, was used instead of Form 35, but in

Morelle Ltd v Wakeling [1955] 2 QB 379 at 381 it was held that although a transfer to a company was not in the prescribed statutory form, it was not for that reason invalidated, having regard to the wide discretion given to the registrar by rules 74, 123 and 322 of the rules of 1925. It followed that the use of Form 19 instead of Form 35 did not prevent the application from being in order. Then as to point (b), rule 259 did not state that a copy of the memorandum and articles of association was to be lodged with the application, merely that it should be lodged. The Land Registry's practice was to requisition for it if it was not lodged, and the fact that a copy was not lodged at the same time as the application for registration did not in his (Plowman J's) judgment mean that the application to register the transfer was not in order. Lastly, as regarded point (c), what was said was that Coyles's application described them as " Frederick Coyle & Company " and not as " Frederick Coyle & Company Ltd." Rule 13 gave the registrar power to correct clerical errors, but that power was not available here, since the transfer was not yet registered. The practice of the Land Registry, however, was to return the transfer for rectification before it was registered. Again, he (his Lordship) was of opinion that the mere fact that it was necessary for the Land Registry to return a document for amendment before it could be registered did not necessarily involve the conclusion that the application was not " in order " in the first instance. In his judgment, the words " in order " could fairly, and should, be construed as " substantially in order " rather than " in perfect order," and the registrar had a discretion to accept applications for registration which were regarded as in order and to enter them in the day list subject to whatever requisitions were considered necessary. That being so, acceptance and entry must be taken as conclusive of the sufficiency of an application in point of form.

Mr Albery had also submitted that the registrar was wrong in cancelling Mr Smith's application for cautions without giving him a hearing. Counsel relied on section 55 (1) of the Land Registration Act 1925, but although that section talked about lodging a caution, the protection given by the Act to a cautioner (namely a right, not a priority, to be heard in opposition to an application to register a dealing) applied only when the caution was entered on the register. Mr Smith's application for cautions was never entered on the register and was cancelled by the registrar because as soon as Coyles lodged their application to register their transfer within the priority period, Mr Smith ceased in effect to have any cautionable interest. In these circumstances, the formal defect points failed. In the result, therefore, both actions failed, and must be dismissed, except that there would of course be an order for specific performance in respect of the sale of the bungalow.

His Lordship ordered the plaintiff to pay Coyles's costs, and Mr Morrison to pay Mr Smith's costs, save those owed to Coyles. A stay of execution was granted pending notice of appeal, notice to be lodged by January 18. An interlocutory injunction restraining dealings in the land was continued.

CHANCERY DIVISION

October 18 1974

(Before Mr Justice MEGARRY)

CRISP v MULLINGS

Estates Gazette February 1 1975

(1974) 233 EG 511

House bought by unmarried couple—Both names on transfer and mortgage, but deposit and outgoings paid by man alone—Woman held to obtain no beneficial interest, despite point as to payments by her for food and clothing—Man must however free her from mortgage liability

By this originating summons Mrs Joyce Violet Crisp sought an order that Mr William George Mullings, of 34 The Green, Morden, Surrey, should concur with her in sale of that property, and a further order for an inquiry as to the beneficial interest in the property vested in each party. The defendant counterclaimed a declaration that he was solely beneficially entitled, subject to a mortgage, and an order that the plaintiff transfer the property to him absolutely.

Mr A H Tibber (instructed by Montague Gardner & Howard) appeared for the plaintiff, and Mr D Gidley Scott (instructed by Lethbridges) represented the defendant.

Giving judgment, MEGARRY J said that the house was registered at the Land Registry with a good leasehold title. The lease was for 60 years to run at a ground rent of £8.50 a year. With slight intermissions, the plaintiff and the defendant cohabited as man and wife, without marrying, from May 1966 until the final rupture in November 1972. The plaintiff was a married woman until there was a divorce some two years ago. The defendant was a widower until he remarried in May 1973. The plaintiff and the defendant first cohabited for over three years at a council house in Thetford, Norfolk. There was then a break of some three months, after which the parties cohabited for a little over a year in a furnished flat in Morden. The landlord of the flat wanted to obtain possession of it, and the parties arranged to buy the house now in dispute. This purchase was assisted by a payment of £250 made by the landlord in order to get possession of the flat, and a loan of £300 by him, secured by a promissory note signed by the defendant and providing for repayment in the two years from November 16 1970. The contract for the purchase of the house was dated December 10 1970, and the plaintiff and the defendant were both parties to it under the description of " the purchaser." The Land Registry transfer was dated December 21 1970, and the plaintiff and the defendant were shown as being the transferees. The transfer was registered on January 14 1971. There was nothing in the contract, the transfer or the entry in the register which stated whether the parties took in equity as joint tenants or as tenants in common. The purchase-money was stated to be £6,250, and the balance of £5,700 required to make up the £550 provided by the landlord of the Morden flat to the requisite £6,250 was provided by a registered charge to a building society. In fact, as the completion statement showed, the vendor made a last-minute reduction of price by £100, but the transfer and land certificate ignored this. The charge to the building society was dated December 21 1970, and the plaintiff and the defendant were both parties to it as borrowers.

Throughout these documents the plaintiff was described as being the defendant's wife and as bearing his surname. It was common ground that the legal estate was vested in the parties as joint tenants, and the dispute was about the equitable interest. It was agreed by counsel that any question as to the quantum of the plaintiff's beneficial interest, if she had one, should be left for an inquiry in chambers or for future proceedings before him (his Lordship). The evidence was conflicting in a number of respects, but many of the conflicts were on matters which, however much they went to credit, either did not affect the issues the court had to decide or else had little bearing on them. The plaintiff, of course, started with the advantage of having the legal estate vested in her jointly with the defendant, though as she was not married to him she could gain no assistance from any presumption of advancement, nor could she rely on any material obligation of support. It was common ground that at all material times it was the defendant who had made all payments under the mortgage and had paid the ground rent, rates, insurance premiums and all gas, electricity and telephone bills. However, Mr Tibber contended that in certain respects the plaintiff had contributed directly to the cost of the house and that in other respects she had made substantial indirect contributions. The first claim to a direct contribution lay in the £550 paid by the defendant's former landlord. He (his Lordship) thought that on balance there was not enough to show that any of the £250 or the £300 (which was repaid by the defendant) was ever the plaintiff's money. Secondly, Mr Tibber pointed to the undoubted fact that both the plaintiff and the defendant had at all times been liable to the building society for the whole of the £5,700 and also for the other obligations of the mortgage. Counsel cited *Cowcher v Cowcher* [1972] 1 WLR 425 for the proposition that he who assumed liability under the mortgage provided the money lent on the mortgage.

A Mr Scott's answer was that whatever the position between the borrowers and the building society, what mattered was the position *inter se* between the borrowers. As already stated, it was common ground that the whole burden of making the payments under the mortgage had been assumed from the start by the defendant. In these circumstances, said counsel, in a dispute between the plaintiff and the defendant, it was quite unreal to regard the plaintiff as having provided half (or any other fraction) of the £5,700 that had been lent, for, as between the parties, she had never assumed, or sought to assume, any of the burdens. He (his Lordship) accepted this submission and held that the plaintiff had not directly provided any of the purchase price. That left Mr Tibber with his contentions as to indirect provision, B which depended in the main upon there having been a division of financial obligations in relation to the house. On any footing, Mr Tibber said, the plaintiff had relieved the defendant of payments for food and some clothing which otherwise the defendant would have had to bear, and this enabled him to make the mortgage repayments, which came to £45.60 a month. He (Megarry J) observed that the defendant did not seem to have been very fortunate in making his repayments, as £5,774 was now outstanding, and this was £74 more than the sum which was lent over 3½ years ago. The payments had thus not even succeeded in keeping pace with the interest due and had not so far paid off any of the capital indebtedness. On any footing, C the plaintiff, as the evidence stood, could not have contributed even indirectly towards discharging any of the mortgage. It was necessary for him (his Lordship) to remember that he was dealing with an unmarried couple and not with man and wife. Mr Scott very properly accepted on the authority of *Cooke v Head* [1972] 1 WLR 518 that this relationship was no bar to the acquisition of a beneficial interest where each had contributed to the purchase in one way or another. The question however, was whether there had been any such contributions; and in considering that, he (Megarry J) must bear in mind that a man owed an obligation to his wife to maintain her and provide a roof over her head which he did not owe to a woman D who was not his wife. In essence, he (his Lordship) could not see that *quoad* property rights there was any real difference whether the two persons were of the same sex, or different sexes, or whether or not there was any sexual relationship between them, provided always that they were not man and wife. He would add that he said nothing about the case where there were any children of the relationship.

In the present case, of course, the property was vested in the plaintiff and the defendant, and the plaintiff could claim the support of the legal title. At first sight it might well be that if the entire beneficial interest was intended to be the defendant's it was a very strange thing for him to have had the legal estate E put into their joint names. As the creation of a trust for sale under which the defendant was the sole beneficial owner seemed pointless, the *prima facie* inference must be that the transaction was intended to pass some beneficial interest to the plaintiff, whether by gift or otherwise: consider *Fowkes v Pascoe* (1875) 10 Ch App 343. The defendant's explanation in his affidavit was that the purchase and mortgage were put in the joint names merely as "a matter of form." Pressed to amplify this, he was unable to give any coherent explanation. He (his Lordship) thought, however, that on other evidence the true explanation emerged. This was that the defendant's earnings were not large enough to enable him to obtain a building society mortgage on his own. On the other hand, the joint earnings of the plaintiff F and the defendant did suffice. The plaintiff was accordingly made a party to the contract, the mortgage and the transfer in order to enable a sufficient loan to be obtained and not with the object of making a gift to the plaintiff of any share of the beneficial interest.

Mr Tibber contended, in effect that the court should not strain unduly to discover what had happened when the property was acquired, but need do no more but see what was a fair result to produce on the state of affairs when the property was sold. What he (his Lordship) had to apply, however, were the equitable principles relating to trusts. In working out those principles in domestic situations it might often be necessary to adopt a somewhat broad approach, but that was quite different G from ignoring those principles and aiming only at a process of rough justice or, for that matter, seeking to award a *quantum meruit* to an estranged wife or a discarded mistress. The question was what interest in property, if any, the claimant was entitled to, and in this the starting point must always be the footing on which the property was acquired, however much this might be explained or modified by subsequent events. The plaintiff alleged that there was an oral agreement between her and the defendant for the defendant to pay the outgoings on the house and for her to pay for the food and certain clothing, and that she carried out this agreement. She also alleged that the agreement was that the house should be owned jointly. H She was very properly pressed in cross-examination on these matters, and he (Megarry J) found her answers unconvincing. There was much difficulty in disentangling what she said had been agreed upon and what she had inferred. He fully accepted her explanation that it was very difficult after all this time to remember what was said; that, however, merely reinforced the uncertainty. He was very far from being satisfied that there ever was any agreement or common intention, or any consensus, that the plaintiff was to have any property interest in the house, either of a determinate amount, or of an indeterminate amount which was to await quantification either in relation to what payments were actually made by each party or otherwise. J

In the end, it seemed to him (his Lordship) that subject to one point, both on the evidence and on the law the plaintiff's claim to any proprietary interest in the house must fail and so must her claim for an order for sale, so that the originating summons would be dismissed. The point referred to was the plaintiff's liability to the building society under the mortgage. He (Megarry J) had dealt with the case on the footing that as between the plaintiff and the defendant the sole liability on the mortgage was the defendant's. Yet although that was the basis on which the parties had proceeded, there had been no formal recognition of this basis. Accordingly he did not think it would be right to make an order in the defendant's favour until he had K heard whether the defendant was offering any formal indemnity for the plaintiff against all liability on the mortgage. Subject to that, the result must be as he had stated. Similarly, subject to anything counsel might have to say, the counterclaim succeeded and the defendant was entitled to the declaration and order claimed.

Mr Scott said that he gave an undertaking that a suitable indemnity would be given; he also accepted that he could not have an order that the plaintiff join in transferring the property to the defendant until she was relieved as between herself and the building society from personal obligation. His Lordship made an order for costs against the plaintiff, who was legally aided, the order not to be enforced without the leave of the court. He adjourned the L hearing of an application by Mr Scott for payment of the defendant's costs from the legal aid fund.

QUEEN'S BENCH DIVISION

January 22 1974

(Before Mr Justice SEBAG SHAW)

LIDDELL v HOPKINSON

Estates Gazette February 1 1975

(1974) 233 EG 513

Former matrimonial home occupied by divorced wife as part of maintenance arrangement—Mortgage instalments kept up by former husband—Oral contract held to have been made by which if wife left house she should have two-thirds of the net proceeds of sale—Contract within section 40, LPA 1925, but enforceable under doctrine of part performance M

This was a claim by Mrs Joan Frances Liddell, of Burnside House, Church Road, Paddock Wood, Kent, against her former husband, Mr Clifford Graham Hopkinson, of 12 Ponsonby Place, London SW1, for £2,864, being two-thirds of the net proceeds of the sale of the former matrimonial home in St Aubyn's Close, Orpington, Kent.

Mrs C M Puxon (instructed by Griffiths, Quinn & Gilmour) appeared for the plaintiff, and Mr C Gray (instructed by Tucker, Turner & Co) represented the defendant.

Giving judgment, SEBAG SHAW J said that Mr Hopkinson left his wife in 1963 after they had been married for 12 years, and the two children remained with their mother in the matrimonial home, a small house with a mortgage in St Aubyn's Close, Orpington. The marriage was dissolved in December 1966, and Mr Hopkinson paid his wife £15 a week maintenance for herself and the children and continued paying the mortgage on the house. Mrs Liddell met her present husband, who was divorced himself and had custody of his four children, but they could not marry, because she would lose her allowance and have to give up the house. Mr Liddell's house was too small for six children, and he could not afford a bigger house. When Mr Hopkinson, who remarried in January 1967, heard of his former wife's attachment to Mr Liddell, he urged them to marry, because if she gave up the house it would relieve him of a financial burden, and according to her evidence he said he would give her two-thirds of the net proceeds from the sale of the house. His story was that he did not agree to pay anything but said he would make some financial contribution, provided assistance in buying a suitable house was needed and provided arrangements were made in a proper form between their respective solicitors. Two-thirds, he said, would be the top limit of the contribution he could give.

Mrs Liddell said that she did not then mention the proposals to her solicitors because she and her former husband were on friendly terms and she thought the arrangement did not need the intervention of solicitors. In April 1970 she married Mr Liddell and went to live in a house in Paddock Wood which he had bought. It cost Mr Liddell more than £2,000 in excess of what the other house had cost him. Mr Hopkinson sold the house in Orpington, but the Liddells received no money. A letter Mrs Liddell wrote to her solicitors about getting some of the money supported her account of what happened. He (his Lordship) felt that Mrs Liddell was a thoroughly honest and reliable witness, and that the squalid truth of the matter was that once Mrs Liddell had left St Aubyn's Close and the defendant had sold the house, he saw no pressing reason to honour the undertaking he had given. He prevaricated on all sides. On the facts, therefore, he (his Lordship) found that there was an agreement with Mrs Liddell that Mr Hopkinson would pay her two-thirds of the proceeds from the sale of the house if she vacated it and got married. That raised the question whether the agreement was a contract. He (Sebag Shaw J) thought that it was. In his opinion it was an oral contract for the disposition of an interest in land within the ambit of section 40 of the Law of Property Act 1925. It was enforceable under the doctrine of part performance by reason of Mrs Liddell's leaving, as she had done, the house she had occupied throughout her married life with the defendant, thus giving up a very substantial pecuniary advantage and lightening the financial burden on him to a corresponding extent. That was quite clearly referable to a contract, and there must be judgment for Mrs Liddell for two-thirds of the £4,296 net proceeds from the sale of the house. That amounted to £2,864, plus 7 per cent interest from the beginning of 1971.

The defendant agreed to pay the money at the rate of £125 a month.

QUEEN'S BENCH DIVISION

November 29 1974

(Before Mr Justice PARK)

BRITISH DREDGING (SERVICES) LTD AND ANOTHER v SECRETARY OF STATE FOR WALES AND ANOTHER; HAYES METALS LTD v SAME

Estates Gazette February 15 1975

(1974) 233 EG 680

Coast protection order within powers conferred by Coast Protection Act 1949 though confirmed by Secretary of State in terms to do with control of rate of sand extraction by commercial firms—True purpose of order protection against erosion, not a reduction in rate of diminution of material resources

These were motions in similar terms by distinct applicants, British Dredging (Services) Ltd and British Dredging (South Wales) Ltd in the first motion and Hayes Metals Ltd, trading as Margam Sand & Gravel Co, in the second, to quash the Borough of Port Talbot Coast Protection Order 1973, made on October 25 1973 by Port Talbot Borough Council, predecessors of the second respondents, the Afan Borough Council, and confirmed on July 12 1974 by the first respondent, the Secretary of State for Wales.

Mr G G Williams (instructed by Holman, Fenwick & Willan, agents for Lean & Lean, of Cardiff) appeared for the applicants; Mr H K Woolf (instructed by the Treasury Solicitor) for the Secretary of State; and Mr G Eyre QC and Mr J Rees (instructed by G Davies, of Port Talbot) for the council.

Giving judgment, PARK J said that for many years the applicants had been extracting sand from the seashore, under licences granted by the freeholders of Margam Sands, and selling it. On October 25 1973 the Port Talbot Borough Council made the order complained of, the effect of which was to make it unlawful for the applicants to continue this business without a licence under section 18 (5) of the Coast Protection Act 1949. The applicants objected to the order, and a public inquiry was held. Having considered his inspector's report, the Secretary of State confirmed the order. The applicants now questioned the validity of the order. They said that it was not within the powers of the Act, and that their interests had been substantially prejudiced by non-compliance with a requirement of the Act. All parties agreed that the court's approach was laid down in *Ashbridge Investments Ltd v Minister of Housing and Local Government* [1965] 1 WLR 1320, where the Court of Appeal had considered a somewhat similar situation which arose under the Housing Act 1957. In that case Lord Denning MR said: "The court can only interfere on the ground that the Minister has gone outside the powers of the Act or that any requirement of the Act has not been complied with. . . . It seems to me that the court should look at the material which the inspector and the Minister had before them, just as it looks at the material before an inferior court, and see whether on that material the Minister has gone wrong in law." Mr Williams submitted that both the Minister and the inspector in the present case had misinterpreted the real purposes of the Act, and that there was no evidence to support the order, which was therefore invalid.

The Act was to amend the law relating, *inter alia,* to the protection of the coast against erosion and encroachment by the sea; "land" included land covered by water, and the definition of seashore was wide enough to include Margam Sands. Section 18 did not provide that any particular activity had to take place, or any particular situation had to arise, before an order could be made under it by a coast protection authority. But whatever the conditions of the seashore, an order could be made only to enable the authority to perform its duty of protecting land against erosion and encroachment by the sea. The words of section 18 appeared to be sufficiently wide to cover every foreseeable situation in relation to the seashore. In his (Park J's) judgment, when the state of the seashore, from whatever cause, was such that it appeared to a coast protection authority to be necessary or expedient for it to acquire the powers granted under section 18 to enable it to perform properly its statutory duty, then the making of a section 18 order would be permissible and would fall within the powers under the Act. An order could not be made merely because there had been a degree of erosion or encroachment by the sea. On the other hand, if an erosion caused a reasonable apprehension of still further erosion it might well appear to a coast protection authority to be necessary or expedient to make an order. There was nothing in the Act which restrained an authority from acting until a catastrophe occurred. Authorities did not have to stand by until a seashore became more and more denuded of sand, until the protection afforded by the sand, for example, to the sea wall, had vanished, so that the wall was breached in the next winter's gale and the land flooded. The powers granted to authorities by section 18 were to enable coast protection authorities to take steps to prevent or mitigate the consequences to their land of the action of the sea in all its moods.

In the present case, the inspector had found that the order was necessary, and there was no dispute that there was evidence to support this conclusion. He said that changes in the coast, including erosion of an underlying stratum of clay, were taking place, owing either to sand extraction, to storms or to alterations in local tide conditions; he was unable to identify which was the real reason. He said that further amenity reasons were not grounds for curtailing the removal of sand, but he saw danger in over-exploitation of the beach's current resources. In the light of those findings, the Minister had ample evidence for confirming the order. The contention that the inspector was seeking to use the Act not to prevent erosion but to protect a material resource was not borne out by the facts. It followed that the contention that the Minister had fallen into error in confirming the order also failed. The Minister concluded that it would be prudent to control the rate of sand extraction. In other words, he considered it prudent to confirm the order. On the inspector's recommendations, he might equally have said that it was necessary or expedient for the authority to acquire section 18 powers, and he could have confirmed the order in those words. There was nothing in *Ashbridge* or the Act to suggest that the Minister could not confirm an order because he thought it prudent to do so. The Minister had neither gone outside the powers of the Act nor wrongly interpreted it, and the motions accordingly failed.

CHANCERY DIVISION

October 29 1974

(Before Mr Justice GOFF)

HARLOW DEVELOPMENT CORPORATION v COX BROS (BUTCHERS) LTD

Estates Gazette February 22 1975

(1974) 233 EG 765

Principle by which breach of a negative covenant is restrained in equity almost as of right does not apply in full force to applications for interlocutory relief—Injunction pending trial to restrain use of Harlow meat warehouse for "freezer centre" suspended for three months to allow for reorganisation of defendants' business

This was an application for an interlocutory injunction by Harlow Development Corporation, plaintiffs in an action against Cox Bros (Butchers) Ltd, of River Way, Harlow, for breach of the covenants contained in a lease of April 10 1972.

Mr G A Lightman (instructed by Wedlake Bell, agents for the corporation solicitor) appeared for the applicants, and Mr G M Godfrey QC and Miss E Gloster (instructed by Jay, Benning & Solomon) represented the respondents.

Giving judgment, GOFF J said that the action was one for breach of a user covenant which read as follows:

"To use the demised premises only for a wholesale warehouse within the meaning of Class X of the Town and Country Planning Act (Use Classes) Order 1963 and for the curing of bacon, cooking of hams and production of sausages PROVIDED THAT in particular but without prejudice to the generality of the foregoing such user shall be conducted in such a manner that there shall be no breach of subclause (II) hereof AND PROVIDED ALSO THAT all necessary licences and consents shall be obtained for any of the foregoing activities."

In April 1972 the defendants caused to be built on the premises a substantial building of which 10,000 sq ft were constructed as an area where their goods could be deposited, stored and processed pending sale, and an adjoining, although separate, 4,000 sq ft were specially constructed for use as a "freezer centre" for the sale of frozen meat and other frozen foods. It appeared from the defendants' evidence that since July 1972 they had been using that 4,000 sq ft exclusively for the purposes last mentioned. The plaintiffs said that such user was user for retail sales and was accordingly a breach of covenant, and for the purposes of the motion that was admitted by the defendants. It followed that in considering what, if any, relief he ought to grant, he (his Lordship) had to consider the defendants as being in the position in which they would have stood had they argued the construction of the covenant before him and failed. The plaintiffs said that the breach was a breach of a negative covenant, and that although what they were seeking at this stage was interlocutory and not perpetual relief, the classic *obiter dictum* of Lord Cairns in *Doherty v Allman* (1878) 3 App Cas 709 at 720 ought to be applied. The principle of that dictum was that once a breach of a negative covenant was established, a court of equity had nothing to do but to say by injunction that which the parties had already said for themselves by covenant, namely that the prohibited thing should not be done, and accordingly an injunction became almost as of course and could be resisted only if special circumstances were shown.

In *Hampstead & Suburban Properties Ltd v Diomedous* [1969] 1 Ch 248 Megarry J decided that the dictum did apply to interlocutory proceedings. He said in conclusion of his observations on that point:

"I see no reason for allowing a covenantor who stands in clear breach of an express prohibition to have a holiday from the enforcement of his obligations until the trial. It may be there is no direct authority on this point; certainly none has been cited. If so, it is high time that there was such an authority; and now there is."

But in *Texaco Ltd v Mulberry Filling Station Ltd* [1972] 1 WLR 814 at 830, the late Ungoed-Thomas J declined to accept that view. He concluded his observations by saying:

"So my conclusion is that the principle of *Doherty v Allman* does not apply to interlocutory injunctions."

It was true that in that case there were questions of fact affecting the validity of the covenants, whereas in the former case the attack on the covenant appeared to have been one of law—the proposition that it was too uncertain. But there the two categorically opposite statements remained. For his part, he (his Lordship) did not think that the dictum could be applied in its full force in an interlocutory matter, particularly having regard to the strong views recently pronounced by the Court of Appeal in the case of *Hubbard v Vosper* [1974] 1 All ER 1023 as to the necessity of preserving the flexibility of the discretion in equity in dealing with interlocutory applications for an injunction. At the same time, in following that direction one should consider all the circumstances; the fact that one had got a negative covenant, a breach of which was admitted as far as the motion was concerned, must be an important factor to be weighed in favour of a plaintiff who said, "Stop, now, and not when the matter comes to trial."

The defendants said that even so he (his Lordship) ought not to grant an injunction, but should stand the matter over until trial, or that the motion should be stood over with the liberty to restore when the decision of the Secretary of State for the Environment on their planning application for change of use was known, and in the last resort that if an injunction were granted it should be suspended for a reasonable time to enable them to close down or reorganise the retail freezer side of their business. They said that there had been long delay on the part of the plaintiffs, and they urged that as a reason why relief should not be granted. They further said that they hoped, at the trial, to make out some case of acquiescence which would disentitle the plaintiffs to relief, but in weighing the force of that, even at this stage, he (Goff J) had to bear in mind that when it got to trial, if a breach of the negative covenant were established, the defendants would be faced with *Doherty v Allman* and would have to make out something very special as an answer to an injunction. The facts as to the planning application were that Harlow Council had given approval for the proposed change of use, but because of the plaintiffs' opposition had been unable to decide the matter, which had been referred to the Secretary of State for the Environment. Even if the Secretary of State confirmed the planning decision, that would not bind the plaintiffs, because while it would be conclusive so far as planning approval was concerned, it would not override the right of the plaintiffs as landlords to insist on the performance of the covenants in the lease. As far as the delay point was concerned, he (his Lordship) had reached the conclusion that there was not sufficient force in the defendants' arguments for him to refuse interlocutory relief on this ground. A letter of September 20 1974 showed that the plaintiffs intended, when they knew the Secretary of State's decision, to consider what they should do. The letter did not say this about the defendants' application alone, but about their actions generally, and added that the plaintiffs would have to consider other leases in the town and the relations between their tenants. The defendants' final point was that the court ought not to grant an injunction which would stop their business, and then possibly, in the end, have a perhaps absurd result, in that the Secretary of State might say yes to planning approval and the plaintiffs go along with it and agree to the use of the premises in the way the defendants wished, but by that time the business would have been brought to a halt. In answer to this, the plaintiffs said that whatever the Secretary of State said regarding the difference of opinion between them and the planning authority, they had the right and duty to consider the matter as landlords, and as the development corporation controlling the new town as a whole they could not be directed to consent to a change of use. While they would review the position in the light of what the Secretary of State said, they insisted in the meantime that the covenants in the lease should be observed. Taking all the circumstances into account, and remembering that there was for the purposes of the motion an admitted breach of covenant, his (Goff J's) conclusion was that there should be interlocutory relief, but that the injunction restraining the defendants from using the relevant part of their business for retail sales should be suspended for three months to give them time to reorganise their operations.

QUEEN'S BENCH DIVISIONAL COURT

December 19 1974

(Before Lord WIDGERY CJ, Mr Justice MELFORD STEVENSON and Mr Justice WATKINS)

SALFORD CITY COUNCIL v McNALLY

Estates Gazette March 15 1975

(1974) 233 EG 1013

An occupier of "patched-up" housing in a clearance area may apply under section 99 of the Public Health Act 1936 for a nuisance order, which must be made on proof of the appropriate facts—Magistrates may however take existence of "patched-up" provisions into account as one of the factors affecting their decision on the terms of the order

This was an appeal by Salford City Council from a decision of the Salford stipendiary magistrate on April 16 1974, granting an application by the respondent, Mrs Brenda McNally, under section 99 of the Public Health Act 1936, for a nuisance order in respect of premises occupied by her at 20 Johnson Street, Lower Broughton, Salford.

Mr R A W Sears (instructed by Sharpe, Pritchard & Co, agents for G F Bannister, of Salford) appeared for the appellants, and Mr B A Hytner QC and Mr F D Hart (instructed by Davis, Hope & Furniss, of Glossop) represented the respondent.

Giving judgment, LORD WIDGERY said that this was the latest of a number of cases which had been before the courts in which the provisions of the Public Health Act 1936 had come into conflict with those of the Housing Acts. The facts were that the property was an unfit house which stood on a clearance site, but the local authority had decided to defer its demolition for at least seven years and to use it, with other such houses, as temporary accommodation for those displaced by demolition. They did this in exercise of their powers under section 48 of the Housing Act 1957 to "patch-up" unfit premises. Under section 99 of the Public Health Act the respondent occupier sought, and the magistrate granted, a nuisance order in respect of the house. The magistrate inspected the premises and found a number of defects existing; there was rising damp, perished plaster, the rear door was rotted and unhinged, there was severe dampness in the first floor and the water closet pipe was cracked and insanitary. By reason of these defects he found that the premises were in a condition prejudicial to health and therefore constituted a statutory nuisance under section 92 of the 1936 Act.

The magistrate did not have before him the full report of *Nottingham City District Council v Newton* [1974] 1 WLR 923, but he was supplied with a copy of a report in *The Times*, which was a comparatively full report, and he concluded that he was bound to make a nuisance order. The situation which arose in the *Nottingham* case was in many respects similar to the present. The decision of the court was that where the essentials of a statutory nuisance had been proved before justices so that they were satisfied that a statutory nuisance existed, they were bound to make a nuisance order. It had, however, been pointed out by the present court that the justices had considerable tolerance in deciding precisely how many of the complaints, and which complaints, required to be remedied and also the time within which that remedy should be supplied. Given that the *Nottingham* case was right (and no one had attacked it in this case) the magistrate's conclusion was entirely right.

Did the resolution under section 48 make any difference? There seemed to him (his Lordship) to be a lot of common sense in the appellants' submissions that it should make a difference. Section 48 was a recognition by Parliament that there would be occasions when people would have to live, perhaps for some time, in houses which were not up to the standards of fitness for human habitation required by the Public Health Act. Nevertheless he had found it impossible to say that the effect of section 48 was to remove the normal Public Health Act standard of fitness in cases where a nuisance within the Act arose. If that had been Parliament's intention it would have been the simplest thing in the world to say so. He was therefore driven to the conclusion that an individual occupier of an unfit house could initiate proceedings under section 99 of the Public Health Act and that the magistrate was bound in law to make a nuisance order once the nuisance alleged had been proved, notwithstanding the fact that the house was one to which a resolution under section 48 of the Housing Act 1957 applied. No doubt the existence of the section 48 resolution was one of the factors which would affect the magistrates' court when dealing with this difficult type of question, but that it was more than a factor and having such weight as the magistrate thought right and no more, he (his Lordship) found it impossible to say. The magistrate took into account all the circumstances before him in reaching his conclusion and the present appeal must be dismissed.

MELFORD STEVENSON and WATKINS JJ agreed, and the appeal was dismissed with costs. The court certified that the case raised a point of law of general public importance, but refused leave to appeal to the House of Lords. Mr Sears told the court that repairs had now been carried out to rectify part of the nuisance—an insanitary water closet and a rotten back door—although treatment for damp had yet to be done. The council was willing to undertake any reasonable work on future defects. The case was said to be a test case, and a further 88 summonses to be awaiting hearing.

On February 13 1975 the Appeal Committee of the House of Lords (Lord Simon of Glaisdale, Lord Edmund-Davie and Lord Fraser of Tullybelton) gave Salford City Council leave to appeal to the House of Lords.

CHANCERY DIVISION

January 25 1974

(Before Mr Justice FOSTER)

BABACOMP LTD v RIGHTSIDE PROPERTIES LTD

Estates Gazette April 19 1975

(1974) 234 EG 201

Contract of sale of house in Mornington Terrace—Purchaser subsequently offers to take subject to tenancies at reduced price—Offer held conditional on swift completion, which did not occur—Deposit repayable, and further sum of £2,500 referred to as "agreed damages" for non-completion awarded—An agreed sum of this kind is not a penalty

This was an action by Babacomp Ltd, of 1-4 Yarmouth Place, Brick Street, London W1, against Rightside Properties Ltd, of Fitzroy Square, W1 for the return of a deposit of £2,500 paid under a contract of sale of 52 Mornington Terrace, NW1, dated July 12 1972, and for agreed damages of £2,500 for failure to complete the sale. The defendants counterclaimed specific performance of an oral agreement for sale dated October 20 1972, or of the July agreement as varied by the oral agreement.

Mr M Mark (instructed by Southall & Co) appeared for the plaintiffs, and Mr R Ellis (instructed by Michael Freeman & Co) represented the defendants.

Giving judgment, FOSTER J said: In this action Babacomp Ltd, which I will call the plaintiff company, seeks an order against Rightside Properties Ltd, the defendant company, for the repayment of a sum of £2,500 paid as a deposit, and for £2,500 agreed damages under an agreement dated July 12 1972, which I will call "the July agreement." The defendant company counterclaims specific performance of an oral agreement made on October 24 1972 ("the oral agreement") for the sale and purchase of 52 Mornington Terrace, NW1, or of the July agreement as varied by the oral agreement. No 52 Mornington Terrace is freehold and is registered at Her Majesty's Land Registry under the title No LN 455052. At all material times the registered proprietor has been, and still is, Mrs Rosa Dora Diviani. On July 3 1972 an agreement was signed by Mrs Diviani and by Mr Moschie on behalf of the defendant company by which she gave an option to the defendant company to purchase the fee simple of the property with vacant possession at the price of £20,000, such option to be exercised within fifteen days. The defendant company gave notice to Mrs Diviani exercising the option by a letter dated July 6 1972. A great deal of evidence has been given as to the subsequent events occurring between Mrs Diviani and the defendant company in regard to the purchase and sale of the property. However, the last letter which is before me is dated January 7 1974, from Conway & Conway (Mrs Diviani's solicitors) to Michael Freeman & Co (the defendant company's solicitors), which is in these terms:

> We are in receipt of your letter of the 2nd instant, and in the circumstances are forthwith instructing counsel to settle the proceedings herein, and will in due course effect personal service thereof upon your client.

What the exact form of those proceedings will be, I do not know, but in view of the impending litigation between the parties in regard to the purchase and sale of the property, I do not express any view except to notice that litigation is pending. I turn now to the July agreement. This is in these terms. The first two recitals read as follows:

> Whereas the vendor is or believes itself to be entitled to an option for the acquisition of a property shortly known as 52 Mornington Terrace, London NW1, and has served notice exercising the same, subject as hereinafter mentioned;
>
> And whereas the vendor has agreed with the purchaser of the said property for sale at the price of £25,000;
>
> Now therefore it is hereby agreed as follows:
>
> (1) subject as hereinafter set forth, the vendor shall sell and the purchaser shall purchase all that the said property for the said purchase price of £25,000;
>
> (2) the purchaser shall on the execution hereof pay to the vendor the sum of £2,500 by way of deposit and part payment of the purchase price.

Paragraph (3) is in respect of office copies of the entry on the register, and is not material.

> (4) The vendor sells as beneficial owner.
>
> (5) Vacant possession of the property shall be granted on completion.
>
> (6) Completion of the sale and purchase hereunder shall take place at the later of the expiration of the period of ninety days from the date hereof or the period of twenty-eight days from the delivery to the purchaser or its solicitors of a notice given by or on behalf of the vendor that it is in a position to complete the contract, and in relation thereto the following provisions shall apply:
>
> (a) the service of such notice shall amount to a warranty on the part of the vendor that it will at the expiration of such notice be in a position to complete this contract;
>
> (b) in the event that such notice shall not be served within a period of six months from the date hereof this contract shall be automatically rescinded and the deposit paid shall be returned without deduction to the purchaser and the vendor shall in addition pay by way of agreed damages to the purchaser the further sum of £2,500.

The remaining paragraphs (7) and (8) are not material. I do not think anything turns on when, where, or how this agreement came into being, though I heard some evidence upon it, and there was no dispute that paragraph 6 (b) came into effect, since no notice was ever served by the defendant company, unless the agreement was subsequently discharged or varied, as is alleged by the defendant company. What happened was that it became apparent that there were a number of tenants in the property and that neither Mrs Diviani nor the defendant company would be able to transfer to the plaintiff company the property with vacant possession. The plaintiff company and the defendant company were concerned with the purchase and sale not only of this property, but also of some four or five other properties, and on

October 24 1972 a meeting took place at the defendant company's then offices at Findlay House, Haymarket. There were present Mr Leslie Ratcliffe, the managing director of the plaintiff company, Mr Julian Robert Smith, a chartered surveyor, who was then employed by the plaintiff company, and Mr Gabriel Moschie, who is the sole director of the defendant company and owns 98 £1 shares of the 100 £1 shares, the issued capital of the defendant company. Both Mr Ratcliffe and Mr Smith gave evidence that a proposal was put to Mr Moschie that the plaintiff company would purchase the property subject to the existing tenancies at a reduced price of £20,000 if completion took place almost immediately, and that as Mr Ratcliffe rose to leave the meeting Mr Moschie agreed to that proposal. Mr Moschie in his evidence said that there was an agreement to reduce the price and make the sale subject to existing tenancies, but that nothing was said about the completion date.

Before I deal with the credibility of the witnesses, it is important to consider the correspondence which ensued. On that same day a letter was written on the notepaper of a company, Sherwood Securities Ltd, an associate company of the plaintiff company, to the plaintiff company's solicitors, Southall & Co. It was compiled jointly by Mr Ratcliffe and Mr Smith and signed by the latter. It dealt with the various properties concerned, but the paragraph concerning 52 Mornington Terrace is in these terms:

> Rightside Properties are unable to sell with the benefit of vacant possession, as they had contracted to do, and they have tried to negotiate a reduced price for the property, subject to the existing tenancies. We indicated that we were prepared to do a deal at £20,000, but if they were not able to sell almost immediately at this figure, we would serve notice on them in order to recover our deposit and the agreed damages of £2,500.

The last paragraph reads:

> We trust that the above résumé of our meeting will lead to speedy conclusions in these transactions.

On October 25, Mr Moschie wrote to the defendant company's solicitors a letter, the first paragraph of which reads as follows:

> We enclose a copy of a letter received from Mrs Brittain regarding 52 Mornington Terrace. At the meeting with Mr Ratcliffe, it was agreed that he would take Mornington Terrace with the tenant at a reduction of £5,000, and we can accordingly put the same proposal to Mrs Diviani.

On October 25, the defendants' solicitors wrote to the plaintiffs' solicitors in these terms:

> We understand that our respective clients met and your client stated that he would be agreeable to purchasing this property subject to the existing tenancy of Mrs Latham, at a price of £20,000, that is a reduction of £5,000 on the contractual purchase price.

On October 26, the plaintiffs' solicitors wrote to the defendants' solicitors in these terms:

> We refer to recent correspondence in connection with the sale of the above property to our clients, and would advise that we have been instructed that our clients are prepared to complete the purchase of the above property, subject to the existing tenancies, at £20,000. It is a condition of our clients' offer that the matter be completed forthwith. Unless we hear from you in early course, we are instructed to serve a notice on you requesting the return of the deposit and the agreed damages of £2,500. We await hearing from you.

And then the next material letter is one dated November 16 1972 from the defendants' solicitors to the plaintiffs' solicitors, which is headed "Re 52 Mornington Terrace," and the first paragraph of which reads:

> We are very pleased to be able to report some progress in the matter on the basis of the suggestion contained in your letter of October 26.

Then there was a little correspondence ensuing in regard to this, until on December 1 1972 the plaintiffs' solicitors wrote to the defendants' solicitors in these terms:

> We enclose requisitions on title, together with a draft transfer for approval, together with copies for your use. May we particularly draw your attention to our requisition requiring vacant possession to be given on completion in accordance with clause 5 of the contract, as it would appear from your letter of November 21 that the property is not vacant.

The defendants' solicitors replied on December 4:

> Are your clients now saying that they are not prepared to complete the purchase subject to the existing tenancy at the reduced purchase price of £20,000? If so, we must of course seek our clients' immediate further instructions, and advise them in the light of the position obtaining under the terms of the contract and read in conjunction with your aforementioned letter.

The plaintiffs' solicitors replied on December 8 in these terms:

> In our letter to you of October 26 we informed you of our clients' offer to complete this matter at the reduced price of £20,000 subject to the existing tenancy, but specifically stated that it was a condition of our clients' offer that the matter then be completed forthwith. Since the matter was not then completed forthwith, we have presumed that our clients' offer has lapsed, but, as requested by you, are taking our clients' further instructions.

On December 18 the plaintiffs' solicitors again wrote to the defendants' solicitors in these terms:

> Further to our letter of December 8, we have now received our clients' further instructions in this matter, which are that they require vacant possession to be given in accordance with the terms of the contract. Please confirm that this will be given, and let us hear from you on other requisitions on title and the draft transfer which we sent to you on December 1.

In the letter of the plaintiff company's solicitors dated October 26 1972, it was stated, as I have read, "It is a condition of our clients' offer that the matter be completed forthwith." This was never objected to by the defendant company or its solicitors, and is in accordance with the evidence of Mr Ratcliffe and Mr Smith. Further, it is clear from that letter that if there was not an early completion the plaintiff company would rely on their rights under clause 6 of the July agreement. In my judgment, Mr Smith and Mr Ratcliffe were honest and truthful witnesses, and I accept their evidence that the offer to purchase at a reduced price, but subject to the tenancies, was subject to a condition that completion should take place almost immediately. In so far as Mr Moschie's evidence is in conflict with that evidence, I reject it. I do not think that Mr Moschie's evidence can be relied on as being honest and truthful.

It is not disputed that completion did not take place within a reasonable time or at all, but it is submitted on behalf of the defendant company that the effect of the oral agreement was to discharge by necessary implication the written agreement, or alternatively that the oral agreement constituted a variation of the written agreement by deleting clauses 5 and 6 of that agreement. I can find no reason why either of those submissions can succeed. If those submissions are right, then there was a new agreement with no completion date agreed at all between the parties, and the defendant company can even today attempt to get specific performance of that agreement, as it in fact seeks by its counterclaim. In my judgment, the effect of the oral agreement was an offer by the plaintiff company to purchase the property, subject to the tenancies, at a reduced price if completion took place almost immediately. This offer was accepted by the defendant company. Completion did not take place. No notice was served by the defendant company, under clause 6 (b) that is, before January 12 1973, and the plaintiff company was entitled to treat the contract as rescinded and claim repayment of the deposit. There is also the question of the further sum of £2,500, described in the written agreement as "by way of agreed damages." The plaintiff company, during the hearing, amended the statement of claim to add a claim that it was entitled to that sum, not only by way of agreed damages but also—and I quote from paragraph 5 of the amended statement of claim—" or otherwise pursuant to clause 6 of the said agreement." For the defendant company it was submitted (a) that as the contract was to be rescinded, the effect was to put the parties in their original position, so that the deposit was returnable, but no claim for damages could be sustained; (b) that in any event, the parties never agreed that the sum of £2,500 represented the true amount of the damages, and indeed, the parties never considered the damage which would ensue if the contract came to an end; and (c) that in any event, the sum was a penalty, and the defendants heavily relied on Mr Ratcliffe's evidence that it was intended to be—and I quote—" a tough penalty, if Mr Moschie tried to sell the property elsewhere."

In regard to (a), it is of course true that if a contract is rescinded, whether by the parties to a contract or by the court, the parties must be restored to their original position, and no question of damages arises. Even if in clause 6 (b) the word " rescinded " is used in this technical sense, however, I do not see why the parties cannot agree that even if there is to be *restitutio in integrum* a sum should also be paid. Alternatively, in view of the reference to damages, the word " rescinded " must, I think, be construed as meaning " brought to an end." In either event, it follows that the plaintiff is not debarred from claiming the sum of £2,500. As regards (b), it is true from the evidence that there was no discussion as to the amount of the damages which might ensue to the plaintiff company. But though it was never discussed, it was inserted in the agreement which was signed by both parties, and as such it was agreed. In my judgment, the claim does not fail because there was no discussions as to the actual damage which might occur, and, as I say later, in fact there could be no claim to damages. As regards (c), I have been referred to a number of cases, most of which are concerned with clauses in hire purchase agreements giving the hirer an option to terminate the agreement on payment of a sum of money. In *Associated Distributors Ltd v Hall* [1938] 2 KB 83 Slesser LJ at p 88 said this:

> He has exercised an option, and the terms on which he may exercise the option are those set out in clause (7). The question therefore whether these payments constitute liquidated damages, or a penalty in the instances mentioned, does not arise in the present case.

In *United Dominions Trust (Commercial) Ltd v Ennis* [1968] 1 QB 54, the Court of Appeal held that the option had not been exercised. Harman LJ at p 67 said this:

> . . . I do not think that there can be a penalty without a breach of contract, for which the penalty may be the result.

In *Sterling Industrial Facilities Ltd v Lydiate Textiles Ltd* (1962) 106 SJ 669 Diplock LJ, as he then was, said this:

> Here there was no question of any breach of obligation, but the defendants were sued in respect of sums payable in a specified eventuality. It had been conceded that there was no case in which it had been held that a payment to be made in a specified eventuality was a penalty, or to be treated as such.

He was not prepared to extend the definition of penalty, and save for certain passages in the speech of Lord Denning in *Bridge v Campbell Discount Company Ltd* [1962] AC 600 there is no authority which suggests that it should be. In the *Bridge* case, apart from the passage in Lord Denning's speech, there are *dicta* to the opposite effect (see Lord Morton of Henryton's speech at p 614). In the present case, neither side suggested that the defendant company was in breach of contract by not serving a notice in time. There cannot therefore be any damages for that breach, however the sum is described. In that event, the sum payable cannot be a penalty, but it is merely a sum to be paid in a specified eventuality. In my judgment, the plaintiff company succeeds in its claim to the second sum of £2,500.

There remains the counterclaim. Even if I am wrong in the view which I have expressed as to the effect of the oral agreement, and the true effect of it was to discharge the written agreement, or at least to vary it by cancelling clauses (5) and (6) of it, the defendant company cannot get specific performance unless (a) it is ready, willing and able to complete and perform the agreement (see paragraph 10 of the counterclaim), and (b) there is a sufficient note or memorandum in writing to satisfy section 40 of the Law of Property Act 1925 of that oral agreement. As to (a), the defendant company cannot complete the agreement, nor is it in a position to compel Mrs Diviani to convey without litigation. It is not now able to perform the contract. As to (b), the defendant company relied on the letter of October 24 signed by Mr Smith and the letter of October 26 from the plaintiff company's solicitors, to which I have referred, and on notes taken by Mr Ratcliffe at the meeting on October 24 1972. As to the notes, Mr Ratcliffe said that he destroyed them after the letter dated October 24 was written. As to the two letters, they constitute a note of an agreement which was subject to completion taking place almost immediately or forthwith. They do not, in my judgment, constitute a note of the agreement alleged by the defendant company. For these reasons, in my judgment, the counterclaim fails. I propose to order that the defendant company repay to the plaintiff company the deposit of £2,500, and in addition pay the sum of £2,500. Both sums will carry interest at 10 per cent per annum from January 12 1973. I propose to dismiss the counterclaim.

An order was made accordingly.

COURT OF APPEAL

February 13 1975

(Before Lord DENNING MR, Lord Justice STEPHENSON and Lord Justice GEOFFREY LANE)

SJC CONSTRUCTION CO LTD v LONDON BOROUGH OF SUTTON

Estates Gazette May 3 1975

(1975) 234 EG 363

Restrictive covenant discharged by Lands Tribunal under enlarged jurisdiction of 1969—Compensation assessed by tribunal on basis of splitting realisable development value 50-50—Nothing wrong with tribunal's approach in a case where there was a loss of amenity, etc, standing to be compensated

This was an appeal by SJC Construction Co Ltd, of Ventnor Road, Sutton, from a decision of the Lands Tribunal dated March 4 1974 determining the compensation payable to the respondents, the London Borough of Sutton, upon modification of a restrictive covenant prohibiting building on land at Cheam Park Way, Cheam, Sutton. The decision of the Lands Tribunal was reported at (1974) 230 EG 906.

Mr G H Newsom QC and Mr B Marder (instructed by A F & F W Tweedie) appeared for the appellants, and Mr P Garland QC and Mr J Mulholland (instructed by Sharpe, Pritchard & Co, agents for W L Barton) represented the respondents.

Giving judgment, LORD DENNING said: This case raises a question of the compensation payable in respect of the removal of restrictive covenants. It concerns an area of land at Cheam in the Borough of Sutton. Originally it was land of about an acre on which there was a house of some historic and architectural interest called Whitehall. It was in a very favoured situation. On the other side of Cheam Park Way there was Cheam Park. This land sloped down towards the park and had extensive views over it. In 1959 the owner of Whitehall sold off one-third of the land at the bottom of

A the area with a restriction that there should be one house only on it—one dwelling-house only, with a garage. That was the restrictive covenant which was put on in 1959. A two-storey house was built on the land at that time; there were then private owners of both Whitehall and the new house. But since that time there have been changes of ownership. The present owners of Whitehall, and the two thirds of an acre with it, are the London Borough of Sutton. The one-third acre was sold off. It is the property of a development company, SJC Construction Co Ltd. The developers have pulled down the house which was originally there. They want to build upon the site six four-roomed flats in a two-
B storey block, six garages and a new access road. That would plainly be a breach of the old restrictive covenant. So the developers wish to have that restrictive covenant removed. For that purpose there has been an inquiry before the Lands Tribunal as to whether the restriction should be removed, and if it is removed, what the appropriate compensation should be. In the old days we used to be very familiar with section 84 of the Law of Property Act 1925. But the provisions have been very much altered by the Law of Property Act 1969. These alterations are of such significance that not much guidance is to be obtained from looking at the old statute or the cases under it. I will not read the section in
C full. It is rather too complicated to do so. It is set out in its amended form in *Wolstenholme & Cherry* and *Preston & Newsom.*

First I will consider the question whether the restriction should be removed or not. The President of the Lands Tribunal, who himself sat upon it, came to some conclusions which I will summarise. First, he found that the proposed discharge or modification did not come within section 84 (1) (a) or (c): that is, the restriction was not obsolete, and its removal would injure the persons entitled to the benefit of the restriction. The only ground which the President
D found to be available was under the new section 84 (1) (aa), that "the continued existence thereof would impede some reasonable user of the land for public or private purposes. . . ." He held that the user which the development company were proposing was a reasonable user for public or private purposes, and that the continued existence of the restriction would impede it. Further, he held that the restriction did secure to the persons entitled to the benefit of it practical benefits of substantial value or advantage to them. But also he was satisfied that the continuance of the restriction in impeding the proposed user would be contrary to the public interest. That is section 84 (1A) (b) of the amending Act.
E And further, he was satisfied that money would be an adequate compensation for the loss or disadvantage which any person entitled to the benefit of the restriction would suffer from its discharge or modification. All those matters were proved. So—summarising it—it is this. The President of the Lands Tribunal was satisfied that the proposed use by the development company was reasonable; that it would be contrary to the public interest to continue the restriction; and that money would be an adequate compensation for the borough council, who were entitled to the benefit of the restriction. So he held that the restriction was to be removed. His findings on that point have not been challenged.

F But now we come to the point before us: what is the compensation to be? The person entitled to the benefit is entitled to be paid "such sum by way of consideration as the tribunal may think it just to award under one, but not both, of the following heads. . . ." The particular one here is, "A sum to make up for any loss or disadvantage suffered by that person in consequence of the discharge or modification." So here the problem before the President was to award such sum as was just to make up to the borough council "for any loss or disadvantage" suffered by them in consequence of discharging the restriction. The President fixed in this particular case the sum of £9,500. He did it

G in this way. He envisaged there might be friendly negotiations between the parties as to the removal of the restriction, and he said, ". . . I think that the most likely outcome of friendly negotiations would have been an agreement to split the development value equally and I so decide." He had evidence from the council's officers as to development value, that is, the value with and without the restriction; and he said the amounts were £35,000 and £54,000 respectively. That gave a realisable development value of £19,000; and splitting it gave £9,500, and he awarded that sum.

Now, Mr Newsom, on behalf of the development company, has submitted that the President was wrong in principle in
H awarding compensation on that basis. He said, in the first place, that the basis was inconsistent with a case decided by the Lands Tribunal in Scotland, *McVey and Robertson v Glasgow Corporation* (1973) 26 P & C R 429. The President was aware of that case. He considered that it was not applicable by reason of the wording of the Scottish Act and the particular provisions of Scottish land law. We have been taken through that case by counsel. It seems to me, as it did to the President, that it is distinguishable on these grounds. First, the Scottish statute is in many respects different from the English. But, secondly, if one examines the facts and circumstances of the Scottish case, it is very
J different. The Glasgow Corporation (who were entitled to the benefit of the restriction) did not suffer any loss or disadvantage from its removal, except that they were deprived of the right or power to obtain money. That appears from page 435. The Lands Tribunal in Scotland held that a right or power to obtain money was not the subject of compensation under the statute, when no other loss or disadvantage had been incurred. That seems to me entirely right. But I think the position would have been very different if there had been any loss or disadvantage, such as injury to the amenities or obstruction of view or increase in noise,
K or anything of that kind. So the Scottish case was decided on its special circumstances, in that there was no such loss or disadvantage at all. The only thing that the Glasgow Corporation lost was the opportunity of extracting money as the price of removing the restriction. Such was not the subject of compensation.

So we come back to the question, what is the basis, or proper basis, of compensation? It is simply compensation to make up "for the loss or disadvantage suffered" by the borough council. There is no method prescribed by the Act by which it is to be assessed. It is essentially a question of quantum. But it is to be assessed for loss of amenities, loss
L of view, and so forth, which are things which it is hard to assess in terms of money. It is similar to the compensation for pain and suffering, which cannot be translated into money terms. So the courts have to arrive at conventional sums which they award. So here the loss was a disadvantage to the borough council, the occupiers of the house Whitehall, and to their tenants. The borough council propose, I gather, to put up many old people's homes on the land they have there. They are entitled to be compensated for the loss of amenities, the loss of view across to the park, and so forth. That has to be translated into money terms. Some calculation has to be made. The President did it by finding the realisable develop-
M ment value, and splitting it equally. I see nothing wrong in his approach. He found some help from the case of *Wrotham Park Estate Co Ltd v Parkside Homes Ltd* [1974] 1 WLR 798. In that case the plaintiff sued for an injunction for breach of a restrictive covenant. Brightman J refused an injunction but granted damages in lieu thereof. He assessed the damages by taking a percentage of the realisable development value. He said at p 815: "I think that the landowner would then reasonably have required a certain percentage of that anticipated profit as the price for the relaxation of the covenant. . . . In assessing what would be a fair percentage I think that the court ought, on the particular

A facts of this case, to act with great moderation." He awarded a percentage of only 5 per cent in that case. The President here looked at it in much the same way. He took a much higher percentage. Instead of 5 per cent he took 50 per cent. He took the realisable development value and split it equally. That was, he said, fair to the parties. It was a method by which he was getting at the loss or disadvantage. I see no error of law in it. The loss or disadvantage is an intangible matter which is incapable of exact calculation in money, and he took a fair and sensible way of assessing it. I see no error of law in the way the President dealt with the matter, and I would therefore dismiss the appeal.

B STEPHENSON LJ: I agree. I too cannot see anything wrong with the method by which the President of the Lands Tribunal has assessed the compensation (or consideration) which he thought just to award the respondents under the amended section 84 of the Law of Property Act 1925. The modification of this restrictive covenant clearly resulted in the respondents suffering the loss of a practical benefit of substantial value or advantage. For that substantial loss or disadvantage they were entitled to substantial compensation. That they got from the President, and by a method which I would not say was the only method, but could not say was unlawful or contrary to the statute or to any authority which
C has been cited to us. I therefore agree that this appeal should be dismissed.

GEOFFREY LANE LJ: I also agree.

The appeal was dismissed with costs.

CHANCERY DIVISION

May 1 1974

(Before PENNYCUICK V-C)

D DAMM v HERRTAGE

Estates Gazette May 3 1975

(1974) 234 EG 365

" Anti-gazumping device "—Supposed oral contract combined with receipt for deposit—On facts, no binding contract ever made—Vice-Chancellor's criticisms of device—" Comes very near sharp practice and could easily be used as engine of fraud "—Receipt could be used as memorandum only against vendor—Further criticism of overstatement of price in draft contract

E This was a claim by Mr Michael Damm, of Sanderstead Road, Sanderstead, Surrey, against Mrs Zena Rosemary Herrtage, of Danecroft, Rose Walk, Purley, Surrey, for specific performance of a contract of sale of Danecroft dated February 8 1972. By counterclaim and by claim in a consolidated action, the defendant sought a declaration that no such contract had been entered into, and an order vacating a caution registered by the plaintiff.

Mr L Price QC and Mr E G Nugee (instructed by Mortlake & Co) appeared for the plaintiff, and Mr G Godfrey QC and Mr D Iwi (instructed by Wilkinson, Kimbers & Staddon, agents for M J Whelton & Co, of Guildford) represented the
F defendant.

Giving judgment, PENNYCUICK V-C said: In this action the plaintiff, Mr Michael Damm, seeks against the defendant, Mrs Zena Rosemary Herrtage, an order for specific performance of the contract for sale by her to him of a property known as Danecroft, Purley, alleged to have been made orally on February 8 1972. The situation is unusual in that Mr Damm, in pursuance of an " anti-gazumping device " recommended by his solicitor, took from Mrs Herrtage contemporaneously with the alleged oral contract a written receipt setting out the terms of the contract. I will state the history of the matter as I find it upon the documents and oral evidence, leaving for subsequent consideration the single G
crucial issue of fact, namely, did Mr Damm and Mrs Herrtage on February 8 1972 enter into a binding oral contract?

Mrs Herrtage is the owner in fee simple of Danecroft, a substantial residence where she has lived for many years with her husband and family. Mr Herrtage had become an invalid, and they decided to sell Danecroft and move into a smaller house. Mr Herrtage, who dealt with all their business affairs, put Danecroft into the hands of two agents, one being the firm of Garrard Smith, who in turn instructed Herring, Daw & Manners, of Croydon, as their subagents. Mr Damm obtained particulars of Danecroft from Herring, Daw & Manners. Mr Damm is of Polish origin and has lived in England since H
1963. He is a man of considerable ability, and now a director of a firm of technical consultants. He speaks English fluently. Early in 1971 he was looking for a house in South-East London. He had previously had some unhappy experiences of " gazumping." It would not be useful to give particulars of Mr Damm's experiences in relation to the purchase of other properties before and after his purchase of Danecroft. In January 1972 he went to Danecroft with his wife and her sister and looked over the property, by which he was much attracted. Then by appointment he called again, accompanied by Mr Doe, a surveyor, on February 8. They were received by Mrs Herrtage, her husband being then in hospital. Mr Doe J
made an inspection and left after an hour or two. Then came the critical interview between Mr Damm and Mrs Herrtage. It is not in dispute that they agreed in principle upon the price of £45,000 plus £500 for certain fixtures and fittings, and upon a deferred completion date.

According to Mr Damm, at the opening of the interview he made clear to Mrs Herrtage that he wished to enter into an immediate binding contract, and again according to him, when they came to terms on the price and completion date they came to those terms on the footing that they were making an immediate binding contract. On that footing, he paid a deposit of £1,000 and she signed a receipt. According to K
Mrs Herrtage, there was no reference to an immediate binding contract, and she received the deposit and signed the receipt on that footing. The receipt was in the following terms: " February 8 1972. Received from Mr M Damm of 79 Dora Road, SW19, a deposit of £1,000 on account of sale of my freehold house 'Danecroft,' Rose Walk, Purley, for the sum of £45,000 [in figures and words] plus £500 for carpets, curtains and other fixtures and fittings. The completion to be agreed mutually and to take place within one year from today. Vacant possession will be given on completion." That was written out by Mr Damm and was then signed " Zena R L
Herrtage, Danecroft, 24 The Rose Walk, Purley." Mr Damm then departed, taking the receipt with him. He posted a copy of the receipt to Mrs Herrtage that evening. I will refer at a later stage in more detail to the evidence as to what took place at this meeting.

Mr Damm proceeded to call on Mr F Coningsby, his solicitor, handed him the receipt and instructed him to act in the purchase. Mr Coningsby's attendance note reads as follows: " Damm. Purley. Clutton, Moore & Lavington, 679 London Road, Mrs Gardiner "—that is the name of Mrs Herrtage's solicitors, Mrs Gardiner being the partner who acted in the matter—" Thornton Heath. Send copy receipt. Not more M
than 5 per cent deposit. Vendor responsible for property. Delay completion one year from exchange, ie move in spring "; and over the page: " (1) Area of land. (2) Solicitor. (3) All buildings in garden included. (4) Confirm carpets, curtains and other fixtures and fittings. (5) Private estate road etc. (6) About 60 years old. (7) Copy of building scheme." The attendance note is undated. There was a conflict of evidence in connection with the date upon which Mr Damm called upon Mr Coningsby. Mr Damm deposed that Mrs Herrtage gave him the name of her solicitors at the end of the meeting on February 8 and that he went the same

A day to Mr Coningsby. The latter said that Mr Damm probably came on the afternoon of the 8th. On the other hand, Mrs Herrtage deposed that she did not know the name of her solicitors until her husband gave it to her on the evening of the 8th. Mr Herrtage confirmed giving the name. Then, according to Mrs Herrtage, she telephoned Mr Damm with the name of her solicitors next morning, the 9th. On this point I accept the evidence of Mrs Herrtage and also Mr Herrtage. It follows that the meeting cannot have been on the 8th but must have been on the 9th. The matter then proceeded between the solicitors and also the estate agents in the same way as if the principals had merely agreed terms
B subject to contract, and as if the solicitors were preparing a written contract to implement those terms. This, as I will explain in a moment, was a feature of the anti-gazumping device.

I should at this stage refer to a few letters which passed between Mr Coningsby, on behalf of Mr Damm, and Mrs Gardiner, the partner in Clutton, Moore & Lavington, on behalf of Mrs Herrtage. On February 9 Clutton, Moore wrote to Coningsby:

We understand that you are acting for Mr Michael Damm on his purchase of the above freehold property from our client, Mrs Herrtage, for the sum of £45,000. We have bespoken office copy
C entries and will send you a draft contract for approval shortly.

On February 10—these letters must have crossed—Coningsby to Clutton, Moore:

We have been instructed by Michael Robert Damm in connection with his purchase of the above freehold property from your client Mrs Herrtage. Our client has handed to us a document a photostat copy of which we enclose

—I interpose to say that is the receipt—

and we look forward to hearing from you in due course. Our client requests that he shall be called upon to pay not more than a 5 per cent deposit in view of the magnitude of the price and that he shall be entitled to interest upon the same in a building
D society deposit. We are instructed to obtain details of the exact area of the land, confirmation that all the buildings within it are included in the price, which also includes all carpets, curtains and other fixtures and fittings. We understand that this property is on an estate in relation to which there is a building scheme and that the property fronts a private road. You will no doubt let us have a copy of any relevant documents in these respects when you reply.

On February 18, Coningsby to Clutton, Moore:

Our client has now written to us stating that his architect estimates that it will be necessary to spend approximately £5,000 in repairs and redecoration of this property. This is a sum which
E apparently your client vendor agrees. Accordingly, our client would prefer the consideration shown in any transfer document to be £50,500 and that an allowance of £5,000 shall be made at completion for the necessary repairs.

On February 28, Coningsby to Clutton, Moore:

Please let us hear from you as soon as possible in relation to this matter. We would say, incidentally, that having perused the receipt dated February 8 1972 this appears to be a satisfactory memorandum of sale, particularly if taken with the cheque for the deposit sum therein mentioned.

March 1, Clutton, Moore to Coningsby:

We enclose herewith a list of the fixtures and fittings which
F are to be sold to your client for the sum of £500.

March 6, Coningsby to Clutton, Moore:

The fixtures and fittings listed are agreed. Please send us office copy entries immediately since this matter was negotiated on February 8 and our client wishes to proceed without delay. If you prefer to cover the matter by the use of a normal contract, please supply the same in duplicate.

April 5, Clutton, Moore to Coningsby:

We now enclose draft contract for approval together with a copy for your use.

The draft contract enclosed is a contract in ordinary form by reference to the Eighteenth Edition of the National Conditions of Sale. The purchase price is stated as £50,000. G
Then the special conditions included this condition:

Notwithstanding that the completion date is fixed by this contract to take place on the (blank) day of (blank) 1972

—in fact the completion date is left blank—

should the vendor desire to do so she shall be at liberty to give to the purchaser six weeks' written notice that she requires completion of the sale and purchase and the purchaser shall be required to complete his purchase six weeks from the date of the receiving of such notice.

On April 13, Coningsby to Clutton, Moore:

We now have pleasure in enclosing some preliminary inquiries H
to which as a matter of courtesy we shall be glad to have your replies. We also return the form of contract prepared by reference to the National Conditions of Sale to which we have no objection other than for the deletion of Special Condition F.

That is the special condition which I have just read.

We think that the position is that there is already an existing agreement for sale dated February 8 1972 providing that completion is to be agreed mutually and to take place within one year from that date when vacant possession is to be given. Accordingly the outside date for completion appears to be February 8 1973 unless mutually agreed before then, and it is upon this basis that the deletion from the contract has been made.
When we hear from you with replies to the enclosed inquiries we J
expect exchange of the contract under the National Conditions of Sale can take place with completion on or before February 8 1973.

Then on April 26, Clutton, Moore to Coningsby:

We now return your preliminary inquiries with our replies and look forward to receiving the part of the contract signed by your client.

May 9, Coningsby to Clutton, Moore:

We enclose the contract signed by our client purchaser in this matter with a further cheque for £1,000. The date for completion is, of course, on or before February 8 1973. K

Then followed a delay of some two months. On July 11 1972, Clutton, Moore to Coningsby:

As arranged with you on the telephone, we return your cheque for £1,000 in respect of the above property. P.S. If your client wishes our client to return the £1,000 paid to Mrs Herrtage we will send you a cheque immediately on hearing from you.

July 13, Coningsby to Clutton, Moore:

Thank you for your letter of July 11 with the return of the cheque enclosed. Is your client going to exchange the contracts prepared or are we to rely on the memorandum of sale, registered on May 8? L

August 3, Clutton, Moore to Coningsby:

Mrs Herrtage has instructed us to write on her behalf to clarify the situation with regard to your client's proposed purchase of her property at the above address. A situation has arisen since your client saw Mrs Herrtage on February 8 1972 which will preclude her from proceeding with negotiations and exchange of contracts with your client. Her husband was at the time seriously ill in hospital and is now unable to move around other than within the confines of his home. It has made it impossible for them to look for alternative accommodation, added to which Mr Herrtage may have to return to hospital for an indefinite period shortly. In view of these unforeseen circumstances, we are returning your client's
initial deposit together with our client's sincere apologies for any M
inconvenience caused. Our client has asked us to say that any reasonable expenses which your client may have incurred will willingly be met by him if necessary. A cheque for £1,000 is enclosed.

Then on August 7, Coningsby to Clutton, Moore:

We thank you for your letter of August 3, upon which we have taken our client's instructions in detail, but he has instructed us to return to you your client's cheque which he is not prepared to accept. The memorandum of sale stands, as does also the caution registered on May 8 in the Land Registry, and completion in the transaction is due on February 8 1973. Perhaps we may add for ourselves that we have every sympathy with your client's position.

A I need not read the subsequent correspondence, which led up to the issue on October 16 1972 of a writ in an action intituled 1972 D 2370. There was in evidence some correspondence between Herring, Daw & Manners and Mr Damm, and also between Garrard Smith and Herring, Daw & Manners. This correspondence, as is the careful habit of estate agents, is copiously interlarded with the expression "subject to contract," but there is no evidence to suggest that Mr Damm himself authorised the use of this expression. The most that could be said against him on this correspondence is that on February 10 1972 Herring, Daw wrote to him, "Further to our recent conversation we understand that you
B are prepared to proceed with the purchase of the above freehold property in the sum of £45,000 subject to contract," and in his reply next day he merely said, "I am pleased to inform you that my offer for the sale of the house has been accepted." I do not think that really goes anywhere.

Pleadings were delivered in that action, 1972 D 2370. Mr Damm, however, for reasons not now material, discontinued that action, and on November 16 issued the writ in the present action, 1973 D 600. I should read some two or three paragraphs from the pleadings in the action. In the statement of claim, endorsed on the writ, paragraph 2:

C By a contract made on February 8 1972 it was agreed that the defendant would sell and the plaintiff would purchase the property [described in paragraph 1] for the sum of £45,000; that the plaintiff would in addition purchase certain carpets and other fixtures and fittings for a further £500; that the contract should be completed on February 8 1973 or on such earlier date as might be mutually agreed; and that vacant possession of the property would be given on completion.

Paragraph 3:

The said contract was made orally between the plaintiff and the defendant at a meeting at the property. At the said meeting the plaintiff paid the defendant the sum of £1,000 by way of deposit on account of the said sale, and the defendant signed and
D gave to the plaintiff a receipt for the said deposit, in which the terms of the said contract were set out.

Paragraph 4:

The defendant alleges that it was a further term of the said contract that the plaintiff would complete at any time the defendant required on being given one month's prior notice by the defendant. The plaintiff is willing to perform the said contract with the addition of such further term as an alternative to the contract pleaded in paragraph 2 hereof.

I need not read the rest of the claim. The defence, paragraph 3:

E It is admitted that a meeting between the plaintiff and the defendant at the property took place on February 8 1972. It is admitted that at the said meeting the plaintiff paid to the defendant by cheque the sum of £1,000 and that the defendant signed and gave to the plaintiff a receipt for such sum. It is denied that any contract was made at such meeting or that such receipt set out the terms of the contract alleged in paragraph 2 of the statement of claim or of any other contract. The said sum was paid on account of any deposit which would be payable under any contract for the sale of the property by the defendant to the plaintiff if any such contract were to be made subsequently. No such contract was ever made.

Paragraph 4:

F The defendant denies that the contract referred to in paragraph 4 of the statement of claim was made but alleges that if such contract was made it contained the further term therein mentioned and the defendant relies on section 40 of the Law of Property Act 1925. The last sentence of that paragraph is not admitted.

I will mention at this stage, in order to get the point out of the way, that I am not satisfied that the parties ever agreed upon an amended date for completion as alleged in paragraph 4 of the statement of claim and paragraph 4 of the defence. I use the word "agreed" without prejudice to the question whether they became contractually bound at all.

The action was consolidated with an action intituled 1972 H 9589 commenced by Mrs Herrtage, in which she claims a G
declaration that no contract had been made and an order for vacation of the caution. The actions came on for hearing on April 8 1974. The principal witnesses were Mr Damm and Mrs Herrtage themselves. Significant evidence was also given on one side by Mr Coningsby and on the other side by Mr Herrtage. Certain other witnesses were called on behalf of Mr Damm, but it would not be useful to refer specifically to their evidence, which in the event threw no light on what took place on February 8. Before I return to the meeting of February 8 1972, I should say something about the anti-gazumping device recommended by Mr Coningsby and put into practice by Mr Damm. The basis of this advice was H
that the purchaser should enter into an immediate binding contract orally with the intending vendor, paying a deposit and taking a receipt which would represent a memorandum sufficient to satisfy section 40 of the Law of Property Act. The matter would then be put in the hands of the purchaser's solicitors, with instructions to proceed in the ordinary way with the preparation of a written contract. Then, if all went well, the respective solicitors would agree the terms of a written contract and the written contract would be signed, the memorandum never being mentioned again. Only if the vendor resiled from entering into a written contract would the oral contract and the memorandum be relied upon. Whatever J
the ethical merits of gazumping, it seems to me that this counter-device comes very near sharp practice and could easily be used as an engine of fraud. I would make it clear that I am not intending to imply that Mr Damm was guilty of any fraud in the present case. A lay vendor and purchaser normally discuss terms in the expectation that if they reach agreement in principle, the terms will be incorporated in a written document with a number of other terms, and that they will not be contractually bound until that written document is signed. It is, of course, possible for them to bind themselves orally, and the purchaser may make it clear to the vendor that that is what he is proposing, but bearing in mind the normal K
practice in this connection, there is obviously room for misunderstanding, and worse, in this transaction at an oral interview from discussion to binding contract. Then again, the transaction is recorded in the document described as a receipt. This receipt would serve as a memorandum against the vendor but not against the purchaser. If indeed the parties intend to bind themselves forthwith, the obvious and straightforward course would be for both of them to sign a document containing the terms of the contract and for that document to be described as an agreement. Mr Coningsby in his evidence described the nature of this device in detail, and gave the interesting reply that it was not particularly tricky or L
unfair.

I must mention one other point, although this point has no direct bearing on the issue in the present action. In the draft contract the purchase price is stated to be £50,000, whereas it was in truth £45,000, the difference representing repairs which the purchaser would expect to do in order to bring the property into proper condition. The object of this overstatement of the price was to facilitate a mortgage. This overstatement of the purchase price was obviously improper, and the solicitors concerned ought not to have lent themselves to it. M

I turn now to the events of February 8 1972. Mr Damm is a youngish man, as I have said, of very considerable intelligence. He was not a man who would lightly permit himself to be overreached, and he had been put on his guard by previous experience of gazumping. He approached the present transaction with the purpose of tying Mrs Herrtage down if he could. This was in itself in no way improper, but the method employed, the anti-gazumping device, was one which calls for careful scrutiny of what took place. Mrs Herrtage is a well-to-do lady, sensible, but without business experience. She was accustomed to rely in all business

A matters upon her husband, who it will be remembered was in hospital on February 8 and is still very much an invalid, although he was able to give evidence from a wheel chair. Mr Damm's account of his interview with Mrs Herrtage on February 8 is as follows. It will be remembered that Mr Doe, the surveyor, left, and Mr Damm and Mrs Herrtage then went together into another room, actually the study, and discussed the intended purchase. Mr Damm says: "I said that provided there was a binding agreement on the spot, I would make a firm offer and pay a non-returnable deposit straightaway and commit myself to the purchase." He said: "I told Mrs Herrtage I was being pressed by my own
B purchaser, and that was a reason for binding myself there and then." Then they discussed the time for completion. Mrs Herrtage said she would be willing to sell at £45,000 plus £500 for fixtures and fittings. He agreed. "We reached a firm agreement at £45,500." She said, as was the fact, that she was the sole owner. Then he identified the memorandum as being in his handwriting and said: "That was intended as a record of our agreement." Then he said he would leave a deposit of £1,000 as part of the agreement. This £1,000 related to late completion. "It would be unreasonable to leave £4,500," that is, 10 per cent. He said: "I insisted on an outside date, and suggested one year."

C Then, according to Mr Damm, Mrs Herrtage took the document out of the room with her, returned and said it was OK to sign, and did sign. He gave her a cheque when she signed the memorandum, and she appeared delighted with the sale. "I would not have handed her £1,000 unless there had been a binding agreement." Then he said: "She made no mention of her solicitors until I was about to leave, and then told me their name." I have already said that I do not accept his evidence on that point. He then goes on to the interview with Mr Coningsby. In cross-examination he adhered to his account. He said: "I told her verbally on February 8 that it was to be binding"—that is, the contract; and he was
D anxious to get on that day a binding agreement. Mrs Herrtage knew it, and nothing was said about her husband in hospital. He denied telling her that the contract would be binding when drawn up; that means a future written contract. He explained his instructions to Mr Coningsby to pay another £1,000 as being by way of goodwill. He also said he asked Mrs Herrtage whether she wanted to speak to her husband and she replied that she was the sole owner. So, according to Mr Damm, he set the scene from the start for the formation of an immediate binding oral contract, and they came to terms as to price and the date for completion on
E that footing.

I turn next to Mrs Herrtage's account. She refers to a discussion as to the date of completion, and she says: "Mr Damm told me he would like to give me £1,000 deposit. There was no mention of a contract. He wrote things down and asked me to sign. Then I said I would prefer to leave it; my husband deals with my business." He said: "I cannot leave £1,000 without a receipt." She then made an abortive telephone call to her bank, where she wanted to speak to the manager. Then she telephoned to the hospital where her husband was. She could not speak to him directly, because he was in a public ward where there was no telephone, but
F she did give a message to a nurse who answered the telephone, asking whether it was all right to sign the receipt, and the nurse came back with the message that it would be all right. That was confirmed by Mr Herrtage. Then, as I have already said, she deposed that she did not give Mr Damm the name of her solicitors because she did not know it. It was not until the next day that she telephoned the name through to him. The expression "subject to contract" was not used in the discussion between Mr Damm and Mrs Herrtage. So Mrs Herrtage's account is that although the discussion was not expressed to be subject to contract, there was no reference to a binding contract; they simply came to terms in principle; and she said she telephoned her husband, through the nurse, G
and got the reply that it was all right to sign the receipt. I need not, I think, refer further to Mr Herrtage's evidence.

I am unable to accept the evidence of Mr Damm in so far as he deposes that at the outset or in the course of the meeting of February 8 he told Mrs Herrtage that he wished to make an immediate binding contract and that she came to terms on that basis. Equally, I accept the evidence to the contrary of Mrs Herrtage. The latter seemed to me to be a reliable witness, and the inherent probabilities favour her account. It must be unlikely to a degree that this lady, with no business knowledge and having to rely on her husband in business matters, would have considered binding herself in this rather H
offhand way, without any advice, to the sale of her extremely valuable property. Again, if that had been her intention, she would naturally have put her signature to a document described as an agreement rather than as a receipt. It is not possible to find a great deal of support for either account, or the reverse, in the correspondence. It is, however, worth pointing out that in Mr Coningsby's attendance note there is no reference to a binding contract having been made. In the first letter which he wrote on February 10 1972 Mr Coningsby made a request that his client should "be called upon to pay not more than a 5 per cent deposit in view of the magnitude of the price." That looks like the negotiation of a term of a J
prospective contract. Then again, in the course of the correspondence Mr Coningsby, although he sent the receipt with his letter of February 10 describing it as "a document," did not until April 13 take specifically the point that there was a written and existing agreement for sale. The nearest he came to that before April 13 was his reference in the letter of February 28 to the receipt being "a satisfactory memorandum of sale," which to my mind is something quite different. I appreciate that Mr Coningsby's reticence as to the existence of a binding contract was part of the anti-gazumping device as explained by him, but it seems to me that a party who plays this sort of double game cannot complain if the K
evidence does not turn out to support the contention he ultimately makes as to what really happened.

As I have said, there is very little in the subsequent correspondence or the documentation which corroborates or otherwise the oral evidence of the two principal parties as to what took place on February 8 1972, and I do not think I can usefully elaborate my finding further. It is perhaps unnecessary to point out that where parties use words of offer and acceptance in connection with a transaction, it depends on the circumstances whether their intention is or is not to bind themselves contractually. *Prima facie,* no doubt, L
unqualified words of offer and acceptance have this result. On the other hand, the parties may intend no more than to agree in principle terms which are to form part of the contract, and this position arises particularly in the case of a contract for the sale of land, where the contract normally includes a variety of terms in addition to price and date of completion. In order to determine the intention of the parties, one must look at the whole of what took place and passed between them. I was referred in this connection to the well-known statement of Lord Cairns in the case of *Hussey v Horne-Payne* (1874) 4 App Cas 311. The observations in this passage are equally applicable to an oral contract or to an M
oral contract with a written memorandum. It is not necessary, although obviously prudent, in this connection that the parties should expressly use the words "subject to contract," or the like, if they wish to exclude the implication of a contractual intention. For the reasons which I have given, I conclude that no binding contract was made between Mr Damm and Mrs Herrtage. I must therefore dismiss this action D 600 and make an order for vacation of the caution as asked in the other action, H 9589.

The defendant was awarded the costs of the consolidated action.

A **COURT OF APPEAL**

May 8 1974

(Before Lord Justice EDMUND DAVIES, Lord Justice STEPHENSON and Sir Gordon WILLMER)

BANBURY BOROUGH COUNCIL v OXFORDSHIRE & DISTRICT WATER BOARD

Estates Gazette May 10 1975

(1974) 234 EG 447

Developer requiring a supply of water to two estates near Banbury held by majority to be liable to contribute towards cost of a 27-in trunk main laid along the 20 miles from B Woodstock to Banbury—Leave to appeal to the House of Lords granted

This was an appeal by the Oxfordshire & District Water Board from a judgment of Mocatta J on May 11 1973 declaring in favour of the respondents, Banbury Borough Council, that the board were not entitled under the provisions of the Water Act 1945 to levy a contribution towards the cost of constructing a 27-in trunk main running for about 20 miles between Woodstock and Banbury and intended in part to supply council houses proposed to be erected by the respondents.

Mr M Mann QC and Mr H Donovan (instructed by Lewin, C Gregory, Mead & Sons) appeared for the appellants, and Mr G E Moriarty QC (instructed by Jaques & Co) represented the respondents.

Giving judgment, EDMUND DAVIES LJ said that in 1971 the plaintiffs were minded to erect a number of council houses on land which they owned within their borough. At Penrose Close they planned to build four bungalows, 12 flats and warden's accommodation, and at Ruscote 16 bungalows. In June 1971 they inquired of the defendants, the statutory body empowered to supply water to the district, about the supply of water to these two sites. In July the defendants said that the estimated cost of laying the mains needed to serve the D Penrose Close development was £1,434, being £1,060 for a main from the development to the existing mains network in the borough and £374 as a contribution towards the cost of a 27-in diameter trunk main between Woodstock and Banbury; and that the corresponding estimated cost in respect of the Ruscote development was £1,227, made up of the sums of £875 and £352. There was no dispute about the plaintiffs' liability for the sums of £1,060 and £875, but a declaration was sought to settle the question whether the defendants could lawfully demand, as they had done, one-eighth of each of the sums of £374 and £352. The items involved totalled a E mere £90.75, but it was common ground that the issue raised was one of considerable importance. Its resolution depended on the construction of section 37 (1) of the Water Act 1945, which provided as follows:

> Where an owner of land proposes to erect thereon buildings for which a supply of water for domestic purposes will be needed, he may require any statutory water undertakers within whose limits of supply that land is situated to construct any necessary service reservoirs, to lay the necessary mains to such point or points as will enable the buildings to be connected thereto at a reasonable cost and to bring water to that point or those points, and thereupon the undertakers shall, subject as hereinafter provided, comply with that requisition: Provided that the undertakers F before complying with a requisition under this subsection—
>
> (a) may require the owner to undertake to pay in respect of each year a sum amounting to one-eighth of the expense of constructing the necessary service reservoirs and providing and laying the necessary mains (less any amounts received by the undertakers in respect of water supplied, whether for domestic or non-domestic purposes, in that year from those mains) until the aggregate amount of water rates payable annually in respect of the buildings when erected and in respect of any other premises connected with the said mains at the rates for the time being charged by the undertakers equals or exceeds such sum as aforesaid or until the expiration of a period of 12 years whichever first occurs; and
>
> (b) except where the owner is a local or public authority, may G also require him to deposit with the undertakers as security for payment of the said annual sums, such sum, not exceeding the total expense of constructing the service reservoirs and providing and laying the mains, as the undertakers may require.

Certainly not later than February 1970, the defendants had planned to construct a 27-in diameter trunk main from the outskirts of Woodstock to an existing service reservoir at Banbury, the distance being 20 miles and the cost £940,000, and by the time the matter came before Mocatta J the necessary works were already under way. The new trunk main was required, as the learned judge had found, to maintain at all times a proper supply of domestic water to H Banbury users existing in 1971 and also to the plaintiffs' two proposed developments. The size of the particular main installed was such that it would in fact be capable of conveying to Banbury additional quantities of water far in excess of the needs of 1971 users supplemented by the plaintiffs' proposed developments; it would accordingly be able to cater for greatly increased demands for water in the borough. The question for decision was whether this trunk main could as a matter of law be a "necessary main" within the meaning of section 37 (1), for if not, the defendants could not require an undertaking from the plaintiffs to pay the £90.75. J

The Act did not define "necessary main," though it contained definitions of "main" and "trunk main." Mocatta J had upheld the plaintiffs' contention that while a main could be either a trunk main or what had been called a distributing main, a trunk main could never be a necessary main for the purposes of section 37 (1). It was clear that the undertaking dealt with by the proviso to the subsection could not exceed a proportionate part of such works as the developer might lawfully require the undertaker to carry out for the purpose of ensuring "a supply of water for domestic purposes . . . needed" by the proposed buildings. Then K could a developer ever require an undertaker to lay a trunk main? It was unsatisfactory to answer only that if such were the intention, section 37 (1) would have said so; it could be argued with more force that wherever the statute intended to exclude trunk mains this would have been made clear. Equally unsatisfactory was the stress laid by counsel for the plaintiffs on the fact that the reference in section 37 (1) to the laying of necessary mains was immediately preceded by a reference to the construction of "any necessary reservoirs." This had led counsel to submit that the undertaking required of a developer could not relate to the cost of any works further back than a service reservoir, and even L to make the unqualified submission that if there was no service reservoir there could be no trunk main. As a general proposition, this was unacceptable; in the present case there already existed a service reservoir, but the fourfold definition of "trunk main" in the Act made clear the untenable nature of the proposition. In a case where no service reservoir existed, how far back might one go for the purpose of determining what works were relevant to the calculation of the amount of the undertaking extracted from the developer? It might be said that one should not go beyond the point at which the distributing mains leading to the point at which the development could be connected to them were themselves M connected to the undertaker's bulk supply. But as counsel for the defendants had pointed out, a new developer might require water to be brought in bulk, and any main constructed for that purpose would be within the definition of "trunk main." Counsel submitted, as he (his Lordship) thought convincingly, that "necessary mains" must include all species of mains and so embrace both trunk and distributing mains. Counsel had argued that the underlying philosophy of the Act was that those requiring water should not be allowed to impose on existing consumers and at no cost to themselves the burden of laying any necessary mains, and that for the

A purpose of section 37 (1) one had to look at the necessity for the main and not at its species. He (Edmund Davies LJ) preferred the submissions of counsel for the defendants, and would allow the appeal.

STEPHENSON LJ delivered a concurring judgment.

Dissenting, SIR GORDON WILLMER said that the question to be decided was restricted to the particular trunk main relevant to the case; and one must ask whether in the context this was a "necessary main." A developer of land could require the water undertakers (a) to construct any necessary service reservoirs, (b) to lay the necessary mains to such point or points as would enable the proposed buildings to
B be connected thereto at a reasonable cost, and (c) to bring water to that point or those points (the "delivery point"). The service reservoirs and the mains which the undertakers could be required to construct must be "necessary" to meet the need in question. The laying of mains from a service reservoir to the delivery point was clearly necessary to meet that need. He (his Lordship) could however find no words in section 37 (1) which enabled the developer to require the laying of mains or other constructional work beyond the service reservoir, or, if there were no service reservoir, beyond the point at which the mains leading to the delivery point could themselves be connected to the undertakers' bulk
C supply. The fact that without the trunk main in question Banbury users, including those in the plaintiffs' new houses, would at times be in danger of having no water went no distance towards showing that the necessary mains to enable the buildings to be connected to the delivery point must be construed as including whatever trunk main was required to ensure the bulk supply of water on which to draw. The definition of "trunk main" in the Act covered (a) a main constructed for the purpose of carrying water from a source of supply to a filter or reservoir, (b) a main constructed for the purpose of carrying water from one filter or reservoir to another filter or reservoir, (c) a main constructed for the
D purpose of conveying water in bulk from one part of the limits of supply to another part of those limits, and (d) a main constructed for the purpose of giving or taking a supply of water in bulk. He (Sir Gordon) was quite unable to see how the laying of "the necessary mains to such point or points as will enable the [proposed] buildings to be connected thereto," as envisaged by section 37 (1) of the Act, could be fitted into any one of these four categories of trunk main. A contrary conclusion could lead to quite absurd results utterly repugnant to commonsense. The owner of the land desiring to erect a building a few yards away from where a trunk main was in course of being laid
E could be required to contribute not only to the expense of laying the necessary main to the point at which his proposed building could be connected thereto but also to the expense of laying the whole trunk main, whatever its size and however long it might be. Any ordinary person would require a good deal of persuasion that Parliament ever intended such a bizarre result. The appeal should be dismissed.

The appeal was allowed with costs above and below, and a declaration made accordingly. Leave to appeal to the House of Lords was granted.

F

HOUSE OF LORDS

March 5 1975

(Before Lord WILBERFORCE, Viscount DILHORNE, Lord DIPLOCK, Lord KILBRANDON and Lord SALMON)

CHERWELL DISTRICT COUNCIL v THAMES WATER AUTHORITY

Estates Gazette May 10 1975

(1975) 234 EG 449

Developer succeeds in appeal to House of Lords from above decision—Important discussion of ambit of section 37 and associated provisions of Water Act 1945—Cases of new mains altogether and improvement of existing mains evidently to be distinguished G

The Cherwell District Council (successors to the Banbury Borough Council) appealed from the decision reported above in favour of the Oxfordshire & District Water Board (now Thames Water Authority).

Mr G E Moriarty QC and Mr A J Anderson (instructed by Jaques & Co) appeared for the appellants, and Mr M Mann QC and Mr H Donovan (instructed by Lewin, Gregory, Mead & Sons) for the respondents.

In a speech with which the remainder of their Lordships agreed, LORD DIPLOCK said: H In 1971, the Banbury Borough Council (the predecessors of the Cherwell District Council) proposed to build, upon land they owned at two sites in the borough, dwellings for which a supply of water for domestic use would be needed. The Borough of Banbury was within the limits of supply of the Oxfordshire and District Water Board (the predecessor of the Thames Water Authority), who were the statutory water undertakers for that area. At neither of the sites were there existing water mains from which a supply of water for the new buildings could be obtained. This would require the laying of new mains. J At that time the source of supply for the borough was water pumped from the River Cherwell into a service reservoir. This source, however, was becoming inadequate for the growing demand. To supplement it the board proposed to extract water from the Thames near Woodstock and convey it by a 27-in trunk main to the existing service reservoir at Banbury. Being desirous of exercising their right under section 37 of the Water Act 1945 to require a water board to lay mains from which a supply of water to the proposed new buildings could be obtained, the council asked the water board how much the board would require them to undertake to pay under the provisions of that section. In reply, the board quoted a figure which comprised two K elements:

(1) the cost of laying the new mains which would be needed in the immediate vicinity of the sites, and

(2) a contribution towards the cost of the proposed new 27-in trunk main conveying water from the Thames to the existing service reservoir for supplying the whole of the borough of Banbury.

The total cost of laying the new trunk main was of the order of £1m. The contribution towards this total demanded in respect of the two sites was small, viz £374 for one and £352 for the other. It appears to have been based upon the L estimated proportion of the total gallonage of water passing through the trunk main which would be supplied to the buildings on those particular sites. The council accepted their liability for the first element, viz the cost of laying the new mains in the immediate vicinity of the two sites, but contended that the water board were not entitled under the section to demand any contribution from them towards the cost of the proposed 27-in trunk main. By originating summons of December 21 1971 the council sought a declaration "whether the defendants" (sc the water board) "can lawfully demand the said sums of £374 and £352 as said contributions in respect of the said 27-in diameter trunk main or M at all." The actual amount at stake was small, but both parties wished to have settled a question of principle as to the construction of section 37 of the Water Act 1945. In the belief that it would facilitate this, leading counsel for the two parties before the commencement of the hearing of the summons before Mocatta J agreed as follows:

We agree that the single issue between the parties is whether a trunk main is in law capable of falling within the phrase "necessary mains" as employed in proviso (*a*) to section 37 (1). We further agree, that, if the answer to that question is "yes," then no further question arises either as to whether the Woodstock-Banbury trunk main is in fact a necessary main or as to whether

A the proportion of the cost attributed to the plaintiff's development is correct. In other words the simple answers "yes" or "no" are each equally conclusive of the action.

In my view, the effect of this agreement is to substitute for the question raised in the originating summons, which was the actual question at issue between the parties, a broader and purely hypothetical question which the court should not have entertained. The result in the instant case has been unfortunate. The agreement assumes that if, upon the true construction of the section, the expression "necessary mains" is capable of including any trunk main at all, the B 27-in trunk main conveying water from the Thames to the service reservoir at Banbury must be included in that expression. It thus diverts the court from a consideration of the meaning of the expression "necessary" as descriptive of the mains to which the section applies, by purporting to exclude as a possible answer to the question raised in the originating summons that although in some circumstances a trunk main may be a "necessary main" within the meaning of the section, the 27-in trunk main from Woodstock to Banbury is not. In the result, the only question argued at the original hearing by Mocatta J and in the Court of Appeal was the hypothetical question: "Aye or nay, does the expression C 'necessary mains,' upon a true construction of the section exclude all trunk mains?" In his reasons for judgment, Mocatta J dealt only with this hypothetical question. He answered it "Aye." In the Court of Appeal, too, the reasons given by the majority (Edmund Davies and Stephenson LJJ) dealt only with the hypothetical question. They answered it "Nay." Nevertheless, the formal order of the Court of Appeal answered the actual question raised by the originating summons. It declared "that in regard to the requirement made by the [Oxfordshire & District Water Board] in respect of the proposed building of the [Banbury Borough Council] at Penrose Drive and at Ruscote in the Borough of Banbury the D 27-in diameter trunk main to be laid by the water board between the outskirts of Wodstock and their service reservoir at Banbury is a necessary main for the purposes of proviso (*a*) to section 37 (1) of the Water Act 1945." It was only Sir Gordon Willmer who directed his mind to the actual question. He was of opinion that the 27-in main was not a "necessary main" within the meaning of proviso (*a*) to section 37 (1) of the Water Act 1945. I think that he was right.

A court that is called upon to construe a statute in order to determine an issue between parties to litigation must not allow itself to be deflected from its task of deciding what the language of the statute means by any agreement between E those parties which involves, expressly or by implication, some assumption as to that meaning. On such a matter the constitutional function of a court as the interpreter of the written law compels it to reach its own unfettered decision. This House would be failing in its duty if it confined itself to the hypothetical question dealt with in the judgments of Mocatta J and the majority of the Court of Appeal, and treated the answer to that question as conclusive of the actual question raised in the originating summons—as was done in the order made by the Court of Appeal. In the view I take as to the true construction of the statute, to do this would provide no useful guidance as to the application of F the Act to other cases. It could only lead to confusion. It might positively mislead. I turn, then, to the question of construction unfettered by any agreement between counsel for the parties as to its meaning, or as to its effect in the particular circumstances of the instant case. The Oxfordshire & District Water Board was constituted by an order made by the Minister under section 9 of the Water Act 1945. The order incorporated, with some immaterial exceptions, the provisions of the third schedule to the Act, which is a modernised version of clauses formerly contained in the Waterworks Clauses Acts 1847 and 1863. The relevant powers and duties of the board are thus to be found in Part IV of the body of the Act (which includes section 37) and in the G third schedule. There is an interpretation section, section 59, in the body of the Act, but a more comprehensive one is contained in clause 1 of the third schedule. (To avoid confusion between sections of the Act which bear the same number as sections in the third schedule, I refer to the latter as "clauses.") The definitions that are relevant to the construction of section 37 appear for the most part only in clause 1, but it is not, and in my view cannot be, disputed that where expressions there defined are used in the body of the Act the same meaning is to be ascribed to them. The only definitions which I find it necessary to cite are "main," "trunk main" and "supply of water in bulk": H

"main" means a pipe laid by the undertakers for the purpose of giving a general supply of water as distinct from a supply to individual consumers and includes any apparatus used in connection with such a pipe;

"trunk main" means a main constructed for the purpose of conveying water from a source of supply to a filter or reservoir, or from one filter or reservoir to another filter or reservoir, or for the purpose of conveying water in bulk from one part of the limits of supply to another part of those limits, or for the purpose of giving or taking a supply of water in bulk;

"supply of water in bulk" means a supply of water for distribution by the undertakers taking the supply. J

"Main" is thus a genus of water pipe of which "trunk main" is a species embracing four sub-species which have the common characteristic that they are pipes conveying water in bulk to some place such as a reservoir from which it is distributed by the undertakers. It is convenient to refer to mains which are not trunk mains as "distribution mains," as has been done in the courts below. The scheme of the Act and schedule as respects the rights of owners and occupiers of existing premises to demand and receive from statutory water undertakers a supply of water for domestic purpose is this: K

(1) Where there are existing mains which are capable of being reached by a service pipe from the premises the owner or occupier is entitled to demand to be connected with the main and to receive a supply of water for domestic purposes, on payment or tender of the water rate. That part of the service pipe that lies between the main and the boundary of the street in which the main is laid is to be laid by the undertakers; the remainder by the owner or occupier of the premises. He must pay the cost to the undertakers of laying their part of the service pipe and connecting it to the main: but no part of the cost of the main itself L can be demanded of him. Even if the capacity of the existing main is insufficient to provide his premises with a constant supply at adequate pressure (as the undertakers are obliged to do by clause 39) they cannot demand any payment towards the cost of replacing the existing main by one of greater capacity or of any other measures required to enable them to fulfil this obligation (see clauses 30, 40 and 41).

(2) Where premises are in an area in which there are no existing mains capable of being reached by service pipes from those premises, the owners or occupiers of such premises may combine together to require the M water undertakers to lay "the necessary mains." The undertakers must comply with this requirement if the requisition is made by owners or occupiers of premises in respect of which the water rates payable annually will, in the aggregate, be not less than one-eighth of the cost of providing and laying the necessary mains (clause 29). This right, which corresponds with a former provision of the Waterworks Clauses Acts, has been supplemented by a new provision contained in the body of the Water Act 1945. In cases where there are not sufficient owners or occupiers of premises

A in the area to qualify them to make a valid requisition themselves, section 36 authorises a local authority to undertake to make good to the undertakers for a maximum period of 12 years the deficiency (if any) between the water rates payable annually in respect of premises in the area and the one-eighth of the cost of providing and laying the necessary mains. Upon such an undertaking being given, the water undertakers are obliged to lay the necessary mains and to bring water to the area. Although when the new mains have been laid the owner or occupier of any premises in the area must pay the cost of laying a service pipe con-
B necting them with the premises, no contribution may be demanded of him towards the cost of laying the new mains or of bringing water to them. His liability is confined to paying the annual water rate levied upon him in common with all other consumers of water for domestic purposes.

Prior to the Water Act 1945 an owner of land had no statutory right to require water undertakers to lay new mains for the supply of water for domestic purposes to a building which he proposed to erect but had not yet erected. All that he could do was to complete the building and hope that when it was completed he, either alone or in combination with
C owners or occupiers of other premises in the same area, would qualify by reason of the aggregate rateable value of the premises to make a requisition upon the undertakers under the provision of the Waterworks Clauses Act corresponding to clause 29 of the third schedule to the Water Act 1945. The Public Health Act 1936, however, made it impracticable to build houses in reliance on this hope alone. Section 137 of that Act compelled a local authority to reject plans for new houses unless they were satisfied that a supply of wholesome water sufficient for the domestic purposes of their occupants would be provided. Section 37 of the Water Act 1945 was a new provision designed to deal with this situation.
D Minor amendments were made to it by the Housing Act 1949, section 46; but I do not think that these affect the meaning of the expression "necessary mains," which was also in the section before it was amended. It is sufficient to cite it only in its amended form:

(1) Where an owner of land proposes to erect thereon buildings for which a supply of water for domestic purposes will be needed, he may require any statutory water undertakers within whose limits of supply that land is situated to construct any necessary service reservoirs, to lay the necessary mains to such point or points as will enable the buildings to be connected thereto
E at a reasonable cost and to bring water to that point or those points, and thereupon the undertakers shall, subject as hereinafter provided, comply with that requisition;

Provided that the undertakers before complying with a requisition under this subsection—

(*a*) may require the owner to undertake to pay in respect of each year a sum amounting to one-eighth of the expense of constructing the necessary service reservoirs and providing and laying the necessary mains (less any amounts received by the undertakers in respect of water supplied, whether for domestic or non-domestic purposes, in that year from those mains) until the aggregate amount of water rates payable annually, in respect of the buildings when erected
F and in respect of any other premises connected with the said mains at the rates for the time being charged by the undertakers equals or exceeds such sum as aforesaid or until the expiration of a period of 12 years, whichever first occurs; and

(*b*) except where the owner is a local or public authority, may also require him to deposit with the undertakers as security for payment of the said annual sums, such sum, not exceeding the total expense of constructing the service reservoirs and providing and laying the mains, as the undertakers may require.

(2) The undertakers shall pay interest at the prescribed rate or, if no rate is prescribed, at 4 per cent per annum on any sum in their hands by virtue of a requirement under paragraph (b) of G
the proviso to the last foregoing subsection, and shall, on the request of the owner of the land, appropriate out of that sum any amount due under the undertaking referred to in paragraph (a) of the said proviso and shall, when the said undertaking is finally discharged, repay to the owner any sum remaining in their hands as aforesaid.

(3) Any question arising under subsection (1) of this section as to the point or points to which mains must be taken in order to enable buildings to be connected thereto at a reasonable cost shall, in default of agreement, be determined by the Minister.

(4) If the undertakers, after receipt of a requisition under subsection (1) of this section and after tender to them of any H
undertaking or deposit which they may require in accordance with that subsection, do not before the expiration of three months (or, where a question has, before that time, been referred to the Minister under the last foregoing subsection, before the expiration of three months from the date when the Minister notifies the undertakers of his decision, if that period expires later) comply with the requisition, they shall, unless they show that the failure was due to unavoidable accident or other unavoidable cause, be guilty of an offence against this Act.

"Mains" is, by virtue of its definition in the Act, a generic term which *prima facie* includes trunk mains as well as distribution mains, unless the context in which it appears makes it clear that it was used in some more restricted mean- J
ing. In section 37 (as also in clause 29) the only express restriction upon its generality is that the mains must be "necessary." This is, in my view, the key word in the section for the purpose of deciding the actual question raised by the originating summons. It poses immediately the question: "Necessary for what?" The section presupposes, first, that there are no existing mains bringing water to points at which it would be practicable at reasonable cost to connect the proposed buildings to those mains by service pipes. If there were, there would be no need for any requisition. Once the building was erected the owner would be entitled to require K
the connection to be made under clause 30, without any liability to contribute to the cost previously incurred by the undertakers in laying the existing mains or to the cost of replacing them by mains of greater capacity to meet the additional demand resulting from the connection of the newly-erected building to them. Secondly, the section presupposes that there will be a supply of water in bulk capable of being brought to the new mains by the undertakers, which (if it is not already available to them) it is their duty to procure without demanding from the requisitioner any contribution to the cost of doing so. Thus the "necessary mains" with which the section deals are confined to new mains to be L
laid by the undertakers in a street or other place where there are no existing distribution mains and to start from a point to which a supply of water is already being brought in existing works belonging to the undertaker; and the answer to the question "Necessary for what?" is: necessary in order to convey the water from that starting-point to points at which it would be practicable to connect the proposed buildings to those mains by service pipes, and for no other purpose.

The section contemplates that as well as "necessary mains" there may be "necessary service reservoirs" to be constructed. *Ex hypothesi* these must lie between the starting- M
point to which a supply of water is already being brought in existing works belonging to the undertakers and the new distribution mains to which the proposed new buildings will be connected. The new main which will carry water from that starting-point to the service reservoir may well fall within the definition of "trunk main"; and if the new trunk main and the new service reservoir will serve no other purpose than to convey water from that starting-point to the new distribution main or mains to which the proposed buildings in respect of which the requisition has been made will be connected it will, in my view, fall within the expression

"necessary mains." For in the context of section 37 I would construe "necessary" as applicable to new mains and service reservoirs provided for that purpose only, and not to mains or reservoirs which will also serve some other purpose of the undertakers in connection with the supply of water to existing consumers or to potential new consumers whose water will not be brought to them through the new distribution main or mains. To ascribe any wider meaning to the word "necessary" where it appears in section 37—or, in a similar context, in clause 29—would in my view conflict with the general scheme of the Act for allocating the cost of works between the proceeds of the annual water rate payable by the general body of consumers of water for domestic purposes and charges additional to the water rate imposed upon particular consumers. Once works have been constructed by the undertakers which bring a supply of water to an existing distribution main, the cost of additions or improvements to any of those works lying between the ultimate source of the supply and service pipes connected with that main, which are needed to meet increased demand for water, must be provided for out of the proceeds of the annual water rate or loans serviced out of the proceeds of that rate. No additional charge may be imposed upon the individual consumers who benefit from particular additions or improvements, even where they are new consumers whose exercise of their right under clause 30 to have their premises connected to an existing main has caused the increased demand. The Act thus appears to accept as a general principle that new consumers as well as old are entitled to the benefit of additions and improvements to existing works rendered necessary by increased demand for water for domestic purposes which they themselves have caused; and that they cannot be required to make any individual contribution to the cost of those additions or improvements beyond what is payable by them by way of the annual water rate in common with consumers who may derive no such benefit.

It would seem to conflict with that principle that an exception should be made in respect of additions or improvements to works lying between the ultimate source of supply and the point to which water had already been brought by the undertakers, merely because the supply to the new consumer involved, in addition to the laying of service pipes, the laying of a new distribution main beyond that point to a point in the vicinity of the premises for which the water was required. Moreover, there are specific indications in section 37 itself that this was not intended. Although in the instant case the board have restricted their demand to requiring the council to undertake to pay annually only a minute proportion of the one-eighth of the total cost of laying the 27-in trunk main from the Thames to the existing service reservoir at Banbury, there is nothing in the section to compel them to do so if that main is included in the expression "necessary main." They could have demanded an undertaking extending to one-eighth of the total cost of the 27-in trunk main, and if the requisitioner had not happened to be a local authority they could have required him to deposit as security a sum equal to that total cost. Again, the method of calculating the sums payable in each year pursuant to the undertaking appears to contemplate that all the works to which the undertaking relates (whether mains or service reservoirs) will serve to supply with water a single identifiable group of consumers liable to pay water rates in respect of water supplied from such of the works as are mains. There is no provision for allocating the total cost of the necessary service reservoirs and the necessary mains between the cost of reservoirs on the one hand and the cost of mains on the other, or between one main and another, so as to enable account to be taken of the water rates payable by one group of consumers in calculating the requisitioner's liability in respect of one part of the works and those payable by another group of consumers in calculating his liability in respect of another part of the works. The calculation envisaged by the section provides only for the assessment of the aggregate cost of works of which it can be predicated that *all* the mains included in them will serve to supply water to the same group of identifiable consumers and none will serve to supply water to other groups as well. In the instant case it is common ground that the 27-in trunk main, when completed, would serve to supply water to the general body of consumers of water for domestic purposes as well as to the proposed new buildings that were the subject of the contemplated requisition. How small a part the necessity to supply water to these buildings played in creating the need to provide the new trunk main is indicated by the minute proportion of the total cost of it that the board thought it equitable to ascribe to the council as requisitioners in respect of those buildings. That the 27-in trunk main would serve such other purposes as well in my view precludes its being a "necessary main" within the meaning of section 37 of the Water Act 1945. I would answer the only question raised in the originating summons accordingly, and would allow the appeal.

COURT OF APPEAL

February 27 1975

(Before Lord Justice CAIRNS, Lord Justice ROSKILL and Lord Justice BROWNE)

CHELTENHAM BOROUGH COUNCIL v A H TAYLOR'S BUTCHERS HIDE SKIN FAT & WOOL CO LTD

Estates Gazette May 17 1975

(1975) 234 EG 531

Nuisance from fat-rendering factory—Numerous extensions to enable company to install machinery, etc, must now cease—Writ of sequestration to issue despite submissions to do with recent difficulties in the animal market

This was an appeal by A H Taylor's Butchers Hide Skin Fat & Wool Co Ltd, defendants in an action brought by the respondents, Cheltenham Borough Council, for nuisance caused by smell from a fat-rendering factory in Arle Avenue, Cheltenham, from an order of Brabin J at Winchester on December 10 1974 directing that a writ of sequestration previously made but ordered to lie in the office should issue on January 2 1975.

Mr P Fallon QC and Mr D Hawkins (instructed by Lewin, Gregory, Mead & Sons, agents for Ivens, Thompson & Green, of Cheltenham) appeared for the appellants, and Mr A Cripps QC and Mr A Fletcher (instructed by Sharpe, Pritchard & Co, agents for the town clerk) represented the respondents.

Mr Fallon said that the case came originally before Brabin J at Exeter in 1971, when the defendants gave an undertaking which was suspended until November 1 1972. Subsequently the council applied for a writ of sequestration on the ground that the defendants were in breach of the undertaking. In October 1973 the judge ordered a writ to issue but to lie in the office until April 30 1974. On April 5 1974 the defendants applied for an extension of time in respect of the writ until August 1 1974. After a further hearing at Lewes on July 26 1974 it was agreed that the writ would lie in the office, not to issue until further application or until discharge of the writ. On December 10 1974 the council sought enforcement of the writ. The defendants contended that they should be granted further time on the ground that by the end of July 1974 they had completed the installation and trials of further equipment designed to prevent smell, and that in view of the dilemma in which they were placed in October 1974 by reason of an unprecedented and unforeseeable increase in the number of cattle killed in abattoirs served by the company, it was unreasonable not to give them a further extension of time in which to test the efficiency of the plant. Brabin J nevertheless ordered issue of the writ.

The defendants were an old-established fellmongering business. They had been in many ways pioneers in the development of a method of dealing with smells arising from the fat-rendering part of their premises. It was accepted that it would be wrong for the company to go on creating obnoxious smells from the factory, but he (counsel) submitted that even so it would be unfortunate to bring the knife down at this time. When the matter came before Brabin J in October 1973 the company had spent nearly £78,000 on remedial equipment. A pilot burner for a catalytic-burner type of air purification system had been installed the previous May. It was thought that installation of the system would cost £16,000, bringing the total amount expended to £93,000. Before the matter came before the judge again in July 1974, he saw all the apparatus that had been installed and was being installed, and it was decided that the writ of sequestration should continue to lie in the office. The council's case when the matter was brought back before the judge last December was that an unprecedented situation had been disclosed the previous October, when six confirmed statutory nuisances had been detected. The council maintained that it had now to be assumed that there was no cure for the smells complained of. While it would certainly be wrong to allow this sort of smell to be perpetuated, there was now a system which appeared to control effectively smells produced from this type of factory. The difficulty had been that at the end of 1974 there was an unprecedented amount of killing of animals. Calves were being sold at 50p each. There was so much material that it was impossible to stop cooking the offal, which had to be dealt with exceptionally quickly. It should be added that throughout the case the judge had formed a very good opinion of the company's managing director, Mr M J Tupper, who had been anxious to dump the offal and had tried to get permission to do so. The simple question was whether or not it was reasonable and right in the circumstances that the company should be given extra time to take them out of a difficult and exceptional period.

Counsel for the respondents were not called on to argue.

Giving judgment, Cairns LJ said that Brabin J had dealt with the case throughout with the greatest care and sympathy for the company, and there were no grounds for saying that he had not properly exercised his discretion. The company now sought extra time until the end of April, but the court had no material before it to justify any such further degree of leniency.

Agreeing, Roskill LJ said that counsel for the company had urged, "Give us to the end of April. If we cannot get it right by then, we surrender." But seven or eight weeks had elapsed since Brabin J had made his order, and the court did not know in any detail what had happened since. He (his Lordship) felt absolutely certain that, had the position really radically improved, the court would have had a mass of affidavit evidence and an application to call fresh evidence. Instead, the court was left for all practical purposes with no more evidence than that given before Brabin J. On that evidence his order was manifestly right. It would have been lamentable if he had not made that order after so long a delay.

Also agreeing, Browne LJ said that he found it impossible to say that the defendants had now solved the problem and that no further nuisances had been or would be caused. It was more than three years since the undertaking was given, and he could see no justification for any further extension of time.

Mr Fallon said that so far as the company was concerned, the fat-rendering part of its plant would now be closed at 12 noon on March 1. As a result of the court's judgment, the company were prepared to offer an undertaking to that effect if the council would not pursue the right of sequestration.

Mr Cripps urged the court merely to dismiss the appeal and not to deal at this stage with the question of an undertaking, which could give rise to complications. The council would not take any step which was not sensible and necessary. If in fact the company shut down their factory satisfactorily, it looked as though that might be an end of the matter.

Cairns LJ said that the appeal would simply be dismissed. If it was necessary to make any application for the discharge of the writ of sequestration, there was no need to come back to the court. The application could be made to Brabin J, who ordered the issue of the writ. The appellants must pay the council's costs.

COURT OF APPEAL

March 25 1975

(Before Lord DENNING MR, Lord Justice STAMP and Lord Justice ORR)

KENT COUNTY COUNCIL v LOUGHLIN AND OTHERS

Estates Gazette May 31 1975

(1975) 234 EG 681

Road charges—Tithe map of 1846 strong evidence against existence of road as a maintainable highway in 1835—No evidence that road had ever been repaired by a public authority—Presumption that arises in such circumstances—Frontagers held liable to pay for the making-up of the road despite decision in their favour in the Divisional Court

This was an appeal by Kent County Council from a decision of the Queen's Bench Divisional Court dated November 19 1974 upholding, on a case stated by Dartford magistrates, objections by the respondents, all frontagers of houses in Fairby Lane, Hartley, to proposals by the council to make up a stretch of the lane at their expense. The decision of the Divisional Court was reported at (1974) 233 EG 759.

Mr R A W Sears (instructed by Sharpe, Pritchard & Co, agents for W G Hopkin, of Maidstone) appeared for the appellants, and Mr T I Payne (instructed by A L Proctor, of Gravesend) represented the respondents.

Giving judgment, Lord Denning said that many old villages were now encompassed by new development, and fields and farms were swallowed up by houses and roads. At the outset the roads were private roads made by the builders and kept up by the residents. Later they were taken over by the local authority and became public roads. When that happened the authority could by statute charge the making-up cost against the frontagers. Thereafter they were maintained by the authority. All that was well known, so that anyone building a house or buying one in a private road should be aware that when the authority took it over he would be liable to pay road charges.

One such old village was Hartley. Its history had been written by its rector 50 years ago. This came into the story because the author drew a map showing the village in late Plantagenet and early Tudor times. It was, he confessed, a piece of imagination; but he showed matters as if there existed 400 years ago a track along the way now called Fairby Lane, today a road with houses on both sides. Basing themselves largely on that map, the householders claimed that the lane was not a private road but a highway repairable at public expense, and therefore that they were not liable to pay the frontage charges. The developers of the houses backed the householders, because they had a considerable frontage; they had agreed to indemnify some of the householders. The question depended on a provision in the Highways Act 1959 which made frontagers liable for the cost of making up a "private street," defined as "a street not being a highway repairable at the public expense." So the question was whether Fairby Lane was a highway maintainable at the public expense. To solve it, the court had to go back 140 years to 1835. If in that year the lane was a highway repairable by the inhabitants at large the frontagers won, and the

A lane should since 1835 have been kept in repair by the authority. By statute the position was fixed in 1835. No new highway coming into existence after that year became repairable by the inhabitants at large unless strict requirements were fulfilled, and nothing had happened since 1835 to make Fairby Lane maintainable at the public expense.

The burden was on Kent County Council to show that in 1835 the lane was not a highway repairable by the inhabitants at large. Of course, no one now living could give evidence about the 1835 position, and the court could only go by what had happened since. In this case the justices found that " neither Kent County Council nor any other authority
B had ever maintained or repaired Fairby Lane." From that evidence it could be inferred that in 1835 the lane was not repairable by the inhabitants at large. The only other thing to go by was maps. There was the map drawn by the rector in his book on Hartley. In the foreword he had said he had been " obliged to fall back, in some degree, upon imagination." He was well endowed with imagination, a writer of fairy tales. The map was useless. It should have been discarded. The county council archivist produced maps between 1769 and 1819. None showed Fairby Lane at all, but they were to so small a scale that they showed only public carriage roads.
C The first large-scale map was the tithe map of 1846 and the schedule annexed to it. The Lord Chief Justice had treated this as of little value, but he (his Lordship) thought it of great value. It was prepared under statutory authority by the Tithe Commissioners, with great care and accuracy, to show all cultivated land, arable and pasture, because tithe was payable on land which produced crops. It also had to show waste land and definite roads which did not produce crops, because tithe was not payable on these. If a road passed over a man's land he would naturally require it to be shown, so as not to pay tithe on it. So on the question whether there was a road at the specific place the tithe map was of much
D importance. Such a map had been so regarded in *A-G v Antrobus* [1905] 2 Ch 188 and *Stoney v Eastbourne Rural District Council* [1927] 1 Ch 367.

The tithe map of 1846 showed no road at all on the line where Fairby Lane now was. It only showed it as a boundary hedge or fence between a cherry orchard and pasture. And in the schedule, those were shown in their full acreage as in the occupation of the local squire, Francis Treadwell, without any deduction for any road. He (his Lordship) regarded that tithe map as sufficient to show that Fairby Lane did not exist as a road in 1846. It might have existed as a footpath, but that was neither here nor there. If there was not
E even a road there in 1835 it was most unlikely that it was repairable by the inhabitants at large; he (Lord Denning) agreed with the statement of Widgery J in *Alsager Urban District Council v Barratt* [1965] 2 QB 343 that " the onus of proof can be satisfied by showing that the highway did not exist even as a road before 1835." The next map was the ordnance survey of 1864-67, made before any development had taken place. It showed a track where Fairby Lane now was, but nothing as to whether the track was a public way or an occupation way. The justices said it appeared " as a highway," but nothing on the map showed that. It could not be a ground for thinking that there was an ancient high-
F way going back before 1835. The court had been shown a map of 1938, with Fairby Lane shown as a made road with houses on the south side and fields on the north. Since the war the whole area had been much developed. It was plain that Fairby Lane had never been repaired or maintained by any public authority and that there was not even a road there in 1835. Those two facts went far to show that it was not then a highway repairable by the inhabitants at large. And there was nothing to the contrary. The justices stated the facts and then said, " We were of the opinion that Fairby Lane was not a private street." That conclusion was not warranted by the evidence. It was not a conclusion to which the justices could reasonably have come. On the facts found G
by them Fairby Lane was not a street maintainable at the public expense, and the appeal should be allowed.

Agreeing, Stamp LJ said that, without hesitation, he had concluded on the balance of probability that had there in 1835 been a road on the site of what was now Fairby Lane it would by the date of the 1846 tithe map have been sufficiently well established to have been indicated on that map. But whether or not the tithe map were regarded as sufficient evidence to show that no road existed, he (his Lordship) agreed with the Master of the Rolls that it ought to be presumed in 1975 that in the absence of any evidence to the contrary, a road found not at any time after 1835 to H
have been repaired by the inhabitants at large or any local authority, and not shown to have existed in 1835, was never so repairable. The map drawn by the rector in his book, published in 1922, must have been taken from the 1864-67 ordnance map, to which it bore a strong resemblance, and the rector must have presumed that the roads shown on that map had been there since time immemorial. He (Stamp LJ) agreed that no tribunal properly approaching the case could have placed reliance on the rector's map and that the justices were wrong in doing so.

Also agreeing, Orr LJ said that on the tithe map the combined effect of the *Antrobus* and *Stoney* cases was that J
an inference might be drawn from the fact that a road was not shown, in as much as a road might involve in total a substantial area withdrawn from agriculture, so that it would be in the landowner's interest to have it shown as untithable. For that reason it would have been wrong for the magistrates to disregard the tithe map. But they did refer to it, and the weight to be given to it was a matter for them. If therefore that matter had stood alone, he (his Lordship) would have hesitated to interfere with their decision. But two other considerations led him to conclude that the appeal should be allowed. The first was that where a road had, on the ordnance-map evidence, existed for at least a century, K
it was a powerful consideration that no evidence had been adduced of any repairs to it at the public expense, and the justices were satisfied that there had been no such repairs. The second was that the magistrates attached undue importance to the map in the rector's book, which the author admitted to have been in part conjectural and which was, if it was evidence at all, heavily outweighed by the six maps between 1769 and 1820, none of which showed the road.

The appeal was allowed. Leave to appeal to the House of Lords was granted.

L

COURT OF APPEAL

November 28 1974

(Before Lord DENNING MR, Lord DIPLOCK and Lord Justice LAWTON)

COURTNEY & FAIRBAIN LTD v TOLAINI BROS (HOTELS) LTD

Estates Gazette June 28 1975

(1974) 234 EG 987

Arrangement as to price in supposed building contract held too uncertain—No binding agreement ever concluded—Contract to negotiate is unknown to the law M

This was an appeal by Tolaini Bros (Hotels) Ltd from a judgment of Shaw J on March 12 1974 holding that a letter of April 10 1969 from the respondents, Courtney & Fairbain Ltd, and the appellants' reply thereto dated April 28 1969, constituted a binding and enforceable contract for the construction of certain works.

Mr D Sullivan (instructed by Wright & Webb, agents for Pollards, of Boreham Wood) appeared for the appellants,

A and Mr J Dyson (instructed by Doyle, Devonshire, Box & Co) represented the respondents.

Giving judgment, LORD DENNING said that Mr Sidney Tolaini, the appellants' managing director, wanted to develop the site owned by the firm in Hertfordshire, and got into touch with Mr Courtney, the respondents' managing director, who was a building contractor but was also well-placed to raise finance for building development. On April 10 1969, after a meeting, Mr Courtney wrote a letter which included the following important words:

> I would be very happy to know that, if my discussions and
> B arrangements . . . lead to . . . a financial arrangement acceptable to both parties, you will be prepared to instruct your quantity surveyor to negotiate fair and reasonable contract sums in respect of each of the three projects as they arise. (These would, incidentally, be based upon agreed estimates of the net cost of work and general overheads with a margin for profit of 5 per cent.)

On April 28 1969 Mr Tolaini replied that he agreed to the terms specified in this letter. Mr Courtney then found someone who provided £200,000 or more to finance the projects, and Mr Tolaini appointed a quantity surveyor; unfortunately, however, there were differences over the price, nothing was agreed, and ultimately the development work was carried out
C by other contractors. He (his Lordship) had come to the conclusion that the letters relied on by the appellants did not constitute a binding contract for works. The words used showed quite clearly that all was left to be agreed in the future, neither a price nor any machinery for ascertaining a price being fixed. Seeing that there was no agreement on something so fundamental, there was no contract to build. But then it was said, might there not still be a contract to negotiate? This subject was touched on by Lord Wright in *Hillas & Co Ltd v Arcos Ltd* (1932) 147 LT 503 at 515, where he said:

> There is then no bargain except to negotiate, and negotiations may
> D be fruitless and end without any contract ensuing; yet even then, in strict theory, there is a contract (if there is good consideration) to negotiate, though in the event of repudiation by one party the damages may be nominal, unless a jury think that the opportunity to negotiate was of some appreciable value to the injured party.

He (his Lordship) thought that a contract to negotiate, like a contract to enter into a contract, was something unknown to the law. Lord Wright's tentative opinion did not seem to be well-founded, because a contract to negotiate must necessarily be too uncertain to have binding force. The appeal should therefore be allowed.

E Agreeing, LORD DIPLOCK said that Lord Wright's dictum (for it was no more) was in his opinion bad law.

LAWTON LJ agreed with both judgments, and the appeal was allowed with costs above and below.

COURT OF APPEAL

December 3 1974

(Before Lord Justice STEPHENSON, Lord Justice ORMROD and Mr Justice THOMPSON)

F JONES AND ANOTHER v RHYS-JONES

Estates Gazette June 28 1975

(1974) 234 EG 987

Purchaser takes conveyance in 1970 subject to restrictive covenant in conveyance of 1961, applies in 1972 for modification—There is no principle to be found in the authorities by which an application of the kind cannot be made so soon after purchase—Indeed, Re Wickins's Application (1962) 183 Estates Gazette 541 shows that it can—Purchaser's application remitted to Lands Tribunal for rehearing and a full decision on the merits

This was an appeal by Mr Robert Keith Jones and his wife, G
Mrs Patricia Iris Jones, of The Hazards, Drysgol Road, Radyr, Cardiff, against a decision of the Lands Tribunal dated June 28 1973 in favour of the respondent, Mrs Joan Margaret Rhys-Jones, of Drysgol House, Drysgol Road, Radyr, on the appellants' application to modify, under section 84 of the Law of Property Act 1925 as amended, a building restriction imposed by a covenant in a conveyance dated December 18 1961.

Mr N T Hague (instructed by Allan Jay & Co, agents for James Morgan & Co, of Cardiff) appeared for the appellants, and Mr J M Henty (instructed by Phoenix Walters & Co, of Cardiff) represented the respondent. H

Giving judgment, STEPHENSON LJ said: This appeal is brought by the appellants, Mr and Mrs Jones, from a decision of the Lands Tribunal (Sir Michael Rowe QC, President) dated June 28 1973, which was annexed, in accordance with the Rules of the Supreme Court, order 61, rule 1 (4), to a case stated at their request on February 18 1974 for our decision pursuant to section 3 (4) of the Lands Tribunal Act 1949. The application was made under section 84 of the Law of Property Act 1925, as amended by section 28 of the Law of Property Act 1969, for the modification of a restriction as to building arising under a covenant imposed on the appellants' J
predecessors in title by a conveyance dated December 18 1961. On objection taken by Mrs Rhys-Jones, the respondent, the tribunal dismissed the application and ordered the appellants to pay the objector's costs. That dismissal is final unless it was erroneous in point of law. The error of law on which the appellants rely is set out in the case stated. The admitted facts, not all of them set out in the tribunal's decision, are as follows. By a conveyance dated December 18 1961, Mrs Osmond, of "Whitehall," Drysgol Road, Radyr, in the County of Glamorgan, sold to a Mr and Mrs Oldham parcels of land in the same road which were identified on a plan as plots 1 and 2. Clause 3 is the covenant with which this appeal is K
concerned. It reads as follows:

> The purchasers with the intent and so as to bind (so far as practicable) the property hereby conveyed into whosesoever hands the same may come and to benefit and protect the estate and property of the vendor being the rest of "Whitehall," Drysgol Road, Radyr, aforesaid and the land enjoyed therewith but not so far as to render themselves personally liable for any breach of covenant committed after they shall have parted with all interest in the property hereby conveyed hereby jointly and severally covenant with the vendor that they the purchasers and their successors in title will at all times hereafter observe and perform the stipulations and restrictions following in relation to the property hereby conveyed namely (a) to fence in and at all times L
> keep fenced in at their own expense and to the satisfaction of the vendor or her agents for the time being the northern eastern and western boundaries of the said land with a wall wooden paling fence or live hedge to be agreed with the vendor or her agents (b) to remove completely to the satisfaction of the vendor or her agents the dilapidated cottages known as "Drysgol Cottages" now situate on adjoining land retained by the vendor such cottages being coloured yellow on the plan annexed hereto (c) not to erect on the land hereby conveyed any dwelling-house or bungalow or other building except in accordance with plans and specifications approved by the vendor.

The Oldhams then built on plot 2 a small dwelling-house M
named (apparently with an eye on the golf-course across the road) "The Hazards," presumably in accordance with plans and specifications approved by Mrs Osmond. By a conveyance dated May 26 1964 the Oldhams sold plot 1 to a Mr Waddilove, and a dwelling-house was built on that plot. At some time before September 1971, the Oldhams pulled down "Drysgol Cottages," the dilapidated cottages standing on the parcel of land which lay between "Whitehall" and plot 2, and which had been identified in the 1961 conveyance's plan as plot 3, and in their place was built a dwelling-house known as Drysgol House. By a conveyance dated April 14 1970, the Oldhams sold plot 2 and "The Hazards" to the applicants,

Mr and Mrs Jones, "subject to the stipulations and restrictions contained in the conveyance"—that is, the conveyance dated December 18 1961—"so far as the same affect the property hereby conveyed and are still subsisting and capable of being enforced." By a conveyance dated September 30 1971, Mrs Osmond sold plot 3 and Drysgol House to the objector, the respondent, Mrs Rhys-Jones, "together with the benefit so far as the same is subsisting and capable of taking effect of the restrictive covenants contained in and reserved by a conveyance dated the 18th day of December 1961. . . ." Before the last sale the appellant, Mr Jones, had applied, on December 7 1970, for planning permission, and obtained it on February 5 1971, to extend "The Hazards" by building another bedroom and bathroom over the double garage; and the appellants wanted to extend the garage to the rear to make a workroom. They gave an undertaking to the respondent, in proceedings for an injunction started on October 4 1971, not to proceed with the extension pending the decision of the Lands Tribunal. At some date before April 1972, Mrs Osmond sold "Whitehall" to the Lord Mayor, Aldermen and Citizens of the City of Cardiff, who have since provided accommodation there for Her Majesty's Judges. On April 18 1972 the appellants applied to the Lands Tribunal for an order that the restriction arising under clause 3 (c) of the 1961 conveyance, which I have read, might "be modified so as to permit the addition of a bedroom and a bathroom to 'The Hazards' by extending the first floor over the existing garage in accordance with the accompanying architect's plans." The application contained these paragraphs:

3. The grounds of this application are that the application falls within paragraph (aa) of subsection (1) of section 84.

4. Particulars of application: (i) the reasonable user of the land which would be impeded by the continued existence of the covenant is as a private dwelling-house, with an additional first-floor bedroom and bathroom; (ii) the restriction in impeding reasonable user does not secure to persons entitled to the benefit of it any practical benefits of substantial value or advantage to them.

The respondent's objection particularised her grounds of objection as follows:

(i) The reasonable user of the land is not as a private dwelling-house with an additional first-floor bedroom and bathroom; (ii) the restriction secures to myself practical benefits of substantial value or advantage; (iii) by reason of the fact that the covenant was only imposed in 1961 it would be improper for the tribunal to exercise its discretion to modify it.

And she claimed £5,000 for the loss of amenity value in the event of the restriction being discharged or modified. Section 84, as amended by the 1969 Act, reads as follows:

(1) The Lands Tribunal shall (without prejudice to any concurrent jurisdiction of the court) have power from time to time, on the application of any person interested in any freehold land affected by any restriction arising under covenant or otherwise as to the user thereof or the building thereon, by order wholly or partially to discharge or modify any such restriction on being satisfied (a) that by reason of changes in the character of the property or the neighbourhood or other circumstances of the case which the Lands Tribunal may deem material, the restriction ought to be deemed obsolete; or (aa) that (in a case falling within subsection (1A) below) the continued existence thereof would impede some reasonable user of the land for public or private purposes or, as the case may be, would unless modified so impede such user. . . .

(1A) Subsection (1) (aa) above authorises the discharge or modification of a restriction by reference to its impeding some reasonable user of land in any case in which the Lands Tribunal is satisfied that the restriction, in impeding that user, either (a) does not secure to persons entitled to the benefit of it any practical benefits of substantial value or advantage to them; or (b) is contrary to the public interest; and that money will be an adequate compensation for the loss or disadvantage (if any) which any such person will suffer from the discharge or modification.

On May 8 1973 there was a full day's hearing by the tribunal of witnesses and counsel on both sides, and on June 28 1973 the tribunal gave its decision dismissing the application on one short ground. After setting out some of the history which I have recounted but omitting the 1970 conveyance, and after referring very shortly to the evidence and arguments, the tribunal stated:

I do not propose to go into this problem, for it seems to me that the general principle enunciated in the judgment of Harman LJ in *Cresswell v Proctor* [1968] 1 WLR 906 applies directly to this case. The covenant was imposed as recently as 1961 and the applicants bought with full knowledge of it in 1971. I do not think it right that purchasers should agree to be bound by a covenant and then seek to modify it within a matter of months. It would of course be a very different matter if those entitled to the benefit of the covenant had waived their rights, but here they do not. I think they are entitled to stand on their rights. I shall therefore dismiss this application.

I read that as meaning that on reflection the tribunal had decided that it need not go into the merits of the application and objection or weigh up the evidence and arguments put before it, and indeed was precluded from doing so, because Harman LJ had enunciated a general principle in *Cresswell's* case which applied to this case and compelled it to dismiss the application. The alleged principle is not clearly formulated, but would appear to be that purchasers who apply to modify a restrictive covenant within a matter of months of having agreed to be bound by it must be refused any modification of it. The applicants were purchasers who had agreed in September 1971 (a mistake for April 1970) to be bound by this covenant and were applying to modify it in April 1972; therefore their application could not be granted.

Mr Hague's argument for the appellants can be bluntly summarised in a sentence: this decision misapplied an undefined and non-existent principle to misstated facts. Mr Henty, for the respondent, contended that the tribunal correctly applied two principles to the facts, and in the alternative that it did decide the case on the merits without considering itself bound by any principle. The alternative he conceded was difficult to argue on the language of the tribunal's decision, and in my judgment the question submitted to this court by the case stated put it beyond argument. For we are asked by the tribunal to say (i) whether a general principle was enunciated in the judgment of Harman LJ in *Cresswell v Proctor*; (ii) if so, whether the tribunal was correct in considering that such principle applied to the facts of the present case; and (iii) whether in dismissing the application the tribunal erred in law. Those questions make it clear to me that the tribunal dismissed the application on the short ground that it was bound to do so by a principle applicable to the facts, and that we have to decide whether the tribunal erred in point of law in holding that there was such a principle, and if there was, that it applied to the facts of this case. Our task is not made easier by the lack of any precise formulation by the tribunal of the principle which it derived from the judgment of Harman LJ. I have already stated the formulation of it which I have culled from the language of the decision. But Mr Henty has extracted two principles which the tribunal certainly did not formulate, but which he formulated thus: (1) where a case concerns an original covenantor and an original covenantee, the tribunal is entitled to require a higher standard of proof or justification than from other applicants: it must be very much clearer that the applicant brings himself within one of the grounds of relief; (2) when a covenant has been comparatively recently imposed the tribunal should exercise its discretion so as to refuse the application.

On examining Harman LJ's judgment I find, first, a decision that the Lands Tribunal was right in exercising its discretion to refuse an application to modify a covenant first imposed only two years before the application. Danckwerts LJ also thought the exercise of discretion right. Winn LJ was not satisfied that it was wrong. But I also find in Harman LJ's judgment characteristically vigorous expressions of disapproval of an application made in such circumstances as he was considering. Those circumstances were briefly that the application

A was made by the original covenantors two years after they had entered into a covenant "not to erect any building whatsoever" on the land conveyed, but to use it as a private garden only. The application was amended and pursued at their instance by trustees appointed by them of a voluntary settlement of which they may have been in effect (as the headnote states), but other members of their family were in name (as the facts found by the tribunal show), the beneficiaries. Although there had been no change in the circumstances of the property, but only a change in their own domestic obligations, yet they had applied about 15 months after the purchase for planning permission to build a bunga-
B low on the land conveyed. Harman LJ found "the idea that you can modify a covenant, voluntarily entered into, within a year or so, without any change in circumstances of the property at all . . . shocking," and modification in circumstances such as those "quite out of the question." Danckwerts LJ considered the application to legalise the breach within such a short period not to be within the true intention of section 84. Winn LJ was less critical of the original covenantors and more critical of the Lands Tribunal. He considered that there might be more excuse for such "a startling prompt attempt to escape from a voluntarily undertaken contractual obligation," and that there were considera-
C tions which to another mind might have overridden the "sanctity of contract" argument which appealed to Danckwerts LJ. But though the effrontery of the application, as it struck two of the Lords Justices, undoubtedly generated enough judicial indignation to drive Harman LJ, at least, into expressions capable of general application, I do not myself find in them the enunciation of any general principle in the form which I think appears in the decision of the tribunal, or in either of Mr Henty's formulations, or indeed in any form which applies to the instant case. If Harman LJ laid down any general principle, it was not necessary to the
D decision of the appeal before him, nor, even if endorsed by Danckwerts LJ, is it binding upon us. I should, moreover, be reluctant to believe that any court had in effect taken away a discretion which Parliament had entrusted to a particular tribunal by directing that it should be exercised in one way only, and I do not think that any member of this court in *Cresswell's* case purported to do that. I agree with the statement of the learned editor of *Preston & Newsom's Restrictive Covenants Affecting Freehold Land* (5th ed, at p 178):

> In *Cresswell v Proctor* the Court of Appeal had to consider an application to modify a very recent covenant made, in effect, at
> E the instance of the original covenantor. Harman LJ and Danckwerts LJ expressed in somewhat strong terms the view that the discretionary powers of the tribunal ought not to be exercised in favour of such an application.

I also agree with the remark at page 187 that they (the two Lords Justices) there "intimated that very new covenants are not lightly to be altered." At most the majority were laying down a guideline that the tribunal should be slow to modify a very recent covenant at the suit of the original covenantor, and it is erroneous in law to confer upon that guideline the style and dignity of a general principle which precludes the tribunal from considering all the circumstances
F of the application before exercising its statutory discretion.

Once the circumstances of the application which the tribunal had to consider in this case are examined, they cannot be said to be so similar to the circumstances of *Cresswell's* application that the guidance given in that case helped, still less compelled, the tribunal to dismiss it. Neither the appellants nor the respondent were original contracting parties to the conveyance of 1961 which imposed the restrictive covenant. The appellants had no relation with the original covenantors except that of purchaser and vendor. At the date of the application the covenant was not two years old, but 10. It was not a covenant prohibiting any building whatsoever, but a covenant not to build except in accordance G with plans and specifications approved by the vendor. In the decade since the covenant was imposed there were changes in the character of the property and the neighbourhood which did not make the restriction obsolete and give a ground for modification within section 84 (1) (a), but which did distinguish this application from *Cresswell's*, and might be very relevant to section 84 (1) (aa), on which the application is founded. Those changes included the building of dwelling-houses on plots 1, 2 and 3. Mr Hague also asked us to bear in mind in addition to those factual distinctions the further consideration that any observations made in *Cresswell's* case might require qualification in the light of H the introduction into section 84 of subsections (1) (aa) and (1) (A) by the 1969 Act. But he did not attach much importance to that, and it is unnecessary to consider whether it has any importance if, as I think, the facts of this case make any principle which can be extracted from *Cresswell's* case inapplicable to them. I also find it unnecessary to decide how far the observations of Russell LJ in *Ridley v Taylor* [1965] 1 WLR 611 at p 623A are in conflict with Harman LJ's opinion in the same case (at p 618B) that "the court should be slow to relieve an applicant of covenants which he himself has entered into." That Diplock LJ in that case agreed with both judgments suggests that the conflict may be more apparent J than real. It is from that judgment of Harman LJ, rather than from his judgment in *Cresswell's* case, that Mr Henty appears to derive his first principle and the Lands Tribunal its practice: see *Preston & Newsom*, pp 186-7.

More relevant, however, to an application like the present by a person who was not the original covenantor is the judgment of Lord Denning MR in *Re Wickins's Application* (1962) 183 EG 541, which was distinguished by the Lands Tribunal in deciding *Cresswell's* case. There the applicant Wickins was not the original covenantor, but a purchaser who bought the property less than a year before applying to modify a covenant which was 27 years old and applicable K only in so far as it affected the property conveyed and was still subsisting and capable of being enforced. The court held that the Lands Tribunal was in error in saying that Mr Wickins had "very recently voluntarily endorsed" the covenant and in dismissing the application on that ground. The Master of the Rolls said:

> "It would be unfortunate if by buying land on those terms, he were to be held to have disabled himself from applying under section 84 to modify the covenant. The beneficial operation of this Act would be entirely defeated by such an interpretation being put on his action. I do not think it is any ground at all L for refusing the application. By being influenced by it, the Lands Tribunal fell into error."

And Mr Wickins's appeal was unanimously allowed and his application granted. That decision is, in my judgment, very much in point and prevents us from accepting Mr Henty's second principle which he derived not, or not directly, from *Cresswell's* case, but from a decision of the Lands Tribunal in *Wynyates Smith Ltd's Application* (1963) 15 P & CR 85. That case is not binding on us, and decides no more than that the fact that a covenant that no more than eight houses should be erected on land was imposed less than 10 years before an application to modify it by allowing the erection of a ninth M house was made by an applicant closely connected with the original covenantor was one factor which could be taken into account in refusing the application. The decision lays down no general principle; it does not justify the statement in the headnote that the covenant was "very recently imposed"; and unless it rightly treated the applicant as the alter ego of the original covenantor, it may be inconsistent with the judgment in *Re Wickins's Application* which I have quoted.

Re Wickins's Application was not brought to the attention of the tribunal, and if it had been I do not think that the tribunal would have fallen into the error into which, on my understanding of its decision, it did fall. In applying the principle that a purchaser who agreed to be bound by a covenant cannot be allowed to modify it within a matter of months (or a year or two) to purchasers who were not the original covenantor, the tribunal was not acting on anything said by Harman LJ in *Cresswell v Proctor* (or in *Ridley v Taylor*), but was acting contrary to what was said by Lord Denning MR, apparently with the agreement of both the other members of this court, who did not give separate judgments, in *Re Wickins's Application.* Without the assistance of authority, I would have thought that the shortness of the time which has elapsed since the burden of a covenant was imposed on an original covenantor or was transferred to a subsequent purchaser was a factor which could properly be put into the scale against modification or discharge whether the application under section 84 be made by an original covenantor (when it would weigh more) or by a subsequent purchaser (when it would weigh less). But the older the covenant, the more time there will have been for other factors such as changes in the property benefited by the restriction to come into the reckoning in favour of modification, and the easier it may be for the tribunal to relieve an applicant of a burden which he has recently shouldered. However, we do not now have to decide whether the judgment in *Wickins's* case permits such an approach or absolutely forbids the tribunal to consider this factor in deciding any and every prompt application by a subsequent purchaser. It is enough for our purpose that according to that judgment the shortness of the time is not a decisive factor which forbids the grant of such an application. And the time in this case is not as short as it was in *Wickins's* case, or as it was stated to be in the decision under appeal. I would accordingly accept Mr Hague's submission for the appellants and answer each of the first two questions put to us by the tribunal in the negative and the third in the affirmative. I would allow the appeal, and remit the matter to the tribunal for reconsideration in the light of our judgments whether in all the circumstances the covenant should be modified, and if it should, to what extent and in what terms.

ORMROD LJ: I agree. This is an appeal from a determination by the Lands Tribunal refusing an application by the appellants to modify a restrictive covenant affecting their property, "The Hazards," Drysgol Road, Radyr, Glamorgan. The matter was heard by the then President, Sir Michael Rowe, on June 28 1973. The way in which his decision is formulated leaves no doubt in my mind that the learned President in effect disposed of the application on a preliminary point, although he had heard all the evidence in the case. He clearly thought that he was bound by a decision of this court in *Cresswell v Proctor* [1968] 1 WLR 906 to reject the application. He said that the "general principle enunciated in the judgment of Harman LJ . . . applies directly to this case." He had previously stated that he did not propose "to go into this problem" for that reason. The principle which the learned President derived from that case is not precisely stated, but the point upon which he relied was that the covenant in question had been imposed "as recently as 1961"; that the applicants bought with full knowledge of it in 1971; and that it was not right that purchasers should agree to be bound by a covenant and then seek to modify it "within a matter of months." The date of the conveyance to the applicants was actually April 14 1970. From the way in which the judgments in *Creswell v Proctor* are expressed, I cannot help thinking that the judges in that case would have been very surprised to learn that their remarks were destined to form part of the law relating to the modification of restrictive covenants under section 84 of the Law of Property Act 1925. As I read the judgments, Harman and Danckwerts LJJ were merely expressing in their own ways their emphatic approval of the decision of the Lands Tribunal in that case to reject an application to modify a restrictive covenant on the ground that it had been recently entered into by the applicant herself (in effect, although not in form) and the respondents themselves.

The decision whether or not to accede to an application for modification of a restrictive covenant under section 84 is, once the requirements of the section have been satisfied, clearly a matter of discretion, and that discretion is now vested by statute in the Lands Tribunal. This court cannot exercise it: it can do no more than review the exercise of it by the Lands Tribunal, according to well-known principles (see *Egerton v Jones* [1939] 2 KB 702 at 709). In my judgment, therefore, this court cannot, or certainly ought not to, fetter this discretion in any way. Accordingly, I would be very reluctant to conclude that any decision of this court laid down any general principle governing the exercise of such discretion, unless compelled to do so by clear and binding authority. It is plain that no question of general principle was under consideration in *Cresswell v Proctor.* Their Lordships, in my judgment, were doing no more than to formulate a proposition of good sense: namely, that where an original covenantor is applying for a modification of a restrictive covenant recently entered into by him, this is one of the matters, and an important one, which the Lands Tribunal can and must take into account and to which it must give due weight in deciding whether or not, in its discretion, to modify the covenant. If the judgments go any further, they go beyond what was necessary for the decision of that case.

The matter, however, does not stop there, for the learned President in the present case has extended what was said in *Cresswell's* case to a subsequent purchaser who buys with knowledge of the restrictive covenant. This was the situation in *Re Wickins's Application* (1962) 183 Estates Gazette 541, in which this court held that such a purchaser was not disabled from applying under section 84 to modify a restrictive covenant. That case is not referred to by the learned President in his decision in the present case, although it is, in my view, conclusive against the view upon which the decision was based. Such a purchaser not only takes with knowledge of the restrictive covenant, but also with the knowledge that Parliament, by section 84, has provided him with the opportunity, if he can bring himself within the section, of applying to the Lands Tribunal to modify it, with prospects of success which will depend upon all the facts and circumstances of the case. In my judgment, therefore, the learned President, in failing to "go into this problem," that is, to take into account all the other factors deposed to in the evidence, has not taken into consideration matters which he ought to have taken into consideration, and accordingly his decision cannot stand. This appeal should therefore be allowed and the matter be remitted to the Lands Tribunal for a rehearing. It is most unfortunate that the parties should be put to this expense, but there is no possible alternative short of compromise at the end of the eleventh hour.

STEPHENSON LJ: The judgments which have been delivered have both been read by Thompson J, and he asks me to say that he agrees with both of them and that this appeal should be dismissed.

An order was made accordingly. Seeking an order for costs below, Mr Hague remarked that the whole of the costs of the hearing had been thrown away; Sir Michael Rowe having retired between the date of the decision and the statement of the case, it would now be necessary to have the evidence heard all over again; and all this had really occurred by reason of the respondents pressing a false argument, in effect as a preliminary point. There was a practice direction of the Lands Tribunal to be borne in mind, as set out in *Preston & Newsom* at page 305, by which an objector in general had a free run without the risk of having to pay the other side's costs, and with a good chance of having his costs paid by the applicant; but what the practice direction did not say, and never had said, was that an objector was allowed two free runs.

A Stephenson LJ: It is a strong point in your favour that the point taken was not one that came out of the head of the President of the Lands Tribunal, but one that was put into his head by Mr Henty.

Mr Hague: It was put into his head by Mr Henty. But, to be fair to Mr Henty, he did not put into the President's head what I may call the *Re Wickins's* point. That was pure invention. It was not referred to in the tribunal, because Mr Henty never suggested for a moment that one could call it a two-year covenant rather than a ten-year covenant.

Ormrod LJ: Somebody complained that he was not referred to *Wickins.*

Mr Hague: It was never suggested by Mr Henty or myself.

B Ormrod LJ: On your side, *Wickins* was conclusive in your favour.

Mr Hague: On that point. But that point was never argued against me.

Ormrod LJ: What point?

Mr Hague: The point that by taking a conveyance subject to a covenant one, so to speak, started time running again. That is what I call the *Wickins's* point. But Mr Henty never argued that. Mr Henty has always indicated to me that that point is quite wrong. So what I may call the *Re Wickins's* point did not arise below. I have made my submissions, my Lord.

Stephenson LJ: Is there anything further that you want to say, Mr Henty? We are proposing to leave the costs of the abortive hearing below to the Lands Tribunal. C

Mr Henty: I would accept that, my Lord. I did not intend to mislead the tribunal. The third ground of my objection was, "By reason of the fact that the covenant was only imposed in 1961 it would be improper for the tribunal to exercise its discretion to modify it." I understand from your Lordships that that was a perfectly reasonable argument. D

Ormrod LJ: Within the discretion, yes.

Mr Henty: As to the costs here, once one reaches that point my learned friend would have to come here. By virtue of the fact that I have been here no extra costs have been incurred by him; he would have had to come to your Lordships' court on appeal to put the tribunal right.

Stephenson LJ: Yes; he had to come. But you need not have supported the tribunal's view. I do not think you can be heard to say that you need not have come here, but as you were here we can ignore the time taken up by your argument in resisting the appeal. E

Ormrod LJ: The short answer to your submission is this: that your solicitors could have written a letter to the other solicitors saying that you could not support this, whereupon it would have been a pure formality.

Stephenson LJ: You took a different course, Mr Henty. We do not blame you, but I am afraid the court is against you. The costs of the appeal will be the appellants', and the costs before the Lands Tribunal will be referred to the Lands Tribunal which completes the hearing. F

TAXATION

CHANCERY DIVISION

July 9 1974

(Before Mr Justice MEGARRY)

PITT (INSPECTOR OF TAXES) v CASTLE HILL WAREHOUSING CO LTD

Estates Gazette June 14 1975

(1974) 234 EG 823

Warehouse company's new access road—Cost held capital outlay for tax purposes, though road built merely as a replacement to avoid local residents' complaints

This was an appeal by the Inland Revenue from a decision of special tax commissioners holding that the cost of providing a new access road to the premises of the respondents, Castle Hill Warehousing Co Ltd, at Sunningdale Estate, Lincoln, was of an income nature, not capital, and was therefore deductible in computing profits for the purpose of corporation tax.

Mr H K Woolf (instructed by the Solicitor of Inland Revenue) appeared for the Crown, and Mr A R Thornhill (instructed by Wedlake, Bell, agents for R A & C P Heptonstall, of Goole, represented the respondents.

Giving judgment, MEGARRY J said that the respondents were a warehousing company at Lincoln. Residents of Sunningdale Drive complained about the disturbance caused by heavy lorries using the road, so the company approached Lincoln Corporation regarding a new access to their site. The corporation agreed to an arrangement whereby the company were to convey a strip of their land to the corporation and build a road, after which the corporation would take the road over. What was now at issue was the nature, for tax purposes, of £5,227 spent by the company on the construction of the road. The Crown contended that it was expenditure incurred on capital account, and relied on the Income and Corporation Taxes Act 1970, section 130. The company argued that the expenditure was incurred on revenue account because it was spent to preserve the company's trade and warehousing property. The special commissioners agreed with the company, and the Crown now appealed from that decision.

The undoubted purpose of the transaction was to preserve the good name and business of the company by stopping complaints from residents. What the company had obtained was an access to their premises which was no better and no worse than they formerly had, but different in direction, location and legal rights. In their decision, the special commissioners had said, "Looking at the matter in a common-sense way, we found that no new capital asset had been created when the company, for the sake of its good name, gave up one road and built another." He (his Lordship) felt some difficulty about two phrases in that sentence—"no new capital asset had been created" and "gave up one road and built another." The first phrase left it uncertain which of the elements of the compound phrase the commissioners were denying. Was it a capital asset, but not new, or was it neither new, nor capital nor asset? It seemed that the commissioners' sentence should have read that no "additional" capital asset had been created and that the company had merely given up one means of access for another.

What, in fact, the company did was to dispose of a tiny piece of their land, so shutting themselves off from access to a highway, and acquire instead a perpetual easement of way in consideration of constructing a new road which gave access to another part of the same highway. The company then had what they never had before, an easement and their own road, part of which they were to retain and part of which they were to surrender when a new ring road came close by. It seemed that by paying once and for all a lump sum of £5,227 the company acquired an asset with enduring qualities intended to be used indefinitely in the manner of a fixed asset of a capital nature, as appurtenant to certain fixed assets, consisting of their warehouses. Counsel for the company contended that the company acquired nothing that could be shown as an asset in the balance sheet, but he (his Lordship) would be surprised to find that problem beyond the wit of accountants. In the end, he could not see how giving up a tiny strip of land and so cutting off access to Sunningdale Drive could transform a capital transaction of that nature into an income transaction. He would be very slow to accept, on those facts, the doctrine of equivalent replacement which seemed to be explicit in the commissioners' decision. The appeal would accordingly be allowed.

TOWN AND COUNTRY PLANNING

A **QUEEN'S BENCH DIVISION**

June 19 1974

(Before Mr Justice MELFORD STEVENSON)

ENFIELD LONDON BOROUGH v SECRETARY OF STATE FOR THE ENVIRONMENT AND ANOTHER

Estates Gazette January 4 1975

(1974) 233 EG 53

Green-belt land near the Ridgeway, Enfield, taken for development—Council's objections—Duty of Secretary of State when considering the matter on appeal from deemed refusal—Duty to "have regard" to development plan does not require B "slavish adherence"—Secretary of State entitled to give permission notwithstanding the fact that the Standing Conference on London and South-East Regional Planning had not yet reported what green-belt land might suitably be taken for development—No implied representation by Secretary of State that he would wait for the report—No estoppel, nor any other breach of natural justice—Council's appeal dismissed

This was a motion by the London Borough of Enfield for an order quashing a decision of the first respondent, the Secretary of State for the Environment, given in October 1973, whereby, on appeal from a deemed refusal of planning consent by the C applicants, he granted the second respondents, Fairview Estates (Enfield) Ltd, outline planning permission for the residential development of 5½ acres of green-belt land situated near the Ridgeway, Enfield.

Mr P M H Mottershead (instructed by the Town Clerk) appeared for the applicants; Mr G Slynn (instructed by the Treasury Solicitor) for the first respondent; and Mr A R Vandermeer (instructed by Vanderpump & Sykes, of Enfield) for the second respondents.

Giving judgment, MELFORD STEVENSON J said: In this case Mr Mottershead, on behalf of the applicants, the London Borough of Enfield, seeks by motion to quash a decision of the D Secretary of State for the Environment which is expressed in a decision letter dated October 8 1973. By that decision the Minister granted permission for residential development of land which is referred to throughout the papers as the "appeal site," that is, some land at the rear of 60 to 64 The Ridgeway in Enfield, and the grounds on which the applicants seek the order that they now pray for is that the requirements of the Town and Country Planning Act 1971 have not been complied with and that their interests have thus been prejudiced. They say that the decision is not only unfair but is also contrary to natural justice, and those contentions are particularised in a number of subparagraphs of the notice of motion. Subparagraph (a) says E that the Secretary of State, contrary to section 29 (1) of the Act, did not have regard to the provisions of the relevant development plan material to the matter before him and that he erroneously had regard to considerations other than the provisions of the development plan. (It is convenient to state that the development plan in question is the Middlesex County Council Development Plan.) It is said that the Secretary of State acted unfairly and in contravention of the principles of natural justice because the decision was founded on a policy or principle newly adopted and applied by the Secretary of State, and that the present applicants had no prior notice of any intention to adopt and apply such a policy. The notice of motion goes on F to say that the Secretary of State had previously represented to the applicant and others that that policy or principle would be implemented only after the receipt of and in accordance with a report to be made by the Standing Conference on London and South-East Regional Planning, to whom the Minister had referred the task of recommending, up to a maximum of 2,000 acres within the green belt, what land might properly be used for residential building purposes, and that at the date when the Minister gave his decision, namely October 8 1973, the standing conference had not yet delivered itself of any decision or recommendation, or identified any area within the green belt which might properly be called on to contribute to the 2,000 G acres for which the Secretary of State was searching. I have been referred to a number of authorities and indeed a large number of pieces of literature called circulars, which are distributed by the Ministry from time to time, containing advice and guidance to local authorities as to the way in which they should exercise their duties as planning authorities. There is no doubt that this body called the standing conference was required by the Minister to undertake this task of investigating throughout the green belt in the south-east of England areas up to 2,000 acres which might properly be used for residential development, having regard to the existing housing pressure which has resulted from the population explosion of which London and the home counties are the victims at the present H time. In effect, Mr Mottershead's argument, as I understand it, amounts to this: that before making the decision of which complaint is advanced the Minister should have waited until the standing conference had completed its task and identified all the bits and pieces of land which in their view might be excepted from the green-belt area, and should have held his hand until he was armed with a report from the standing conference embodying all that information.

The history of this particular matter is that in 1972 application by the intending developers, called the Fairview Co ("Fairview"), was made to the applicant borough for outline planning permission to develop the appeal site, which is clearly J identified on a large plan I have been shown outlined in red. The applicant council did not announce any decision within the statutory period, whence it followed that the application was deemed to have been refused, and against that refusal was brought the appeal which the Minister had to consider and upon which his decision is the subject of this case. It was agreed in the early stages that the matter should be dealt with on the basis of written statements, and indeed the developers, the Fairview Co, set out their case in a document, I think it is WDD3, which is exhibited in an affidavit by the town clerk of the applicant borough. And in October 1972 the Secretary of State asked for objections by residents in addition to the K council concerned, requesting that those representations might be made in writing. That request generated a large bundle of correspondence which forms a not inconsiderable part of the papers in front of me. I am not going to read all the letters which were written by indignant and apprehensive residents of Enfield expressing apprehension, and indeed bitter resentment, at the prospect of the invasion of the green belt which the development I am now considering would have involved. Mr Mottershead's argument is based on what he says is a misconception or misconception of the Minister's duty under

A section 29 of the Town and Country Planning Act 1971. Section 29 of that Act lays down, in effect, directions to be followed by planning authorities who have to deal with applications for planning permission, and it is not disputed, indeed it is so provided by a later section, that the Minister in dealing with the kind of appeal which he dealt with in this case has to follow those same directions as they appear in section 29. Among the relevant passages in a long verbose section appear these words: " . . . where an application is made to a local planning authority for planning permission, that authority, in dealing with the application, shall have regard to the provisions of the development plan, so far as material to the application, and
B to any other material considerations, ..." Subsection (3) (a) of the same section provides that the authority " in determining the application, shall take into account any representations relating thereto which are made to them, before the end " of the relevant period, and subsection (4) provides that " in determining any application for planning permission . . . the local planning authority shall take into account any representations relating to the application which are received by them before the periods " which are specified elsewhere and do not matter, I think, for the purpose of this present case.

The contentions presented to me can, I think, be fairly summarised in the proposition that the requirement to have
C regard to the provisions of the development plan achieves a mandatory effect which compels the Minister to adhere implicitly to the terms of the development plan. The other side of the coin is that the decision of this appeal of course also has to take account of other material considerations, and the other material consideration in this case is the undisputed pressure for housing accommodation which is a feature of the whole of the south-east of England, and in particular the metropolitan area. And it is said that although the Minister was armed with a large bundle of objections from residents and others who feared they might be affected, he ought, before deciding in any way to depart from the development plan in this instance, not only to have awaited the result of the report
D of the standing conference, but also to have given to everybody concerned yet another opportunity to make representations to him and to advance arguments in support of their views. I have also been referred to a White Paper [*Widening the Choice: The Next Steps in Housing*, Cmnd 5280], issued I think in April 1973 by the Secretary of State, which again contains a mass of impressive prose, but the most important part of it is paragraph 16:

" The Government considers, as did the GLDP Panel, that there are limited areas of green-belt land where development could start quickly because services are readily available and which could be used for housing without damage to basic green-belt policies, and
E indeed to their advantage. Such areas could be identified by careful examination by those best qualified for the job, the local authorities. The Government is therefore asking the Standing Conference on London and South East Regional Planning, with the authorities concerned, to undertake this task urgently. From these areas the Government is looking for some 2,000 acres to be made available quickly, which might, after going through the normal planning procedures, be found suitable for housing. These releases of green-belt land will be matched by extensions or improvements elsewhere. In the light of the studies referred to in paragraph 15, local authorities will be encouraged to propose additions to green-belt areas in their development plans. They will also be given encouragement and advice on ways of enhancing the amenity of green-belt land."

F Those views are extended and particularised in a number of circulars to which Mr Slynn, on behalf of the first respondent to this motion, has drawn my attention.

I think it is sufficient, so far as I am concerned, to look at the language of the decision letter to which I have referred, namely the letter of October 8 1973, and examine the language of the relevant paragraph in which the Minister sets out the substance of his decision. After a preamble referring to the fact that the written representations made in support of the appeal and those of the council and other interested parties had been considered, and to a visit by an officer of the department to the site, paragraph 3 is as follows, and this is addressed to the agents for the developers' firm:

" In your representations on behalf of your client you stress the G
urgent need to make available land for housing, and the appeal proposal has been examined with that need in mind to see whether an exception to the normal presumption against development in the approved green belt would be justified in this instance. The appeal site is on the edge of the built-up part of Enfield and has a residential development on two sides. The physical situation of the site is such that it is not thought that its loss from the green belt would be unacceptable, and its development residentially is not seen as an outward spread of the built-up area of Enfield. Because of the slope of the land away from the ridge, it is not thought that houses on the site would be prominent when seen from the open land to the west. They would be seen against a background of existing development to the east and south and against houses along the Ridgeway. After
careful consideration of all the facts and representations, including H
those put forward by local people and other third parties, the conclusion has been reached that in the particular circumstances of this case the housing gain from development of the site outweighs other considerations. It is noted that reference was made by the local planning authority to the decision issued in August 1972 on an appeal about residential development of a small area of green-belt land to the south of and adjoining the present appeal site. The circumstances of that decision have been taken into account, but do not alter the view that the present appeal should be allowed."

That referred to a decision refusing development on the green outlined area on the plan to which I have already referred. I do not think it necessary to go into any detail about that. I
find myself wholly unable to extract from the history of this J
case anything that could be regarded as an implied representation by the Minister that he would not depart from the original development plan until the standing conference had delivered itself, or indeed anything which deprived or estopped the Minister from exercising the unfettered discretion vested in him to decide any such appeal as was before him at the time on its merits as he saw them, and I am wholly unable to extract from the evidence in this case anything which could possibly be regarded as a breach of the principles of natural justice.

I have been referred to a number of very helpful authorities. One in particular I think is especially worth referring to. That is a case in Scotland, *Simpson v Edinburgh Corporation* 1960 SC
313, and it appears from the headnote that a section of the K
Town and Country Planning (Scotland) Act 1947, section 12, which is in similar language to section 29 in the present case, had to be considered. The headnote quotes a part of the section of the Act of 1947, saying:

" Subject to the provisions of this and the next following section, where application is made to the local planning authority for planning permission, that authority may grant permission either unconditionally or subject to such conditions as they think fit, or may refuse permission; and in dealing with any such application the local planning authority shall have regard to the provisions of the development plan, so far as material thereto, and to any other material considerations." L

What happened in that case was that the University of Edinburgh obtained planning permission for the construction of some modern university buildings in George Square, Edinburgh, which involved the demolition of two sides of the square and the destruction of its characteristic Georgian appearance, the sort of situation that always inflames preservationists. The owner of a dwelling-house in George Street objected to the university's proposed building operations on the ground that the purported planning permission was contrary to the city's development plan and was *ultra vires,* and he sought what in Scotland they call a declarator to that effect. Lord Guest, who
was then Lord Ordinary, decided that the local resident had M
no right to sue, and that in any event the local authority had power to grant planning permission contrary to the development plan. There is an interesting passage on pages 318-319 of the report in which the position and significance of the development plan are dealt with by Lord Guest. He says:

" Section 12, which has already been quoted, obliges the local authority in dealing with applications for planning permission to have regard to the provisions of the development plan so far as material thereto and to any other material considerations. It was argued for the pursuer [that was, the local resident] that this section required the planning authority to adhere strictly to the development plan. I do not so read this section. To ' have regard to ' does not in my view

A mean 'slavishly adhere to.' It requires the planning authority to consider the development plan, but it does not oblige them to follow it. In view of the nature and purpose of the development plan, to which I shall refer later, I should have been surprised to find an injunction on the planning authority to follow it implicitly, and I do not find anything in the Act to suggest that this was intended. If Parliament had intended the planning authority to adhere to the development plan, it would have been simple so to express it. If a glossary of the phrase 'to have regard to' is to be sought in the Act, it is to be found in sections 4 and 19, where it is plain that 'having regard to' in that context does not mean 'to follow implicitly,' but means 'having in view.' It was pointed out that if the phrase was mandatory, then the addition of the words 'to any other material considerations' in section 12 (1) of the Act would, if
B other material considerations were inconsistent, face the planning authority with an impossible task of reconciling the two. In my opinion the meaning of section 12 (1) is plain. The planning authority are to consider all the material considerations, of which the development plan is one."

Coming back to the decision letter in this case, and referring again to paragraph 3 of that letter, which I have already read, I can find nothing to support the inference that/the Minister had here failed to have regard to all material considerations and to apply his mind thereto. In particular I find that he applied his mind to the presumption that the green-belt principle must be strictly observed, indeed the green belt must not be loosened, and
C that he decided that like other belts the state of congestion required that it should be relaxed. Thus he was careful to set out all the details that follow from the examination of this site, the proposed development, the slope of the land, and other things, which indeed convinced me that whoever drafted this decision was deeply conscious of addressing his mind to the question of whether the presumption that the integrity of the green belt should be observed if it could be should be the dominant factor in considering whether in the particular circumstances of this case, this particular development should be permitted. Having reached that conclusion, I cannot find any representation, anything in the nature of an estoppel, still less any
D breach of what I understand to be natural justice. The papers, as I have already indicated, are burdened with many, many letters from indignant residents, who are angry about what was proposed, but the whole material is there, and it was agreed that the matter should be decided by written statements on both sides, and so it was, and I think it was right. It follows that I must dismiss this motion.

Each respondent was awarded costs against the council.

E

QUEEN'S BENCH DIVISIONAL COURT

July 16 1974

(Before Lord WIDGERY CJ, Mr Justice BRIDGE and Mr Justice MAY)

WHITFIELD v GOWLING

Estates Gazette January 11 1975

(1974) 233 EG 147

Enforcement notice—Plea of no case, since no evidence adduced in support of allegation that respondent was owner and occupier of land in question—Proof that he was owner and occupier six months before service is good enough unless he can produce evidence of some change in the situation

F This was an appeal by Mr T J R Whitfield, clerk to the Cumberland County Council, from the dismissal by magistrates sitting at Penrith on October 11 1973 of a complaint against the respondent, Mr Joseph Alan Gowling, of Whinney Hill, Catterlen, Penrith, charging him with failure to comply with an enforcement notice, contrary to section 89 (1) of the Town and Country Planning Act 1971.

Mr M Rich (instructed by Sherwood & Co, agents for T J R Whitfield, of Carlisle) appeared for the appellant, and Mr J Trench (instructed by Doyle, Devonshire, Box & Co, agents for T H Campbell Wardlaw, of Newcastle-upon-Tyne) represented the respondent.

Giving judgment, LORD WIDGERY said that Cumberland G County Council served an enforcement notice on the respondent on December 1 1971 on the basis that he was owner and occupier of land at Whinney Hill, Catterlen, Penrith. The notice required him within 28 days to remove certain buildings, including an oil storage tank, from the land in question and restore it to its previous condition. The respondent appealed against the notice, but after an inquiry held in October 1972 the notice was confirmed and limited to come into effect on January 3 1973. The respondent having failed to comply with the terms of the notice, he was duly summoned. Before the magistrates he contended that he had no case to answer, because there was no evidence to show that he was the owner and occupier of the H land on the date of service of the notice. The local planning authority was then allowed to put in further evidence and produced a certificate under section 16 of the Town and Country Planning Act 1962 which was signed by the respondent and admitted that he was the owner and occupier of the land in question when he made an application for planning permission in June 1971. Also submitted was a statement signed by the respondent to the effect that he was the owner and occupier of the land pursuant to section 215 of the Town and Country Planning Act 1962, and that no other person had any interest in the land.

No authorities were cited to the magistrates, and it was for them to decide whether the proof of ownership subsisting in J June 1971 could be carried through to December 1971. There was a dictum of Fry J relating to the proof of ownership in *Governors of Magdalen Hospital v Knotts* (1878) 8 Ch D 709. In that case it had been submitted that there was no presumption that in 1783 the hospital remained the owner of land which had been conveyed to it in 1763, but the learned judge determined against that submission, saying that when there was evidence that land was conveyed to a particular person there was, in the absence of anything to the contrary, a presumption that the land remained the property of the person to whom it had been conveyed. In the present case there was evidence that the respondent was the owner and occupier of the land six months K before the enforcement notice was served. The magistrates should have considered the matter on the footing that having been satisfied about ownership and occupation in June 1971, they should have presumed that that ownership and occupation continued until December 1971, unless there was some ground for thinking that that was not the case. The magistrates should therefore not have acceded to the submission of no case to answer, and the matter should be sent back with a direction to continue the hearing.

BRIDGE and MAY JJ agreed, and had nothing to add. The costs of the appellant were ordered to be paid by the respondent

L

QUEEN'S BENCH DIVISIONAL COURT

November 4 1974

(Before Lord WIDGERY CJ, Mr Justice BRIDGE and Mr Justice SHAW)

LTSS PRINT & SUPPLY SERVICES LTD v HACKNEY BOROUGH COUNCIL AND ANOTHER

Estates Gazette January 11 1975

(1974) 233 EG 147

Enforcement notice—Secretary of State's mistaken general statement as to planning law probably did not affect his decision, but case remitted to him for "whatever adjustment he might think necessary" M

This was an appeal by LTSS Print & Supply Services Ltd against a decision of the second respondent, the Secretary of State for the Environment, upholding an enforcement notice served by the first respondents, Hackney Borough Council, by which further use of premises at 138 Kingsland Road, London E2, for the retail sale of furniture was prohibited.

Mr A Dawson (instructed by Kingsley Napley & Co) appeared for the appellants, and Mr H K Woolf (instructed by the

A Treasury Solicitor) represented the second respondent. The first respondents took no part in the appeal.

Giving judgment, LORD WIDGERY said that the notice recited that the sale of furniture was a change of use for which planning permission had not been given. The planning history of the premises was that in 1961 they had been used by a company dealing in veneers and plywoods, mainly on a wholesale basis. Afterwards the premises had been used for the retail sale of furniture. There was no contest that the established use of the premises up to 1969 was that of a wholesale warehouse, and it was not surprising that the Secretary of State had upheld the enforcement notice and the local authority's contention that there had been a material change of use. The Secretary B of State, as he was entitled to, had then gone on to consider whether planning permission for the offending use should be granted. The argument for the appellants before the inspector was that, within the definition of "warehouse," it would be open to the appellants to use the premises for the purpose of a cash-and-carry grocery business. Since, it was said, that would be more detrimental to the environment than the sale of furniture, planning permission should be granted. In considering whether or not furniture-selling would be better than cash-and-carry grocery, the Secretary of State was entitled to consider what were the alternatives. Had he asked himself that question, he would have referred himself to section 23 (9) of the Town C and Country Planning Act 1971, which provided that where an enforcement notice had been served planning permission was not required for the use of the land for the purpose for which it could lawfully have been used if the development had not been carried out. If a notice was served the landowner or occupier could revert to any use which would have been lawful under the terms of the 1971 Act. The Secretary of State had not, however, taken that course. Instead, he had said that there was no right to revert to the original use once a new use had been established.

That was an error. It was highly improbable that the mistake had made any difference to the Secretary of State's decision. The inspector, in recommending refusal of permission, had obviously D contemplated the traffic which would have been attracted to the premises, and had gone on to give strong views for saying that the land was unsuitable for the present use. It was unlikely that the Secretary of State, in considering the inspector's report, had been influenced by the error of law he had made earlier. He had stated that he did not propose to grant permission because of the inspector's reasons. However, it was right for him to be given an opportunity of making whatever adjustment he might think necessary to his decision, and the case should be remitted to him for further consideration.

BRIDGE and SHAW JJ agreed, and an order was made accordingly. Leave to appeal was given.

E

QUEEN'S BENCH DIVISION

July 30 1974

(Before Mr Justice MILMO)

BATES v SECRETARY OF STATE FOR THE ENVIRONMENT AND ANOTHER

Estates Gazette January 11 1975

(1974) 233 EG 157

F **Tree preservation order—No force in complaints about council's procedure—Criticism of "classification" of trees, even if established, no ground for quashing order—Trees to be preserved adequately identified by description, and in particular by reference to numbers marked on each tree**

This was an application by Mr Edgar Bates, of 2 Seymer Road, Romford, Essex, to quash Tree Preservation Order No 8/69 (Dudbrook-Navestock), made by the second respondents, Essex County Council, in May 1969 and confirmed, subject to modifications, by the Minister of Housing and Local Government, predecessor of the first respondent, the Secretary of State for the Environment, on October 31 1969.

The applicant appeared in person. Mr J M Sullivan G (instructed by the Treasury Solicitor) appeared for the Secretary of State, and Mr D R P Mole (instructed by the council solicitors) represented Essex County Council.

Giving judgment, MILMO J said that on April 28 1969 the applicant entered into a contract with the owners of Dudbrook Hall, Navestock, Essex, for the felling of 60 trees covered by a Forestry Commission licence, dated April 24 1969, permitting felling. These trees constituted the total subject matter of the tree preservation order which it was now sought to quash. On April 25 1969 the chairman of the landscape and conservation subcommittee of Essex County Council had received from an officer of the county planning department a request for a tree H preservation order to be made in respect of the 60 trees. On April 29 the chairman gave written authorisation for the making of a tree preservation order in respect of the trees, and the order was made under the seal of the council on May 20 1969. At a meeting of the landscape and conservation subcommittee on the following day the action taken by the chairman was recorded and confirmed, and the relevant resolution was entered in minute 53 of the subcommittee for that date. The minutes of the meeting were duly signed by the chairman of the subcommittee at its next meeting on June 18 1969. By letter dated May 23 1969 notice was given to the applicant of the making of the tree preservation order, and the notice complied with all the J statutory requirements and regulations. By letter dated May 26 1969 to the Minister of Housing and Local Government the applicant appealed for a public inquiry so that his objections to the order might be heard. The Minister, as he was fully entitled to do, decided not to hold a public inquiry, and by letter dated July 11 1969 he notified the applicant that he could take his decision on the basis of written representations and on inspection of the trees and their surroundings by one of his officers in company with a representative of the council and the applicant or his representative. On October 7 1969 the Minister's representative visited the site and made an inspection of the trees. He did so in the presence and company of the applicant and duly made K his report to the Minister. On October 31 the Minister confirmed the order subject to modifications, the effect of which was to reduce the number of trees covered by the order from 60 to 20. The applicant sought to quash the order so confirmed.

The applicant's first ground of appeal was an unparticularised allegation that the confirmation by the Minister was not within the powers of the Town and Country Planning Act 1962. The applicant said he did not rely in support of that ground upon any matters going beyond the allegations in his fourth ground, which amounted to the statement that the council's order was not authorised in the prescribed manner and not recorded in the council's minutes as required by the relevant provisions of the L Local Government Act 1933. He (his Lordship) would deal with that point in a moment. He found no evidence whatsoever to support the charge made in ground 2, that while the Minister accepted that over 40 of the 60 trees concerned should be cut down, he failed to take into consideration the resulting effect on amenity, and further failed to consider the correctness of the classification of the trees in the schedule to the order. On the amenity point, all the evidence pointed the opposite way. The reference to the "correctness of the classification of the trees" appeared to be a criticism of the species of trees listed in the schedule of the order. He (his Lordship) was not satisfied that there was any misdescription, and even if there had been a wrong M identification of the species of tree, he did not consider that it would have invalidated the order in any way, as the trees to be preserved were adequately identified by the rest of the description given, and in particular by the number marked upon each tree. Further, the applicant was present when the inspector examined the trees and raised no objection to the inspector's opinion as to the species. Besides, of the six trees alleged to have been wrongly described, five were eventually excluded from the order. The third ground of appeal was the submission that the schedule to the Minister's decision letter contained material errors in the specification of six of the trees forming the subject of the order, and that turned out to be no more than a repetition, equally

A without substance, of the applicant's complaints as to supposed misdescription. As for ground 4, he (Milmo J) found as a fact that, as already stated, the making of the order was properly and adequately recorded in minute 53 of the meeting of the council's landscape and conservation subcommittee on May 21 1969, which minutes were duly signed at the next meeting of the subcommittee on June 18 1969. For the reasons appearing in the affidavit of the clerk to the council, he (his Lordship) was satisfied that the order was made in the prescribed manner and that the subcommittee's chairman was duly authorised by the council to make the order on its behalf.

Plainly, the applicant was under the impression that the hear- B ing before the High Court would be a rehearing of the matters in respect of which the legislature had given to the Minister the duty of decision. Basically the applicant's complaint was that the Minister had based his decision on the report of the inspector and that the applicant's expert disagreed with the inspector's opinions. This was not a ground upon which the High Court would be entitled to quash the Minister's order. In the result, there were no grounds for quashing the order, and the application had to be dismissed with an order that the applicant pay the costs of both respondents.

C

COURT OF APPEAL

October 17 1974

(Before Lord DENNING MR, Lord Justice ORR and Lord Justice BROWNE)

HARWICH HARBOUR CONSERVANCY BOARD v SECRETARY OF STATE FOR THE ENVIRONMENT AND OTHERS

Estates Gazette January 18 1975

(1974) 233 EG 263

Planning permission—Scheme for marina for 500 yachts in D Harwich harbour called in by Secretary of State—Inspector says scheme would create hazards in the operation of the port, Secretary of State says Department of Trade's consent concludes that issue—Not so, Department of Trade concerned only with a narrower question—Order quashing decision confirmed on appeal

This was an appeal by the Secretary of State for the Environment and Stour River Estate from a judgment of MacKenna J in the Queen's Bench Division on December 3 1973 (reported at 230 EG 225) quashing a decision under section 35 of the Town and Country Planning Act 1971 by which the Secretary of State granted planning permission for the development of a 500-berth E yacht marina and associated facilities at Erwarton Bay, Shotley, near Harwich. The respondents to the appeal were the Harwich Harbour Conservancy Board.

Mr J Priest QC, Mr H K Woolf and Mr P Vallance (instructed by the Treasury Solicitor) appeared for the Secretary of State; Sir Michael Havers QC and Mr A R Grove (instructed by Joynson-Hicks & Co) for Stour River Estate; and Mr K Jupp QC, Mr M G V Harrison and Mr D Robins (instructed by Sherwood & Co) for the board.

Giving judgment, Lord Denning said that in order to implement their proposal for a yacht marina, the developers, Stour River Estate, had to get permission from the Secretary of State for Trade and Industry, who came into the matter because of F two private Acts dealing with Harwich harbour which were passed in 1863 and 1865, and a public Act, the Coast Protection Act 1949. Apart from that, the developers had to get planning permission under the planning Acts, and their proposal was called in by the Secretary of State for the Environment under section 35 of the Act of 1971 for his own decision. He ordered an inquiry, and appointed an inspector who, although putting on one side many grounds of objection, was much impressed with the difficulties which would arise in the future in the user of the harbour. The inspector heard evidence from the masters of vessels and stated: "Harwich Harbour has seen a fantastic growth of commercial tonnage over the last 10 years. . . . The application is for 500 large yachts capable of going to sea to be G moored opposite Parkeston Quay, a very busy part of the harbour, and I think that this substantial number of pleasure vessels might well have a detrimental effect on the commercial operation of the port and the safety of those who use it." On that ground the inspector recommended that planning permission be refused. The Secretary of State, having considered his inspector's report, took a contrary decision and granted planning permission. He said, in effect, that there were no sufficient grounds for refusing it. There was no doubt that he was much influenced by a decision of the Department of Trade. He thought that the responsibility in respect of traffic and navigation in Harwich Harbour could safely be left in the hands of that H department, and he need not worry himself about that aspect. Harwich Harbour Conservancy Board complained about the Secretary of State's decision, and the judge upheld their case. The other parties now appealed from the judge's order.

Under the Harwich Harbour Act 1863, section 17, it was enacted that it was not lawful for any person to construct any work below high-water mark if such work would be injurious to the navigation of the harbour or the entrance thereof or would tend to obstruct access thereto. The main objective of that section was to ensure that the works themselves, such as a jetty put out in the harbour, would not obstruct or tend to obstruct the free passage of vessels up and down, or cause silting J to take place. It was not in the least directed to the number of vessels which might use the harbour. The Harwich Harbour Act 1865, section 26, stated that a person who wanted to erect a pier or wharf, or any other erection, had a right to do so, unless in the opinion of the conservancy board, or the Board of Trade on appeal, the construction would be injurious to navigation. There again, it was the works themselves which were to be considered—whether they would be injurious to navigation—and not the user consequent on them or the number of vessels which might use them. It appeared, therefore, that under those private Acts the Department of Trade was not to consider the effect of building works on the amount of traffic in the harbour, but the effect of the works in terms of actual physical hindrance of the K vessels going to and fro. The Coast Protection Act 1949, section 34 (1), said that "no person shall without the consent in writing of the Minister of Transport . . . (a) construct, alter or improve any works . . . so that obstruction or danger to navigation is caused or is likely to result." That seemed wider at first sight than the private Acts, but the question was whether it was limited to a narrow consideration, whether the works themselves or their operation were an obstruction or danger to navigation, or whether it extended to a much wider consideration, whether the user of such things as a marina or jetty, or the number of vessels going to them, would be likely to cause interference, obstruction or danger to navigation. L

It seemed to him (his Lordship) that there was no doubt that the Department of Trade, in correspondence with the conservancy board, led them to believe that the department were taking a narrow view of their powers under the provisions of the 1949 Act. At the inquiry, the board made the positive statement as part of their case under the 1865 Act: "The department [of Trade] were unable to take into account the effect of the increase in traffic resulting from the proposals." That interpretation was never contradicted, and the inspector quite clearly thought that the narrow view was being taken and that the consent of the department was only a consent so far as the narrow view was concerned—obstruction by the works themselves, and not what M effect the user would have on the use of the harbour by vessels. He (Lord Denning) thought that that was the correct view. The wording of the 1949 Act was in relation to whether, in the operation and the construction of the works, an obstruction to navigation was caused or was likely to be caused. It was the same as in the construction of the 1865 Act. One looked at whether the operations themselves caused obstruction or danger to the vessels using the harbour, and not at the wider position—what was the effect of the increasing number of vessels on navigation. That was how the inspector understood the matter. He went on to consider the effect of increased traffic and came

A to the conclusion that the proposal might well have a detrimental effect on the commercial operations of the port.

The real point in the case was whether the Secretary of State for the Environment had misdirected himself in law. Having regard to the specific legislation, he took the view that the exercise of planning powers would be inappropriate, and he was accordingly satisfied that planning permission ought not to be refused on this ground. His decision letter stated:

> "The Secretary of State has taken note of the inspector's conclusion that there might well be a detrimental effect on the commercial operation of the port, arising from the presence of a large number of yachts. Having regard to the responsibilities of the Secretary of B State for Trade and Industry with regard to possible 'obstruction or danger to navigation' the Secretary of State does not consider, on the material before him and using his best judgment, that there are any grounds for refusing planning permission for this reason."

Sir Michael Havers had submitted that the Secretary of State had brought his own judgment to bear on the material before him. He (his Lordship) could not so interpret the decision letter. The Secretary of State was considering that the Department of Trade had responsibility in the matter and that if they consented to it, as he seemed to think they had done, that was good enough. In that he was in error. The Department of Trade were limited to the impact of the works themselves on the vessels passing to and fro. At the planning inquiry, the evidence of C masters of ships using the harbour was analysed, and the inspector went into it most carefully. With the Department of Trade, there was no such inquiry required, with all the careful investigation that took place at the planning inquiry. In other words, the hindrance to traffic and detrimental effect on the port seemed a very proper matter to be considered. This was a matter for the inspector, and on his reference the Secretary of State ought to have taken the responsibility of deciding it himself and not have fallen back on the previous matters dealt with by the Department of Trade. If he had considered it himself, he (Lord Denning) thought he would have been very slow to interfere with the recommendations of his own inspector.

D ORR and BROWNE LJJ agreed, and the appeal was accordingly dismissed.

QUEEN'S BENCH DIVISIONAL COURT

November 19 1974

(Before Lord WIDGERY CJ, Mr Justice MAIS and Mr Justice CROOM-JOHNSON)

KENT COUNTY COUNCIL v LOUGHLIN AND OTHERS

E

Estates Gazette February 22 1975
(1974) 233 EG 759

Magistrates entitled to ignore negative evidence of tithe map and find that a lane to be made up by the council was already a public road—Case a "very close thing," but frontagers hold decision quashing council's provisional apportionment

This was an appeal by the Kent County Council from a decision of Dartford justices on February 19 1974 holding that Fairby Lane, Hartley, Kent, was a public road and F quashing a provisional apportionment of the cost of making it up served by the council on the respondents, Mr E J Loughlin and 13 other frontagers.

Mr R A Sears (instructed by Sharpe, Pritchard & Co, agents for the solicitors to the council) appeared for the appellants, and Mr R A Henderson (instructed by A L Proctor, of Gravesend) represented the respondents.

Giving judgment, LORD WIDGERY said that the council, who wanted to make up part of Fairby Lane and felt it was a private road, sent a provisional apportionment to the appropriate frontagers so that they should contribute to the expense. The 14 people with property fronting on to the lane objected on the ground that it was not a private road, and G this contention was upheld by the magistrates.

As often happened in similar cases, there was no positive evidence as to when the lane became a public highway. The council were thus thrown back on the tactics of endeavouring to prove that the highway came into existence after the year 1835. The relevance of that date was that under an Act of that year, highways dedicated to public use before that date were to be maintained at public expense. This did not automatically apply to roads which came into existence after that date. As it was not possible to prove the status of the road in 1835, the county council set out to prove that it did not even exist as a road in 1835. If so, it would automatically H follow that it could not have been a public highway in 1835, and this would have settled the issue. The council placed a number of maps before the magistrates. The first three in order of date, those of 1769, 1779 and 1794, did not show the road, and neither did an ordnance survey map of 1799. A map by a Captain Mudge, who surveyed the roads of Kent in 1801, also did not show the road, nor did an ordnance survey map of 1819 or a tithe map of 1846. The first map featuring the road was an ordnance survey map of 1864. The frontagers put forward a book by the Reverend Geoffrey Bancks, a former vicar of Hartley, published in 1927, which contained a map of the area in late Plantagenet and early J Tudor times. This showed some kind of road or way following the line of Fairby Lane. In the foreword to the book, *Hartley Through The Ages,* the author pointed to some difficulties with the map. He said: "The maps I have endeavoured to make as truthful as possible, but I have been obliged to fall back to some degree on imagination." The magistrates in their conclusions said, "We are of the opinion that Fairby Lane is not a private street." This was a conclusion of fact with which the Divisional Court could not interfere. The court could only overrule the magistrates' decision if they were satisfied that it was perverse, or one which no reasonable bench of magistrates, properly instruct- K ing themselves, could have reached.

He (his Lordship) did not think the magistrates should have been significantly affected by the Rev Bancks's work. His map was intended to show late Plantagenet and early Tudor times, and those times were not of concern in the present case. A map which was a reconstruction could not be given anything like the authority and weight of a contemporary map made by a professional map-maker. He (the Lord Chief Justice) did not think it was putting it too high to say that *Hartley Through The Ages* was an unreliable guide. With the ordnance survey maps, all the council L could prove was that Fairby Lane came into existence between 1801 and 1864. This left the tithe map of 1846, which did not show the road. Tithe maps were produced to illustrate the various areas where tithes could be levied. They were not intended to be used as a kind of ramblers' guide by people seeking a route from A to B. The position of roads was often of little importance. The fact that the tithe map omitted the road was of far less value than if it included it. The man who made the map could have thought the road was irrelevant for his purposes. If the road was shown on the map, it would be important, because if it was shown it would mean that it definitely existed. The ultimate M question in the case was whether the magistrates' decision was perverse. Were they entitled, as reasonable people, to take the view that the tithe map was unreliable? If so, the council would be left with no means of saying that the lane came into existence after 1835. Tithe maps should be approached by reasonable people with care. The case was a very close thing, but he (his Lordship) had reached the conclusion that the magistrates were entitled to take the view they did and to refuse to place reliance on the tithe map. In his opinion, therefore, the appeal failed.

MAIS and CROOM-JOHNSON JJ agreed, and the appeal was dismissed with costs.

A

QUEEN'S BENCH DIVISIONAL COURT

November 12 1974

(Before Lord WIDGERY CJ, Mr Justice MAIS and Mr Justice CROOM-JOHNSON)

HARRIS SIMON & CO LTD v MANCHESTER CORPORATION

Estates Gazette February 22 1975

(1974) 233 EG 761

Trader's objection before a jury to corporation's decision to stop up street—Finding in trader's favour can be upset only if perverse—Verdict one which 12 reasonable men could reach—Corporation's appeal fails B

This was an appeal by Manchester Corporation from a decision of Manchester Crown Court on November 13 1973 allowing an appeal by Mr Cecil Simon, of Harris Simon & Co Ltd, Jubilee Street, Manchester, against the stopping-up of two streets giving access to the rear of his premises.

Mr S J Fawcus (instructed by Church Adams Tatham & Co, agents for Cobbetts, of Manchester) appeared for the appellants. The respondent did not appear and was not represented.

Giving judgment, LORD WIDGERY said that the appeal
C arose out of a somewhat unusual jurisdiction. By virtue of the Manchester General Improvement Act 1851, power was given to Manchester Corporation to stop up roads. Under the Act, anyone could appeal to quarter sessions if he thought he would be aggrieved by the stopping-up of any street, court, alley, way or passage, and quarter sessions were to empanel a jury of 12 disinterested people to decide whether the street, etc, was unnecessary or might beneficially to the public be stopping-up of the streets. Counsel for the corporation was or aggrieved thereby. Manchester Corporation had recently decided to stop up Brewery Street and Briddon Street, Manchester. These two streets, approaching each other at right
D angles, formed alternative means of access to a private right of way enjoyed by Mr Simon leading through to his furniture-manufacturing premises, which fronted on to Jubilee Street. Therefore Mr Simon had two routes whereby he could approach his private right of way. He objected, as he was entitled to do. At the hearing, the corporation raised the question whether he should or should not be regarded as a person who would be injured or aggrieved by virtue of the stopping of the streets. Counsel for the corporation was endeavouring to play down the consequences which the stopping-up of the streets would have on Mr Simon, and Mr Simon was resolutely asserting that he would be aggrieved. What
E had been proposed was that on his losing his public right of way over the two streets he should be given a private right of way by agreement in lieu, and there were arrangements of a complicated nature as to the provision of keys, either personally to Mr Simon or in the hands of a janitor who would be in attendance on the gates to ensure they could be opened whenever Mr Simon wished his vehicles to go through. The corporation asserted that any injury to Mr Simon would be very small indeed: Mr Simon, on the other hand, said that he would be seriously inconvenienced by the gates and fences. He asked the jury if they would like to have to unlock gates before they could get into their own homes. He said that
F the two streets were the only access to the ground floor of his premises. His firm had enjoyed free access to their premises for more than 75 years and he wanted the position to remain so. He maintained that in any event his firm would be worse off, because the arrangement with the gates and keys would be an encumbrance or, at all events, an obstacle.

The judge, summing up to the jury, left them with three questions to answer: (a) whether Brewery Street and Briddon Street were unnecessary for public thoroughfare; (b) whether the stopping-up of these streets would be beneficial to the public; and (c) whether Mr Simon would be injured or aggrieved by the stopping-up of the streets. The jury answered yes to question (a), gave no answer to question (b), G
which they had been told arose only if their answer to (a) was negative, and answered yes to (c). As they decided that Mr Simon would be injured or aggrieved by the stopping-up of the streets, his appeal was allowed by the Crown Court. The corporation had a right of appeal to the Divisional Court if they could show that the decision below was wrong in law or in excess of jurisdiction. The right was exercisable by way of case stated. The case here did not include any evidence from the court below, but the appellants *de bene esse* were allowed to read some of the evidence. Counsel contended on their behalf that the jury's verdict should be treated in the same way as the judgment in a civil case. He invited the court H
to say that the real test was whether the jury's conclusions, particularly on question (c), were conclusions which reasonable men might come to. He (his Lordship) did not accept that that was the right approach to the problem. The right of appeal by way of case stated was not one which involved a rehearing. There was no power in the Divisional Court to concern itself with the matter at all unless it could be said that the decision of the court below was wrong in law or in excess of jurisdiction.

The situation appeared to be unique, in the sense that the court could not think of a case where a jury verdict was open to appeal by way of case stated on a point of law. He (Lord J
Widgery) thought that there were only two ways in which an error of law could be said to arise in such a way as to justify the interference of the court under the powers of section 10 of the Courts Act 1971: if the judge when addressing the jury had himself made an error in law, or if it could be said that the verdict was perverse in the true sense, that was to say, was a conclusion which no reasonable jury could have reached. The latter test was the one applied by the Divisional Court in appeals by way of case stated from magistrates' courts, and he (his Lordship) thought it would be wrong to apply some looser test, because the whole con- K
cept of appeal by rehearing was foreign to appeal by way of case stated. If the normal test were applied, and the question was "Was the conclusion of the jury on question (c) a conclusion which no reasonable jury could have arrived at?" the answer, as he thought, was no. It seemed that the 12 disinterested people on the jury might well have been impressed by Mr Simon's argument that he was losing something, an unrestricted right, and receiving a restricted right. He (the Lord Chief Justice) was not prepared to say that no reasonable jury could have come to the conclusion that Mr Simon was aggrieved.

MAIS and CROOM-JOHNSON JJ agreed, and the appeal was L
accordingly dismissed.

QUEEN'S BENCH DIVISIONAL COURT

October 22 1974

(Before Lord WIDGERY CJ, Mr Justice MILMO and Mr Justice ACKNER)

R v SECRETARY OF STATE FOR THE ENVIRONMENT AND ANOTHER, *EX PARTE* HOOD

Estates Gazette March 8 1975

(1974) 233 EG 931

"Road used as public path" may be reclassified as a mere footpath on special review of definitive map despite the presumption as to use as a bridleway enshrined in section 32 (4) (b) of the National Parks and Access to the Countryside Act 1949—Dissenting judgment of Ackner J in case of importance to horsemen M

This was a motion by Mrs Margaret Cynthia Hood, of McNab Piggeries, Shuart Lane, St Nicholas, Birchington, Kent, for an order of certiorari to quash a decision of the first respondent, the Secretary of State for the Environment, contained in a decision letter dated August 22 1973. The decision confirmed a proposal of the second respondents,

A Kent County Council, that a highway shown on the definitive map prepared under Part IV of the National Parks and Access to the Countryside Act 1949 as a "road used as a public path" should be reclassified as a "footpath."

Mr L Hoffmann (instructed by Cripps, Harries, Hall & Co, of Tunbridge Wells) appeared for the applicant, and Mr H K Woolf (instructed by the Treasury Solicitor) represented the first respondent. The second respondents did not appear and were not represented.

Giving judgment, Lord Widgery said that the highway in question was a short and unimportant length of path but the application raised an important principle. By section 27 of
B the National Parks and Access to the Countryside Act 1949, county councils were required to produce draft maps of their area, showing on the map a footpath or a bridleway as might appear to the council to be appropriate wherever, in their opinion, such a right-of-way subsisted, or was reasonably alleged to have subsisted, at the relevant date. By section 27 (2) a map prepared in accordance with the Act was also to show any way which, in the opinion of the authority carrying out the survey, was, or was reasonably alleged to be, a road used as a public path at the relevant time. Section 27 (6) defined the various ways to which earlier subsections had made reference. Thus "footpath" meant a highway over which
C the public had a right-of-way on foot only other than such a highway at the side of a public road; "bridleway" meant a highway over which the public had the following (but no other) rights-of-way: viz, a right-of-way on foot and a right-of-way on horseback or leading a horse, with or without a right to drive animals of any description along the highway; "public path" meant a highway being either a footpath or a bridleway; and, finally, "road used as a public path" meant a highway other than a public path used by the public mainly for the purposes for which footpaths or bridleways were so used. The conception of a "road used as a public path" was
D a slightly unusual one. To comply with the definition, the road in question had to be a highway where the rights of the public were not precisely defined. If it was clear that the rights of the public were confined to those embraced by a footpath or bridleway, and thus the path was a public path within the several definitions, the classification had to be footpath or bridleway, as appropriate. It was only where the public rights were not clearly confined to footpath or bridleway that resort was had to the imprecise definition of the phrase "used as a public path."

The 1949 Act contained complicated provisions whereby the draft map produced by the county council was made open to
E inspection, and landowners and other interested parties were able to make representations as to its accuracy. Eventually a definitive, or final, map was produced. By section 32 (4) a definitive map and statement prepared under section 32 (1) were conclusive as to the particulars contained to the following extent: (a) where the map showed a footpath, the map was conclusive evidence that there was at the relevant date specified in the statement a footpath as shown on the map; and (b) where the map showed a bridleway or a road used as a public path, the map was conclusive evidence that there was at the said date a highway as shown on the map, and that the public had thereover at that date a right-of-way
F on foot and a right-of-way on horseback, or leading a horse. The presumption created by section 32 (4) (b) was unexpected. In order that a road could be classified as a road used as a public path it sufficed that the rights of the public should be more extensive than a footpath or bridleway. It was not necessary, to come within the definition, that the road should be shown as a bridleway. It was sufficient if the public rights were more extensive than a footway without being more precisely defined. By virtue of the presumption, however, if a road was shown on the map as a road used as a public path the effect was to establish conclusively that, at the relevant date, a right-of-way on foot *and* a right-of-way on horseback existed over the path. Accordingly, any road G
classified as a road used as a public path produced in practice a consequence that the public could claim a right of bridleway over that road.

In section 33, as amended by the Countryside Act 1968, a comprehensive procedure was laid down for the revision of the maps. Thus, on the occasion of any review, amendments could be made to the maps. In the present case the sole reason for altering the classification was because the category "road used as a public path" had been abolished by provisions contained in the Act of 1968, Part III of the third schedule. The appellant's interest was in preserving the path as a path which could be used by the public with horses, and H
it was her contention that with the removal of the classification "road used as a public path" there should be substituted either "a byway open to all traffic" or "a bridleway." The county council, supported by the Secretary of State, took the view that the proper new classification was as a "footpath." The third schedule to the 1968 Act provided by paragraph 9 (1):

> In the special review the draft revision, and the definitive map and statement, shall show every road used as a public path by one of the three following descriptions: (a) a "byway open to all traffic," (b) a "bridleway," (c) a "footpath."

Paragraph 10 provided: J

> The considerations to be taken into account in deciding in which class a road used as a public path is to be put shall be (a) whether any vehicular right-of-way has been shown to exist; (b) whether the way is suitable for vehicular traffic having regard to the position and width of the existing right-of-way, the condition and state of repair of the way, and the nature of the soil; (c) where the way has been used by vehicular traffic, whether the extinguishment of vehicular rights-of-way would cause any undue hardship.

No doubt the first question which arose when the special review was carried out, and consideration turned upon a road K
previously classified as a "road used as a public path," was whether the correct new classification should not be "by-way open to all traffic," and for this purpose he (his Lordship) thought the paragraph 10 tests sufficient and comprehensible. There was however nothing in the paragraph to assist in deciding whether the classification should be "bridleway" or "footpath" if it was not to be "by-way open to all traffic." Accordingly it seemed that in such circumstances considerations besides those set out in the paragraph must come into play. The applicant said that since, under section 32 (4) (b) of the 1949 Act, the road when classified as a road used as a public path was conclusively presumed to be a road over L
which the public had rights of footway and bridleway, the new classification should reflect that position and be a bridleway. The respondent argued that section 32 (4) (b) merely raised an evidentiary presumption arising out of the fact that the map for the time being classified the way in question as a road used as a public path. Once that classification was removed, it was contended, there was nothing left in section 32 (4) (b) upon which the evidentiary presumption could bite. Accordingly, the question of "footpath or bridleway" had to be decided in the same manner as such questions were decided by the county council when preparing the original draft maps under the 1949 Act. In other words, in M
deciding whether the new classification should be "footpath" or "bridleway," the county council were to make such determination as might appear appropriate. It was further said that no prejudice to the public would result, because substantially similar provisions for reconsideration of the county council's reclassification would exist under the Countryside Act 1968 as under the Act of 1949. In his (Lord Widgery's) judgment, the argument for the Secretary of State and the county council was correct. For reasons already stated, he could not accept that the only matters to be taken into account were those set out in paragraph 10 of the third

A schedule. Nor could he accept that a presumption relating to a bridleway which flowed from the fact that that way was properly described as a " road used as a public path " could be in any sense conclusive of the proper classification of the way under the 1968 Act between " bridleway " and " footpath." For those reasons the application should be dismissed.

MILMO J agreed with the judgment of Lord Widgery.

Dissenting, ACKNER J said that the third schedule to the Act of 1968 dealt with the inadequacy of the review provisions in the Act of 1949 by important amendments to section 33. The section was so amended that in carrying out a review the appropriate authority were required to have regard to the B discovery by the authority of any new evidence, or of evidence not previously considered by the authority, showing that there was no public right-of-way over land shown on the map as a public path or as a road used as a public path. Further, special provision was made to remedy the imprecision of the definition of " road used as a public path." The first review carried out by any authority after the coming into force of the Act of 1968 was to be a special review involving reclassification of roads used as public paths. It was clear from paragraph 10 that the test for reclassification was of an extremely limited nature, since it related only to whether the *first* of the three descriptions, namely " by-way open to all traffic," was or C was not the appropriate description, and provided no test whatsoever for differentiating between a " bridleway " or " footpath." That seemed to him (his Lordship) to suggest quite clearly that roads used as public paths were not to be down-graded beyond the status of bridleway unless there was evidence that could be properly considered under section 33 of the 1946 Act as amended by the schedule. Nothing could have been easier than to have provided in paragraph 10, if Parliament had so intended, that Part III of the third schedule was to authorise a free-ranging inquiry; that in deciding whether or not a " road used as a public path " should be D classified as a bridleway or footpath there were to be no restrictions upon the evidence that could be received; and that the " conclusive evidence " provisions of section 32 (4) (b) were no longer to operate in this regard. That that had not been done in clear terms was incontestable, and there appeared to him (Ackner J) to be no warrant for filling the lacuna by implying words—and many words would be required —which the draftsman had considered unnecessary. If Parliament wished substantially to increase the powers of authorities to interfere with or reduce public rights, it should do so in clear and unequivocal terms.

What the inspector here did, as he was entitled to do, apply- E ing the three tests for reclassification provided in paragraph 10, was to conclude that the route ought not to be included on the draft revised map as a " by-way open to all traffic." That conclusion left unresolved whether the path should be classified as a bridleway or footpath. The inspector had gone on to conclude, while reserving to the Secretary of State the legal issue as to whether the provisions of section 32 (4) (b) of the Act of 1949 had been affected by the third schedule to the Act of 1968, that it was " unlikely, in the context of this route's general situation and relationship to other rights-of-way, that bridle rights ever existed, except perhaps in a general way over all the land included in Herne Common." In his opinion, F the route should therefore be reclassified on the revised map as a footpath. His recommendation to that effect, which was accepted by the Secretary of State, was thus based neither on the criteria of paragraph 10 nor on the basis of evidence acceptable or admissible within the terms of section 33 of the 1949 Act as amended by the Act of 1968. He (his Lordship) concluded, accordingly, that the Secretary of State was not entitled to reclassify the road otherwise than in accordance with the conclusive presumptions created by section 32 (4) (b) of the 1949 Act, that was to say as a bridleway, and he for his part would therefore have quashed the decision.

The application was dismissed with costs.

QUEEN'S BENCH DIVISIONAL COURT G

October 9 1974

(Before Lord WIDGERY CJ, Mr Justice MILMO and Mr Justice ACKNER)

JOHNSTON AND ANOTHER v SECRETARY OF STATE FOR THE ENVIRONMENT AND ANOTHER

Estates Gazette March 15 1975
(1974) 233 EG 1015

Enforcement notice—Planning unit—Forty-four garages originally purpose-built for owners' taxi-cab fleet—Unit is each garage, not the block—Further point on landlords' inability to control tenants' actions—" Structure of enforcement provisions of Town and Country Planning Act 1971 misunderstood " H

This was an appeal by Mr Sidney Albert Johnston and Mrs Elsie Winifred Johnston, owners of 44 garages in Salisbury Road, Harringay, London N4, from a decision of the first respondent, the Secretary of State for the Environment, upholding five enforcement notices served by the second respondents, the London Borough of Haringey, requiring the appellants to discontinue the use of certain of the garages for motor-vehicle repairs.

Mr J H Hames QC and Mr J Denniston (instructed by J Craigen, Wilders & Sorrell) appeared for the appellants; Mr H K Woolf (instructed by the Treasury Solicitor) for the first respondent; and Mr M Spence (instructed by D E Wood) for the second respondents.

Giving judgment, LORD WIDGERY said that the 44 garages in question, which were built in a courtyard, were originally intended to stable taxis owned and operated by the owner of the property. Over the years individual garages were let to tenants; some tenants took two or more adjacent garages using them as a single unit, and in some cases garages were used for repairs as opposed to mere garaging purposes. The K council's object in serving the five enforcement notices was to restore the original conception that the garages were for garaging motor cars *simpliciter*. In settling the enforcement notices the council had obviously come to the conclusion that the planning unit was not the entirety of the 44 garages but that individual planning units could be discerned in garages or groups of garages in separate occupations, and they had not attempted to allege a material change of use in respect of the premises as a whole. On behalf of the appellants, Mr Hames now contended that all the garages should have been treated as an entirety. The first step, in these circumstances, was to look at the planning unit. The most valuable contri- L bution to a solution of the problem of what was the proper planning unit was the judgment of Bridge J in *Burdle v Secretary of State for the Environment* [1972] 1 WLR 1207 at 1213, which suggested that one should start with the " unit of occupation," in other words that *prima facie* the planning unit was the area occupied as a single holding by a single occupier. He (his Lordship) would not for a moment wish to suggest that the rule was absolute, admitting of no exceptions, but it was clearly right as a first step. In deciding what the appropriate planning unit was, the unit of occupation was of great, if not of predominant, importance. It was a consideration which the Secretary of State or the local M planning authority might take into account in deciding what was the appropriate unit. In argument in the present case, the premises were likened to a purpose-built block of flats, and Mr Hames contended that in consequence the block should be regarded as a single planning unit. He (Lord Widgery) would have thought that in almost every case of a block of flats, with the flats let separately to different tenants, the planning unit would be the flat in question. It was not to be overlooked that it might be in the interests of the occupier himself to have the planning unit confined to the land in his own occupation. If it were so confined he was at least in control of everything happening in the planning sphere which might adversely affect him, whereas if he were compelled to share a planning unit with another or others he might find enforcement action being taken against him when the true fault was that of one of the other occupiers of the unit.

The second point taken by Mr Hames concerned one particular enforcement notice relating to garages Nos 42 and 43. Prior to 1970 garage No 44 had had the same occupier as 42 and 43 but had been used for a long time for repair work, whereas 42 and 43 had been used simply for garaging purposes. After 1970 the occupier proceeded to use all three garages for repair purposes, and the council sought to enforce the cessation of the use of 42 and 43 for those purposes, although they recognised that enforcement in respect of 44 in isolation could not be effective. The inspector in his report had said in effect that before 1970 there was a composite use of the single planning unit—part used for garaging and part used for repair work—but after 1970 there was a single use of all three garages for repair work. In those circumstances, treating the three garages as the planning unit, it was a perfectly proper conclusion for the inspector to reach that there had been a material change in the planning unit comprising the three garages, because what had formerly been a dual or composite use had become a single use for repair purposes—what was often described as an intensification of the repair use of such a character and extent as to result in a material change in the use of the planning unit as a whole. That seemed to him (his Lordship) to be absolutely right.

Mr Hames also took the point that since the garages were let to tenants who had a measure of security under the Landlord and Tenant Act 1954, it was impossible for the owners to comply with the notice and obtain the discontinuance of the use within 28 days or within any foreseeable time. Counsel's argument was that where an owner had leased the property to a tenant and was unable to recover possession, any enforcement notice should be expressed to apply either at no particular time or at some indefinite time, because it could not be foreseen at what precise moment possession would be obtained and the opportunity acquired to have the use changed. He (his Lordship) thought that the structure of the enforcement provisions of the 1971 Act had been misunderstood. Enforcement notices were served not only on the owner of the land; they were also served on the occupier. So far as concerned the occupier, all he had to do to comply with the direction in the notice was discontinue the use of the garage for repairs. When one came to the landlord and asked what the consequences would be if the use were not discontinued, the answer was in section 89 (5) of the 1971 Act. Under the section, it was only a person who was in a position to determine the unlawful use and who, notwithstanding the notice, either continued to use the land in defiance of it or caused or permitted some other to do so, who was at risk. In his (Lord Widgery's) judgment the period for compliance, highly relevant and appropriate for the occupier of the land, did not really affect the owners, because unless they themselves did something which could be said to amount to a using of the land for the unauthorised purpose or causing or permitting the tenants so to do, they were not at risk and not concerned with the compliance provisions of the notice. Any change such as Mr Hames suggested would render the notice completely ineffective against the person who ought to comply with it, namely the occupier. It was effective and proper against the occupier and did not put the owner at risk. The appeal should be dismissed.

MILMO and ACKNER JJ agreed, and an order was made accordingly.

QUEEN'S BENCH DIVISION

July 26 1974

(Before Mr Justice WILLIS)

SHEPPARD v SECRETARY OF STATE FOR THE ENVIRONMENT AND ANOTHER

Estates Gazette March 29 1975
(1974) 233 EG 1167

Purchase notice—Secretary of State may apparently obtain unlimited time for consideration of his decision by notifying owner that he does not propose to confirm—On facts, Secretary of State had no power to refuse to confirm a notice demanding acquisition of a paddock adjoining Hangleton Manor, Hove—Reference to Noble v Armitage [1962] EGD 534

This was an application by Mrs Freda Sheppard to quash a decision of the first respondent, the Secretary of State for the Environment, dated January 30 1973 refusing to confirm a purchase notice served by the applicant on the second respondents, the Borough of Hove, demanding the acquisition of land adjoining Hangleton Manor, Hove, Sussex.

Mr C S Fay (instructed by Boxall & Boxall, agents for Edwin Boxall & Kempe, of Brighton) appeared for the applicant, and Mr K H T Schiemann (instructed by the Treasury Solicitor) represented the first respondent. The second respondents were not represented.

Giving judgment, WILLIS J said: This is an application under section 245 of the Town and Country Planning Act 1971 (hereinafter referred to as "the 1971 Act") to quash a decision of the Secretary of State for the Environment dated January 13 1973 refusing to confirm a purchase notice relating to land adjoining Hangleton Manor, Hove, Sussex. The circumstances in which the purchase notice came to be served by the applicant, Mrs Sheppard, are as follows. On February 10 1967 planning permission was granted to the trustees of the Knole Estate for conversion into flats of Hangleton Manor, a house of architectural distinction which was graded II on the statutory list and had been used for some years previously as an hotel, and for the erection of 17 houses on land adjoining. This included permission for the erection of four houses and garages on the land the subject of the purchase notice (hereinafter called "the paddock"). On the same day application was made by a firm called SCB Contractors Ltd, of which both the applicant and her husband were directors, for permission to recondition Hangleton Manor as a single-dwelling unit and for the erection of 13 dwelling-houses on land adjoining. This application did not include the paddock as a site for four houses and garages, and the paddock was shown uncoloured on the plan attached to the application. In the meantime Hangleton Manor house, with certain adjoining land including the paddock, was transferred by the trustees to the applicant on March 17. The application of SCB Contractors was granted, subject to conditions, on March 28 1967. Later events can be summarised as follows. The company built four houses under the permission granted to them, and the remaining 9 houses under revising permissions. The house was not converted into a private house, but was sold by the applicant and reconditioned in order to revert to its earlier use as an hotel. The paddock remained undeveloped in the ownership of the applicant. In 1970 the applicant sought permission to build four houses on the paddock, as had been allowed to the trustees, but planning permission was refused, and the refusal was upheld on appeal on April 14 1971. She therefore served on the local authority on August 5 1971 a notice under section 129 of the 1962 Act (now section 180 of the 1971 Act) requiring them to purchase the paddock as being land which had become incapable of reasonably beneficial use in its existing state or of being rendered capable. The local authority declined to buy the land, and on October

29 1971 they transmitted the notice to the Secretary of State in accordance with section 181 (1) (c) and (3) of the 1971 Act.

It is necessary now to refer to some of the sections of the 1971 Act which are relevant to this application. Section 182 sets out what might be called the "procedure before action." On receipt of a purchase notice from the local authority the Secretary of State is required by subsection (1) to *consider* whether he will confirm the notice or "take other action under section 183 of this Act," but whatever action he *proposes* to take, he is required by subsection (2) to notify the server of the notice and the appropriate authorities. Any notification under subsection (2) must give the recipients the opportunity of being heard by a person appointed by the Secretary of State before he confirms the notice or takes any other action under section 183. Subsection (4) in fact allows the Secretary of State to change his mind after a hearing and take action different from that which he had at first proposed. Section 183 provides in subsections (1) to (4) for the various courses of action which the Secretary of State can take. These are: (1) confirm the notice, if he thinks that the conditions of section 180 are complied with; (2) grant planning permission or revoke or amend conditions, in lieu of confirmation, in order that the land can be used beneficially; (3) secure the beneficial use of the land by directing the grant of permission if application is made, again in lieu of confirmation; and (4) confirm the notice, but substitute another acquiring authority from that appearing in the notice. Finally by subsection (5) it is provided:

> "In section 182 of this Act, any reference to the taking of action by the Secretary of State under this section is a reference to the taking by him of any such action as is mentioned in subsections (1) to (4) of this section, or to the taking by him of a decision not to confirm the purchase notice either on the grounds that any of the conditions referred to in subsection (1) of this section are not fulfilled or by virtue of section 184 of this Act."

Section 184 provides as follows:

> "(1) This section shall have effect where, on an application for planning permission to develop any land which has a restricted use by virtue of a previous planning permission, permission is refused or granted subject to conditions and an owner of the land serves a purchase notice under section 180 of this Act. (2) For the purposes of this section, land is to be treated as having a restricted use by virtue of a previous planning permission if it is part of a larger area in respect of which planning permission was previously granted (and has not been revoked) and either "—(a), which is the material paragraph here—" it remains a condition of the planning permission (however expressed) that that part shall remain undeveloped or be preserved or laid out in a particular way as amenity land in relation to the remainder. . . . (3) If a copy of the purchase notice is transmitted to the Secretary of State under section 181 (3) of this Act the Secretary of State, although satisfied that the land has become incapable of reasonably beneficial use in its existing state, shall nevertheless not be required under section 183 (1) of this Act to confirm the notice if it appears to him that the land ought, in accordance with the previous planning permission, to remain undeveloped or, as the case may be, remain or be preserved or laid out as amenity land in relation to the remainder of the large area for which that planning permission was granted."

Section 186, so far as material, provides:

> "(2) If, before the end of the relevant period, the Secretary of State has neither confirmed the purchase notice nor taken any such action in respect thereof as is mentioned in section 183 (2) or (3) of this Act, and has not notified the owner by whom the notice was served that he does not propose to confirm the notice, the notice shall be deemed to be confirmed at the end of that period."

Subsection (3) prescribes the relevant period, and subsection (4) provides:

> "Where the Secretary of State has notified the owner by whom a purchase notice has been served of a decision on his part to confirm, or not to confirm, the notice (including any decision not to confirm the notice in respect of part of the land to which it relates, and including any decision to grant any permission, or give any direction, in lieu of confirming the notice, either wholly or in part) and that decision of the Secretary of State is quashed under the provisions of Part XII of this Act, the purchase notice shall be treated as cancelled, but the owner may serve a further purchase notice in its place."

The Secretary of State took a provisional view that the purchase notice should not be confirmed, and he wrote (*inter alios*) to the applicant a letter dated January 24 1972 which he intended to be a notification of his proposed action. The relevant paragraphs read as follows:

> "5 From the evidence at present before him the Secretary of State takes the view that the land has become incapable of reasonably beneficial use in its existing state, but in view of the provisions of section 32 of the 1968 Act he does not consider that he would be justified in confirming the notice as he feels that the site ought, in accordance with the planning permission granted on March 28 1967, reference BP 30996/M/12791/67, to be laid out as amenity land in relation to the remainder of the larger area for which that permission was granted.
>
> "6 He therefore proposes, unless notified by February 25 1972 that the county council, the borough council or the server of the notice desires an opportunity of being heard by a person appointed by him for that purpose, *not* to confirm the purchase notice."

The applicant required an opportunity to be heard, and a hearing took place at which the applicant, the local authority and others made representations. The inspector recommended, and the Secretary of State accepted, that although the land was incapable of reasonably beneficial use in its existing state and that the conditions in section 183 (1) were met, confirmation should be refused because the land had a restricted use in accordance with section 184 by reason of a condition imposed on the paddock under the planning permission of March 28 1967 and ought to be land laid out as amenity land in connection with the larger area for which the permission was granted on March 28 1967. This condition read as follows:

> "10 Additional trees shall be planted as shown on the deposited plan and the remainder of the land within the curtilage of the site shall be laid out and planted with trees, shrubs and/or grass within one year from the completion of the works and shall be so maintained to the satisfaction of the borough council."

The reason was stated to be "to safeguard the visual amenities of the locality." It is the imprecision of the language used in this condition which has given rise to the principal difficulties in the case and has involved much argument whether it was effective to impose on the paddock a restricted use under section 184. If it was not, it is common ground that the Secretary of State had no power to refuse to confirm the purchase notice on the grounds which he gave.

Before considering condition 10 and counsel's submissions upon it, it is convenient to deal with a preliminary point taken by Mr Fay. By section 186 (2) and (3) of the Act, a time-limit for action under section 183 is imposed on the Secretary of State which requires him to take the relevant action within six months of receiving the purchase notice from the local authority, or nine months from its receipt by the local authority, whichever expires first. If that action is not taken within the limitation period, the notice is deemed to be confirmed. The decision of the Secretary of State refusing to confirm the purchase notice in reliance upon section 184 was dated January 30 1973 and was outside the limitation period. Mr Fay's submission is that on a true construction of section 186 (2) the purchase notice must be deemed to have been confirmed before January 30 1973, namely on April 29 1972 and that the purported decision in January was *ultra vires* and should be quashed. Submissions on this difficult point were made to the inspector at the hearing on behalf of the applicant and the local authority, and the convenient course was taken of reducing them to writing so that they were before the Secretary of State when he reached his decision, as they have been before me. Mr Fay's submission, in brief, is that within the relevant period the Secretary of State had "neither confirmed the purchase notice nor taken any such action in respect thereof as is mentioned in section 183 (2) or (3)," and had not "notified the owner by whom the notice was served that he does not propose to confirm the notice." Counsel submits that there is no question but that the Secretary of State had neither confirmed nor taken either course referred to in section 183 (2) and (3), and that, despite the use of the word "propose" in the fourth contingency, the refusal contemplated in that event is just as much a final decision as the courses of action contemplated earlier in the subsection. In particular it is not a reference to the notification of tentative or provisional action contemplated by section 182 (2). He (Mr Fay) further submits that if, contrary to his main submission, the notification of the proposal is referrable to section 182 (2), then the letter of the Secretary of State dated January 25 1972 has not, by reason of its conditional phrasing, achieved its intended purpose of notifying the applicant that the Secretary of State had reached a provisional conclusion not to confirm the notice.

Mr Fay submits that the scheme of the Act and of earlier legislation makes it clear that the Secretary of State has been placed under a strict time-limit for the taking of any of the courses open to him under section 183, including by subsection (5) a decision not to confirm, and that it would be wrong to construe section 186 (2) in such a way as to enable the Secretary of State to evade the limitation merely by giving a notification under section 182 of his proposal not to confirm, whether or not there follows a hearing and whatever may be his final decision. Section 186 (2) has provided, says Mr Fay, for a final decision, whatever it is, to be given during the limitation period, and section 186 (2) does not provide via section 182 (2) what would seem to be wholly arbitrary, and in many ways a capricious escape-route, because the word "propose" is used in both subsections. Mr Schiemann submits that there is only one course of action which is caught by the limitation period, and that is confirmation. Any other course, whether refusal to confirm *simpliciter*, or action in lieu of confirmation, is properly labelled a "refusal to confirm," even if there is what counsel calls a "consolation prize" in the form of (for example) planning permission. In other words, if the Secretary of State proposes to take any course other than to confirm, he only has to give notification that he does not propose to confirm in order to have unlimited time thereafter to come to his final decision. If this was Parliament's intention, I confess I find it difficult to understand why it chose such an elaborate and obscure way of expressing it. It only needed to provide that if the Secretary of State proposed initially to confirm he must come to a final decision, whatever it might be, within, say, nine months. Although it may be that during the years since the original Act of 1947, with its amending Acts and Acts of consolidation following at regular intervals, the expression of a simple idea in simple language has become increasingly elusive, I cannot think that if that had been the intention the subsection would not have said so. I think that if the Secretary of State proposes to confirm or take action in lieu of confirmation, he must notify the owner of his intention under section 182 and give a final decision, including a decision not to confirm, within the limitation period. Why then should there be an apparent exception to this procedure deriving from the words "and has not notified . . . that he does not propose to confirm?"

Mr Fay says the exception is apparent and not real, and that when the matter is closely examined, the apparent proposal represents a final decision to refuse confirmation, and so all courses of action are parcelled within the same limitation period. This is a neat construction, and I would like to be able to construe the subsection in this way. However, I find it difficult to accept, particularly when there are found in the same section the phrases "not propose to confirm" and "decision not to confirm." *Prima facie* there is an element of futurity short of finality in the one which is not present in the other. But when I ask myself why Parliament provided for time not to run when the Secretary of State gives notice of one proposed action, namely not to confirm *simpliciter*, rather than any of the other possible courses of action open to him, particularly when a hearing can follow from the notification of any one of them, I cannot find an answer which seems to me convincing. The possible time-scale in the various contingencies provides no clue. I cannot think Mr Schiemann's suggestion that a proposal to confirm is the only course which gives the owner what he calls the "green light," and therefore the Minister must hasten, is a likely explanation, particularly since the notification may bring about a hearing after 28 days' notice. In any event, many owners would think a proposal to grant planning permission was an even greener light. In the end, despite the most helpful arguments from counsel, I find myself unable to find a satisfactory rationale for Parliament to have provided an exception to the limitation period in section 186 (2). Equally I am not convinced by Mr Fay that to have done so produces results so absurd and capricious that I feel impelled to accept his construction of the section. I think that the relevant words in section 186 (2) do look back to section 182 (2), and construing them as best I can from their context, I have come to the conclusion that the limitation period does not run after the Secretary of State has given notice under section 182 (2) that he does not propose to confirm. Nonetheless, if those who may have to consider this section share my difficulty, it may be thought right at some convenient time to clarify the meaning.

Has the Secretary of State given such a notice in this case? It is conceded in the decision letter of January 30 1973 that the letter of January 24 1972 could have been better expressed, and by Mr Schiemann that, although it is a letter in standard form, it does not fall within the strict construction of section 182. On the face of the letter, the Secretary of State said that he proposed not to confirm *unless* a hearing was required. The hearing was required and held, and thus, so runs the argument of Mr Fay, there never was a true section 182 notification in the events which happened. I doubt whether such a notification should depend for its validity on so stringent a construction. It is clear from paragraph 7 of the letter that it was intended to be a notice under section 182, and although that, of course, would not be enough to validate a true invalidity, and the use of the word "unless" was strictly inaccurate, the sense seems to me to be clear that the server was notified of the proposed course of action and given an opportunity of a hearing. I propose therefore to uphold the letter as a valid notification under section 182, and Mr Fay's submission on this point also must accordingly be rejected. But it may be thought that in future the phrasing should be changed.

I pass to the main issue in the case, namely whether, by reason of condition 10 of the planning permission of March 28 1967, the Secretary of State was entitled to refuse to confirm the purchase notice in pursuance of section 184. Mr Fay has taken a number of points in support of his submission that condition 10 cannot be relied on by the Secretary of State. First he says that the condition, on its true construction, does not apply to the paddock at all. It applies, he says, to the curtilages or gardens of the 13 houses proposed to be built, that is the remainder of the gardens of each individual site after the planting of any additional trees. Although almost any construction is possible with a condition so obscurely phrased, this is not one of which I should have thought it was susceptible. Mr Fay, however, suggests that a clue to this construction is to be found in the wording of conditions in the later revising permissions for the plot

A development, where lay-out conditions in identical language were imposed on the front gardens of the remaining houses. Unless his construction is accepted, says Mr Fay, the four houses built under the first permission would escape the lay-out restriction on their gardens. I doubt if it is permissible to try to construe condition 10 by reference to later permissions, but even if it is I do not think it is possible to give it the construction for which Mr Fay contends. Assuming did not fairly and reasonably relate to the development he is wrong on this point, Mr Fay says that condition 10 permitted, or if it did so relate at the time it was imposed, it cannot be said so to relate to any development which B remained an effective subject of planning permission when the Secretary of State made his decision. If the condition related to anything, says Mr Fay, it related to that part of the permission which was concerned with the development of the manor house by changing its use and not with the house development, and that since the manor development did not, and now will not, take place there is no effective development to which the condition can fairly and reasonably relate. Finally Mr Fay submitted that for the purposes of section 184 a condition does not remain restrictive of land in so far as it relates only to land which at the date of the permission and at all times thereafter was never in the owner- C ship or under the control of the owner or developer of the only land which remained the effective subject of planning permission, that is to say in this case the dwelling-houses, in which the applicant has never had an interest.

Mr Schiemann, faced with what is on any view an ambiguous and badly-drafted condition, invites me to approach its construction the broad way, having regard to the commonsense of the transaction and the real intention of the parties: see *Crisp from the Fens v Rutland County Council* (1950) 1 P & CR 48, per Bucknill LJ at page 55, approved in the case of an ambiguous condition in *Miller-Mead v Minister of* D *Housing and Local Government* [1963] 2 QB 196. Adopting this approach, counsel submits, the paddock was included in the land subjected to condition 10 and so remained; the question of ownership of the land at the time the condition was imposed was irrelevant, having regard to the appropriate statutory provisions; and in the end, the real issue was whether in the circumstances of this case, giving the words of section 184 their natural and ordinary meaning, the Secretary of State was entitled to refuse to confirm. He (counsel) conceded that in this particular case the section operated harshly, but he asserted that it was one of the hard cases which Parliament must have had in mind when enacting E section 184 to deal with the particular situation, and, as he put it, "to catch a variety of fish." I find the task of construing this condition extremely difficult. The draftsman has created quite unnecessary difficulties for anyone in the position of the Secretary of State or the court by using the word "curtilage," which has a meaning for a lawyer, but which he could not have used in any context recognisable by a lawyer in this case. What is "the site?" When that has been identified, what is "the remainder of the land within the curtilage of the site" after additional tree planting has been allowed for? *Quot homines, tot sententiae.* The inspector's view of the condition was this:

F It appears to me that the condition applies to the whole of the comprehensive application site area edged red, being land which included the purchase notice site which for very many years had quite clearly been an integral part of the curtilage of the manor house, closely linked to the buildings, yard and grounds by the several access ways through the surrounding flint walls.

In his decision letter the Secretary of State says:

It is agreed that this could have been better phrased, but it is considered that "curtilage" may be taken as a reference to the curtilage of the manor house, which includes all the land edged red on plan E and used for the comfortable enjoyment of the house, and that "the remainder of the land" clearly refers to that land other than the curtilages of the 13 houses.

Mr Schiemann submits that the condition covers land G other than the curtilages of the houses, the retained items such as the dove-cote, and the areas of existing trees and additional tree-planting. I have already referred to Mr Fay's construction, which I regret that I am unable to accept, and I refrain from referring to the construction put forward by the corporation. There is no doubt, from the reasons given for condition 3 and from passages to which I was referred in the inspector's report, that the council were very anxious that Hangleton Manor should be restored to a family home, and it seems that they persuaded the company to combine an application to change the use of the manor with the application for building 13 houses on part of the manor land, H which was all that the company was interested in. The majority of the conditions imposed by the permission seem to me to be clearly designed to enable the new development to be undertaken consistently with the restoration of the manor as a house. Condition 10 does not seem to me to have any purpose, if it can be given a meaning, other than to enhance the setting of the manor house. I do not think it could be said to relate fairly and reasonably to the development of the 13 houses. If the paddock can be said to be covered by condition 10, therefore, it could only be related to permission for the development of the manor by its change of use. My mind has wavered, but in the end, despite the J ambiguous phraseology of the condition, I think it was intended at the time it was imposed to provide for the laying-out of some land belonging to the house in order to create an attractive setting for it, and that it would have been understood by both parties at the time to have included the paddock.

I have reached the conclusion that condition 10 covered the paddock with considerable hesitation, but if I am right the question arises whether "it remains a condition of the planning permission that [the paddock] shall be laid-out," etc. If I am right in taking the view that condition 10 related K only to that part of the permission affecting the change of use of the manor, the question arises whether the permission can be considered as divisible and looked at in parts. It is clear from authority that it is proper to do so in certain circumstances (*F Lucas & Sons Ltd v Dorking & Horley Rural District Council* (1964) 62 LGR 491). Further, it is clear from a number of cases dealing with enforcement notices that a planning authority cannot succeed for breach of condition unless the development to which it was attached by the planning permission has actually been carried out (*Mounsdon v Weymouth & Melcombe Regis District Council* [1960] 1 QB 645 and *Noble v Armitage* [1962] EGD 534, per L Lord Parker CJ). It may be that in accordance with *Lucas's* case it is right to regard the 13 houses as each subject to separate permissions. But whether that is so or not, I am of the opinion that it is right to consider condition 10 as relating solely to the development of the manor house, and that that development should be treated as subject to a separate permission. That development has not been carried out, and if proceedings had been brought in respect of an alleged breach, they would, in my view, have been bound to fail. If, therefore, condition 10 would have been ineffective to found enforcement proceedings, can it nonetheless remain effective to found a decision unfavourable to the applicant under M section 184? I think not. Although I would have thought that a condition could remain effective under section 184 despite the permitted development not having been carried out, there must at least be a possibility that the development to which the condition is attached will take place. Here there was none. When the Secretary of State reached his decision, Hangleton Manor had become an hotel and not a private home, and nothing was left of the permission to which the condition was attached. As in the enforcement cases, at the relevant time there was nothing except the condition itself, and in my view a condition in suspended animation, as it

A were, is ineffective to bring it "within the planning net" (see per Ashworth J in *Mounsdon's* case, *supra*). I have reached this conclusion roughly by the route plotted for me by Mr Fay, but I think Mr Schiemann's rather simpler approach leads me to the same result. In the circumstances of this case, and giving the words of section 184 their ordinary and natural meaning, did there remain a condition which justified the Secretary of State in refusing to confirm the purchase notice? Put a little more specifically, and adapting Mr Fay's words, if the Secretary of State had asked himself the question "Does condition 10 now remain a condition which fairly and reasonably relates to a development which B is still an effective subject of the planning permission?" I think the answer in each case should have been no. I am glad to have been able to reach this conclusion. It seems to me that this issue under section 184 is likely in most cases to depend on the particular facts, and in the rather exceptional circumstances of this case, as Mr Schiemann acknowledges, any other conclusion would clearly have worked hardship on the applicant.

This is sufficient to dispose of the main issue in the applicant's favour, and it is not necessary to express any view on Mr Fay's submission on the question of the ownership of the paddock. C In the event that this case goes to appeal, it is right that I should record that Mr Fay also criticised the decision of the Secretary of State as failing to comply in a number of respects with the requirements of the Act, but it is not necessary that I should deal with them having regard to the conclusion I have reached.

Mr Fay: My Lord, in those circumstances I ask for an order that the decision of the Secretary of State be quashed, and I ask for an order that the costs of this matter should be paid by the respondent.

Mr Schiemann: My Lord, I have nothing to say on the matter of costs. May I mention one matter: in the early part of your Lordship's judgment your Lordship said that it was a common D ground that the Secretary of State had no power to confirm the purchase notice if the condition was bad. Your Lordship will recall that I expressly reserved (and I say this in case the matter goes further) the position. And it would certainly be helpful, and I think my learned friend would not object, if that reservation were referred to in case the point were raised.

Willis J: I had your reservation quite clearly in mind, Mr Schiemann. I had not thought that the way I put it, as it were, put you out of court on that in subsequent proceedings. The way I expressed it was: if it was not effective to impose on the paddock a restricted use, it is common ground that the Secretary of State had no power to refuse to confirm the purchase notice. What I had in mind in that was that the Secretary of State having E concluded that, but for section 184 he would have been bound to confirm.

Mr Schiemann: I think if your Lordship were to add "on the grounds that he gave," that would cover the point.

Willis J. Yes, certainly.

QUEEN'S BENCH DIVISION

April 26 1974

(Before Mr Justice WILLIS)

F WALTER HERMANNS & SONS LTD v SECRETARY OF STATE FOR THE ENVIRONMENT

Estates Gazette April 5 1975

(1974) 234 EG 47

Planning permission refused on appeal for new premises for existing poultry-dressing business—Decision proceeded on erroneous basis that business should be transferred to more suitable site—Owner's appeal allowed

This was an appeal by Walter Hermanns & Sons Ltd, of East Markham, Retford, Nottinghamshire, against a decision of the respondent, the Secretary of State for the Environment, G dated May 29 1973, confirming the refusal by Nottinghamshire County Council on March 8 1972 of planning permission for construction of new premises for the appellants' poultry-dressing business.

Mr A de Piro QC and Mr N F B Jarman (instructed by Shaen, Roscoe & Bracewell, agents for Clay, Allison & Clark, of Retford) appeared for the appellants, and Mr I D L Glidewell QC and Mr G Slynn (instructed by the Treasury Solicitor) represented the respondent.

Giving judgment, WILLIS J said: This is an appeal against a decision of the Secretary of State for the Environment under section 245 of the Town and Country Planning Act H 1971, dated May 29 1973 and dismissing an appeal by Walter Hermanns & Sons Ltd against a refusal of planning permission dated March 8 1972 by the Nottinghamshire County Council. The decision is challenged on the usual grounds, that it was not within the powers of the Act and that the relevant requirements were not complied with, but before me Mr de Piro conceded that ground 2 of the notice, which asserted that rules 12 and 13 of the Town and Country Planning (Inquiry Procedure) Rules 1969 had not been complied with, was not pursued; and further, that in ground 5 the allegation that the inspector took no, or no proper, note of the evidence J called by the applicants was abandoned. It is plain, as Mr de Piro agrees, that there is no possible criticism which can be made either of the conduct of the inquiry or of the inspector's report so far as the rules are concerned. The gravamen of the complaint is that the decision of the Secretary of State was made contrary to the rules or requirements of natural justice. I shall return to that later.

Application was made to the local planning authority for permission for the erection of a single-storey building for poultry dressing and storage in connection with an established business on the adjoining land, the floor area of the building being some 5,700 sq ft, for which an industrial development K certificate had already been granted. It is to be observed, without going into detail, that among the grounds for refusal of the application was, as it was stated, that the application was for a major extension of an industrial type of operation. The facts of the matter can be stated quite shortly, and involve some historical notice of the activities of Mr Hermanns and his son over the last few years. Mr Hermanns senior was a refugee from Nazi Germany before the war, and came to this country and settled in the area of East Markham, which is an attractive village in the countryside of Nottinghamshire, as an agricultural worker at first. L About 12 years or so ago he started a poultry business on land adjoining the appeal site, and again without going into details, over a period of years between 1961 and 1969 he built one house and occupied another, and he got permission for certain buildings and uses in connection with this poultry business which he built up over the years. Briefly the position was that starting in 1961 with two 16-year-old girls in a very small poultry business, by 1969 he had 40 employees with a large turnover and a higgledy-piggledy collection of buildings and uses which necessitated his employees working in very cramped and unsatisfactory conditions. So in 1971, when it was apparent that more working space under cover was M needed, he made an application for an extension of the killing and dressing building by an area of some 1,200 sq ft. Before that could be dealt with, he was able to buy the field which contained the appeal site, of about four acres, which was used intermittently for rough grazing.

The appellants' case before the inspector was that, contrary to the way in which the local planning authority expressed themselves in the first ground of the refusal, the proposal was not for an extension of the business in a new building, but for a reorganisation of the existing jumble of buildings and uses. What was proposed was that the machinery, which was inadequately housed and operated in

A an existing building and which involved some 20 of the 40 employees working in the open at times, should be put under cover in the new building on the adjacent site, the existing buildings being used for storage and packing, all of which would effectively concentrate the operations of the company on one side of the road. The position on the ground was that there was a complex of buildings and uses comprising this undertaking, which carried on the collecting, killing, plucking and dressing of poultry. Within the area of business activities which was being considered by the inspector were buildings and uses, some of which were permitted under planning permission, some of which had not
B been authorised and some of which might or might not have been deemed to be authorised if the matter were tested. At all events, there had been no enforcement proceedings in respect of anything which the appellants had done upon the land up to the time of the appeal, although the local planning authority indicated, dependent upon the result of the appeal, that they would decide whether or not, and to what extent, any enforcement proceedings might be taken at a later date. It was quite clear that between the application and the hearing of the appeal the local authority became increasingly concerned at what they found had occurred between the first permission and the situation that they were
C considering at the time. It was apparent that there were upon this site, or collection of sites, a very vigorous and enterprising father and son, who from very small beginnings had built up a thriving business employing largely female labour. Their attitude, I think, can be judged from paragraphs 25 and 26 of the inspector's report. In view of the submissions which have been made I think it right that I should record these in full:

> 25. The site occupies almost a quarter of the field of which it is a part and this field has always been in agricultural use, proposals for other forms of use having been refused planning permission. The proposed development would adversely affect
> D the appearance of the area and would expose the remainder of the field to successive pressures for similar industrial activity. The adverse effects of the existing factory due to noise, smell, flood-lighting and the generation of traffic would be likely to be increased in a manner quite unsuited to a rural residential village and on a scale which could no longer be termed "small," "light," or "cottage" industry.
>
> 26. The proposed development may be welcomed by the public health inspector as an improvement, but the appellants' manager says that conditions were not very good when he came three years ago. A major expansion and capital investment has taken place since, when thought should have been given to rational-
> E isation and the prevention of the present unsatisfactory situation. The energy and enterprise of the appellants is admirable and there is no suggestion of forcing them out of East Markham, but employment in their business has expanded from two in 1961 to over 40 in 1971. At present employees come from quite a wide area, and because of its subregional rôle the enterprise could operate with equal success and could expand more appropriately if it were to be located at Tuxford or at Retford, where there is suitable land allocated for industrial use and presently available. A large degree of unauthorised use by the existing concern has emerged during consideration of the current proposal and the appeal, and this disregard of planning legislation, while it may have been due to ignorance rather than wilfulness, does not
> F diminish concern at the possibility of further encroachment. Any enforcement action will be dependent upon the result of the appeal.

It seems to me that the case for the authority was being presented not only on amenity grounds but on the likelihood of further applications to expand the business if the appeal were to be allowed. To eliminate that risk, it was argued that although it was not the council's intention to compel the appellants' removal from East Markham, it would be better for the business to be established in an industrial area elsewhere where it would be appropriate and could expand as required.

I come, therefore, to the findings of fact, to the earlier G
paragraphs of which I should make a brief reference before I come to the all-important paragraph 36. The inspector, Miss Albery, refers to the attractiveness of East Markham. She finds that the site can hardly be seen from any roads except Beckland Hill, which divides the property owned by the appellants; that the present enterprise employs between 50 and 60 people and has a throughput of some 20,000 birds a week collected from 100 to 150 customers; that the employees come from a number of areas, from East Markham, from Tuxford, from Ollerton, Elkesley and from Retford; that collections are made from areas within a radius of 60 miles, and deliveries are made to Nottingham and H
Derby, and in some weeks to Smithfield. That finding, of course, justifies the categorisation of this as a subregional business applied to it by the local planning authority. In view of the submissions of Mr Glidewell, which I shall refer to in a moment, it is necessary, I think, that I should refer in more detail than otherwise I would to the following paragraphs in the findings of fact:

> 8. In 1960 planning permission was granted for a private garage now used as a cold store. In 1961 planning permission was granted for a shed for handling and dressing Mr Hermann's own poultry in connection with a one-acre smallholding owned by him, which was used at the time of the appeal for the preparing and packing J
> of poultry. In 1963 part of the house was being used as an office and permission was obtained for conversion of an integral garage to a dining-room and for the erection of a new garage on to which sanitary accommodation has since been built in addition to an unauthorised canteen building. In 1966 and 1968 permission was obtained for the erection and extension of a shed about 40 ft by 45 ft for storing crates ancillary to the existing use, and that has always been used to house the killing and processing machinery.
>
> 9. Making-up of packing-cases and storage in conjunction with the appellants' poultry business now take place on their smallholding land, where they own a second house, on the west side of Beckland Hill. About eight of their employees park cars there, K
> and no application for planning permission has been made in respect of these uses.
>
> 10. The site is about 200 ft wide and varies from 150 to 200 ft in depth, and the appellants have laid down on it a concrete slab and erected a cold store, intended to form part of the building subject to appeal, and no planning permission for this has been granted.

In paragraph 14 there is a reference to the view of the parish council, and I only refer to it to point out that that council put forward as one of their grounds of objection the probability of expansion. There is a finding in paragraph 15 that there is suitable land allocated for industrial use and L
available at Tuxford and Retford. That brings me to the crucial paragraph in this case and ultimately to the crucial words. Paragraph 36 is headed "Inspector's Conclusions" and reads:

> Bearing in mind the above findings of fact, I am of the opinion that if suitably designed the proposed building would not materially affect the attractive appearance of East Markham, on account of its low-lying and relatively unobtrusive situation, but the scale and intensity of activity combined with the intrinsic nature of the appellants' business are already unsuited to its location in a residential area on the outskirts of a quiet country village. Nuisance from noise and smell would be somewhat reduced by the removal of the machinery to the proposed new M
> building, but the amount of traffic generated would not be reduced, and I am not persuaded that there would be no likelihood that the business would continue to grow. I conclude that alleviation of the present unsatisfactory working conditions should be achieved by removal to a suitable site elsewhere.

In giving his decision the Secretary of State has dealt with the matter concisely, and in paragraph 2 attaches a copy of the inspector's report, refers to the facts set out in paragraph 35 of the report, some of which I have referred to, and quotes verbatim the inspector's conclusion in paragraph 36. The decision letter goes on: "The inspector recommended that the appeal be dismissed. The Secretary of State agrees with the inspector's conclusions and accepts her recommendation." Since, therefore, the Secretary of State has based himself, so far as one can see, entirely upon the findings of fact in paragraph 35 and the conclusion in paragraph 36, it is important to consider what seem to be the bases for those conclusions.

Mr de Piro submits that so far as the first part of paragraph 36 is concerned, he has a finding in his favour on the merits; that contrary to the local planning authority's submissions, the inspector (and so the Secretary of State) took the view that there would not be any amenity objections to a suitably-designed building, because it would hardly be seen; that it would in fact reduce the noise and smell coming from the present activities; and that although the amount of traffic generated would not be reduced, it at all events would not be increased. Therefore, from a planning point of view, Mr de Piro submits that his clients should have succeeded in the appeal. He says that the inspector, very understandably, has fallen into error in two ways. He draws attention first to the phrase, "I am not persuaded that there would be no likelihood that the business would continue to grow," and submits that this is really putting the onus the wrong way round. It is not a submission which he presses very heavily, and I do not think it is sustainable as an objection on its own. Secondly, and this is the nub of his complaint, he criticises that part of the conclusion which reads, "alleviation of the present unsatisfactory working conditions should be achieved by removal to a suitable site elsewhere." He concedes, of course, that alleviation of the present unsatisfactory conditions could be achieved by removal to a suitable site elsewhere, but submits that the inspector has so expressed herself as to base the decision to dismiss the appeal upon the conclusion that the whole enterprise should be removed to a suitable site elsewhere. What Mr de Piro submits, in a nutshell, is this, that the inspector, who is not to be criticised on any other grounds at all, has fallen into error in taking the view that the proposal to alleviate the existing unsatisfactory conditions by erecting a new building upon the land which had been acquired was being compared with an ideal situation, where it was to be assumed that the existing enterprise was not there at all.

If that is what the inspector has in fact done, Mr Glidewell concedes that she should not have done so. He further concedes, as I understand him, that it would not have been right to take into account that applications to expand at East Markham in the future might be encouraged if the present appeal was to be allowed. In support of his submission, Mr de Piro has referred me to a decision of the Secretary of State in an appeal in the case of a development in the Easthampstead Rural District. The decision is reported in the 1970 *Journal of Planning & Property Law* at page 547, and it was a case where the application was for an extension by way of an additional floor to office accommodation in Bracknell. It would have been part of a depot which had a number of planning permissions given by the local planning authority, who had taken no steps to try to secure its removal, although they took the view that it was unfortunate that the building was there at all. The inspector took the view that one aspect of the matter was whether "it would undesirably intensify or accentuate the activities carried on in the depot." The Minister, while agreeing with a great deal of his inspector's conclusions, thought that it would be wrong in principle for the permission to be refused on the sole ground that it would tend to consolidate an existing depot. The report goes on: "The suggestion that the undertaking is basically in the wrong location and should move to a new site raises a separate issue which the Minister has no jurisdiction to consider in the context of the appeal." This is not of course a decision which is binding upon me, but it is one which I find helpful in approaching this particular matter, although Mr Glidewell endeavoured to distinguish it from the facts of this case.

Mr Glidewell seeks to support the Minister's decision in this way. He says that what was relevant at the inquiry was expansion, but expansion looked at in two different ways; one, future expansion, which was in the inspector's mind, and the other, what one might call present expansion, which quite plainly was not in her mind but is the basis now put forward for supporting the decision of the Secretary of State. It is said by Mr Glidewell that parts of the company's activities were carried on in breach of planning control, or at all events without prior planning permission; that there has thus taken place some unauthorised expansion of the business over the years; that the permission sought was for a building to house and consolidate this unauthorised expansion, and that this in itself, if granted, constituted expansion. If this view of the matter be right, Mr Glidewell submits that the Secretary of State was entitled to dismiss the appeal albeit that such a concept was not present to the mind of the inspector, or, so far as I can see, of anyone else at this inquiry. The extent to which there had been infringements of planning control was not and could not have been determined at the inquiry, which was not concerned with enforcement procedure, nor could there have been, as it seems to me, anything but the most approximate equation between infringement and the new building. I think the short answer to Mr Glidewell's argument should be that if the appeal has been determined against the appellants on a view of the facts which was never a ground of refusal, never discussed at the inquiry, except by way of historical reference or in relation to removal of the business elsewhere, and never apparently present to the inspector's mind or a factor in her conclusions, there has been unfairness to the appellants. Nevertheless, I recognise that policy requirements can, and if necessary should, affect a decision, and that there are of course many cases in which the Secretary of State is entitled in his discretion to draw inferences and conclusions from the words which are used in the inspector's conclusions and findings of fact which may not be obvious at first sight. It has been said many times that inspectors' reports are not to be construed strictly as if they were Acts of Parliament or legal documents, but equally, it seems to me, the courts should not be astute to find a ground for supporting a decision involving a construction which seems to it to be over-subtle or smacking of semantics.

With great respect to Mr Glidewell, I do not think it would be right to support this decision, which on the face of it, as it seems to me, has proceeded on an erroneous basis and which the Secretary of State has explicitly accepted, by what I regard as an artificial gloss on perfectly clear words which really admit of no other interpretation than the one contended by Mr de Piro, and which I understand Mr Glidewell concedes involves an erroneous approach unless his interpretation is found acceptable. Since I reject Mr Glidewell's interpretation, it follows either that the decision was based upon conclusions which involved fundamental error, or if that were not so, upon a view of the facts which was not an issue in the appeal nor, on the face of it, a factor in the conclusion and so, as I think, irrelevant. In either event, therefore, it seems to me, that the rules of natural justice require that the decision of the Secretary of State should be quashed.

The appellants were awarded costs.

A

QUEEN'S BENCH DIVISION

July 31 1974

(Before Mr Justice WILLIS)

MOLTON BUILDERS LTD v WESTMINSTER CITY COUNCIL AND ANOTHER

Estates Gazette April 12 1975

(1974) 234 EG 115

Enforcement notice—Suite in Whitehall Court used as offices—Consent of Crown Estate Commissioners to operation of notice not a derogation from grant—Commissioners in any event fulfilling a public duty and therefore unable to fetter
B **the exercise of their discretion by contract**

This was a claim by Molton Builders Ltd, of 32 Old Burlington Street, London W1, against Westminster City Council for a declaration that an enforcement notice served upon them restraining use as an office of a suite in Whitehall Court, London SW1, was void and of no effect, and secondly and alternatively, against the Crown Estate Commissioners for damages for derogation from grant.

Mr J Harman QC and Mr L H Hoffmann (instructed by Beer, Dunnett & Co) appeared for the plaintiffs; Mr I D L Glidewell QC and Mr A B Dawson (instructed by the city's solicitor) for the first defendants; and Mr P Oliver QC and
C Mr G Slynn QC (instructed by the Treasury Solicitor) for the second defendants.

Giving judgment, WILLIS J said: This case raises an interesting point under section 199 of the Town and Country Planning Act 1962. In form the plaintiffs claim that an enforcement notice served on them by the first defendants, the council, is void and of no effect, and further or alternatively, damages against the second defendants. The condition precedent of the service of an enforcement notice by the first defendants was the granting of consent to the service by the second defendants, and it is claimed by the plaintiffs
D that the consent was a derogation from the grant of the subject premises under lease by the commissioners. Mr Harman has abandoned the claim that the giving of consent was also a breach of the implied covenant for quiet enjoyment, and by consent I am absolved from the duty of assessing any damages to which the plaintiffs may be entitled.

The subject premises consist of a suite at Whitehall Court, London SW1. On May 9 1931, Whitehall Court was demised by the Commissioners of Crown Lands, the predecessors of the second defendants, to Whitehall Court Ltd for a term of 99 years from January 5 1930. The user covenant in clause 9 prohibited use of the premises for the purposes of
E any trade, manufacturing, business, lodging-house, private hotel or subletting in furnished or unfurnished apartments, except as mentioned in subclause (c), and by (c), the vital clause in this case, the premises were to be used for "private dwelling-houses, first-class clubs or residential or professional chambers or offices only." There was a provision granting consent to use of the cellar for maintenance machinery, subject, however, to the rights and powers of all local or other bodies or authorities. On June 5 1968, the lessees, then called Whitehall Court (Holdings) Ltd, granted an underlease for the term, less the last 10 days, to Clabon Developments Ltd. The user covenants were in identical terms to those referred
F to above. On September 29 1970 Clabon granted to the plaintiffs a subunderlease of a suite of offices on the fifth floor of Whitehall Court, consisting of 11 office rooms, etc, formerly a flat, for a term of 61 years from December 24 1967. At all material times thereafter the suite was occupied by Mr Jack Dunnett as solicitors' offices. The relevant covenants in this subunderlease were, so far as material: "(L) (i) Not without the previous licence in writing of the managers"—Whitehall Court (Holdings) Ltd—"to carry out or permit to be carried out any development to the suite within the meaning of the Town and Country Planning Acts"; "(ii) To give to the managers particulars of any notice or proposal issued to the lessees by a planning authority and without delay G
to take all necessary steps to comply with any such notice"; and "(P) To use the suite as a suite of offices or for the purpose of a private residence in one occupation only and not to use or permit the suite to be used for any illegal . . . purpose." By reason of the terms of the user covenants cited, the plaintiffs claim—and it is not disputed—that the parties to the lease and underlease respectively contemplated that the subject premises would be used in accordance with one or other of the recited users, including offices or professional chambers. It is not disputed that there is no substantial difference between those two users.

Section 199 (2) of the 1962 Act reads, so far as material: H
"Except with the consent of the appropriate authority (a) no order or notice shall be . . . served under any of the provisions of . . . section 45 of this Act"—the power to serve enforcement notices—" . . . in relation to land which for the time being is Crown land." The meaning of "Crown land" is "land in which there is a Crown interest or a Duchy interest": "Crown interest" means "an interest belonging to Her Majesty in right of the Crown or belonging to a Government Department," and so on, and "Duchy interest" means "an interest belonging to Her Majesty in right of the Duchy of Lancaster, or belonging to the Duchy of
Cornwall." The "appropriate authority" for the purposes of J
section 199 (2), in the case of land belonging to Her Majesty in right of the Crown and forming part of the Crown Estate, means the Crown Estate Commissioners. The Civil List Act 1952 provides by section 1 that the hereditary revenues of the Crown during the present reign shall be paid to the Exchequer and made part of the Consolidated Fund, in return (by section 2) for the annual payment of the Queen's Civil List. The position of the commissioners is regulated by the Crown Estates Act 1961. It is provided, so far as is material, by section 1 that the commissioners shall be
"charged on behalf of the Crown with the function of K
managing and turning to account land and other property, rights and interests, and of holding such of the property, rights and interests . . . as . . . cannot be vested in the Crown . . . and the property, rights and interests . . . shall continue to be known as the Crown Estate." Subsection (2) vests in the commissioners powers of management and improvement of the Crown Estate. By subsection (3), "It shall be the general duty of the commissioners . . . to maintain and enhance its value and the return obtained from it"—the Crown Estate—"but with due regard to the requirements of good management." By subsection (4), provision
is made for the commissioners to comply, after consultation, L
with directions given to them by the Chancellor of the Exchequer or the Secretary of State, who are to have regard, before doing so, to subsection (3). Section 3 requires the commissioners to obtain the best consideration for their land transactions which can reasonably be obtained, subject to the Crown's monopoly position.

In June 1971 the council were desirous of serving an enforcement notice on Mr Dunnett requiring him to discontinue use of the suite as an office. They served requisitions under section 215 of the Act of 1962 upon (*inter alios*) the plaintiffs and the second defendants for information
as to interests held in the premises. These were followed by M
a request from the council to the commissioners for their consent to the service of an enforcement notice on Mr Dunnett. This was granted by letter of July 27 1971 in which the commissioners concluded that there were no grounds on which they could properly refuse the council's request. Thereafter the notice was served, and Mr Dunnett lodged a notice of appeal to the Secretary of State which is held in abeyance pending a final decision in these proceedings. Mr Harman submits first of all that the Crown Estate is private land held for the private purposes of the Crown and is analogous rather to the Duchy land than, for example, to an

airfield held in trust for the Crown for defence of the realm. If in such cases as the present it is the private interest of the Crown which is engaged, Mr Harman submits that there is no reason why the ordinary obligations of landlord and tenant should not apply. He concedes that the Crown cannot by contract fetter powers entrusted to it to be exercised for the public good (*Commissioners for Crown Lands v Page* [1960] 2 QB 274), but he says that section 199, confers no powers on the commissioners, but rather imposes fetters on a planning authority in relation to Crown land. Therefore the Crown, like any other landowner, is not permitted to derogate from its grant. It has nevertheless done so, says Mr Harman, by granting consent to service of the enforcement notice.

Mr Harman has referred me to a number of authorities, through which he has traced the development of the doctrine of derogation, and to *Megarry & Wade's Law of Property*, 3rd ed, p 684, where the effect of the cases is summarised. It is there stated: "To constitute a derogation from grant there must be some act rendering the premises substantially less fit for the purposes for which they were let." Mr Harman submits that the maxim is one which has been gradually extended, and he takes as what he calls his sheet anchor the words of Warrington LJ in *Harmer v Jumbil (Nigeria) Tin Areas Ltd* [1921] 1 Ch 200 at 223 and 224:

> Undoubtedly this is the first case in which this particular question has arisen, but for myself I confess that I can see no reason why the principle should be limited in the way contended for by the defendants. The premises have become unfit for the purpose for which they were demised, and this because it is no longer legal for the plaintiff to use them for that purpose by reason of the acts of the defendants; and I cannot see why the acts of the defendants should not have the same effect in law whether the change brought about by them is a physical change in the condition of the demised premises or a change in their condition brought about by the fact that it has become illegal to use them for the purpose intended by the demise.

Mr Harman submits that the maxim has developed beyond the relationship of landlord and tenant, and extends to sale and to the grant of a *profit-à-prendre* or a licence. He adopts as part of his argument the passage on p 817 of *Megarry & Wade, supra*:

> In so far as this doctrine restricts the grantor's freedom to use any of his neighbouring land which he may have retained, it is really part of the law of property, since the rights which it creates bind not only the grantor but also all who claim title through him; so that, in effect, the grantee and his successors in title have a proprietary interest of a special kind against the grantor's land, into whosesoever hands it may pass.

Finally, as I understand his submissions, Mr Harman acknowledges that despite the development of the maxim, it has not yet reached the point where he can claim that he is covered by authority; but he invites me to extend the maxim still further and bring the giving of consent under section 199 within its scope.

Mr Oliver, for the commissioners, submits in the first place that the commissioners manage the hereditary revenues as the Crown Estate for the public and that in all their dealings with the estate they act in a public capacity. The discretion given them in section 199 to give or withold consent is in their public capacity as managers of a public fund, and when exercised, it is exercised in a matter of public interest under a public statute. The exercise of such a discretion, he says, cannot be fettered by contract (see the case of *Page, supra*). Assuming, however, that the commissioners were acting in this matter for the Crown in a private capacity and, contrary to his submission, that the maxim of nonderogation can apply (1) where no retained land is in issue, and (2) in favour of a person claiming under the grantee, he submits that there are two questions which must be answered: one, what is the scope of the grant which the plaintiff claims and from which he alleges derogation, and two, does what has been done amount to a derogation from that grant? Counsel accepts that as between landlord and lessee it was contemplated that the premises could be leased for any of the purposes prescribed in the lease, but he says, as is conceded by Mr Harman, that that imports no guarantee by the commissioners that the premises can always be used for any of the specified purposes. The grant, he says, cannot be construed otherwise than as one for use of the premises in accordance with the law at any particular time. He submits that as a matter of construction, the right to use must be subject to the qualification that the purposes are permitted by the general law, and that nothing which the commissioners do to facilitate the operation of the general law can amount to a derogation from grant (*Pwllbach Colliery Co Ltd v Woodman* [1915] AC 634). The general law has changed since 1931, so that by the time of the present action permission was required for making any material change in the use of premises (section 13 (1) of the Act of 1962). Further, service of an enforcement notice under section 15 of the Act of 1968 achieves nothing of itself, says Mr Oliver, because the person served can appeal to the Minister under section 16, and by subsection (7) an appeal operates as a deemed application for planning consent which may or may not be granted. Thus the giving of consent does not frustrate or prevent the user of the premises. If an enforcement notice follows, it puts the person concerned in a position of having to seek planning permission which, under the general law, he should have sought in the first place for a change of user. The commissioners in 1931 cannot have implicitly granted what they could not grant expressly, that is, a right to use the premises otherwise than in accordance with the general law. Mr Oliver submits that to amount to a derogation, the grantor must authorise or actively participate in the act done (*Malzy v Eichholz* [1916] 2 KB 308); that the grantor is under no obligation to take steps to prevent the interference by a third party (*Master v Hansard* (1878) 4 Ch D 718); and finally, by parity of reasoning, that where a landlord has power to give or withhold consent to the doing of an act by a third party which may restrict the tenant's enjoyment, he is under no obligation to withhold consent.

Mr Glidewell, for the council, is concerned only if affirmative answers are given to the two questions, one, would consent have involved a derogation towards the plaintiffs if the commissioners had been a private person, and two, were the commissioners exercising a private right as opposed to a public function? Counsel adopts Mr Oliver's argument on these questions, but submits that if they do not avail the commissioners, the position is not that the consent is ineffective or unlawful simply because the giving of it involves a derogation from grant. The consent is lawful, he says, but the plaintiffs have a remedy in damages against the commissioners. He concedes that to succeed on this point would be likely to yield him only a Pyrrhic victory, since a large number of office users have been established in Whitehall Court without the obtaining of planning permission and the commissioners could hardly be expected to invite claims for damages by giving consent to the service of further enforcement notices.

It will be convenient to consider first whether the giving of consent under section 199 amounts to a derogation of grant actionable at the suit of the plaintiffs. I defer for the moment a consideration of the questions whether the plaintiffs as subunderlessees are entitled to maintain such an action and whether an action will lie where no retained land is in issue, and assume that both questions are answered in Mr Harman's favour. The rule, as has been said, is clear; the difficulty, as always, is in its application. That difficulty is increased when the facts of the case under consideration differ greatly from those of any case which has hitherto been decided. It has been variously and authoritatively stated: "A grantor having given a thing with one hand is not to take away the means of enjoying it with the other"

A (*Birmingham Dudley & District Banking Corporation v Ross* (1888) 38 Ch D 295, per Bowen LJ at 313); "If A lets a plot to B, he may not act so as to frustrate the purpose for which in the contemplation of both parties the land was hired" (*Lyttelton Times Co Ltd v Warners Ltd* [1907] AC 476, per Lord Loreburn LC at 481). The doctrine bestows a right wider than an easement, and the act for which it can be invoked need not be actionable as a tort (*Aldin v Latimer Clark, Muirhead & Co* [1894] 2 Ch 437). It was applied to an incorporeal hereditament (*Peech v Best* [1931] 1 KB 1). An extension of the doctrine resulted from the decision in *Harmer v Jumbil, supra,* a case in which the lessor granted B a lease of land for the purpose of an explosives factory and later granted a mining lease of adjoining land. The lessee of the latter land put up buildings at such a distance from the magazines as to imperil the factory's licence. He was restrained from doing so, since he stood in the shoes of the grantor, who in the circumstances would have been impliedly obliged to do nothing which would bring about the forfeiture of the licence. At p 219 Lord Sterndale MR said:

> The question is whether in the circumstances of this case, doing what the defendants did, doing, that is to say, that which *ipso facto* as soon as it was done caused the forfeiture of the licence and made it illegal to carry on the business in the C premises, was a derogation from the grant of the lessor. I agree with the learned judge that the question has never arisen under circumstances of this kind, and I have found it a very difficult case to decide, and although I feel myself bound to differ from the conclusion at which the learned judge in the court below [Eve J] arrived, I do so with great hesitation, because I think that this is a case which presents novel circumstances.

Then Younger LJ on p 226 says:

> For the obligation laid upon the grantor is not unqualified. . . . The obligation, therefore, must in every case be construed fairly, even strictly, if not narrowly, it must be such as, in view of the surrounding circumstances, was within the reasonable contem- D plation of the parties at the time when the transaction was entered into, and was at that time within the grantor's power to fulfil. But so limited, the obligation imposed may, I think, be infinitely varied in kind.

Thus there can be a derogation even though there is no physical interference with the plaintiff's enjoyment. The act of the lessor must be one which renders the premises unfit, or substantially less fit, for the purpose for which they have been demised. On the other side of the line, if the grantor's action merely reduces the amenity of enjoyment or involves the grantee in greater expense without preventing or substantially interfering with the user, no action will lie for E derogation of grant (*Browne v Flower* [1911] 1 Ch 219; *O'Cedar Ltd v Slough Trading Co Ltd* [1927] 2 KB 123).

It is common ground in the present case that at the time the lease was granted in 1931 both parties contemplated that the premises could be used for the specified purposes. At the time of the 1970 lease to the plaintiffs from Clabon, the purposes were reduced to a residence or suite of offices, but it is the use of flat 139 as offices which the council proposes to prohibit, and that use is common to all three leases. What was the scope of the grant? It is common ground that there cannot be implied any warranty that any of the prescribed users may be continued during the currency of the lease F (*Hill v Harris* [1965] 2 QB 601). In 1931 there was no effective planning legislation by which the user of the subject premises could be controlled. In particular, no permission was needed to convert a residential flat into an office or vice versa. Effective control was introduced by the Town and Country Planning Act 1947, and has so continued as to prevent material change in the use of land taking place without planning permission. Had these premises not been Crown land, no question of consent to the service of an enforcement notice would have arisen; but the submission of the plaintiffs is that by reason of the 1931 lease, which prescribed the uses that would be permitted, the commissioners are precluded from consenting to the operation of the ordinary law of the land upon the G premises, because such consent might lead to interference with the particular use prescribed and would thus amount to a derogation of grant.

If that submission is correctly stated, it would, I think, present novel circumstances and involve a very considerable development of the maxim to hold that the act of consent of itself frustrated or prevented the use of the suite as offices or substantially interfered with its enjoyment. I do not think Mr Glidewell exaggerates when he says that if that submission is correct it would effectively substitute the Crown Estate Commissioners for the local planning authority in all such cases as the present and the private interests of the Crown H Estate for the public purposes of the development plan. Indeed, Mr Harman does not shrink from accepting that there would thereby be created a privileged class of lessees and tenants who would effectively be controlled in their use of the premises only by the commissioners, acting within the user covenants in the lease and guided principally, at all events, by a consideration for the good management of and revenue from the estate. If this conclusion is correct, I find it very surprising, but I do not think it is. I think that Mr Oliver is correct when he says that the right to use the premises for the purposes set out in the 1931 lease must be J construed as subject to the qualification that such uses must not be otherwise than in accordance with the general law. Is the Crown to be liable for damages if the discretion vested in the commissioners to give or withhold consent is exercised in favour of enabling the general law to take its course and have the question whether or not development has taken place without permission and, if so, the consequences, determined according to law? I think not. Consent can be withheld, and if so no proceedings under section 45 of the 1962 Act can be taken. But it does not seem to me that the commissioners should be inhibited against exercising their discretion by the spectre of a claim for damages for deroga- K tion of grant if they consent to the general law being applied in a particular case. As Mr Oliver says, the individual is only being put in the position of having to obtain permission which he should have applied for in the first instance.

Whether or not the grant should be construed in this qualified way, does the act of consent fall within the maxim, or can the maxim properly be extended to embrace such an act in the circumstances of the present case? The use as offices could continue, notwithstanding the giving of consent, until an enforcement notice was served and for a period of at least 28 days thereafter (section 15 (7) of the 1968 Act). The local planning authority can withdraw an enforcement L notice at any time before it takes effect. More important, as it seems to me, a right of appeal to the Minister is given by section 16 on the grounds set out in subsection (1). The appellant can have a hearing before an inspector (subsection (2)), and by subsection (3), the enforcement notice is to have no effect until the final determination or withdrawal of the appeal. When an appeal is brought, the appellant is deemed to have made an application for permission for, in this case, the office user (subsection (7)), and by subsection (5), the Minister can quash or vary the notice in the appellant's favour or can grant planning permission for such user with or without conditions. Thus it is apparent that the service of M an enforcement notice does not in this or any other case *ipso facto* lead to the prohibition of the user unless the person served decides not to appeal, nor *a fortiori* does the giving of consent under section 199. I do not think that this position of uncertainty as to the final result of the giving of consent is met by Mr Harman's submission that if, contrary to the facts here, the plaintiffs had known of the commissioners' intention to give consent, they would have succeeded in *quia timet* proceedings in restraining them from putting it into effect. I have very considerable doubt that an injunction would or could have been granted in such circumstances. I do not think that the giving of consent by the commissioners can possibly be said to have frustrated or prevented one of the purposes for which it was contemplated the premises could be used or to have rendered it impossible for the premises to be legally used for office purposes. If that is right, this case is outside the scope of the maxim as laid down in the authorities hitherto, and I can find no justification for extending its scope in the circumstances of this case. The commissioners have exercised a discretion given to them under the general law, the result of which may, in unpredictable circumstances, involve cessation of office user in the future. In other words, the giving of consent which in all probability would be followed by an enforcement notice puts the continuation of the office user at risk and no more. In my judgment this does not involve a derogation from grant.

I pass to the question whether the commissioners were acting in a private capacity or for public purposes in the exercise of a public duty. This is not an easy question to answer in the case of such a unique body, but it is not, I think, susceptible of much elaboration. I do not think that a comparison with the Duchies of Lancaster and Cornwall is of assistance, or that there is any necessary analogy to be drawn from their juxtaposition in section 199. The only guide to the position of the commissioners under the law is to be found, it seems to me, in the Acts to which I have been referred, and I think that the proper inference to be drawn is that in managing the Crown Estate the commissioners are acting in a public capacity in relation to the Crown revenues which are surrendered and form part of the Consolidated Fund. By section 199 it seems to me that the commissioners are given a discretion to give or withhold consent in a public capacity as managers of a public fund and that the exercise of that discretion is a matter of public interest under a public statute. If this is the correct view of the position of the commissioners, then they cannot by contract fetter their power to give consent (*Commissioners of Crown Lands v Page, supra*), and the plaintiffs must fail on this ground also. The interesting questions whether the maxim of derogation from grant can apply where no retained land is in issue or is capable of assisting persons claiming under the original grantee do not, therefore, arise for decision by me, and with no regrets, I do not therefore attempt any opinion on matters which properly belong to another division.

Orders for costs were made in favour of both the council and the commissioners.

QUEEN'S BENCH DIVISION

July 9 1974

(Before Mr Justice WILLIS)

RUGBY SCHOOL v SECRETARY OF STATE FOR THE ENVIRONMENT AND ANOTHER

Estates Gazette May 3 1975

(1974) 234 EG 371

Bloomsbury conservation area—Council's refusal to consider an application by the Rugby Estate for permission to redevelop two houses—Refusal based on agreement by estate governors with council on need for a " master plan " for area—Refusal supported, having regard to " very special conditions " of the application—Discussion whether governors' co-operation in master plan proposals puts them into less favourable position than the ordinary applicant for planning permission

This was an application by the Governors of Rugby School to quash a decision of the first respondent, the Secretary of State for the Environment, dismissing an appeal under section 36 of the Town and Country Planning Act 1971 from the refusal of the second respondents, the London Borough of Camden, to give a decision on an application for planning permission for development of a site in Lambs Conduit Street, Bloomsbury, London.

Mr R Carnwath (instructed by Iliffes) appeared for the applicants, and Mr G Slynn (instructed by the Treasury Solicitor) represented the first respondent. The second respondents took no part in the appeal.

Giving judgment, WILLIS J said: This is an application made under the Town and Country Planning Act 1971 by the Governors of Rugby School to quash a decision of the Secretary of State for the Environment by which he dismissed their appeal to him under section 36 of the Act against the failure of the Camden Council to give a decision on an application for planning consent under section 29 of the Act. The application was made on December 29 1972 and proposed the uniting of the upper floors of 53 and 55 Lambs Conduit Street on the governors' estate in Bloomsbury to provide three four-person flats. The council declined a decision upon the application in circumstances which will appear, and the governors thereupon appealed to the Secretary of State under the default procedure of section 37 of the Act. It was agreed to dispense with a formal hearing, and the matter was referred to an inspector with plenary powers under the Ninth Schedule. After viewing the site and the area, and considering the representations, the inspector dismissed the appeal.

The Rugby Estate includes a considerable number of houses occupied in the main by tenants of long standing in single floors. A number of the houses are listed as being of architectural or historic interest, and so they present considerable problems in modernisation. Therefore the policy of the governors as a first step, and in agreement with the Camden Council, has been to embark upon a scheme of modernisation and improvement to provide up-to-date flats in non-listed houses. The houses under appeal are in this category. In their statement on the appeal the Camden Council point out that No 55 was vacant and No 53 had tenants on two floors. The buildings, which are unlisted, are in the Bloomsbury conservation area, and in a residential area at 200 persons per acre. The council's case, in brief, for declining to deal with the application and inviting the Minister to dismiss the appeal was that piecemeal development in such an area was bad planning practice, and that following certain ad hoc applications in 1970 and 1971 the governors had agreed with the council's officers to prepare a master plan setting out the aims and objects of the estate management proposals. The discussions were well advanced, and the plan should, said the council, be finally agreed as an overall strategy before individual applications were dealt with. The governors accepted that the residential part of the appeal premises was programmed under stage II of the overall plan so far as it had gone, but contended that the letting of the ground-floor shop now would help to finance the essential improvement to the residential accommodation and listed buildings. In his letter of decision dated January 4 1974 the inspector states in paragraph 2:

> It seems to me that a decision in this case rests primarily on whether it is fair and reasonable for the council to demand the production of a master plan covering reconditioning and improvements to the properties within the Rugby Estate, before considering individual applications affecting only some of the property.

In paragraph 3, after setting out the proposals designed to bring dilapidated premises up to modern standard, he states:

> I consider that any such proposals should be welcomed but must be related to the wider issue of improvement works within the Rugby Estate as a whole.

In paragraph 4, after supporting the council's view that approval should be withheld from the current proposal in advance of a comprehensive scheme, the inspector said:

In these circumstances I do not consider it would be unreasonable for your clients to press ahead and complete their proposals so that a programme of work can be agreed upon with the council. Such a procedure would, in my view, be in the best interests of the Rugby Estate as a whole, and of those at present living within it in particular.

In paragraph 6, the inspector said:

I have considered the other matters raised in the written representations but I am of the opinion that they are not sufficient to outweigh the considerations which have led to my decision.

He dismissed the appeal. Counsel for the applicants takes a short point. He says that by section 29 of the Act, which by reference governed the application, the scene is set by its form. The applicant decides the nature of the application and is entitled to have it dealt with in one or other of the ways provided under section 29 (1). Where he is acting on appeal, whether under the normal or default procedure, the Secretary of State is required, says counsel, to follow the same course. The applicant formulated a detailed application which the council declined to consider, and the inspector, by adopting the council's reasoning, has by purporting to give a considered refusal likewise failed to determine the appeal as required under section 36, or has taken into account collateral or irrelevant considerations. Counsel submits that any other view involves the applicants being unfairly prejudiced compared with an individual owner simply because they are estate owners and in this case because they have shown their willingness to try to achieve an agreed master plan with the council. In short counsel submits that the application and appeal were never dealt with as required by sections 36 and 29. Mr Slynn, for the respondent, accepts that if the decisive matter had been the council's refusal to consider the application in pursuance of their declared policy, their action would have been contrary to the well-known statement of the law by Bankes LJ in *R v Port of London Authority, ex parte Kynoch Ltd* [1919] 1 KB 176 at p 184, and could not be supported; equally if the Minister had done likewise and declined to give a decision. The Minister, says Mr Slynn, has not done likewise, and the inspector's report looked at as a whole, shows that although he supports the council's policy of not dealing with individual applications in advance of agreement between the council and the governors on a master plan for the whole estate, he has considered whether there are circumstances affecting this particular application which would justify treating it as a special case and granting permission. He has concluded there are not, and has therefore dismissed the appeal. The Secretary of State, says Mr Slynn, is entitled to have a policy and act upon it "provided that the existence of his general policy does not preclude him from fairly judging all the issues which are relevant to each individual case as it comes up for decision" (see per Cooke J in *Stringer v Minister of Housing and Local Government* [1970] 1 WLR 1281). Finally Mr Slynn stresses the very special circumstances of this case, where the policy followed by the council and adopted by the Minister had been acquiesced in by the governors until the present application, success in which would bring some immediate financial gain.

This case has given me some concern. I would have thought it difficult to justify a policy which would entail refusing to consider, and on appeal refusing, applications to modernise houses owned by third parties, for example on either side of Nos 53 and 55, simply because such a policy was in line with that adopted for the Rugby Estate although no criticism could otherwise be made of the applications. If that be right, why, as counsel for the applicants asks, should the Rugby Estate be put in a less favourable position merely because they are estate owners and are co-operating with the council? If that question accurately represents the position here, it requires an answer. I think in the end the decision of the Secretary of State cannot be said, by its endorsement of the council's policy, to be itself a failure to exercise his statutory duty upon appeal. It would not be helpful to speculate when a negative decision following a refusal to consider might amount to a failure to act within the statute. It is sufficient for the purposes of my decision in this case to say that I think that the question posed above by counsel is not strictly accurate, and that there are very special conditions present upon the Rugby Estate which explain the council's policy and the Secretary of State's approval of it. It is not without importance that the governors themselves support that policy in principle and have been active in furthering it. It would thus require special circumstances, as it seems to me, to justify permitting an individual application to proceed in advance of agreement on a master plan if the restrictive policy being followed met in principle with the approval of the Secretary of State. When the decision letter is read as a whole, it seems to me that the approach of the inspector has been (a) to consider the fairness of the policy; (b) to approve it in principle for the reasons he gives; and finally (c) after viewing the area and considering all the relevant facts to conclude that the application should not be treated exceptionally. If that be the correct view of the inspector's decision letter, it seems to me to accord with the guidance to be found in the decision of Cooke J in *Stringer's* case. For the foregoing reasons, therefore, and in the special circumstances of this case, I am satisfied that the Secretary of State properly exercised his statutory duty in deciding to dismiss the appeal. The motion fails and must be dismissed with costs.

QUEEN'S BENCH DIVISION

November 25 1974

(Before Mr Justice CANTLEY)

BEACONSFIELD DISTRICT COUNCIL v GAMS

Estates Gazette June 7 1975 and February 28 1976

(1974) 234 EG 749 and (1974) 237 EG 657

Permission granted for bungalow in green belt on condition that existing farmhouse should be demolished—Obligation of a contractual nature enforceable by injunction—Developer must either demolish house or leave bungalow unoccupied

This was an application by Beaconsfield District Council for an injunction restraining Mr William Gams and his wife, Mrs Gladys Ellen Gams, from occupying or permitting occupation of a bungalow built on green belt land at Iver Heath, Buckinghamshire, until a farmhouse on the land had been demolished as required by a condition subject to which planning permission for the bungalow had been granted.

Mr J Grove (instructed by Lovell, White & King) appeared for the plaintiffs, and Mr M Tennant (instructed by E D C Lord) represented the defendants.

Giving judgment, CANTLEY J said that Mr Gams was a builder and builders' merchant. In 1966 he bought Church Farm, Iver Heath, a farmhouse and 25 acres of land situated in the green belt. The area was primarily agricultural, and the intention of the plaintiffs as local planning authority for the district was then, and was still, to keep it so. Building land in the green belt in the rich country in Buckinghamshire was an extremely valuable commodity, and Mr Gams realised there were great commercial potentialities in the development of the 25 acres of Church Farm, if such development were ever permitted. His (Cantley J's) impression of Mr Gams was that he was an astute businessman and a candid and honest witness. He said in evidence that when he bought Church Farm he knew that the land had good potential. He knew that applications for building development had already been refused in the vicinity, and he knew that these applications went on until, in the course of time, someone breached the dam and development was permitted. Mr Gams, with a building business of his own and some property, was prepared to wait, although not, perhaps, to wait patiently. In the meantime, he had bought the farm and he carried on pig-farming there.

In 1967 he let the farmhouse to a US Army officer, and he himself lived in a house in Hayes, Middlesex. In 1972 he conceived the idea of erecting a bungalow on the land. He was 65, and the years went by all too quickly at that age. It seemed to him that if he was going to live on Church Farm a bungalow would be more suitable for him and his wife. In addition, at that time his mother-in-law lived with him, and she could only climb stairs with great inconvenience, so that a bungalow would be much better for her. It might also have occurred to Mr Gams that a bungalow would be a very good piece of property to build on this land in the game of Monopoly he was playing. He therefore wrote to the planning authority on March 22 1972 saying that for medical reasons it was imperative that he, his wife and her mother lived in a bungalow. He applied for permission to build the bungalow, saying, "If I am given permission to build the bungalow, I will demolish the farmhouse." Permission was granted on August 11 1972, and the planning authority wrote the undertaking offered them into the conditions appended. They stipulated that the bungalow was not to be occupied by anyone not in employment on the farm or in local agriculture, and that it was not to be occupied at all until the existing house was demolished to their satisfaction. They gave as one of the reasons for imposition of these conditions the need for conformity with green-belt policy. The bungalow was built, and Mr Gams entered into occupation nearly 18 months ago, but the farmhouse was still not demolished. Further, it was at least doubtful whether Mr Gams was employed in agriculture locally. If he was, it would not be for much longer. He had let his 25 acres to an ex-manager of his, who carried on the farm but lived in a house in Slough.

The situation today, therefore, was that one had a man living in the bungalow who was probably not employed in agriculture locally, a man who was engaged in agriculture locally but lived in a house in Slough, and a farmhouse which was still not demolished. While the bungalow was being built, Mr Gam's 25-year-old daughter got married. Her husband had his own electrical-goods business, and they set up house at Ickenham, some five miles away from her parents. There was nothing wrong with the farmhouse's structure; he (his Lordship) had seen pictures of the property, and there was nothing in its appearance to suggest that it was ripe for demolition. Indeed, like the bungalow, it was a very nice building. Mr Gams thought it would be most advantageous if his daughter and son-in-law came to live in the farmhouse rent-free. The house had four large bedrooms, a sitting room, a dining room, a kitchen and a study. In February 1973 Mr Gams's solicitor wrote to the planning authority: "It does appear to our client that it is a terrible waste of good accommodation for the farmhouse to be demolished." He asked the council for permission for the house to remain, at least for a limited period, to be let to someone who was homeless. The council refused, and pointed out that the demolition of the old farmhouse was the main basis under which permission for the building of the bungalow was granted. Mr Gams appealed against the council's refusal to the Secretary of State, and his appeal was awaiting decision. There were some further matters he (his Lordship) felt he should mention. Mr Gams had made a number of applications in respect of this land for residential development in 1970. In 1971 he applied for permission to erect two filling-stations. Again in 1971 he applied for residential development. All applications were refused. He had had another bungalow on the coast for six or seven years, but the family lived in it only at weekends.

The failure to demolish seemed to him (his Lordship) to be a calculated and flagrant breach of a contractual obligation stated in plain terms and well understood by Mr Gams, who had frankly admitted that he knew in 1972 he would get permission for the bungalow only on the basis of the condition as to demolition. To get a benefit for himself, he had entered into a bargain, his obligation under which he had flouted from the outset. The council now applied for an injunction preventing the occupation of the new bungalow until the existing farmhouse had been demolished; in other words, they applied for the enforcement of Mr Gams's covenant. This was not a capricious restriction; it was a restriction imposed on behalf of the public to preserve the green belt. The farmhouse was on agricultural land which had not been classified in any other manner. If Mr Gams preferred to live in a bungalow built on the land, he accepted that he would have to remove the other house. He was not put under any direct pressure. He understood these matters very well, and therefore had opened negotiations by offering to make the agreement which he had since continued to flout. Paying substantial damages would be a good proposition to him if he was left with two attractive houses in the green belt. He was a businessman. There was nothing unlawful or disgraceful in hoping to turn a farm of 25 acres into a housing estate, provided it was done with planning permission, and there was nothing disgraceful in making a profit out of it. But what Mr Gams could not do was defy the planning authority by blatantly seeking to retire from the bargain he had made. There would be an injunction restraining the defendants from permitting the occupation of the bungalow until either the existing farmhouse was demolished or the covenant had been discharged or modified, either by the Minister, by successful application to the Lands Tribunal, or by agreement between the defendants and the plaintiffs. The injunction would, however, operate very harshly and cruelly if it came into effect too soon. Mr Gams was awaiting the outcome of his appeal to the Minister, as to which he (his Lordship) said nothing and decided nothing. Though Mr Gams had lost this round, there was no reason why he should be dealt with harshly. The order would not be enforced for two months.

BEACONSFIELD DISTRICT COUNCIL v GAMS

In this case, a report of which appeared in our issue of June 7 1975 (234 EG 749), the plaintiffs were successors to the Eton Rural District Council, to whom in March 1972 the first defendant had applied for planning permission for construction of a bungalow on some 25 acres of green belt land. Eton RDC deferred consideration of the application pending the making of an agreement under seal in pursuance of section 12 of the Buckinghamshire County Council Act 1957, subsections (1) and (2) of which provide as follows:

(1) Every undertaking given by or to [an empowered council] or by the owner of any legal estate in land and every agreement made between the council and any such owner being an undertaking or agreement—

(a) given or made under seal on the passing of plans or otherwise in connection with the land; and

(b) expressed to be given or made in pursuance of this section;

shall be binding not only upon the council and any owner joining in the undertaking or agreement but also upon the successors in title of any owner so joining and any person claiming through or under them.

(2) Any such undertaking or agreement shall be treated as a local land charge for the purposes of the Land Charges Act 1925.

The contemplated agreement under seal was made on August 9 1972, and by its terms the first defendant undertook (1) that the bungalow should not be occupied otherwise than by a person locally employed in agriculture, and (2) that the bungalow should not be occupied until a farmhouse which stood upon the 25-acre site had been demolished.

On August 11 1972 Eton RDC granted permission for construction of the bungalow, imposing conditions in substantially the same form as the undertakings contained in the agreement under seal. As our report goes on to state, the first defendant built the bungalow and had been in occupation of it for some 18 months, though it was at least doubtful whether he was employed in agriculture; in addition, the farmhouse had not been demolished.

We have been informed of the position with regard to the 1957 Act and the agreement of August 9 1972 through the kindness of Mr A D Walker of Lovell, White & King, solicitors for the plaintiff council. Mr Walker comments that the first defendant was seeking to say, among other things, than an agreement such as had been made by him with the local planning authority could be overruled by the Secretary of State for the Environment on appeal. Subject to the points made above, our account of the case (in particular, the paragraph at the end of the report in which the core of Cantley J's judgment in favour of the plaintiffs is expressed) appears to be reliable. The flaw in the report is the statement in the second paragraph of the judgment to the effect that the planning authority had merely written the undertakings offered them into the conditions imposed on the grant of planning permission. This was incorrect.

QUEEN'S BENCH DIVISIONAL COURT

January 27 1975

(Before Lord Justice WIDGERY CJ, Mr Justice ASHWORTH and Mr Justice MICHAEL DAVIES)

COLLIS RADIO LTD AND ANOTHER v SECRETARY OF STATE FOR THE ENVIRONMENT AND ANOTHER

Estates Gazette June 21 1975

(1975) 234 EG 905

Enforcement notice prohibiting change of use from warehouse to cash-and-carry store—Planning officer's opinion of supply of land may be treated as evidence though unsupported by documentary sources—Impact of cash-and-carry business on other shops etc in the district is a relevant consideration for the planning authority or Secretary of State—Owners' appeal dismissed

This was a motion by Collis Radio Ltd, of St John's Square, Wolverhampton, and Eclipse Radio & Television Services Ltd of Park Lane, Halesowen, Worcestershire, to remit a decision of the first respondent, the Secretary of State for the Environment, dismissing an appeal by the two companies against an enforcement notice served on them by the second respondents, Dudley Borough Council.

Mr I Glidewell QC and Mr R J Moshi (instructed by Simpson, Silvertown & Co, agents for Betesh & Co, of Manchester) appeared for the appellants, and Mr H K Woolf (instructed by the Treasury Solicitor) represented both respondents.

Giving judgment, LORD WIDGERY said that the appellants owned and occupied land in Halesowen covering an area of 18,000 sq ft. There was a large building on the land, and the surrounding ground was used as a car park. The site was on land allocated for industrial development. The appellants had had planning permission to use the building as a warehouse since 1967, and they used it to store electrical stock. In 1972 they applied for permission to change the use of the warehouse to a retail cash-and-carry business. The retail customers would look at the electrical goods at the warehouse, pay cash for them and carry them away. This selling method would allow the appellants to sell their goods at a significantly lower price than ordinary retail shops. There was some muddle over this application of 1972; the local planning authority said that it had been refused, but the appellants claimed that they had received no notification of any refusal. In any event, the appellants had now been using the site as a cash-and-carry warehouse for the last 18 months, and the local planning authority served an enforcement notice requiring the warehouse to be returned to its former use. The appellants appealed to the Secretary of State, who, after a public inquiry, dismissed the appeal. The appellants now came to the Divisional Court, where of course they had a right of appeal only on points of law. What they said was that the Secretary of State's decision letter was erroneous in two respects. First, the Secretary of State found that there was a shortage of land for industrial development in the area, and that what there was would soon be used up. The appellants said that this was unsupported by evidence. Secondly, the Secretary of State had found that a proliferation of cash-and-carry warehouses would affect other commercial interests in the district, and the appellants said that this was not a material consideration.

So far as the first point was concerned, the borough council's planning officer had given evidence at the inquiry, and expressed the view that there was only a small amount of land available for industrial use, which land was not likely to remain available for long. His opinion was unsupported by plans, schedules or any documents, and counsel for the appellants argued that his evidence was merely an unsupported assertion. He (his Lordship) was unable to accept this argument. The planning officer lived with the situation day after day. He knew whether the supply of industrial land was drying up or not. He could speak from his special knowledge of the area. If he said that there was very little industrial land left and that this was likely to be used up in the near future, then this was some evidence on which the Secretary of State could base his decision. What weight he attached to that evidence was entirely up to him, and the court, acting on well-known principles, was unable to interfere with his decision on the ground put forward.

The second issue was the Secretary of State's finding that a proliferation of development like cash-and-carry warehouses in the area would be likely to have a severe effect on the Halesowen shopping centre. The inspector thought that if planning permission were given for shopping purposes to one firm in the industrial development area, similar applications would follow from other firms and these would be difficult to refuse once one application had been granted. This view was supported by evidence that enforcement orders had been served by the council on two similar premises in the area which had been changed to retail cash-and-carry warehouses. Counsel for the appellants argued that the consequences to other sites if planning permission were given on one application was not a proper consideration for the planning authority. He accepted that granting planning permission to site A might affect site B when there was a close relationship between them, but he did not accept that the prospect of proliferation of similar use in this case was a proper consideration for the local planning authority or the Secretary of State. He (his Lordship) was quite unable to agree. Planning was something that dealt with locations, not with individual sites. It was of the greatest importance when considering a single planning application to take into account what the consequences, or "side-effects," of granting it would be. An application for planning permission for site A had to be judged according to the consequences to sites B, C and D; this was the most elementary planning procedure.

ASHWORTH and MICHAEL DAVIES JJ agreed, and the appeal was accordingly dismissed.

VENDOR AND PURCHASER

CHANCERY DIVISION

May 17 1974

(Before Mr Justice WHITFORD)

FEDERATED HOMES LTD v TURNER

Estates Gazette March 1 1975

(1974) 233 EG 845

Contract for sale of development land conditional (1) upon planning consent being obtained within five years, (2) upon a right of access being obtained for the benefit of all land covered by the contract, including land which in the event might be retained by the vendor—Provisions not inserted solely for the protection of the purchaser, so that he cannot waive them and claim specific performance on his failure to obtain planning consent within the five-year period

This was a claim by Federated Homes Ltd, of Bucklersbury House, London EC4, against Mr James Turner, of Hempton Court Farm, Almondsbury, Gloucestershire, for specific performance of a contract entered into in 1968 for sale of land forming part of the farm.

Mr A de Piro QC and Mr A C W Hordern (instructed by J Eaton & Co) appeared for the plaintiffs, and Mr J A R Finlay QC and Mr D Hawkins (instructed by Lee, Bolton & Lee, agents for Cooke, Painter & Co, of Bristol) represented the defendant.

Giving judgment, WHITFORD J said: In 1968 the plaintiffs in this proceeding and the defendant entered into an agreement under which the defendant was going to sell and the plaintiffs were going to buy certain land. It was part of a farm, Hempton Court Farm, at Almondsbury in the County of Gloucester. The terms which were agreed between the parties were set out in a composite document dated June 27 1968, consisting in part of what appears to be a standard set of terms issued by the Bristol Law Society and in part of certain typewritten additions thereto which were bound up with the standard conditions. There was a plan forming part of the agreement to which reference is made, and that plan shows the area to be sold marked with a red boundary. I should mention in passing that within the red land there is an area marked with a green boundary. The green area is in fact the farmhouse and the land immediately adjacent to the farmhouse. There were certain special provisions with regard to this part of the red land which are of no significance for present purposes, and I need not mention them further. I should however add that one finds additionally upon the plan an area of land delineated in blue which was no part, directly, of the agreement entered into between the plaintiffs and the defendant. The blue area was in fact an area of land also owned by the defendant, and before ever this agreement was entered into there had been some negotiation relating to the proposed acquisition of the blue land for the purposes of the building of a school, or an addition to a school, by the county council. In fact, the matter was in substance concluded, so far as the acquisition of the blue land is concerned, by the time the agreement for the sale of the red land was entered into, save only for this, that the price to be paid for the blue land had not been finally determined.

The agreement of June 27 sets out in clause 5 the price which it was agreed should be paid, and the price was "£7,000 per acre of the red land." The land is described as "agricultural land," which indeed it was at the time in question. Having regard to the price which the plaintiffs were prepared to pay, it is of course at once apparent that they were not intending to pursue the purposes for which the land had previously been used, namely agriculture. They hoped that upon an appropriate application they would get permission to build. It was quite plainly envisaged at the time when the agreement was entered into that planning permission might not be granted. That is apparent from the terms of the agreement, because certain provisions were inserted in the agreement, to be found in the typewritten sheets, and these provisions deal, or purport to deal, with the eventualities which were to ensue if planning permission were not obtained. The whole of the dispute in this action is about these particular provisions, to which I will shortly have to turn. As a matter of history, however, it is convenient to say that after the agreement had been signed, the plaintiffs took all appropriate steps to make their application for planning permission for building of the type envisaged in the agreement for sale. There is no reason to suppose that they, on their side, did not pursue this application with all reasonable diligence. In fact they have been initially refused permission, but an appeal is on foot. The appeal has not yet been heard and is not due to be heard until later this year. No doubt part of the difficulties which have undoubtedly arisen could have been avoided if it had been possible to get these matters dealt with rather more expeditiously, but it was not possible, so we are left with the situation in so far as it is covered by the terms in the agreement. The original agreement set a term of five years. Without going into the details of it, what basically it provided was that there should be a five-year period during which the plaintiffs would be able to pursue their application for planning permission and at the end of that period certain eventualities were to follow if planning permission had not been obtained. By the end of, or towards the end of the five-year period, it was quite apparent to the plaintiffs that they were not going to get planning permission within the five-year period, and it is plain that at this stage they considered where their best interests lay and what it would be most suitable to do. A letter was written on their behalf by their solicitors as long ago as April 4 1973 in anticipation of the event that planning permission would not have been received by the expiry of five years from the date of the agreement, and in that letter, which was addressed to the defendant's solicitors, an indication is given in these terms:

> "My clients have now indicated that I should shortly write you formally exercising the option in respect of the whole of the land on the contract terms (in which event there would be no retained land), and adding that they wish to take up the conveyance one month thereafter, at which time they would pay £175,000. The balance of purchase-money would then be paid in three equal annual instalments."

The exact importance of that paragraph can only be understood when we come to consider the relevant terms of the

A agreement, but it does, I think, show the way in which the plaintiffs were approaching the matter at this time. They apparently took the view that the agreement was in the nature of an option agreement, and that they were entitled, in certain circumstances, to exercise their option to acquire the property, and they were proposing to do this. What can perhaps be described as a formal notice was sent on April 17 1973. Under this the plaintiffs' solicitors wrote to the defendant's solicitors informing them again that they were proposing to proceed with what I think they were still thinking was an option agreement, upon the basis that the plaintiffs would purchase the land, waiving any conditions relating
B to the securing of planning consent and certain other conditions which I think it is suggested really only make sense if they are considered as being part of the conditions relating to the obtaining of the planning consent. At any rate, here, before the expiry of the five-year period, was a notice sent saying that the plaintiffs wanted to proceed and were prepared, in order to proceed, to waive certain conditions, and indeed yet another letter was sent in May of the same year, following an exchange of counsel's opinions as to whether or not the plaintiffs could exercise some valid right, be it by way of option or otherwise, to have the agreement specifically performed. In this further letter, dated May 30 1973, it was
C again stated that the plaintiffs were proposing to waive any conditions set out in two clauses of the agreement to which I will shortly turn, clauses 18 and 19. In this letter it was said that although the view was taken on the plaintiffs' side that in the circumstances they were entitled to ask for completion on the basis that they paid for the first 25 acres at the agreed price, which would result in a capital payment of £175,000, and were thereafter entitled to pay the balance in three equal annual instalments, they were nonetheless prepared, if it would suit the defendant better, to pay the whole of the agreed price for the whole of the land, which comprised some 17 acres, namely £490,000. The defendant, for
D reasons best known to him, took the view that he did not want to complete upon these terms. He had been advised that the plaintiffs were not entitled to call, in the circumstances, for a completion of the contract upon the terms which they set out in any of their letters, or indeed upon any terms. It appears that he had been advised that if at the expiry of the five-year period no planning permission had been secured, the agreement would terminate. However, to make doubly sure, shortly after the expiry of the five-year period in question, notice was sent from the defendant's side to the plaintiffs' solicitors by a letter dated June 28 1973, stating that the
E defendant was treating the contract as determined by the expiry of five years from its date pursuant to the provisions of clause 18.

The whole of this case really turns upon what in essence may appear to be a relatively simple point. Everybody is agreed that at the beginning there was an agreement which envisaged the sale of the whole of the red land at the agreed price, £7,000 per acre. Everybody is agreed that by clause 18 of the agreement a condition was written into the agreement which really came to this: that the agreement was conditional upon the plaintiffs obtaining in respect of the red land outline planning permission. On the plaintiffs' side,
F what basically is said is that the whole of this provision to be found in clause 18, and indeed certain other provisions to be found in clause 19, were inserted solely for their benefit as purchasers, and to some extent this does, no doubt, seem a commonsense sort of approach. They were, after all, buying agricultural land at a price which I suppose in 1968 was very much in excess of 10 times the value of the land as agricultural land, and in those circumstances, understandably enough, they might want to provide an escape route if at some stage when the application was being put forward it became quite plain that the land would never be permitted to be used for anything other than agricultural purposes. If that be right, then the basic contention on the plaintiffs' side G is that the whole of these provisions relating to the obtaining of planning permission being conditions which were inserted for their sole benefit, they were entitled to waive the conditions if they were so minded and can now call for specific performance of the agreement. That is just what they do in this action.

Before I come to consider the relevant clauses of the agreement, I think it may perhaps help if I say one word about the law. This is not a point which is or can, in any sense, be said to be new. There have been cases in the past where vendors have sought to escape from an obligation which they have entered into to sell property upon the sort of H basis which arises here, and there is no doubt that in the past the courts have taken the view, in relation to contracts for sale, where difficulties in completion may have arisen because of the impossibility of or an objection to the performance of a condition on one side, if that condition benefits only the other party, then in fairness he ought to be entitled to waive it, because it would not be right to allow the vendor to escape from his obligations upon a point which the purchaser ultimately decides is really, taking matters by and large, one which is of not sufficient substance or importance, so far as he is concerned, for him to desire to insist on it. The first of the cases to which I was referred (and I was referred J to quite a number, though I do not propose to deal with them all in detail) was the case of *Hawksley v Outram* [1892] 3 Ch 359. That was a case of a sale of a business which had been carried on under the name of Benjamin Outram & Co, and under the contract for sale the purchaser was going to be entitled to carry on the business under this name, and there was a restrictive covenant in the contract binding the vendors not to carry on a business of this character within a certain area. It was a case where complication arose because one of the vendors was not going to be available at the time when the agreement was to be concluded, and had given a power of attorney to the other, who signed the sale agreement on his K behalf. The defendants defaulted upon completion, and on proceedings being brought for specific performance the question was raised as to whether or not the power of attorney authorised the giving to the plaintiff purchaser the right to use the name Benjamin Outram & Co and as to whether or not it authorised the attorney to subject the company selling to a covenant in restraint of trade. I need only refer to the judgment in the Court of Appeal given by Lindley LJ, who, having dealt with the facts, at page 375 of the report continues in these words:

"The purchaser, however, says, 'A doubt being raised whether L the power of attorney authorises the giving the purchaser power to carry on the business in the name of the old firm, which might expose the old partners to liability, or authorises the binding the vendors not to carry on the same business, I will waive those stipulations.' That appears to me to remove all difficulty, because it is quite obvious that those two clauses are inserted simply and purely for the benefit of the purchaser; and if there is any doubt whether they are binding upon the vendors, and the purchaser waives them, what have the vendors to complain of? What conceivable difficulty remains if that is done? I can see none. Of course, if those clauses were so inextricably mixed up with other parts of the transaction that they could not be severed, there might be a difficulty; but they are not. They are M perfectly severable. The attorney has got power to sell the business, and says, 'I agree to sell the business.' He also says, 'I agree further that the vendors shall not carry on business in competition with you within a radius of 50 miles.' The validity of that further agreement is doubtful. The purchaser says, 'If that is doubtful, I give it up.' Where is the difficulty in enforcing specific performance of the other stipulation, which is not tainted with any doubt or affected by any difficulty? It appears to me that the plaintiff is right. It is obvious that we should be doing grievous injustice if we were to decide in any other way, because the plaintiff has, on the faith of this agreement, entered into possession and paid considerable sums of money. If he cannot get specific performance he will never get his money back, and

A it appears to me that there is neither principle nor authority which can or ought to prevent us from making a decree in his favour."

This they did. I cite the whole of that passage because I think it is always important in these cases to have in mind the exact circumstances in which the decision was given. This was quite plainly a case where the court took the view that if an order for specific performance were not made, a grievous injustice would be inflicted upon the party seeking the order, and furthermore it was a case where the decision was taken upon the basis that it was, in the words of the Master of the Rolls, quite obvious that the clauses upon
B which the question of a waiver arose were being "inserted simply and purely for the benefit of the purchaser."

A number of other authorities were referred to of respectable antiquity, but a very much more recent case which was drawn to my attention was the case of *Heron Garage Properties Ltd v Moss* [1974] 1 WLR 148, which was decided by Brightman J this year, and was a case in which purchasers were buying a part of a site, and the agreement was expressed as being conditional upon the purchaser obtaining planning consent within a specified time for the redevelopment of the property for certain purposes. It was a case where the agreement provided that if the consent was not obtained within
C the specified period, either party could determine the agreement by giving notice in writing. That was an express term of the agreement, in which the relevant clause, clause 7, started off by saying, "This agreement is expressly conditional upon the purchaser obtaining detailed town planning consent for the redevelopment of the property . . . " and so on, and continued at a later stage, "In the event of town planning consent in such terms as above not being granted within six months from the date of the first local town planning authority meeting after the date hereof (or within such extended time as the parties hereto may agree in writing) then after the expiration of such period (or such extended
D period) either party may by notice in writing to the other party determine this agreement, whereupon the deposit paid hereunder . . . " and so forth. So that in some ways, the circumstances were not unlike the circumstances which I have to consider. Again, in this case, the question which was canvassed was whether, in fact, the purchasers had a right unilaterally to waive the conditions of clause 7. It was a case in which Brightman J came to the conclusion that they had no such right. In his judgment he reviews the decision in *Hawksley v Outram* to which I have already referred. He refers to another case of *Bennett v Fowler* (1840) 2 Beav 302,
E which is not, to my mind, quite so much in point, and he refers to a decision given by Kekewich J in *Lloyd v Nowell* [1895] 2 Ch 744, where again the question arose as to whether a particular stipulation was for the sole benefit of the party seeking to waive. In rejecting the contention put forward by the purchaser in the case before him, Brightman J puts it in this way. As I understand it, he accepts the position that there may be circumstances in which a contracting party seeking specific performance can waive a stipulation, but the only circumstances which Brightman J envisages might lead to this right were these, that there might be some express term in the agreement giving the party the right to waive, or
F the circumstances must be such that there must be an inevitable implication, taking the terms of the agreement as a whole, that the condition sought to be waived is for the benefit, and for the benefit only, of the person seeking to waive it.

I think now I must turn to the agreement, and it is necessary for me to read in full those clauses of the agreement which were referred to in argument. Clause 18, which is the first and perhaps most important of the clauses, is in the following terms:

"This contract is conditional upon the purchasers obtaining in respect of the red land outline planning consent free from conditions or subject to conditions acceptable to the purchasers G
for the erection of units at an average density of not less than 10 units per acre within five years of the date hereof. The purchasers shall use their best endeavours to obtain such outline planning consent in respect of the red land and shall be entitled to call upon the vendor for assistance (but not monetary assistance) to this end. At all times the purchasers shall notify the vendor of the issue of any consent relating to the red land or any part thereof within 48 hours of the receipt of such consent. If such planning consent be not granted within the said five-year period then this contract shall become null and void except as hereinafter provided. In the event of outline planning consent being granted within the five-year period for the erection of units at the said average density upon an area of at least 25 acres then this H
contract shall become binding in respect of that area of land and the purchasers shall be required to purchase and the vendor shall be required to convey that area of land at the price of £175,000 for the first 25 acres thereof and at the rate of £7,000 per acre for the remainder and the purchasers shall be required to complete the purchase of 25 acres of that area of land within one month of the granting of such outline planning consent. This contract shall remain conditional in respect of the remainder of the red land not comprised in the area which shall have received planning consent as aforesaid until the termination of the said period of five years. In the event of outline planning consent for the erection of units at the said average density being granted in respect of an area of less than 25 acres in the first instance then J
the vendor may decide that this contract shall become binding in respect of that lesser acreage by giving within one month of the issue of such consent notice in writing to the purchasers and the purchasers shall then be required to purchase that lesser acreage at a price calculated at the rate of £7,000 per acre and to complete that purchase within one month of the receipt of such written notice but the vendor shall not be required to convey an area of less than 25 acres in the first instance unless he has given such written notice to the purchasers of his decision to do so. In the event of outline planning consent being granted within the said five-year period for the erection of units at the said average density on a total area greater than 25 acres but less than the total area of the red land either in the first instance or as a K
result of subsequent grants of planning consent then this contract shall become binding in respect of the area of land in excess of 25 acres only if the vendor is reasonably satisfied that provision has been made by the purchasers in consultation with the local planning authority for the development of the balance of the red land, then remaining and not having the benefit of outline planning consent, with units at not less than the same said average density per acre. The vendor, when satisfied with the above (for which purpose he shall be allowed one month), shall give notice in writing to that effect to the purchasers and this contract shall thereupon become binding in respect of the whole area in respect of which the said outline planning consent has been granted and the purchasers shall be required to purchase and the vendor to L
convey the said area of land at a price to be calculated at the rate of £7,000 per acre, and unless the vendor shall signify the contrary in writing within the said period he shall be deemed to be satisfied. In the event of this contract becoming binding in respect of an area in excess of 25 acres at any one time either in the first instance or subsequently the purchasers will be required to complete their purchase within one month of the date on which the contract becomes binding only in respect of 25 acres and to pay the sum of £175,000, and the balance of the purchase moneys shall be paid in three equal annual instalments the first of which shall be paid one year after the said sum of £175,000 falls due and the vendor shall convey such part or parts of the land in respect of which there is a binding contract as the purchasers may M
require but so that the vendor shall not be required to convey a greater area of land at any time than is covered by the purchase moneys then paid calculated at the rate of £7,000 per acre. If at any time during the fifth year of the said period of five years the vendor shall wish to make his own planning application in respect of all or any part of the red land which shall not then have received consent he shall be entitled to do so and to call upon the purchasers for particulars of any application they shall have made in respect thereof. In such case any consent received by the vendor shall be treated for the purpose of this contract as a consent received by the purchasers and the provisions herein contained as to purchase of land with the benefit of an overall average density of 10 units to the acre shall apply."

There is then a definition of units, which is not material for the present purposes. The next relevant clause is clause 19, which is in the following terms:

> "This contract is also conditional upon the purchasers being able to obtain for the benefit of the whole of the red land right of access in accordance with the requirements of the local planning authority to the public highway across the land over and to the south-west of the railway. The costs (if any) of obtaining this right of access shall be borne in full by the purchasers up to a total sum of £10,000 and in the event of these costs exceeding the said sum the excess costs over and above £10,000 shall be shared equally between the vendor and the purchasers. To the extent that the cost of constructing the access road (the cost of which is the purchasers' liability) shall be greater than the normal costs at that date of constructing a normal access road of similar width resulting from it having to be constructed across the railway tunnel, then such additional costs shall be deemed to be a part of the costs of obtaining the right of access."

Clause 20 is of no particular materiality; it merely provided that the vendor was going to get £5,000, which would be non-returnable. But clause 21 must also be considered. It is in these terms:

> "It is a condition of this contract and shall be a clause in any subsequent conveyance or conveyances that the vendor shall reserve a full and free right of way for all purposes of a width sufficient to permit a roadway to be constructed to the satisfaction of the local planning and highway authorities across any land conveyed, for the benefit of and for the purpose of obtaining access to all the retained land; this right of way to follow the route of the roads to be constructed on the land conveyed and the purchasers to ensure that such new roads are routed and constructed in a manner which will permit such satisfactory access. The vendor shall further reserve the rights for the benefit of the retained land to connect to and make use of the services and utilities in the nature of sewerage and surface water drainage systems water electricity and gas supplies as may in due course be provided on the land conveyed and the purchasers shall ensure that these said services and utilities which are so provided on the land conveyed are of a capacity sufficient so far as the rights of way are concerned to serve the development that may reasonably be expected to take place on the retained land and so far as the sewerage and surface water drainage systems water electricity and gas supplies are concerned to serve that part of the red land which shall not be or have been conveyed to the purchasers and further so far as water electricity and gas supplies are concerned so far the purchasers are able. For the purposes of this clause the 'retained land' shall mean the land shown on the plan attached hereto and thereon edged blue and the portion of the red land not conveyed as aforesaid."

It is perhaps to be observed just in passing that so far as clause 21 is concerned, it appears to me, on the plain language used in the clause, and upon the definition of the "retained land," that so far as any question of access and any access road is concerned, the benefit of that might enure to the blue land, but the benefit of any arrangements as to sewerage, surface water drainage systems, water, electricity and gas supplies was going to enure only in respect of any red land which, for the purposes of clause 21, came to be "retained land." Perhaps the most important words in these three clauses are the very opening words of clause 18. For the purposes of argument clause 18 was divided into a number of subclauses by counsel on behalf of the plaintiffs, and his numbering was followed in the course of his address by counsel for the defendants. Clause 18, as can be seen, starts off by making the contract conditional upon the obtaining in respect of the red land of outline planning consent for the erection of accommodation at a certain density. The next provision, which was referred to as 1 (a), is the provision in clause 18 under which the purchasers had to use their best endeavours to secure the requisite consent. The following provision, which was referred to as 1 (b), is in these words, which I must repeat: "If such planning consent be not granted within the said five-year period then this contract shall become null and void except as hereinafter provided." The plaintiffs' case, as I understand it, really proceeds first upon this basis, that really one can stop at this point. The whole of this provision was, it is said plainly, and I think there is some sense in it, quite plainly inserted to safeguard the purchasers. They wanted to escape if they did not get the planning permission, and it was for that reason and that reason alone that provision was made in that part of clause 18 which was referred to as 1 (b) for the contract becoming null and void if planning permission were not granted within the five-year period. So far as the defendant is concerned, what he says first of all is that as a mere question of construction, if one looks at the language, there is not really any doubt about what the parties had agreed at the time when they entered into this agreement; they had agreed that there was to be a period within which planning permission might be sought. At the end of that period, subject to the exceptions to which I will shortly have to come, if planning permission for the red land had not been obtained there was to be an end of the matter. On the plaintiffs' side it is said, "Well, it may well be that this agreement so states, but what does it matter to the defendant whether planning permission is obtained or not? What he wanted was his £7,000 per acre for his 70 acres, and that we are prepared to pay him. He entered into this agreement to sell this acreage to us at that price. It may be that a safeguard was inserted to safeguard our position; we are now prepared to waive it. Why should the contract become null and void?"

I think one has got to be very careful, in relation to agreements of this character, in approaching the question of waiver on conditions. I do not think, myself, that it is fair to consider this solely upon the basis of the provisions of clause 18 up to the end of that part of it which has been designated as clause 1 (b), because quite plainly at the beginning what had been envisaged was the possibility of situations rather more complex than the situation in which everything was going to stand or fall with the application in respect of the whole of the red land. Thus, even if planning permission were not to be obtained within the five-year limit, the agreement was not necessarily going to become null and void, and there are elaborate and sometimes not very easily reconcilable provisions which follow thereafter and deal with the situation which might arise if, say, planning permission were secured for 25 acres or more than 25 acres or for a lesser acreage. It is to my mind quite plain, upon the provisions of clause 18 alone, that a situation could conceivably arise in which the purchaser was going to get planning permission for an area of land which would entitle him to enforce his rights to purchase, but that there might be left to the vendor other areas of land in respect of which no planning permission was going to be obtained, which would be "retained land." Clause 21, in terms, constitutes a recognition of the fact that at the end of the day, quite apart from the question of the blue land, there might be a certain portion of land which would by then have to be regarded as "retained land," in respect of which the vendor was going to secure certain rights. It must also be remembered that the agreement was not conditional merely upon the securing of planning consent. It was also conditional upon the purchasers being able to obtain for the benefit of the whole of the land the rights of access referred to in clause 19, the rights of access which might be of benefit, so far as the vendor was concerned, under clause 21 in relation to retained red land, or indeed in relation to the blue land. What was submitted was that so far as the blue land was concerned, that really need not enter into consideration at all, because the agreement for the purchase of the blue land was concluded before ever this agreement was entered into, and only a consideration of price remained outstanding. However that may be, the possibility of rights relating to access might, I suppose, at least notionally affect the price which the council was prepared to pay for the acquisition of the blue land, and indeed the right of access in favour of the blue land contained in clause 21 might be a negotiable right of value to the vendor.

I think, however, that basically this case turns upon what seems to me to be a relatively simple and straightforward question of construction. It may well be that the terms of an agreement of this character might be so drafted that it was made quite apparent in terms on the face of the agreement that a condition relating to the acquisition of planning permission with a certain period was a condition which could be waived by the purchaser, and if anybody is seeking to enter into an agreement of this character in the future, and in fact wishes to retain a right to waive conditions relating to the obtaining of planning permission, I would have thought they would have been very sensible so to do. It may well be that the terms of particular agreements may be such that one must be able to come to the conclusion that the conditions being waived are conditions which are favourable only to the person seeking the waiver. The express terms of this agreement go to this: no planning permission within the five-year period and the contract becomes "null and void except as hereinafter provided." It is common ground that the subsequent exceptions do not save the situation, because there was no planning consent in respect of any part of the red land, and the exceptions are only really dealing with the situation which arises in relation to partial consent. This is very strong language to use. It would have been very easy if one party or the other were to have the right to take some decision as to what was to happen to insert express words indicating this. The possibility that a point might arise where somebody might make a choice was plainly envisaged by those who were concerned to draft these provisions. One finds in clause 18 itself, for example, the expression "the vendor may decide," and at a later stage, "unless the vendor shall signify the contrary." On the face of it, it seems to me that there really is no doubt as to what the intention of the parties was at the time when the agreement was signed, if one really takes the language of the document, and to my mind it would have required a very strong case indeed to establish that this provision, as clear as any provision could be in its terms, must be subject to this: that taking the agreement as a whole, this and the subsequent provisions are provisions which were really only inserted for the protection of the purchaser and which he was, in fact, entitled to waive. Setting aside, however, any question of subsequent conditions, and merely considering what I think is the strongest way in which the case was put upon the plaintiffs' side, I have come to the conclusion that if you take the agreement down to the end of that part of 18 which was designated as 1 (b) alone, there is no sufficiently strong implication that this provision was for their benefit alone, no implication sufficient to override what, as I have said, seem to me to be very clear words. As I have indicated, I do not think it is right to take it alone, and once one brings into consideration the possible alternative events envisaged in the remainder of clause 18, not, I am afraid, always without difficulties of interpretation, and in clauses 19 and 21, one must reach the conclusion which was reached by Brightman J in *Heron Garage Properties Ltd v Moss,* namely, that this is not a case where it is obvious on the face of the contract that the stipulations which the plaintiffs sought to waive by those letters to which I have already referred were for their exclusive benefit and that, in all the circumstances of this case, they ought to be entitled to have their waiver accepted and to be granted an order for specific performance. In the result the action fails.

COURT OF APPEAL

February 5 1975

(Before Lord Justice CAIRNS, Lord Justice LAWTON and Mr Justice MACKENNA)

STRUTT v WHITNELL

Estates Gazette April 26 1975

(1975) 234 EG 285

Vendor fails to give vacant possession of house with sitting tenant—Purchaser need not mitigate damage by accepting offer to repurchase made by vendor—Entitled to keep house and sue for difference in value attributable to tenant's presence

This was an appeal by Mr Norman Whitnell and his wife, Mrs Christine Marjorie Whitnell, of Cottage Grove, Clacton-on-Sea, Essex, from a decision of the district registrar at Colchester on April 3 1974 assessing at £1,900 the damages due to the respondent, Mr John Alan Strutt, of Vicarage Gardens, Clacton-on-Sea, upon his claim for breach of a contract of sale of 22b Bockings Grove, Clacton.

Mr A Scrivener (instructed by Gotelee & Goldsmith, of Ipswich) appeared for the appellants, and Mr M J Hyam (instructed by Elwes, Wittey & Co of Colchester) represented the respondent.

Giving judgment, CAIRNS LJ said that the contention raised on behalf of the defendants before the registrar was that certain events had occurred as a result of which the plaintiff could have mitigated his damage, and that if he had done so he would have suffered no damage at all. Consequently he was entitled to no more than nominal damages in the action. The plaintiff was a property developer, and the defendants were friends of his. On April 26 1969 they purchased from him 22b Bockings Grove for £4,650. They let the house to a tenant who became a protected tenant under the Rent Act. In 1970 structural defects became apparent in the house. The defendants were in the position that they could not sue the builder in respect of these defects because they had no privity of contract with him. It was therefore arranged in a friendly manner between them and the plaintiff that he would buy back the house so that he could make a claim on the builder. Accordingly the plaintiff did buy back the house, the conveyance taking place on March 3 1971. The sale was a sale with vacant possession, according to the terms of the contract. It was thought at that time that the tenant would be willing to leave. But unfortunately he did not do so, and there was no ground under the Rent Act on which an order for possession could be obtained against him. There was a meeting in the late spring of 1971 when the matter was discussed between the plaintiff and the two defendants, and the plaintiff said he might have to sue them for damages for breach of the term as to vacant possession. The first defendant, Mr Whitnell, said, "Don't do that, we will buy it back." The plaintiff said he could not do that because he had already sold the house in a portfolio with 11 other properties. That was a totally untrue statement which at the hearing he denied making. The registrar disbelieved him, accepting the defendants' account of the conversation. Mr Whitnell, not deterred by that attitude of the plaintiff at that interview, made at least one further offer in similar terms, that offer being made, according to Mr Whitnell's own evidence, when he got the letter before action claiming damages.

The plaintiff duly sued for breach of contract, the damages being quantified as the difference in the value of the house with vacant possession and the value without it. The defendants recognised that to that claim they had no defence. They therefore simply appeared in the action but served no defence, and judgment was accordingly entered for the plaintiff for damages to be assessed. The issue of damages was referred to the registrar. There was evidence from surveyors on each side as to the value of the house both with and without vacant possession. The conclusion at which the registrar

A arrived was that the difference was £1,900, and at that sum he assessed the damages. The contention of the defendants before the registrar and in the Court of Appeal had been that the plaintiff could have mitigated his damage so as to reduce the defendants' liability from £1,900 to nothing, simply by accepting the offer which the defendants had made to him, there being no good reason for refusal. What was said by Mr Scrivener, on behalf of the defendants, was that this was a simple case of mitigation of damage on principles which had been laid down in a number of cases, and in particular in the decision of the Court of Appeal in *Payzu Ltd v Saunders* [1919] 2 KB 581. Certainly if one took passages
B from the judgments in that case, particularly the judgment of Scrutton LJ, there were statements of general principles in relation to mitigation of damage which, if they were treated as of universal application, might be said to apply to the facts of this case. Nevertheless the court had been referred to no case which was really parallel in its facts and in which the rule as to mitigation of damage had been applied.

He (his Lordship) had reached the conclusion that the rule could not be applied to a case where the buyer of either goods or real property had a cause of action in damages for defects in the goods or the property or for breach of covenant in the contract, and where some offer had been made to the buyer
C by the sellers which might be said to amount to providing him with an alternative remedy. One of the cases to which the court had been referred was a decision of Devlin J in *Heaven & Kesterton Ltd v Etablissements Francois Albiac & Cie* [1956] 2 LLR 316. At page 321 the judge said: "That is tantamount to depriving a buyer of his right to reject altogether and saying that, notwithstanding that he is properly rejecting, he is to be put in precisely the same position as if he had taken up the goods and as if his only remedy was to be compensated for the damage on the basis of defective quality." The circumstances here were that the only remedy
D in law which the plaintiff initially had in respect of this breach of contract was a right of action in damages. If it were to be said that an alternative remedy was offered to him by the defendants by way of selling the house back to them, then he (his Lordship) would say that, just as in the case before Devlin J, the plaintiff was not bound to choose between the two remedies. He would add that it did not appear to him that there was any such offer made by the defendants as could put the plaintiff in the position that he was bound to accept it in mitigation of damage. Nothing appeared to have been said as to the price at which the defendants would repurchase. Even if the offer had been in definite terms, the
E defendants had failed to establish that it was unreasonable for the plaintiff to have refused it or that he had no good reason for refusal. The registrar came to the right conclusion, and he (Cairns LJ) would dismiss the appeal.

Agreeing, LAWTON LJ said that the question was whether, if the law gave a man a remedy, he could be said to act unreasonably if he elected to keep that remedy and not accept some other remedy which might be available to him in law, or alternatively might arise, as in this case, out of an offer made by those who were in default. The registrar had approached the problem on the basis that the law gave the plaintiff a remedy, that he had elected to keep his remedy
F and that in those circumstances it could not be said that he had acted unreasonably. The registrar was saying that the plaintiff had a right to keep the house and sue for damages. That was what he did. Like the registrar, he (his Lordship) could not say he was wrong.

Also agreeing, MACKENNA J said that there was no evidence of any definite offer by the defendants to buy back the property at the contract price. The evidence did not establish that the plaintiff had unreasonably refused such an offer. At the worst he had refused to enter into negotiations with the defendants, giving, as the registrar had found, an untrue reason for his refusal. That was not enough, on his (MacKenna J's) view of the law, to disqualify him from G
recovering substantial damages. It looked as if the plaintiff had behaved badly towards the defendants, who were his friends, and had earned the description of "Shylock" which the registrar had given him, together perhaps with an additional epithet which would be appropriate to an untrue witness. If that were so, it was a pity that he must recover substantial damages. But it was no ground for deciding his case otherwise than in accordance with the principles of the law of damages, which were well settled in his favour. With that much reluctance, he (his Lordship) agreed that the appeal failed.

The appeal was dismissed. Leave to appeal to the House H
of Lords was refused.

CHANCERY DIVISION

May 10 1974

(Before Mr Justice WALTON)

RIGHTSIDE PROPERTIES LTD v GRAY

Estates Gazette June 14 1975

(1974) 234 EG 819 J

Notice to complete sale of land not complied with, and vendor elects to accept supposed repudiation and treat contract as discharged by non-performance—Notice in fact insufficient—"At least 21 days' notice" means 21 days excluding day of service and date for completion—Further points concerning receipt by post and validity of service on a solicitor outside normal working hours—Vendor's election itself a repudiation available for acceptance (and ultimately accepted) by purchaser—Extended discussion of relationship between law and equity in a case of the kind—Purchaser need not show himself ready, able and willing K
to complete—Instead of preparing for a completion which ex hypothesi will not take place, he may simply accept the repudiation and sue for damages

This was a claim by Rightside Properties Ltd, of 56 Haymarket, London W1, against Mr Bernard Gray, of 5 Salem Road, Paddington, London W2, for damages for breach of a contract of sale of 53 Bassett Road, North Kensington, London W10, for £50,000.

Mr R Ellis (instructed by Randall, Rose & Co) appeared for the plaintiffs, and Mr N C H Browne-Wilkinson QC and Mr M Blythe (instructed by Allen & Son) represented the L
defendant.

Giving judgment, WALTON J said that on April 25 1972, in consideration of £50, the defendant gave the plaintiffs an option to purchase 53 Bassett Road for £50,000. This option was exercised by a letter from the plaintiffs dated April 30 1972, and it was common ground that the effect of that letter was to bring into existence a contract by correspondence within the meaning of section 40 of the Law of Property Act 1925, and that that contract was accordingly governed by the Statutory Form of Conditions of Sale 1925. By condition 1 of that form, the date for completion was fixed at the first day after the expiration of seven weeks from April 30 1972, or M
(as that was a Sunday) the next following working day, namely June 19 1972. Completion did not in fact take place on that date, and by letter dated June 23 1972, enclosing a notice therewith, the defendant's solicitors wrote to the plaintiffs' then solicitors as follows:

"We have discussed this matter with our client and in the circumstances have agreed that your clients should, despite their actions so far, be given the opportunity to complete the contract. Accordingly we enclose herewith by way of service upon you a notice to complete the transaction in accordance with No 9 of the Statutory Conditions of Sale."

Condition 9 provided as follows:

(1) If the purchaser shall neglect or fail to perform his part of the contract the vendor may give to the purchaser or to his solicitor at least 21 days' notice in writing specifying the breach and requiring the purchaser to make good the default before the expiration of the notice.

(2) If the purchaser does not comply with the terms of the said notice (a) the deposit money, if any, shall, unless the court otherwise directs, be forfeited to the vendor, or, in the case of settled land, to his Settled Land Act trustees; (b) the vendor may resell the property without previously tendering a conveyance or instrument of transfer to the purchaser; and the following provisions shall apply.

(3) Any resale may be made, by auction or private contract, at such time, subject to such conditions, and in such manner generally, as the vendor may think proper, and the defaulting purchaser shall have no right to any part of the purchase money thereby arising.

The notice enclosed was dated June 23 1972, which was a Friday. It was signed by the defendant's solicitors, and required completion " within 21 days from the date hereof." The plaintiffs did not comply with the notice, and after the expiration of the period thereby limited, namely on July 20 1972, the defendant's solicitors wrote to the plaintiffs' then solicitors making it perfectly clear that he regarded the contract as ended. In substance, he purported to accept a repudiation by the plaintiffs of the contract, and his solicitors added that it was not reasonable for Rightside to expect the date for completion to be extended beyond the expiry of the notice to complete. If, of course, the notice given to Rightside was for any reason bad, the letter of July 20 1972 was *prima facie* itself a wrongful repudiation of the contract by the defendant, a repudiation which the plaintiffs were entitled, if they were so minded, to accept. They did not, however, do so at once. Indeed, they had on June 29 1972 entered into a contract to resell the property to one Leslie Ratcliffe for £53,000. By a coincidence, also on July 20 1972, Mr Ratcliffe's solicitors had written to Rightside's then solicitors asking for authority to inspect the register, a request which placed Rightside's solicitors, in the circumstances, in a near impossible position. The defendant himself was also in a difficult situation; he had counted on receiving the proceeds of sale of 53 Bassett Road in time to enable him to complete a subsequent auction purchase into which he had entered, and the delay in completion had embarrassed him. Neither side knew of the difficulties in which the other had been placed.

The present proceedings were begun by Rightside on December 15 1972, claiming specific performance of the contract or alternatively damages for breach. The statement of claim was delivered only after there had been an application to strike out the action for want of prosecution, but thereafter matters progressed normally. The defence and counterclaim, as finally amended, pleaded the terms of condition 9 of the statutory conditions of sale, and then continued: " By letter dated June 23 1972, the defendant's solicitors gave notice to the plaintiffs pursuant to the said condition 9 requiring the plaintiffs to complete the said contract within 21 days. Wrongfully and in breach of contract the plaintiffs failed to comply with the said notice." And the conclusion was stated that Rightside were not entitled to the relief claimed, or alternatively to equitable relief. Rightside's reply and defence to counterclaim, as finally amended, was basically that the notice actually given did not give Rightside 21 days' notice as required by condition 9, and further or alternatively that the period of 21 days was in the circumstances unreasonably short. On opening the case for Rightside, counsel at once elected to accept the repudiation of the contract contained in the letter of July 20 1972, and to abandon the claim to specific performance. He indicated, in relation to the disputed condition 9 notice, that he would argue simply that the requirements of the condition had not been complied with, and that he did not intend to pursue the allegation that 21 days' notice was unreasonably short. As the case proceeded, therefore, he (counsel) called only one witness, Rightside's former solicitor, Mr Freeman. A cautious and impressive witness, Mr Freeman was unable to say whether the notice dated June 23 1972 was received by his firm in the course of the sole post delivered on Saturday June 24 or by either of the two posts delivered on Monday June 26; the original of the letter did however bear what appeared to be his firm's date stamp of June 26 1972. The defendant's solicitor, Mr Lavender, proved that the letter was signed by him on Friday June 23, and that had the settled routine of his firm been followed, the letter would have been put into the post with a first-class stamp late on Friday evening.

He (his Lordship) thought that the onus of showing that Rightside's solicitor received the notice at any earlier date than June 26 1972 rested on the defendant, and that it had not been discharged. The only proper inference to be drawn from the evidence was that the letter in question, with its enclosure, would not have been delivered on the Saturday morning in the ordinary course of post, and was therefore delivered on the following Monday, June 26. If that were so, then there could be no pretence that the notice was a valid notice. Even making the assumption most favourable to the defendant, moreover, namely that the notice was delivered on June 24, the length of the period specified, when reckoned from that date, could only have been 20 days. Hence, the notice was, on any footing, too short by at least one day. Counsel for Rightside had additionally submitted that it did not matter when the letter enclosing the notice was delivered; what mattered was when, as a matter of ordinary routine, it could be expected that it would be opened by the solicitor. It would be absurd, he said, if service on a solicitor (as expressly envisaged by condition 9) could be validly effected out of normal working hours; and Saturday was not a normal working day for most solicitors. He (Walton J) was inclined to think that that submission was correct: see *Papillon v Brunton* (1860) 5 H & N 518. It would seem the height of absurdity if a notice, whose function was always to draw some fact to the attention of another party for action or information, could be served at a time when it would, as a matter of ordinary routine, be utterly impossible for it to convey any information to anyone. Counsel also said that in any event a notice requiring something to be done " *within* 21 days " did not give the period of " at least 21 days " referred to in condition 9: the formula " at least " indicated that the period allowed must be exclusive both of the day of service and of the day of expiry of the notice. Here again, this appeared to be correct: see *Young v Higgon* (1840) 6 M & W 49 at 54, and *In re Railway Sleepers Supply Co* (1885) 29 Ch D 204.

It was thus plain that the notice enclosed with the letter of June 23 1972 was wholly invalid for any purpose. The defendant could not attempt to say that the notice was given under the general law, and not under condition 9 at all. He must be bound by his own solicitor's description of the notice they were serving on his behalf, and otherwise the notice would be a " tricky " notice, and effect would not be given to it in equity at all. In any case, the contractual time for completion was June 19 and the notice was sent only four days thereafter. He (his Lordship) was unaware of any case in which a notice under the general law served such a short time after the expiration of the contractual date for completion had ever been held to be good. It was not merely delay in performing the obligation of completion, but unreasonable delay which under the general law entitled the other party to make time of the essence of the contract. He (Walton J) did not consider four days' delay to have been unreasonable on any view of the case. There was also the point that condition 9 did exist and plainly precluded service of a notice of shorter length under the general law. Setting the notice aside,

A therefore, counsel for Rightside said that this was a simple case in which time never became of the essence of the contract, nor was there ever any such unreasonable delay on the part of Rightside as would amount to repudiative delay on their part. Hence, the contract was still in being when, on July 20 1972, the defendant repudiated it; such repudiation discharged Rightside from any obligation which might otherwise have rested on them to perform any conditions precedent on their part, such as proving their ability and readiness to pay the purchase money; they had finally accepted the repudiation, which had not in the meantime been withdrawn, as it might have been; and accordingly they were now entitled
B to damages for fundamental breach by the defendant of the contract.

Counsel for the defendant said, however, that this was far too simple a way of looking at the matter. He began by submitting that at law, as distinct from the position in equity, time was of the essence. On this basis, his analysis of events was as follows: (a) by their failure to complete on June 19 1972 Rightside had repudiated their obligations under the contract; (b) by the terms of his letter of July 20, the defendant accepted that repudiation, and thus brought the contract to an end; but (c) that this was subject to Rightside being able
C to show that they were entitled to specific performance—presumably as at the date when the letter of July 20 was sent—in which case it would be the defendant and not Rightside who had repudiated the contract. And that involved, said counsel, Rightside showing that either on June 19 or within a reasonable time thereafter, they were ready, willing and able to perform their part of the contract. This did not seem to square with section 41 of the Law of Property Act 1925, which was the same as section 25 of the Judicature Act 1873 and read:

> "Stipulations in a contract, as to time or otherwise, which according to rules of equity are not deemed to be or to have become of the essence of the contract, are also construed and have
> D effect at law in accordance with the same rules."

Nevertheless counsel relied on two cases which he said established his proposition. The first was *Howe v Smith* (1884) 27 Ch D 89, a purchaser's action for specific performance which became one for the return of a deposit. It was found that the purchaser had by persistent delay repudiated the contract. Fry LJ said at p 103:

> The 25th section of the Judicature Act 1873 enacted that stipulations in contracts as to time, which would not before the passing of the Act have been deemed to be of the essence of such contracts in a court of equity, should receive in all courts the same
> E construction and effect as they would theretofore have received in equity. The effect of this clause is, in my opinion, that the purchaser seeking damages is no longer obliged to prove his willingness and readiness to complete on the day named, but may still recover if he can prove such readiness and willingness within a reasonable time after the stipulated day; and the inquiry therefore arises whether the purchaser in the present case could aver and prove such readiness and willingness within a reasonable time.

Fry LJ's words were approved by Lord Atkinson in *Stickney v Keeble* [1915] AC 386 at 404, an action for the return of a deposit by a purchaser who had served a valid notice to complete. After commencement of the action, the vendor had put it out of his own power to give specific performance by
F selling elsewhere, though specific performance was not requested. Lord Parker of Waddington observed that this fact would formerly have been fatal to any claim of the vendor's to relief in equity from the consequences of the repudiation at law, and that section 25 of the Act of 1873 was not to be interpreted as justifying any other result. He (Walton J) thought that statement of Lord Parker's apt in the circumstances of *Stickney v Keeble*, but the doctrine could not be wrenched out of context. As Lord Parker himself had said, section 25 (now section 41) meant that where equity would formerly have disregarded a stipulation as to time and restrained an action at law based on the breach thereof, the courts constituted by the Act of 1873 were to disregard it in like manner. G Applying that principle, could there be any equitable reason why, in a case where, at a time before the purchaser became in unreasonable delay beyond the date fixed for completion—the case here—the vendor repudiated the contract, the purchaser should not simply accept such repudiation as discharging him from all further performance under the contract, including that of preparing for completion, which *ex hypothesi* would then never take place? In such a case equity must simply follow the law, and could not be so esoteric as to place on the purchaser the burden of showing that, *ex hypothesi* at a time after repudiation, he was in a position to complete the contract which it was then quite H certain would never be completed. In his (Walton J's) judgment, in equity as well as at common law, the wrongful repudiation by one party of his obligations under the contract entitled the other to accept such repudiation, and thereby put an end to the contract, and such other was, as a consequence, discharged from performing any conditions precedent which it would otherwise fall on him to discharge. Accordingly, Rightside were never at any time under an obligation to show that they were in a position to perform their part of the contract. Even if this were wrong, it was surely only at the "material" time or times that a plaintiff J must show this. One of those times must have been the date for completion, and it appeared sufficient for the plaintiffs to point to the fact that they had entered into a valid contract for a subsale whose time for completion spanned the time for completion of the contract now in question.

Even this was not the end of the matter, because counsel for Rightside submitted that even at common law, time was not of the essence of the date for completion. On the true construction of the contract here, that was to say, the date for completion was not of the essence. The true inquiry was as to the effect which the parties had agreed a failure to complete on the day fixed for completion was to have. K Was it a fundamental term of the contract, subject only to the saving hand of equity, or was it to be a merely venial breach, not in any sense fundamental? He (his Lordship) thought counsel for Rightside was correct in submitting it was the latter. The provisions of condition 9 were intelligible only on such a basis, for they did no more than provide for the consequences of a fundamental breach after—and only after—the service of such a notice as was thereby envisaged. There never was such a breach by Rightside as would have been equivalent to a fundamental breach of the contract, even apart from the doctrines of equity. It followed that Rightside were entitled to the damages claimed for the defendant's L wrongful fundamental breach of contract by repudiation.

Judgment was entered for Rightside and an inquiry into damages ordered. The plaintiffs were awarded three-quarters of their costs.

COURT OF APPEAL

October 17 1974

(Before Lord Justice RUSSELL, Lord Justice CAIRNS and Sir John PENNYCUICK) M

MOUNTFORD AND ANOTHER v SCOTT

Estates Gazette June 14 1975

(1974) 234 EG 819

Specific performance—Six months' option to purchase house for £10,000 obtained by purchaser's agent for £1—No reason in law or equity why option should not be exercised and contract so constituted specifically enforced—It is the contract of sale that is specifically enforced, not the option agreement

This was an appeal by Mr Calvin Scott, of 49 Enmore Road, London SE25, from a judgment of Brightman J in the Chancery Division on May 24 1973 awarding the respondents, Mr Frederick Mountford and his wife, Mrs Hilda Mountford, of Woodbastwick Road, London SE26, specific performance of a contract of sale of 49 Enmore Road constituted by the exercise on March 29 1972 of an option contained in an agreement of December 12 1971.

Mr H G Narayan (instructed by Suriya & Co) appeared for the appellant, and Mr I H Maxwell (instructed by Hancock & Willis) represented the respondents.

Giving judgment, RUSSELL LJ said that the plaintiffs were nominees of a small property development and building company called H & L Cronk Ltd, and Mr Mountford was the accountant. The defendant, a West Indian housepainter, was totally illiterate, although he spoke and understood English well. His Italian wife could not read English. At a meeting on December 12 1971 with the plaintiffs' agent, the defendant signed an agreement giving the plaintiffs a six months' option to purchase his house for £10,000 in return for £1. The agreement was read over by a friend of the defendant who was present at the meeting. A month later the defendant purported to withdraw the offer, but the plaintiffs' solicitors told him he was not entitled to do so, and two months later, on March 29 1972, the option was exercised. The defendant's first argument was that the option agreement was vitiated by the plaintiffs' agent's representation that he (the defendant) could withdraw from the agreement up to six weeks after signing it. This was a hopeless contention, as the judge had come to the clear conclusion that no such representation had been made. Next it was said that the agreement was an unconscionable bargain. But there was no evidence that £10,000 was not a reasonable price, and there was nothing remotely approaching the circumstances in which equity was prepared to intervene. Finally, the startling proposition that the law would not regard £1 as valuable consideration for the option appeared to be based on a misreading of a Canadian case, *Gilchrist Vending Ltd v Sedley Hotel Ltd* (1967) 66 DLR (2d) 24, which suggested only that possible future obligations avoidable by payment of $1 were illusory as consideration. The position was, therefore, that the option agreement here was valid and effective and that a contract of sale was constituted when the option was exercised. There was no justification for the proposition advanced on behalf of the defendant that the contract so created ought not to be specifically enforced, but that the purchaser should have only been awarded damages. He (his Lordship) would add that in circumstances of the kind the court was asked to order specific performance of the contract of sale, not of the promise in the option agreement not to withdraw the offer for the period in question. Provided the option was valid and for valuable consideration and duly exercised, it was irrelevant to the issue of remedy under the contract of sale that the valuable consideration could be described as a token payment. So also if the option agreement was under seal with no payment, which was what he (his Lordship) took Brightman J to be referring to when discussing a "gratuitous option." While, therefore, he (Russell LJ) agreed that a valid option to purchase constituted an interest in the land affected, he did not think, as the judge appeared to have done, that that fact was necessary to their common conclusion as to the appropriate remedy. For all these reasons, he would dismiss the appeal.

CAIRNS LJ delivered a concurring judgment in which he observed that in so far as there was a difference in approach between the trial judge and Russell LJ, he agreed with Russell LJ. SIR JOHN PENNYCUICK agreed with both the judgments delivered. The appeal was dismissed with costs.

CHANCERY DIVISION

March 18 1974

(Before Mr Justice WALTON)

LAKE v BAYLISS AND ANOTHER

Estates Gazette June 21 1975

(1974) 234 EG 907

A vendor who sells to a third party holds the purchase price as trustee for the original purchaser, who can accordingly trace it and recover it—Dicta concerning the position of a vendor who enters into a contract and then receives better offers—Semble, these should in an appropriate case be passed on

By this interpleader originating summons, Mr Joel Seth Lake, solicitor, of Cardiff, asked the court to determine what he should do with the sum of £50,000 received by him on behalf of his client, Miss Mervyn Victoria Bayliss, on sale of some 3¼ acres of land at Overndale Road, Mangotsfield, Bristol, to a purchaser other than Dr Henry Andrew Mullen, plaintiff in an action against Miss Bayliss for breach of a contract to sell the land in question to him.

Mr P J Millett QC and Mr C I Howells (instructed by Field Fisher & Martineau, agents for Hallinan, Blackburn, Gittings & Hambleton, of Cardiff) appeared for Dr Mullen, and Miss S Cockburn (instructed by Seifert, Sedley & Co) represented Miss Bayliss. The plaintiff took no part in the argument.

Giving judgment, WALTON J said that on December 27 1968 Miss Bayliss entered into a written agreement to sell Dr Mullen certain land at Mangotsfield, Bristol. The consideration was unusual, and it was that, in a sense, which had given rise to the present proceedings. The consideration was primarily to withdraw two writs which Dr Mullen had issued against Miss Bayliss and to assume certain liability in respect of a planning application. Having entered into that agreement, Miss Bayliss sold the land elsewhere, notwithstanding that Dr Mullen had issued a writ on June 13 1969 claiming specific performance of his agreement. There was either a muddle in a proper search by the blameless purchaser, or the purchaser described the land in a slightly different way from that in which it was described in the application for the registration of the estate contract. Miss Bayliss sold the land for approximately £50,000. The proceeds were received by her solicitor, Mr Lake, who, having received the proceeds, began to have qualms about accounting to Miss Bayliss for the money. Ultimately, by the present summons he applied for directions as to what he ought to do with the money. An order was made in the proceedings by which an issue was to be tried "whether the funds . . . should be paid forthwith to the defendant in the issue or should remain in court pending the determination of the action now proceeding in the Cardiff District Registry of the Chancery Division of this court between the parties to the issue, the short title of which is *Mullen v Bayliss*. . . ."

What now ought to be done with the money? Counsel for Dr Mullen urged that although the relief sought was unusual, if one went back to first principles there was really no difficulty, as it was one of the standing doctrines of the court that on a contract for the transfer of property being entered into, the vendor or intending transferor became a trustee for the purchaser or transferee. Admittedly, it was a qualified trusteeship. It did not have all the usual incidents of a bare trusteeship for a cestui que trust, and the most glaring departure from such normal incidents was that the trustee had his own interest, the receipt of the purchase money, to protect and a lien on the property until he had been paid his purchase money in the normal course. Nevertheless counsel said that basically, in regard to the land contracted to be sold or transferred, no matter what might be the case in relation to other collateral matters such as moneys paid

A under a contract of insurance or receipts under derequisitioning procedures, there was no doubt that the property was held as by a trustee. That being the case, if the contract was completed and the vendor wrongfully—as was alleged—sold the property, then the purchaser was entitled to say that what had happened was that the vendor had sold trust property, and he was entitled under those circumstances to follow the trust property and say when it came to the final performance of the contract that he would take the proceeds of sale instead of the property. There was a dearth of authority on the point, but counsel suggested that the reason was that four matters must concur before it was worthwhile for a purchaser to B pursue this remedy. First, the vendor must have sold the property twice over, and that was rare, because most people honourably fulfilled their obligations; secondly, for some reason the registration of an estate contract by the purchaser must either not have taken place, or if it had taken place it must have been for some rare reason, such as an error in searching, of no effect; thirdly—and this was most vital—the purchaser must be in time to intercept the purchase money, to trace the purchase money, for once it had been paid into an overdrawn bank account or something of that nature, which was the only likely occasion on which these prior circumstances would concur, it was too late to follow it; and C lastly, the purchaser must prefer to pursue the tracing of the purchase money rather than to pursue his remedy in damages. Where the vendor was solvent there would be little in it, and again, where the vendor had a lien on the trust property for a purchase price which would normally be roughly the amount of the sale price, there would not be much in it, and probably it would be hardly worthwhile in these days seeking to trace the proceeds of sale rather than simply taking the remedy in damages. In the present case, however, the consideration to be provided by Dr Mullen was rather special, and if he could not trace the purchase money in the way D suggested, he would be forced into a completely different contract from the one he intended to enter into, because he would be left with the rights which he was going to give up in exchange for the land, and he would get nothing but damages. That would be a very extraordinary result as far as he was concerned.

Authority was silent on the point at issue save for one case, a decision of Lord Eldon LC in *Daniels v Davison* (1809) 16 Ves Jun 249, where he said at 254:

> "My judgment on that point [after dealing with other points in the case] lays out of consideration the question, whether, taking Cole not to be affected with notice, Davison, the vendor, is to E be considered in equity as holding the money, derived from the second purchase, viz, the difference between the prices, in trust for the person, to whom he had first agreed to sell the estate. The estate by the first contract becoming the property of the vendee, the effect is, that the vendor was seised as a trustee for him; and the question then would be, whether the vendor should be permitted to sell for his own advantage the estate, of which he was so seised in trust; or should not be considered as selling it for the benefit of that person, for whom by the first agreement he became trustee; and therefore liable to account."

It was true that Lord Eldon put it in the form of a query, but his words were to be understood in the sense that he F would have given, if pressed, as answer to that query, that that indeed represented the law. Counsel for Dr Mullen had also found out that Sir George Jessel, when Solicitor-General, arguing the leading case of *Shaw v Foster* (1872) LR 5 HL 321 in the House of Lords, had stated flatly at 327,

> "*Daniels v Davison* shows that after a contract for the sale of an estate, if the vendor sells to another person for valuable consideration, he is accountable for the money as a trust."

He (his Lordship) considered that that was in line with authority and represented the law.

Counsel for Miss Bayliss took a number of points on her G behalf, but her main point was that in the sort of circumstances where the vendor entered into one contract and then resold the property elsewhere, the purchaser was restricted to damages. The trusteeship, she said, was as if it never had been, and the purchaser must accept that accordingly. He (Walton J) thought it would be *pessimi exempli* if a vendor was entitled to shed the character of a trustee by a wholly wrongful act on his or her part. Once one had undertaken the rôle of trustee it was a rôle which, unless discharged by some external circumstance, one must carry out to the bitter end if so required by the other party to the contract. The vendor could not be heard to say that because of her wrongful H act in reselling the property she never was a trustee. She remained a trustee right down to the moment of resale, and accordingly was bound to hold the purchase price as trust property to transfer to the purchaser on the purchaser completing the obligations on the purchaser's part. Counsel for Miss Bayliss had raised a very good point, which was how far this concept of trusteeship went. Did it follow, for example, if counsel for Dr Mullen were right, that if a vendor obtained better offers for the property after having entered into a contract for sale, he would be under an obligation to pass them on? If so, said counsel, that would be an extraordinary result. J He (his Lordship) did not think it was such an extraordinary result, but he also did not think that that sort of question—good question though it was—could be asked *in vacuo*. One must have a concrete case in which that sort of thing happened before one knew what the duty of the vendor was. In most cases, one knew that the purchaser was purchasing the property for his own occupation. Under those circumstances, unless the subsequent offer was astronomical, there would be no real reason for thinking that it would be of any interest to the purchaser. In other cases the vendor might well know that the purchaser was buying for speculation, and although he (his Lordship) did not say that it would K necessarily be the case he could see that under certain circumstances it might very well be the case that the vendor ought to pass the offers on.

Another point taken on Miss Bayliss's behalf was that by the pleadings in the action for specific performance (which had been referred to as "the Cardiff action," because it was commenced in the Cardiff District Registry) Dr Mullen ought not to be allowed to amend the pleadings in that action, because the amendment relating to the following of the proceeds of sale would be something new and different which had happened subsequent to the issue of the writ, and there- L fore a new writ and a new proceeding would be called for. He (Walton J) would be very sorry to think that the Rules of the Supreme Court were so ossified and stylised that it was impossible to pursue an action when it was changed somewhat into a protean shape by reason of the conduct of the defendant. But in any event it seemed that there was an implied term, as it were, in the statement of claim following on the allegation of the written agreement, "and thereby the defendant became a trustee of the property for the plaintiff"; and if one read that in, which was implied anyway, the transition, if the defendant had wrongfully disposed of the trust property in exchange for much fine gold which Dr M Mullen sought to trace, became a simple and clearly allowable amendment. Of course, the Cardiff action was still being very bitterly fought, and at the end of the day there might be no such contract as alleged by Dr Mullen, but on the basis that there was, the money should remain in court to abide the outcome of the Cardiff action. There could be no grounds for an order that Miss Bayliss pay the money to Dr Mullen, because in fact she had not got it. It was a trust fund under the jurisdiction of the court, and all Dr Mullen would need was a declaration that he was entitled to it.

LANDS TRIBUNAL

CAPITAL GAINS TAX

A MARSON (H M INSPECTOR OF TAXES) v HASLER

(Ref/79/1974)

(R C WALMSLEY FRICS)

March 4 1975

Estates Gazette March 29 1975

(1975) 233 EG 1183

First capital gains tax reference to tribunal—Newsagent's, tobacconist's etc shop with living accommodation in Enfield —Value of freehold subject to 21-year lease expiring in 1972—Property offered to tenant for £10,000 in 1965, and B after negotiations sale completed for that sum 2½ years later—Taxpayer's valuer's use of differential rate of return "unacceptable"—Tribunal's comments on opinion evidence and analysis of comparables—Market value in April 1965 held to be £8,000

Giving his decision, MR WALMSLEY said: This is the first reference to the Lands Tribunal arising out of the capital gains tax provisions of the Finance Act 1965. The parties are Mr J F Marson, HM Inspector of Taxes, Chelmsford ("HM Inspector of Taxes" or "the Revenue"), and Mr F W Hasler ("the taxpayer"). The event which has given C rise to these proceedings is the disposal, on September 29 1967, of a property, 602 Hertford Road, Enfield, Middlesex.

Relevant Statutory Provisions

The relevant provisions of the statute are as follows:

Section 19 (1)

Tax shall be charged in accordance with this Act in respect of capital gains, that is to say chargeable gains computed in accordance with this Act and accruing to a person on the disposal of assets.

Section 22 (1)

All forms of property shall be assets for the purposes of this D Part [Part III] of this Act.

Paragraph 25, Schedule 6

If the person making a disposal so elects (and the taxpayer has, in this case, so elected), . . . it shall be assumed . . . that the assets disposed of . . . being assets which were in the ownership of the said person on April 6 1965, were on that date sold, and immediately re-acquired, by him at their market value on the said April 6 1965.

Section 44 (1)

. . . in this Part of this Act "market value" in relation to any assets means the price which those assets might reasonably be expected to fetch on a sale in the open market.

E *Section 44 (6)—now re-enacted as section 47 (1) of the Taxes Management Act 1970.*

If and so far as the question in dispute on any appeal against an assessment to tax (whether capital gains tax or corporation tax) . . . is a question of the value of any land . . . then . . . the question shall be determined on a reference to the Lands Tribunal.

The subject property, which belonged to the taxpayer's wife, comprises a shop with living accommodation and premises, on the west side of Hertford Road, Enfield, just south of the junction with Turkey Street. The trade carried on was (and still is) that of newsagent, tobacconist, confec- F tioner and stationer. The interest to be valued is the fee simple subject to a lease for 21 years from March 25 1951 at a rent of £150 pa exclusive for the first 7 years, £160 pa for the next 7 years and £180 pa for the last 7 years. The lessee was responsible for all repairs and insurance of the property.

In July 1965 discussions took place between the lessee, a Mrs Groom, and her landlord, Mrs Hasler, as to the former purchasing the freehold. Mrs Hasler's solicitors, Hill & Abbott, of Chelmsford, sought valuation advice. They wrote to a firm of chartered surveyors, W Goodchild & Co, of Enfield, asking for a report and valuation. On July 30 1965 G Goodchilds submitted a detailed report ("the 1965 report"), the concluding sentence of which was:

"In our opinion the current market value of this freehold property, subject to the lease to Mrs V L Groom expiring on March 25 1972, at the present rent of £180 per annum exclusive, is £10,000 subject to there being no unusual charges or restrictions."

The property was accordingly offered to the tenant in August 1965 at £10,000. Against this figure Mrs Groom offered £6,000. In 1966 and again in 1967 Mrs Groom expressed her continued willingness to purchase; and in May H 1967 her solicitors came forward with an increased offer of £9,000. On receiving this, Hill & Abbott wrote to Goodchilds for advice as to the 1967 value of the freehold reversion, "as it may well be that Mrs Hasler should receive more than the figure of £10,000 originally quoted in 1965." After making a further inspection Goodchilds replied, on June 13 1967, "if the lessee is prepared to make an offer of £10,000 for the property then we would still advise your client to accept." That advice was adopted, and three months later the sale to the tenant was completed at £10,000.

In the above circumstances the taxpayer contends that the value of 602 Hertford Road in April 1965 was £10,000, the J same figure as had been professionally advised in July 1965 and as had been realised some 2½ years later. The Revenue contend that the value in April 1965 was no more than £8,000. The question for determination by the tribunal is whether the market value of the subject property on April 6 1965 was £8,000 or £10,000, or some other figure and if so what.

At the hearing Mr Raymond Sears (instructed by the Solicitor of Inland Revenue) appeared for the Revenue and called evidence from Mr R S Stubbs ARICS. Mr Malcolm Spence (instructed by Hill & Abbott, of Chelmsford) appeared for the taxpayer and called evidence from Mr P A Nicholas K FRICS. Mr Stubbs has had over 26 years' professional experience, the last 20 of which have been in the district valuer's office at Enfield, where he holds the position of senior valuer. Mr Nicholas has been retired since September 1967, prior to which date he had been a partner in W Goodchild & Co for nearly 20 years in charge of their Enfield office: the 1965 report and the subsequent correspondence relating to 602 Hertford Road had been his personal responsibility.

A The respective valuations, slightly adjusted for purposes of comparison, were as follows:

		By R S Stubbs ARICS for the Revenue			By P A Nicholas FRICS for the taxpayer	
Unexpired term						
Net rent		£180			£180	
Y.P. for 7 years	@ 5%	5.79	£1,042	@ 5%	5.79	£1,042
Reversion, on March 25 1972						
Estimated net rental		£1,125			£1,250	
Y.P. in perpetuity	@ 9%	11.11		@ 8%	12.50	
		£12,500			£15,625	
Deferred for 7 years	@ 9%	0.547	6,840	@ 7%	0.623	9,734
B			£7,882			£10,776
			say £8,000			say £10,000

As can be seen, there are three points on which the valuers disagree and which I will deal with in turn:

(a) As to the net rental value.

(b) As to the rate per cent for capitalising that estimated rent.

(c) As to whether the rate per cent for deferring the capital value should be different from that adopted for capitalising the estimated rent.

C

Net Rental Value

Mr Nicholas' figure of £1,250 pa estimated rental value was a spot amount based essentially on his experience, but bearing in mind also the rents agreed for certain shops (including 491, 493 and 497 Hertford Road) in a parade newly built in 1965 and let on 21-year leases.

Mr Stubbs' figure of £1,125 pa had been arrived at after analysing *inter alia* the sale prices and lease rents involved in transactions relating to three nearby shops in Hertford Road, Nos 590, 592 and 598. These shops had been bought by a single purchaser in 1964-65 and were resold by him in 1967. Brief details of the transactions are as follows:

D

No 590, bought Aug 1964 for	£3,500		
No 592, bought Nov 1964 for	£6,500		
	£10,000	Resold Sept 1967 for	£9,500
No 598, bought Jan 1965 for	£5,000	Resold June 1967 for	£6,950
	£15,000		£16,450

No 590 had been bought with vacant possession, but was resold subject to a seven-year lease from Sept 1964 @ £450 pa.

No 592 was bought, and resold, subject to a seven-year lease from Sept 1963; three years @ £450 pa and four years @ £500 pa.

No 598 had been bought with vacant possession, but was resold subject to a seven-year lease from Jan 1965 @ £800 pa.

E Mr Stubbs' devaluations were all in terms of a Zone A of 15 ft depth and a Zone B of 25 ft for shop space, with appropriate allowances for residential accommodation and for other ancillaries. From his devaluations Mr Stubbs had derived figures which he had then applied to the component floor areas of the subject property.

Having studied Mr Stubbs' analyses, and having inspected, I accept his opinion that the net rental value of No 602 in April 1965 cannot be put any higher than £1,125 pa.

F

Rate per cent for Capitalising the Estimated Net Rent in Reversion

Mr Nicholas' rate of 8 per cent was based solely on his experience. Mr Stubbs' figure of 9 per cent was based on his experience applied to his analyses of the sale prices of 590, 592 and 598 Hertford Road as above, and also of the prices realised for two other nearby shops, 537 and 613 Hertford Road (referred to again later). I find, on studying these analyses, nothing to justify as low a rate of return as 8 per cent, and ample evidence to support a rate of at least 9 per cent. I note that the transactions relating to 590, 592 and 598 Hertford Road point to the market having risen by about 10 per cent between August 1964 and September 1967.

G

Differential Rate per cent for Deferring the Capital Value

When questioned about his reason for capitalising his estimated rental of £1,250 pa on the 8 per cent table but deferring the resultant capital value on the 7 per cent table, Mr Nicholas answered: "I always use this method when the reversion of a shop exceeds five years and when I think the rent is bound to increase." (In this context, "increase" refers to an expectation that the rental value at the date when the current lease expires will prove to be higher than its rental value as estimated at the date of valuation.) There were good grounds for Mr Nicholas' expectations of such an "increase." H By 1965 the London Borough of Enfield had put in hand very considerable development on both sides of Hertford Road, comprising over 700 residential units. Although building works had apparently not progressed far at the date of the 1965 report, the local authority's intentions were well known, and the development was duly carried out over the succeeding years.

The above answer by Mr Nicholas is consistent with the following comment included by him in his firm's 1965 report:

> "In our opinion the trading position of the shops in this part of the Hertford Road will improve in the next few years and the general nature of this trading centre will tend to change J from a good second-class position to an almost first-class shopping position. These changes will result in higher rents being obtained for the property and higher capital values."

Although I recognise that there is some logic in Mr Nicholas' use of differential rates per cent as explained above, his valuation approach is not the normally-accepted method of valuing a reversion. Nevertheless, if Mr Nicholas had produced analyses of his comparables, devaluing these on the same basis, then the multipliers used in his valuation might well have proved helpful. But Mr Nicholas had not prepared analyses of his comparables either on this basis or on any K other basis, and I therefore find his use of a differential rate of return unacceptable. A similar valuation approach was put before the tribunal in *Selkirk and Another v Birmingham Corporation* (1969) 210 EG 995 and was similarly rejected.

Looking at the component parts of the two valuations, the figures put forward by Mr Stubbs seem to be supported at every stage. By contrast the figures put forward by Mr Nicholas (where they differ) appear to be wholly unsupported.

Two Schools of Valuation

In upholding, as I must, the valuation put forward for the L Revenue, it may be helpful to add a general comment. Mr Nicholas strongly criticised Mr Stubbs' reliance on analyses, and described the process of analysis as being "a mere juggling with figures." He regarded Mr Stubbs as belonging to the "Analytical School" of surveyors, and he readily accepted that he himself could be regarded as belonging to the "Forty-years-man-and-boy School." The tribunal has every respect for able and practical surveyors who belong to the latter "school," but the fact should be recognised that when a member of this "school" finds himself unable to agree values with an equally able and practical member of the "Analytical School," then on a reference to the Lands M Tribunal the latter surveyor is apt to have the easier passage.

The reason for this is quite simple. The tribunal reaches its decisions on the evidence presented, and although it does, of course, draw on its own skill, that skill is applied not to the valuation of the subject property but to weighing the evidence given—a process conveniently described as "valuing the valuations" (and which may involve also "valuing the valuers"). Opinion evidence, if it is to be certain of carrying weight, needs not only to be based on factual evidence but also to be *demonstrably* so based. Factual evidence bearing on the value of any land commonly takes the form of comparables; and the purpose of analysing the comparables

A is to enable *unlike* features to be identified and distinguished, and to enable *like* features to be compared. In the present case the procedure adopted by the taxpayer's surveyor was to produce a factual schedule of transactions *simpliciter*. The surveyor for the Revenue, for his part, as well as producing analyses of his own comparables had also analysed the comparables put forward for the taxpayer. On studying the analyses as a whole I find that the taxpayer's set of comparables provide support not for his own case but for the Revenue's.

The market value of the freehold of 602 Hertford Road on April 6 1965 is determined at the sum of £8,000.

B This disposes of the reference, but there are three further matters to which I may refer:

1. It is a characteristic of a freehold reversion that—all factors remaining equal—its value may be expected progressively to increase by mere effluxion of time. This factor, even apart from the apparent rise in the market between 1964 and 1967 already noted, makes it unlikely that the freehold interest subject to the same lease would have the same value both as at 1965 and as at 1967.

CGT Agreements as Comparables

C 2. Nos 537 and 613 Hertford Road have been mentioned. These two shops were sold in 1967-68 and they fell to be valued for capital gains tax. No 537 was sold with vacant possession in May 1967 for £11,500; and the district valuer reached agreement with surveyors that its value at April 1965 had been £10,000. No 613—in a poorer trading position than either No 537 or No 602—was sold with vacant possession in October 1968 for £8,500; and the district valuer reached agreement with surveyors that its value at April 1965 had been £8,000.

I should interpolate here that the above two "comparables" were brought forward not by Mr Stubbs but by Mr Nicholas. D Mr Stubbs told me that he had been relying on market transactions only; he regarded capital gains tax agreements as being secondary evidence only; such agreements were in any case confidential and could not be disclosed without the permission of the taxpayer (which permission, in these two cases, he had sought and obtained after receiving the taxpayer's schedule of comparables).

As regards the weight attaching to this kind of evidence, I agree with Mr Stubbs that a *figure* of 1965 value agreed for capital gains tax purposes cannot be regarded as primary evidence; but I find the above information in respect of 537 and 613 to be helpful, because it shows some consensus of E professional opinion that shop values in Hertford Road were indeed increasing after 1965.

3. The third matter is one which I refer to out of courtesy to the taxpayer, because it was stressed by Mr Spence in his closing speech. The taxpayer, said Mr Spence, considered it a scandal that these "oppressive proceedings" had been insisted on by the Revenue: it had been quite wrong of the district valuer not to have accepted the valuation made by Mr Nicholas in 1965; the amount of tax involved was very small, certainly less than £600, yet the taxpayer had been faced either with incurring the risk of costs or with paying F too much tax.

The above criticism must be looked at against the background of the events leading up to the reference. What apparently happened was as follows.

In or about June 1969 the particulars of the sale of 602 Hertford Road in 1967 at £10,000 were referred to the district valuer by HM Inspector of Taxes. The district valuer, having in due course made his valuation, wrote to the taxpayer's accountants on July 3 1969 inviting their agreement to a figure of £7,500 for capital gains tax purposes. The accountants in reply referred the district valuer to their client's solicitors, Hill & Abbott. That firm, having been G contacted, referred the district valuer on July 31 1969 to W Goodchild & Co and informed the district valuer of the report and valuation made by that firm in 1965. On August 26 1969 the district valuer wrote to Goodchilds and invited *their* agreement to his £7,500 valuation figure. In response Goodchilds sent to the district valuer a copy of their firm's 1965 report. Negotiations thereafter with Goodchilds were protracted, being conducted first with one of their partners Mr Rudd and later with another partner Mr Thomas—both these partners being members of the Royal Institution of Chartered Surveyors. Agreement was eventually reached between the district valuer and Goodchilds in July 1971 at a H figure of £8,000 (the same figure as the Revenue contended for in the present proceedings). On July 22 1971 the district valuer reported this agreement to HM Inspector of Taxes. There the matter rested for nearly six months, but on January 4 1972 the district valuer was informed by HM Inspector of Taxes that the figure agreed was not acceptable to the taxpayer. Mr Spence told me that, in this matter, the taxpayer had been acting throughout on advice.

Mr Nicholas' evidence to me was that the solicitors had been unhappy about the £8,000 figure; they had ascertained that Goodchilds were unwilling to reopen the matter ("Mr Thomas washed his hands of the case"); on December 1 J 1971 Mr Nicholas had been asked if he, himself, would give evidence, and to this he had agreed, albeit reluctantly; subsequently he had called on his former office to collect his own file covering the period 1965 to 1967, but he had not been in touch in any way with either Mr Rudd or Mr Thomas on the matter.

Not only, it appears, had Mr Nicholas refrained from making any contact with the present partners of his former firm; he had refrained also from making any contact with the district valuer. Nor had the district valuer at any time received a request from Hill & Abbott to make contact with Mr Nicholas. K

The circumstances as described demonstrate, in my opinion, that the taxpayer's criticism of the conduct of the Revenue is wholly groundless. It is clear that, under the law and in order to reach finality, the Revenue had no alternative but to refer the matter to the Lands Tribunal for determination.

The taxpayer will pay the costs of HM Inspector of Taxes, such costs failing agreement to be taxed by the Registrar of the Lands Tribunal on Scale 4 of the County Court Scales of Costs, with certificate for increased fees to expert witness.

LANDS TRIBUNAL

COMPENSATION

A W & S (LONG EATON) LTD v
DERBYSHIRE COUNTY COUNCIL

(Ref/76/1974)

(DOUGLAS FRANK QC (President) and R C WALMSLEY FRICS)

November 22 1974

Estates Gazette January 25 1975

(1974) 233 EG 403

Purchase notice—Residential development land at Mickleover, Derby, required for school—Entry not taken by acquiring authority nor possession given—Date for assessment of com-
B **pensation held to be date of tribunal hearing in October 1974 —Fall in value since acceptance of purchase notice in August 1973 held to be 20 per cent**

This reference concerns the compensation payable by the Derbyshire County Council (as successors to the Derby Borough Council) on the acquisition of 1.445 acres of land lying to the north-east of Matthew Way, off Haven Baulk Avenue, Mickleover, Derby. The claimants are W & S (Long Eaton) Ltd and the interest being acquired is the freehold in possession.

The reference land was, until the end of 1972, part of a block of about 23 acres intended for residential development. The borough council, however, had given an indication that the 1.445
C acres was needed for school purposes, and accordingly it was excluded when a planning application was lodged in respect of the remainder of the block. Planning permission for the 21½ acres was granted on September 26 1972 and this area is currently in course of development by a company, David Charles Ltd.

To put the matter of the reference land on a proper footing, a separate planning application was lodged on March 19 1973 in respect of the 1.445 acres; and permission was duly refused on May 24 1973, on the ground that "the site is required for the provision of a school." On July 17 1973 the claimants served on the town clerk of Derby Corporation a purchase notice under
D section 180 of the Town and Country Planning Act 1971, requiring the corporation to purchase their interest in the land. On August 14 1973 the town clerk wrote to the claimants' solicitors accepting that notice in the following terms:

> ". . . the council are prepared to accept the purchase notice of 17 July submitted by you to the council on behalf of the above company. You may take this letter as formal notice of acceptance. I enclose claim form in duplicate. Please return one copy, duly completed, to the Director of Estates and Development who will then commence negotiations with you."

Negotiations proved abortive, and on March 22 1974 the claimants lodged a reference with the Lands Tribunal to deter-
E mine:

> "The value of the land if sold in the open market by a willing seller, as at the date of determination of the reference or (if possession should be taken by the corporation before the reference is determined) at the date of the taking of possession by the corporation."

The hearing before us was opened on October 22 1974. Mr M Burke-Gaffney (instructed by S J B Thomas & Co, of Nottingham) appeared for the claimants and called evidence from Mr S H Moss FSVA [of Nelson H Sutton & Co, of Long Eaton]. Mr C J Chapman, solicitor, appeared for the acquiring authority and called evidence from the county estates officer, Mr D E F
Birley ARICS.

"Important and Difficult Question of Law"

Before we come to the evidence it is necessary to determine an important and difficult question of law, namely the date by reference to which compensation is to be assessed. The acquiring authority have not entered upon the land, nor has possession of the land been given. Mr Chapman, relying on *Birmingham Corporation v West Midlands Baptist (Trust) Association (Inc)* [1970] AC 874, says that the date of the hearing by the Lands Tribunal is the appropriate date. Mr Burke-Gaffney, on the other hand, argues that in so far as the speeches in the G
House of Lords in the *Birmingham* case referred to rule (2) of section 5 of the Land Compensation Act 1961, they were *obiter*, and that in any event they were no authority for the proposition that, in cases where entry has not been made, values must necessarily be ascertained as at the time of the hearing.

This question, as to the date of assessment, is not an academic one in the present case, because (as will be seen) it is common ground between the surveyors that there had been a marked fall in values between August 1973 and October 1974.

The compensation in the *Birmingham* case fell to be assessed by reference to rule (5) of section 5 of the Act of 1961, that is to say the cost of equivalent reinstatement. In that case the H
earliest date on which rebuilding might reasonably have been expected to be begun was some 14 years after the date of the notice to treat, and it was manifestly unjust that the cost of equivalent reinstatement should be ascertained by reference to prices so long out of date. The argument in favour of the notice to treat date rested upon what had commonly been assumed to be law in rule (2) cases.

"Birmingham" Case Opinions on Rule (2) Followed

Even if we were to accept Mr Burke-Gaffney's argument that the opinion of their Lordships, as to the irrelevance of the notice to treat date in rule (2) cases, is *obiter*, that opinion is of such persuasive authority and was expressed so decisively and unani- J
mously that we feel bound to accept it. Yet that is not the end of the matter, for it does not necessarily follow that the date of the hearing before the Lands Tribunal is the correct date in all cases; that much is clear from their Lordships' speeches.

It seems to us that we should conclude from the *Birmingham* case that there are at least three different events, the date of occurrence of which may determine the date of assessment. They are:

(1) entry by the acquiring authority;

(2) the making of an enforceable contract; and

(3) the hearing by the Lands Tribunal. K

In the instant case, as we have indicated, the first event has not occurred. Has the second event, the making of an enforceable contract occurred? Section 180 of the Act of 1971 enables an owner to serve a purchase notice, ie a notice requiring the council of the area concerned to purchase his interest. Under section 181 the council must serve a notice stating that either they are willing to comply with the purchase notice or they are not willing. In the former event the council are deemed to be authorised to acquire the interest compulsorily in accordance with the relevant provisions and to have served a notice to treat

A (which by section 208 cannot be withdrawn) on the date of service of the council's notice. The " relevant provisions " include by reference the usual means for the ascertainment of the amount of the purchase money.

In our view the two notices, coupled with the statutory procedures for determining the price, do constitute a contract for sale. But is the contract, at that stage, an enforceable one? We think not. It is well settled that the court does not order specific performance of a contract until the price is fixed. Nor do we find any authority in the *Birmingham* case for adopting the date of a contract unless that contract has become enforceable: see for example the speech of Lord Reid at page 899 and that of
B Lord Morris at page 907.

" Hesitation and Reluctance "

Accordingly, we find that the correct date by reference to which compensation is to be assessed in this case is the date of the hearing before us. We have come to this conclusion with some hesitation and reluctance, because there is little doubt that the parties assumed, at the time when they served their notices, that the price payable would fall to be ascertained by reference to values then prevailing.

Turning now to the evidence, both Mr Moss and Mr Birley were in the difficulty that they had no knowledge of any sales of residential land in Derby later than 1973. Their valuations
C had accordingly been prepared on the evidence available in 1973, and they were able to offer only *ipse dixit* opinions as to the subsequent drop in the level of value by October 1974.

What we propose, therefore, is to consider their valuations as prepared and put forward, and to make a finding as to the amount of compensation which would have been payable if we had decided that the relevant date for our purpose was the date in August 1973 when the claimants' purchase notice was accepted. That figure will then be adjusted, on the further evidence given by the surveyors, to a value as at the date of the hearing.

The valuation put forward by Mr Moss for the claimants may be summarised as follows:

D

16 plots for detached houses at £4,500 a plot ...	£72,000
8 plots for semi-detached houses at £4,200 a plot ...	33,600
	£105,600
Less cost of road (including sewers) 360 ft run at £20	7,200
	£98,400

During the course of the hearing it became agreed that, although some particular purchaser might well have in mind the development of the reference land with differing-sized plots as above, an equally reasonable assumption would be an intended development with 22 equal-sized plots suitable for the erection
E of " Aquarius " detached houses, this being the typical house on the adjoining land under development by David Charles Ltd. We intend therefore, for convenience, to regard the valuation for the claimants as representing:

1.445 acres (= 22 plots at £4,800)	£105,600
Less cost of road and sewers	7,200
	£98,400

The valuation put forward by Mr Birley for the acquiring authority was quite straightforward:

1.445 acres at £25,000 per acre	£36,125

F

The disparity between the two valuations, one at £98,400 and the other at £36,125, is remarkable, bearing in mind that the subject-matter of the reference is wholly uncomplicated. What is being valued is the freehold interest in a rectangular plot of land measuring about 200 ft by rather more than 300 ft, which is agreed to be ripe for immediate residential development and requiring only the provision of an internal road at a cost which is accepted to be £7,200.

Comparables Analysed

Mr Moss relied for his opinion of value on two comparables, which we shall call " A " and " B." Mr Birley relied on two other comparables, which we shall call " C " and " D." It is G necessary to consider each of these comparables in some detail, because they provide between them the only available evidence as to what was the value of the reference land in August 1973.

Comparable A comprises 4,660 sq yds of land off Haven Baulk Avenue, Mickleover, sold by auction on May 29 1973 for £31,000 with outline planning permission for nine houses, and ripe for development. Some 425 ft of roadworks was required for its development (some over an existing sub-standard private road), the cost of which was estimated by Mr Moss, and was accepted by Mr Birley, to be £7,650. This comparable may be analysed thus:

H

0.963-acre (= 9 plots at £3,450)	£31,000
Add cost of road and sewers	7,650
Cost of land, when serviced (= 9 plots at £4,300) ...	£38,650

Comparable B comprises 1.11 acres of land in Farley Road, off Burton Road, Derby, sold by auction on November 28 1972 for £56,000 with outline permission for 12 units and ripe for development. Some 365 ft of roadworks was required for its development, the cost of which was estimated by Mr Moss, and was accepted by Mr Birley, to be £7,300. This comparable may be analysed thus:

J

1.11 acres (= 12 plots at £4,675)	£56,000
Add cost of road and sewers	7,300
Cost of land, when serviced (= 12 plots at £5,275) ...	£63,300

Comparable C comprises 11.2 acres of land off Haven Baulk Avenue, Mickleover, purchased by agreement by Derby Borough Council for educational purposes in October 1973 for £105,000. We find that 3.2 acres of this comparable had agricultural value only, and that the other 8 acres was residential development land. Prior to starting this development, however, a purchaser would have had to acquire from adjoining owner(s) access K facilities for roadworks and sewer construction. The parties agreed that one-third of the increase in value, consequent on the securing of the necessary easements, would have had to be set aside for their acquisition. Mr Birley put in an analysis of the sale price on this basis, but his analysis was challenged by Mr Moss in a number of respects. In the light of the opinions expressed, and having made our own inspection, we consider it likely that a reasonable analysis of this comparable would be:

8 acres residential land (= £26,750 acre)		£214,000
Less cost of constructing off-site access and sewer connection		3,100
Value with benefit of easements for access and drainage		£210,900
Less value as agricultural land, 8 acres at £400 acre		3,200
Increase in value due to acquisition of easements ...		£207,700
Allow for the purchase of the easements, one-third		69,200
		£138,500
	Defer 3 years at 10%	0.75
	Say	£103,700
3.2 acres agricultural land at £400 acre	Say	1,300
	Purchase price	£105,000

L

M

Comparable D comprises two plots of land, one 6.193 acres and the other 3.83 acres, off Sinfin Avenue, between Allenton and Chellaston, Derby, purchased by agreement by Derby Borough Council for educational purposes in June 1973 for £115,000. The 6.19-acre plot is suitable for residential development, but it includes about half an acre which is sterilised by the presence of overhead high-tension electricity cables. The 3.83-acre plot is unsuitable for development. As with Comparable C Mr Birley put in an analysis of the sale price, but that calculation

	Reference Land		Comparables			
	Claimants (per Mr. Moss)	Authority (per Mr. Birley)	A	B	C	D
Per house plot						
Without internal roads or sewers	£4,475	£1,650	£3,450	£4,675	£1,675	£1,425
Cost of providing above services	325	325	850	600	—	—
Plot, serviced	£4,800	£1,975	£4,300	£5,275	£ —	£ —
Assumed number of plots	22	22	9	12	128	91
Average gross plot size	300 sq yds.	300 sq. yds.	530 sq. yds.	470 sq. yds.	300 sq. yds.	300 sq. yds.
Per acre						
Without internal roads or sewers	£6,800	£25,000	£32,200	£50,450	£26,750	£22,750
Cost of providing above services	£5,000	£5,000	£7,800	£6,550	—	—
Land, serviced	£73,000	£30,000	£40,000	£57,000	£ —	£ —
Net area	1.445 acres	1.445 acres	0.963 acre	1.11 acres	8 acres	5.7 acres

was challenged by Mr Moss. On the evidence, and after inspection, we consider that a reasonable analysis of Comparable D would be:

Plot of 6.193 acres		
5.70 acres residential land (= £22,750 acre) ...		£129,700
0.49-acre sterilised by high-tension cables		Nil
		£129,700
Less cost of sewage pumping station		8,000
		£121,700
Defer 1 year at 10%		0.91
		£110,750
Plot of 3.83 acres		
3 acres at playing field value £1,000 acre	£3,000	
Less for diversion of distribution cables	1,000	
	£2,000	
Defer 18 months at 10%	0.87	
Say	£1,750	
0.83-acre with derelict cottage, farmhouse and farmbuildings	2,500	
		4,250
Purchase price		£115,000

In order to apply the above analyses of the comparables it is necessary to express all four in terms both of "£s per house-plot" as used by Mr Moss, and in terms of "£s per acre" as used by Mr Birley. As regards Comparables C and D, we are concerned only with the values attributed to the areas of residential land. The table opposite sets out the essential details. It will be seen that, for ease of comparison, we have attributed to Mr Birley as well as to Mr Moss, the unchallenged figures of cost for the provision of internal roads on the reference land and on Comparables A and B.

Council's Comparables "Of No Assistance"

We have, with reluctance, come to the conclusion that Comparables C and D are of no assistance. This reluctance stems from our recognition that the areas of residential land within these two comparables are physically more closely similar to the reference land than are Comparables A and B; Comparable C, indeed, shares a common boundary with the reference land and is suitable for development at a similar density of 16 houses to the acre (as is also Comparable D). Three considerations, however, combine to render Comparables C and D unhelpful.

First, because Mr Birley was not in a position to speak at first hand to these, his own comparables: the negotiations on behalf of the purchasing authority had been handled by the borough estates officer, now retired; Mr Birley, as county estates officer, had simply inherited the Derby Corporation files and was not fully conversant with background matters, including the planning circumstances.

Secondly, because the prices per acre (and consequently the prices per house-plot) shown for these two comparables in the above table cannot be regarded as very dependable; they emerge from contentious analyses which include significant variables, such as periods of deferment.

Thirdly, because we have no evidence as to the cost of servicing Comparables C and D with internal roads and sewers; prospective purchasers of the reference land would, in our opinion, think primarily in terms of £s per house plot, fully serviced, when deciding on their bids, and would use Mr Birley's test of £s per acre, prior to the provision of internal roads and sewers, only as a check on their primary figure. We accept that the valuation put forward by Mr Birley is supported by the figures shown in the above table for his two Comparables C and D, but the test of correctness lies elsewhere—in the figures for the other two Comparables A and B, which have the added advantage of being open-market transactions.

As to Comparable A, this is a site quite near the reference land and of much the same quality, but not so economical in shape.

Comparable B seems to us a special site; it is in a high-class residential locality and has an exceptional view, and yet is within 20 minutes' walk of the town-centre. Although this site was sold with planning permission for 12 units, it has been developed with nine superior detached houses. Mr Birley told us that land in Derby was often worth more per acre for development in larger house-plots than for houses on smaller plots. This opinion was not shared, however, by Mr Moss, who thought there was no difference at all—"the one balances out the other." For our part we are assuming that, for residential land in Derby in 1973, much the same price per acre would be obtained for residential land, whether the purchaser intended a development with modest houses at, say, 16 to the acre or with bigger houses at, say, 10 to the acre.

It is clear from the table that Mr Birley's figures are well below the level of value demonstrated by the open-market sales A and B. But it is equally clear that Mr Moss's figures are well above that level, because plot values taken from Comparables A and B cannot be applied to the reference land without first making adjustments for the difference in plot size. If the plots on Comparables A and B had averaged only 300 sq yds each, the prices obtained for these sites at auction, instead of representing £4,300 and £5,275 per plot, would have represented £2,430 and £3,360 per plot respectively.

Value at August 1973

Looking at the evidence as a whole we put the value of the reference land at August 1973 at £58,500. This amount may be regarded as derived primarily from a plot calculation:

22 house-plots fully serviced @ £3,000	..	£66,000
Less cost of road (including sewers) ..	..	7,200
		£58,800

the above figure of £58,800 being rounded down after applying an acreage calculation as a check:

1.445 acres @ £40,000 per acre	..	£57,800

The figure of £58,500 requires adjusting for the change in values between August 1973 and October 1974. Before we pass to that question, however, it seems only courteous to refer to a residual valuation which was put in by Mr Birley, and on which much of the hearing time was spent. Mr Moss put in his own version of this form of calculation, and Mr Burke-Gaffney called evidence on building costs from a Mr S R Ellison who has been all his life in the building industry. Towards the end of the hearing a document was handed in, headed "Amended residual valuation with agreed construction costs and plot numbers." The parties had agreed (a) that the construction cost of a typical house on the reference land, including road and sewers and a builder's profit at 15 per cent, would be £6,565 and (b) that the land could be regarded as accommodating 22 house-plots. But there still remained differences on three matters:

(1) what selling price should be assumed for the typical house,

(2) what allowance, if any, should be made for developer's profit and risk additional to the builder's profit on the house, and

(3) what allowance should be made for interest on capital.

It transpired, however, that although Mr Birley had produced the original residual valuation, he was not really relying on it. "I do not," said Mr Birley, "attach a great deal of weight to residuals in any form." And Mr Moss, when putting in his own residual calculation, made it clear that he did not regard the exercise as likely to be a helpful one. Suffice it for us to say that, having considered all the figures used in these various calculations, we find in them nothing to cast doubt on our findings as to the value of the subject land at the time when the notices were served.

Value at October 1974

Turning finally to the question of the value at the date of the hearing, Mr Moss expressed the opinion, in answer to a question from the tribunal, that "there should be perhaps a 20 per cent difference between 1973 and October 1974." Mr Birley thought there had been a greater fall. His own valuation had not been made until May 1974; he considered that values had remained stable between August 1973 and May 1974, that they had then fallen by 20 per cent to the early part of October 1974, but that there had been a further fall of 20 per cent during the fortnight immediately preceding the date af the hearing. In preferring the evidence of Mr Moss we intend no disrespect to Mr Birley, whom we accept as an able and experienced valuer; but we have in mind that Mr Moss was basing his opinion on over 50 years' professional experience, and we attach more weight to his *ipse dixit* opinion than to that of the county estates officer.

Applying a reduction factor of 20 per cent to the ascertained August 1973 value produces a figure of £46,800.

The acquiring authority will pay the claimants the sum of £46,800, to which will be added a surveyor's fee based on Scale 5 (a) of the Scales of Professional Charges of the Royal Institution of Chartered Surveyors, also any legal costs properly incurred from the date of the acceptance of the purchase notice up to the date of the notice of reference.

Having read the decision in this matter and having then opened the sealed offer of compensation lodged by the acquiring authority, we find that the amount thereof is less than the amount of our award; accordingly the acquiring authority will pay to the claimants their costs of this reference, such costs, unless agreed, to be taxed by the Registrar of the Lands Tribunal on the High Court Scale. For the avoidance of doubt we would add that Mr Ellison does not qualify as an expert witness.

CLEGG v SHEFFIELD CORPORATION
(REF/125/1972; REF/181/1972)

(V G WELLINGS QC)

December 16 1974

Estates Gazette February 1 1975
(1974) 233 EG 528

Dwellings in Sheffield clearance areas—Cleared-site values—Claimant appearing in person and calling no expert evidence produces schedule of settlements effected by him with acquiring authority at substantially higher amounts—Tribunal rejects valuer's assurance that council has no other records to support higher figures—Registrar's letter to council inviting production at further hearing of all completion statements and records of similar properties acquired by them at site value at relevant periods—Evidence that investigation of 3,500 such transactions would take 5-6 months and bring "housing clearance programme to a standstill" for that time—Tribunal's criticism of council's valuers' investigation of settlements—Their evidence treated with "the greatest caution"—Award of higher amounts than council's valuations

Giving his decision, Mr V G Wellings said: These are references to determine the amount of compensation payable for the compulsory acquisition by the acquiring authority under Part III of the Housing Act 1957 of:

(a) In the case of Ref/125/1972, 115, 117, 119 and 121 Fawcett Street and 24 and 26 Mushroom Lane, Sheffield, pursuant to the City of Sheffield (Netherthorpe) Compulsory Purchase Order (Nos 790-792 and 794) 1965;

(b) in the case of Ref/181/1972, 17 and 19 Carlton Street, Sheffield, pursuant to the City of Sheffield (Attercliffe) Compulsory Purchase Order (No 910) 1967.

In order to suit the convenience of the parties the two references were in effect heard together, although some of the evidence was necessarily heard separately.

Ref/125/1972:

Possession of the six reference properties was taken by the acquiring authority on different dates between March 6 and September 25 1967. Nothing appears to turn on the precise date in that year on which possession was taken. The reference was made by the acquiring authority on July 25 1972 and named the claimants as being Mr John Henry Clegg and Mr C E Styan, who was alleged therein to be a mortgagee. The acquiring authority received no claim from Mr Styan and indeed no notice to treat was served on him. Notice to treat was served on Mr Clegg and he submitted a claim in respect of the reference properties in the sum of £625 plus fees. Mr Clegg appeared before me, but Mr Styan did not appear and was not represented.

The statement of reasons for the compulsory acquisition which accompanied the 1965 order stated (in effect) that the 411 dwellings shown coloured pink on the map to the order had been represented as unfit for human habitation and that the area had been declared to be a clearance area under Part III of the Act of 1957; that the dwellings comprised in the clearance area were in a multiplicity of ownership and that by bringing the land into one ownership by means of a compulsory purchase order the acquiring authority proposed to make the cleared site available for development for residential purposes.

A **Evidence on May 1 1974 (first day of hearing) as to Ref/125/1972:**

Mr Clegg said that he had held a single lease of all the reference properties for a term of 99 years from September 29 1881 at an apportioned ground rent of £13.54. He had taken an assignment of the lease in 1953, at which time it included five other dwelling-houses. From memory the price paid by him had been £500 or £600. The reference properties were the subject of a mortgage in favour of Mr Styan. The principal sum was £1,000 with interest at 4 per cent. None of the principal sum had been repaid. He claimed compensation for the reference properties in the sum of £625 plus fees. They were worth more than that
B amount, but he so limited his claim in order to accommodate the acquiring authority. He understood that the acquiring authority was going to allege that the compensation to which he was entitled was £155. That sum was insufficient, as other transactions between himself and the acquiring authority (to which I shall later refer and which I shall call "Mr Clegg's comparables") would show.

Mr G B Woofitt ARICS, a chief housing valuer of the acquiring authority with eight years' experience, gave evidence as follows:

(1) The reference properties were to be valued in accordance with the provisions of section 59 of the Act of 1957, the second schedule to the Land Compensation Act 1961 and, by virtue of
C the acquiring authority's proposals, on the basis that the sites were available for residential development in accordance with the provisions of sections 14 and 15 (1) of the Act of 1961;

(2) his valuation was:

Site value (ie, site cleared of buildings and available for development)

510 sq yds at 10p per sq yd pa	£51.00 pa
Less ground rent	£13.54 pa
	£37.46 pa
YP for 13½ years at 14% and 3% (Tax 41.25p) ...	4.1
	£153.59

D

Therefore site value compensation applies, say £155

(3) he took an interest rate of 14 per cent because of the lack of security due to the unfitness of the properties; this was a figure at which he arrived from his experience;

(4) he produced a schedule of four alleged comparable settlements effected by the acquiring authority in respect of other properties compulsorily acquired together with supporting documents, which schedule and documents on their face justify his figures of 10p a sq yd. The residues of the leasehold interests referred to in the schedule vary between 10 years and 15 years
E and the interest rates applied to the ground rents vary between 10 per cent and 3 per cent (tax 41.25p) and 20 per cent and 3 per cent (tax at the appropriate rate);

(5) he had no knowledge whatsoever of Mr Clegg's comparables (see below). It was possible that when the details had been ascertained those comparables might be of assistance in valuing the 13½ years' residue of the relevant term;

(6) he had completed a transaction with Mr Clegg in respect of 26-38 Latimer Street, Sheffield, which property was subject to the 1965 CPO in which the acquiring authority had on April 25 1969 paid £636 for 636 sq yds or £1 a sq yd. The property was, however, freehold. It was in a poorer condition than that
F of the reference properties.

Ref/181/1972:

Possession of the properties in Carlton Street was taken as to No 17 on May 4 1970, and as to No 19 on July 27 1970. Again nothing appears to turn on the precise date in that year on which possession was taken. This reference was made by the acquiring authority on November 10 1972 and named Mr Clegg (only) as the claimant. He, after service of notice to treat on him, submitted a claim for compensation in the sum of £500 plus fees for the two properties. The statement of reasons accompanying the 1967 CPO stated that the 177 dwellings shown coloured pink on the map to the order had been represented as unfit for human
G habitation and that the area had been declared to be a clearance area; that the 177 dwellings comprised in the clearance area were in a multiplicity of ownership, and by bringing the land into one ownership by means of a compulsory purchase order the acquiring authority proposed to make the cleared site available for industrial purposes.

Evidence on first day of hearing as to Ref/181/1972:

Mr Clegg said that in his experience as a property owner in Sheffield the two properties would in the open market at the relevant date together have fetched £2,000 as a site available for industrial development. He had sold to the acquiring authority by private treaty a number of vacant sites (Mr Clegg's compar-
H ables: see below) for which he had been paid much larger sums than the sum of £135 which, he understood, the acquiring authority was offering by way of compensation. He provided the addresses of such properties but could provide few other details. He limited his claim to £500 plus fees in order to accommodate the acquiring authority. As soon as 17 and 19 Carlton Street had been pulled down their sites had been used for the storage of steel girders by the occupier of an adjoining factory. Indeed Carlton Street as a whole had become an industrial area and was near a ring road.

Mr D Airey ARICS has been employed as a chief housing valuer in the acquiring authority for four years. His evidence
J was as follows:

(1) The sites were to be valued in accordance with the provisions of section 59 of the Act of 1957 and the second schedule to the Act of 1961, and by virtue of the acquiring authority's proposals the valuation had to be on the basis that the sites were available for development for industrial purposes in accordance with sections 5, 14 and 15 (1) of the Act of 1961.

(2) His valuation was as follows:

17 Carlton Street:

Site value (ie, site cleared of buildings and available for development)

100 sq yds at £1 per sq yd		£100 K

Therefore site value compensation applies

19 Carlton Street:

Freehold subject to a lease for 800 years at a ground rent of £3 pa

Ground rent	£ 3 pa	
YP in perpetuity	11	
		£ 33
Total		£133

Value of claimant's interest £135

L

(3) He produced schedules of alleged comparable settlements effected by the acquiring authority as follows:

(a) six settlements effected by the acquiring authority with local surveyors in respect of properties comprised in the 1967 CPO, which settlements on their face justify £1 per sq ft for No 17;

(b) two settlements effected by the acquiring authority with local surveyors in respect of long leaseholds comprised in the 1967 CPO, which settlements on their face justify a multiplier of 11 of the ground rent in respect of No. 19;

(c) five transactions relating to long leases sold at auction, which transactions on their face justify a multiplier of 11 of the ground rent in respect of No 19. M

(4) He had no knowledge of Mr Clegg's comparables (see below) but would endeavour to investigate them and provide details when the hearing resumed.

Evidence on July 23 1974 (second day of hearing) as to Ref/181/1972:

No further oral evidence as to Ref/125/1972 was given on this day. The facts, completion statements and other documents with respect to Mr Clegg's comparables had been ascertained and an agreed schedule and agreed supporting documents were produced.

A The schedule referred to the following transactions between Mr Clegg and the acquiring authority:

(1) A leasehold interest of 800 years from March 25 1857 at an annual ground rent of £11.37½ in sites in Egerton Street and Dene Lane, Sheffield (one mile south-west of city centre; 2½ miles south-west of Carlton Street; 2½ miles south of Fawcett Street) was sold on June 10 1966 by Mr Clegg to the acquiring authority for a price (after deduction for ground rent) of £2,700; representing for 572 sq yds: £5 per sq yd.

(2) A freehold site at 173-183 Gleadless Road, Sheffield (1½ miles south of city centre; three miles south-west of Carlton Street) was sold by Mr Clegg to the acquiring authority on September 21 1972 for residential development for £1,400: representing for 660 sq yds:
B £2.12 per sq yd.

(3) A freehold site at 216-222 Gleadless Road (same distances as before) was sold by Mr Clegg to the acquiring authority on May 8 1967 for residential development at £1,250; representing for 857 sq yds: £1.50 per sq yd.

(4) A freehold site sold in 1957 for road-widening purposes and to which I need not further refer.

Mr Airey, continuing his evidence, said, by reference to p 20 of the written statement of the City and County Borough of Sheffield Development Plan, that Carlton Street was in 1970 zoned as a residential area, which, however, was destined, by 1972, to become an industrial area. No express planning per-
C missions had been granted in respect of the reference properties. He said that the site of 17 Carlton Street was too small to be developed as a dwelling-house and that for that reason planning permission for residential development of it could not be expected. Moreover there would be no demand for a dwelling-house on the site. He thought that the site was capable of being developed for the erection of a small workshop but that there would be no demand for it. He thought that Mr Clegg's comparables were no useful guide. The site at Egerton Street was much larger and more obviously capable of development for industrial purposes. If that site had been sold in 1970 he would have expected that a price representing £10 per sq yd (that is to say, double the 1966 price) would have been obtained. Moreover
D Egerton Street was nearer to the city centre, where sites command higher prices. The sites in Gleadless Road were not comparable because they were more readily capable of development, albeit for residential purposes. He said that when he prepared his valuation he was not aware of any of Mr Clegg's comparables. He made a search for comparable properties but he did not come across them. He said to me in respect of 17 Carlton Street: "I can assure you that there are no records at the council other than those now before you which would support a higher figure than £1 per sq yd." He agreed that the Carlton Street area is now an area of general industry: there are steel rolling mills within 200 yds of No 17.

E With respect to the assurance which Mr Airey gave to me I asked Mr Dawson, solicitor appearing on behalf of the acquiring authority, whether I could accept that assurance. His reply was: "You cannot be sure that there are no other properties which might be relevant to this inquiry." Because of the importance of the matter and in order to make sure that I had not misunderstood Mr Dawson's reply I repeated the question. I received precisely the same answer.

Events following second day's hearing:

Following Mr Dawson's reply mentioned above, I caused a letter to be sent by the Registrar of the Lands Tribunal to the acquiring authority (with copy to Mr Clegg) inviting the acquir-
F ing authority at a further hearing on a date to be fixed to produce to me all their completion statements and other records relating to transactions at site value whether or not under compulsory powers to which they were parties:

(1) In respect of leasehold land at ground rents possession of which was taken in 1967;

(2) in respect of freehold land with vacant possession, possession of which was taken in 1970;

(3) in respect of freehold land subject to long leases, possession of which was taken in 1970,

other than the completion statements and records already produced. The letter also stated that at the hearing I would expect to receive assurances from the acquiring authority's witnesses G
that the evidence was complete and that they would no doubt explain which of the new material (if any) was relevant and which was irrelevant and to what extent their valuations were affected thereby.

Evidence on third day (October 16 1974) of hearing with regard to both references:

Mr Airey, recalled, said that the number of transactions to which the above-mentioned letter referred was of the order of 3,500. To investigate the question whether any of those transactions had relevance to the inquiry in these two references would take five to six months and would involve several departments of the acquiring authority and would bring its housing H
clearance programme to a standstill for the months involved. Before producing his valuation on the first day of hearing he had not made a search of all the possible material. But in his experience, he thought that the material put forward by him was the most useful material available. He had had to make a search in order to find the comparable transactions upon which he had relied in his schedule. He had not dealt with them personally. He produced a further schedule of seven properties in the vicinity of the subject properties, which schedule, on its face, supports a valuation of £1 per sq yd for 17 Carlton Street. The sites in question varied between 62 sq yds and 220 sq yds in area, all of them, according to Mr Airey, being, because of their small size, incapable of development. There was nothing in the J
schedule or in Mr Airey's evidence to indicate that the sites were acquired for the purpose of being developed as industrial premises.

Mr Wooffitt, recalled, said that the details set out in his schedule produced on the first day of the hearing had not been prepared by him. They were not transactions personally known to him. He, however, checked them before giving evidence and came to the opinion that these were the best comparables available. He did not make any investigations to see if there were any other properties which might be relevant. The schedule produced on the first occasion had in fact been prepared two years earlier by a valuer who had now left the employment of the acquiring K
authority. He thought that none of Mr Clegg's comparables had any bearing upon his valuation. They represented either freehold or very long leasehold transactions in respect of sites which were much more readily capable of development and much more in demand. He produced two further schedules each in respect of two properties which he thought were relevant in respect of his valuation and which on their face supported it. He had no knowledge of these properties when first giving evidence.

Observations on the evidence of Mr Wooffitt and Mr Airey:

In my view it was their duty, in so far as they intended to support their opinions by their authority's own settlements, to L
make a full and proper investigation of those settlements, such as would have enabled them to assure the tribunal that the settlements on which they did rely were truly representative and that there were no other settlements which might be more favourable to the claimant than to the acquiring authority. Mr Wooffitt made no such investigation and that made by Mr Airey was inadequate. Having regard to the dates on which the references were made, they both had ample time in which to make a full and proper inquiry without risk of causing any administrative inconvenience to the acquiring authority. I agree with Mr Dawson's very proper comment with regard to the assurance which Mr Airey gave me on July 23 1974. I reject that assurance. In the circumstances I shall treat the evidence of M
Mr Wooffitt and Mr Airey with the greatest caution. If I had received on behalf of Mr Clegg expert evidence which I could have preferred to their evidence I would have rejected it completely. Since Mr Clegg called no expert evidence I cannot adopt that course, since, otherwise, more particularly in the case of REF/125/1972, it would be difficult or impossible for me to fulfil my duty to determine the amount of compensation payable.

Decision with respect to Ref/125/1972:

In my judgment 10p per sq yd for the 13½ years' residue of the lease of the 510 sq yds of the site is too low. I propose to increase that price by 25 per cent. I also propose to substitute for Mr

Wooffitt's multiplier 4.5 YP (12 per cent and 3 per cent—tax at 41.25p). I am not satisfied that the security of what is in effect a ground rent of the site is affected by the unfitness of the buildings (if they be unfit: this was not proved before me) to the extent suggested by Mr Wooffitt. The calculation therefore becomes:

510 sq yds @ 12.5p per sq yd per annum	£63.75 pa
Less ground rent	£13.54 pa
	£50.21 pa
YP for 13½ years @ 12% and 3% (Tax 41.25p)	4.5
	£225.94
Say	£225

That sum is below the ceiling value (£704.07 according to Mr Wooffitt).

I award under Ref/125/1972: compensation in the sum of £225 together with surveyors' fees (if any) on scale 5 (a) of the Royal Institution of Chartered Surveyors and any legal fees incurred by Mr Clegg before the date of the reference. I make no award in respect of the mortgage alleged to be vested in Mr Styan.

Decision with respect to Ref/181/1972:

I agree with Mr Airey to the extent that if there had been no compulsory purchase order the site of 17 Carlton Street might reasonably have been developed by the erection on it of a small workshop. It appears to me also that there would have been some demand for such a small workshop. This is supported by the industrial development which has in fact taken place in Carlton Street and by the user (admittedly in the nature of storage in connection with a factory) of the site itself which has in fact occurred. It is nevertheless a difficult matter to assess precisely how much demand there would have been.

I am struck by the difference between £1 per sq yd suggested by Mr Airey for No 17 and £10 per sq yd which, he says, is the 1970 equivalent for the Egerton Street site in Mr Clegg's comparables. Even after making allowances for differences in size of the respective sites and their locations, so large a disparity has not been explained by Mr Airey to my satisfaction. I suspect that his valuation of £1 per sq yd for No 17 is, notwithstanding his evidence, one more appropriate to a site available for development for residential purposes: consider the settlement relating to 26-38 Latimer Street.

Having regard to all the circumstances and to the assumptions which, by statute and on the authorities, I am required to make (including no doubt that if there had been no compulsory purchase order there would have been some prospect, having regard to the relevant development plan, that Carlton Street would or might become an improving area, even if it is otherwise to be assumed to consist of unfit houses: such unfitness was not proved before me), I propose to substitute for Mr Airey's valuation £2.50 per sq yd (for 100 sq yds) or £250. That sum is less than the ceiling value of No 17 (which was said by Mr Airey to be £316).

I have no means of judging whether the multiplier of 11 of the ground rent of 19 Carlton Street is sufficient. I suspect, however, that since there would be no prospect of increasing the ground rent so that it kept pace with inflation, a multiplier of 11 is correct, if not generous. The compensation in respect of No 19 will therefore be £33 and the total compensation payable under REF/181/1972: £283: say £285. I award that sum together with surveyors' fees (if any) on scale 5 (a) of the Royal Institution of Chartered Surveyors and any legal fees incurred by Mr Clegg before the date of the reference.

The claimant appeared in person; the Solicitor to the Sheffield Corporation for the acquiring authority.

HICKMOTT v DORSET COUNTY COUNCIL

(Ref/127/1974)

(J D RUSSELL-DAVIS FRICS)

February 6 1975

Estates Gazette March 15 1975

(1975) 233 EG 1009

First case under Part I of Land Compensation Act 1973—Claim for compensation for depreciation due to public works—Tribunal finds freehold value of 16th/17th-century house in Dorset village not measureably depreciated by "physical factors" of noise and vibration from increased traffic allegedly due to widening of bend in class B road—Use of road by more traffic at greater speed not encouraged by "one small, isolated and local improvement"—Claimant to make contribution to costs

Giving his decision, MR RUSSELL-DAVIS said: This is the first reference to come before the Lands Tribunal arising out of Part I of the Land Compensation Act 1973, which provides for compensation for depreciation caused by the *use* of public works. It concerns certain road improvements to the B3162 road in the small village of Drimpton in the County of Dorset. The claimant is Mr John George Russell Hickmott, a retired heating engineer, who is the owner and occupier of a private house known at Hillcrest, formerly Drimpton Farm House, adjacent to the highway; he claims the sum of £8,000. The respondents, Dorset County Council, while admitting the claimant to be qualified, contend that there is no depreciation; hence no compensation is due.

The claimant was represented by Mr R J Soper of counsel [instructed by Clarke, Willmott & Clarke, of Crewkerne], who called evidence from Mr A L Jayne FSVA, an experienced valuer and a member of the firm of Wyatt, Powell & Co, of Yeovil, and from the claimant himself. The county council was represented by Mr P K Harvey, the deputy county solicitor, who called evidence from Mr Rex Dixon ARICS ARVA, the county valuer since 1969.

Provisionally-listed Building

The interest claimed to be depreciated (section 1) is the freehold in possession of the claimant of the two-storey stone and tiled house, having a frontage of 52 ft 3 in to the B3162 road, a return frontage and garage access to the C43 road, and a total area of about one-eighth of an acre. The house dates back to the 16th and 17th centuries and is included in the Dorset County Council's provisional list of buildings of architectural and historic interest (with associations with the cavalier, Colonel Nathaniel Whetham) under group III. It contains three-four living and five-six bedrooms with appropriate offices, a car port and very small garden.

The "public works" consist of the improvement of the carriageway of the B3162 at a bend in the centre of the village, the subject house standing on the outside of the bend. The works entailed the increase in width of the carriageway from 15 to 22 ft, and consequential realignment. The "physical factors" complained of as depreciating the value of the interest are, in particular, noise and vibration, and to a lesser extent, smell, fumes and smoke from the increased traffic using the highway; the last three factors were not proceeded with.

The claimant also maintained, and evidence was given, that the danger, and apprehension of danger, from traffic had measurably increased. It is, however, appropriate for me to say at this stage that these are not "physical factors" nor are they expressly covered in section 1 (2) of the Act.

The B3162 is the Bridport-Chard road, and Drimpton lies some 2½ miles from Broadwindsor. Hillcrest's return frontage is to the C43, a spur road leading to the B3165, thence to Crewkerne.

A The relevant date under section 9 (2) (a) is the date on which the public works were first used after completion of the alterations, which is agreed to be August 1971. The claim was made within the time prescribed under section 3 (2), on June 16 1974. Compensation has to be assessed under section 4 by reference to prices current on the first day of the claim period, which is 12 months from the date on which the highway was first used after the completion of the improvements; the valuation date for this purpose is therefore August 1972. Under subsection (2) account shall be taken of any intensification that may then reasonably be expected of the use of the works in the state in which they
B were on the valuation date.

No relevant works had been carried out to the subject house under sections 23 and 27, and it is agreed that none could reasonably be expected to have been carried out; for example, noise insulation (section 4 (3)). Under section 7, compensation shall not be payable on any claim unless the amount exceeds £50.

In evidence Mr Jayne stated that due to the road widening at this point there had been a considerable increase in traffic noise and vibration from vehicles from both the Bridport and Chard directions, but he was at a disadvantage in that
C there were no statistics available of traffic-flow before the works. The road could originally have been described as a country lane. Now traffic from the Bridport direction, proceeding downhill, tends to pull out towards the house in order to see around the bend, while traffic from the other direction, uphill, tends to accelerate in order to take the hill. Since August 1971 cracks had appeared in the ceilings by the beams in the two front reception rooms, and in a cement fillet to the north-east gable wall which had had twice to be renewed. The houses was built without foundations in the modern sense (not surprisingly for a house of this age), and he felt that the stability would in time become affected.

D He had observed intensive use by all kinds of private cars as well as by lorries, especially "supermarket containers," etc. Both Bridport and Chard, as well as Crewkerne, had experienced recent growth. The B3162, he said, was used as an alternative to avoid congestion on the two east-to-west main roads, the A35 (coastal road) and the A30, as well as, presumably, the north-to-south A3066 Yeovil-Bridport road. Word was getting around that the B3162 was a good, safe route; he admitted, however, that in parts the road was twisting and narrow. Because the corner was safer, parking outside the house had increased by the customers of the
E adjacent general store and the small draper's shop next to that, the latter in the occupation of the claimant's daughter.

£8,000 Depreciation Claimed

He estimated the value of the residence immediately before the works at the sum of £22,000, and its value afterwards at £14,000; and the depreciation therefore at £8,000. He had taken into account some physical damage to the structure due to the impact of lorries against the flank when the sharp corner into the C43 was negotiated.

I apprehend this to be, in the main, opinion evidence derived from his experience. He is responsible in his firm
F for negotiating the sales of such properties but it was not supported by any comparable transactions as at either date. He admitted that he had not been acquainted with conditions before the works, and he had been in practice in the area only since mid-1971, a period of 4½ years.

Mr Hickmott's evidence in the main corroborated Mr Jayne's from his personal experience of conditions both before and after the works; he had lived at Hillcrest since December 1965. He had had correspondence with the county council over a period, and he had proposed the formation of a 5-ft footpath as part of the works in order to keep vehicles further away from the house, but this had not been acceded to. He described the cracking already apparent, and G
as to the foundations he had ascertained that these were in effect large stones bearing on to a clay and gravel soil. He described the parking outside the house; the effect would be to deter any would-be purchaser.

Previously he estimated the speed of traffic around the corner at 20-25 mph; it now averaged 40-50 mph; there is no "slow" sign. He had carried out an unofficial traffic count for short periods at different times of the day over three days, Thursday to Saturday, September 12-14 1974, which averaged from 1.1 to 1.6 vehicles per minute in either direction; he had made no such count pre-1971.

In cross-examination, he agreed that the centre line of H
the road was 5-6 ft further away from the house; but he averred that traffic did in fact attain a speed of 40 mph uphill and 40-50 downhill—he had timed it over ¼-mile; he did not claim this was a safe speed, nevertheless it was frequently made, especially by "locals," who also often turned across the traffic into the C43 at speed without proper visibility of oncoming uphill traffic.

Mr Dixon gave evidence of the acquisition of the land for the road improvement, which provides for the thoroughfare an extra 12-ft width for carriageway, and verge and retaining wall on the inside of the bend; it had the effect of J
taking traffic further away from Hillcrest. He declined to commit himself how far faster speed caused greater vibration, and he claimed no special knowledge of the effect of vehicle speed on foundations.

Tests had been made by the county council, he said, and noise had been measured *outside* the front of the house on the L 10 scale at 60 dBA, well within the limit of toleration. A traffic count in October 1974 gave 1,086 vehicles in the 18-hour period from 0600 hours to midnight, an average of about one per minute, of which some 15 per cent of the vehicles exceeded 30-cwt capacity. A projection to 1980 would give about 2,000 vehicles over the same period; the K
factor for the holiday month of August compared with October would give about 1 to 0.8. This evidence was not deposed to first hand by a member of the county highways department, but was not objected to by Mr Soper.

He considered that the value of Hillcrest both before and after the works was £18,000; he cited the sales at mid-1972 of two not dissimilar properties in Beaminster, which he did not claim as truly comparable, but they gave a wide bracket into which a valuation of Hillcrest could be slotted after completion of the works.

L

"Frightening Repercussions" of Allowing Claim

In his concluding address Mr Harvey claimed that Mr Hickmott had not proved his case of depreciation, and it was unrealistic. To allow the claim would, by opening the door to the payment of compensation in respect of such minor road improvements, create "frightening repercussions."

At the conclusion of the hearing, between 15.45 and 16.15 hours, I drove the nine miles along the B3162 road from Bridport to Drimpton, and beyond the village as far as the crossroads with the B3165. During the whole journey I passed or overtook no more than six moving vehicles after M
leaving the Bridport outskirts. On the next morning, in fine weather, at about 11.00, I inspected Hillcrest and spent nearly an hour observing conditions from the front living-room. Between 14.15 and 14.45 on the same day I parked a short distance from and in view of the house, during which time 20 vehicles passed in either direction, including those turning into the C43.

Although this was a January day I had no difficulty in coming to the conclusion that traffic using the road is light, and no more at any season than is to be expected on many unimportant B roads.

A I also drove as far as Winsham, about four miles towards Chard. I found the whole length of the road between Bridport and Winsham to be twisting and in many places distinctly narrow, not exceeding some 12 ft in width, so that it is difficult for a car and commercial vehicle to pass in safety or comfort without one having to pull into the road verge. I tried to drive at speed downhill around the inside of the bend and had to brake from more than 35 mph to nearly 20 mph in order to achieve it in safety. I believe that Mr Hickmott's estimate of 40-50 mph is, no doubt innocently, widely exaggerated, even in the case of the most foolhardy driver.

B I noticed that vehicles do park in front of the house, as well as in the space immediately in front of the two small shops, but I am not convinced that this necessarily has anything to do with the works; if it is more than previously, it could be due partly to the increased population of the village. In any event it is an open question whether such could be included in the physical factors enumerated in section 1 as giving rise to a claim of depreciation in value.

It has not been proved to my satisfaction that the cracking in the plaster and to the fillet complained of is not due to other factors than traffic vibration; nor has expert evidence C been given as to the likely effect of such traffic volume on the stability of the house.

The claimant has lived in the subject house for nine years, during which time it is perhaps easy to make a subjective comparison between conditions as they existed and now. But the test is *depreciation in market value*. I do not think that a prospective purchaser in mid-1972 would, in the same market conditions, have reduced an offer, which he would have made before the improvements. It could be argued that the *appearance* of greater road safety at that point would have enhanced the value.

From the evidence before me and with the benefit of my D fairly detailed observations (although restricted to periods over two days), I conclude that the nature and general condition of the B3162 do not encourage more traffic to use it, or at appreciably greater speed, because of one small, isolated and local improvement to a sharp bend. If there is an increase in the weight and amount of traffic, it is more likely to be consistent with the increase in volume on all classified roads in the south-west of England over the last four years or so.

I viewed externally the two houses in Beaminster, five miles from Drimpton, mentioned in Mr Dixon's evidence. Beaminster is a small market town, and, although it is perhaps not E now relevant, I found neither to provide much assistance. But I come down nearer to Mr. Dixon than to Mr Jayne that the mid-1972 value of the subject house is about £18,000, and I am more persuaded by his evidence than I am by Mr Jayne's that its pre-works value is about the same figure.

Accordingly, I find that there is no measurable depreciation in the value of the subject property due to the physical factors caused by the use of the public works, as enumerated in section 1 of the Act, and I have no alternative but to dismiss the claim.

At the reading of this decision Mr Harvey said very fairly F that as this was the first case under Part I of the 1973 Act to have been heard, the county council wished to be lenient in the matter of costs. The claim was not made frivolously and the claimant believed firmly in his cause. However, the council's staff have put a great deal of time into the preparation of their responses; accordingly I think it is just that the claimant should make a contribution of £20 towards their costs, and I so award.

G

DHN FOOD DISTRIBUTORS LTD AND OTHERS v TOWER HAMLETS LONDON BOROUGH

(Ref/36-38/1974)

(DOUGLAS FRANK QC, President)

January 30 1975

Estates Gazette March 22 1975

(1975) 233 EG 1109

Compulsory acquisition of Bow, London, premises of three companies one of which wholly owned, and had the same directors as, the other two—Claim for disturbance by two H companies for extinguishment of their businesses—Legal ownership of freehold of reference properties the sole asset of third company, which tribunal holds not the "agent or tool" of, or trustee for, the parent company—Question of implied rescission of seven-year-old contract for transfer of freehold interest to parent company—Compensation limited under section 20 of Compulsory Purchase Act 1965 to occupation from date of entry until earliest date for determination of tenancies

Mr D M W Barnes (instructed by Asher, Fishman & Co) appeared for the claimants; Mr G N Eyre QC and Mr A D Dinkin (instructed by the solicitor to the London Borough of J Tower Hamlets) for the acquiring authority.

Giving his decision, Mr Frank said: These are applications by (1) DHN Food Distributors Ltd, (2) Bronze Investments Ltd, and (3) DHN Food Transport Ltd (all in liquidation) for determinations of preliminary points of law, which, put shortly, are whether the first- and third-named claimants are entitled to compensation for disturbance (amounting to extinguishment of their businesses) as a result of the compulsory acquisition by the acquiring authority of the premises in which those businesses were carried on. I shall henceforth refer to the first-named claimants as "DHN" the second as "Bronze," and the third as "the K Transport Company."

The reference properties were situated in Malmesbury Road, Bow, London E3, and were included in a compulsory purchase order confirmed by the Minister of Housing and Local Government on October 12 1970. Notices to treat and notices of entry were served on October 30 1970 and possession was taken on March 17 1971. The freehold interest in those properties was conveyed on May 10 1971 for £360,000, but it was agreed that sum did not include any disturbance or any other claim under rule (6) of section 5 of the Land Compensation Act 1961 and that any such claims should be left for subsequent determination. DHN and the Transport L Company claim that they are entitled to compensation for the total extinguishment of their businesses, but the acquiring authority say that they had no interest in the land acquired entitling them to compensation for disturbance.

History of the Claimant Companies

Prior to 1963 there existed three entirely separate companies all engaged in the same kind of business and in that year they agreed to amalgamate. Hence DHN was formed and became registered on April 10 1963 as a company carrying on business in the wholesale and cash-and-carry grocery trade. They then decided to buy the properties the subject M of this reference, they having no other, and entered into an agreement with the Palestine-British Bank in order to provide the necessary finance for the acquisition. The essential terms of the agreement with the bank were as follows:

(a) The Palestine-British Bank were to pay for the acquisition of the properties. The properties were then owned by Stephen Smith & Co Ltd. They were to be sold by the vendors at a price of £115,000.

(b) The properties were to be transferred to Bronze, which was then a company controlled by the bank. It was registered on October 25 1962 and its shares were wholly owned by the bank.

(c) Bronze was to enter into a contract to sell the properties to DHN for £120,000. Completion under that contract was to be one year after the completion of the purchase by Bronze. £20,000 of the purchase price was to be paid upon exchange of contracts and the balance of £100,000 upon completion.

(d) DHN were to pay interest at 12 per cent per annum on the balance of £100,000.

(e) The bank were to have an option to acquire a 12½ per cent equity interest in DHN for cash consideration at par on the completion date.

(f) DHN were to be entitled to full occupation of the properties as from the date of acquisition by Bronze.

The freehold interest in the properties was transferred to Bronze on January 6 1964 and DHN entered into occupation about that date and commenced to trade from them soon afterwards. By a contract dated May 27 1964 DHN were to purchase the property from Bronze with completion to take place on or before January 6 1965. The contract was subject to the National Conditions of Sale (17th ed) and special conditions A to K endorsed thereon. It may here be observed that condition 7 provides:

"if the purchaser (not being already in occupation as lessee or tenant at a rent) is let into occupation of the property before the actual completion of the purchase, then, as from the date of his going into occupation and until actual completion, or until upon discharge or rescission of the contract he ceases to occupy the property, the purchaser shall (1) be the licensee and not the tenant of the vendor. . . ."

"Completion Never Took Place"

On December 14 1964, Bronze agreed to the postponement of their completion for a further year upon consideration of a payment, but in the event completion never took place. By an agreement dated February 8 1966 DHN agreed to purchase the shares in Bronze owned by the Israel-British Bank (successors to the Palestine-British Bank) for £3,597 5s and further undertook that upon completion it would procure that Bronze would pay to the bank £116,402 15s, being the balance of the £120,000 payable under the agreement. The shares in Bronze were acquired by DHN pursuant to that agreement.

At all times from February 8 1966 until the voluntary winding-up of the companies, DHN owned the whole of the issued share capital of Bronze. The directors of the two companies were at all times the same. Bronze had no employees nor a bank account. It carried on no trading or other operations. The only asset of Bronze was its legal ownership of the freehold of the reference properties. All interest due under the above mortgage was paid by DHN. Rent in respect of the reference properties, which were subject to leases, was collected by agents on behalf of DHN and paid by the agents to DHN.

At all times from January 6 1964 until May 10 1971, Bronze was the registered proprietor of the freehold interest in all the reference properties. The accounts for Bronze from January 7 1964 show rent receivable £28,150 in respect of the period from January 7 1964 to July 31 1966 and thereafter at the rate of £13,500 per annum, which has been charged as an expense in the accounts of DHN.

The Transport Company was registered on April 10 1968 and at all times the whole of its issued capital was owned by DHN. The directors of the two companies were the same. The Transport Company had no bank account. It owned vehicles which were used wholly and exclusively in connection with the business carried on by DHN, worked for no other person and carried on no other business or activity.

Two Preliminary Issues of Law

In 1971 and 1972 all the three claimant companies went into voluntary liquidation, and it may be taken, at least for the purpose of this determination, as an inevitable consequence of the compulsory purchase. Accordingly the preliminary issues of law as expressed by the claimants and acceded to by the acquiring authority are as follows:

1. Whether DHN and the Transport Company are confined to such compensation for disturbance as can be allowed under section 20 of the Compulsory Purchase Act 1965.
2. If so, whether their compensation for loss or injury under section 20 has to be assessed on the basis that they could expect to remain in occupation of the subject land for only such period as would extend from March 17 1971 to the earliest date on which their tenancy could have been determined by a notice to quit served on that date.

I should add that Mr Eyre, for the acquiring authority, did not concede that the claimants were even entitled to compensation under section 20 of the 1965 Act. That section provides:

"If any of the land subject to compulsory purchase is in the possession of a person having no greater interest in the land than as tenant for a year or from year to year, and if that person is required to give up possession of any land so occupied by him before the expiration of his term or interest in the land, he shall be entitled to compensation for the value of his unexpired term or interest in the land, and for any just allowance which ought to be made to him by an incoming tenant, and for any loss or injury he may sustain."

He conceded, however, that if section 20 of the 1965 Act applied and if (which was not the case) entry had taken place after 1973, then the claimants would have been entitled to have had taken into account the benefits of the Landlord and Tenant Act.

Mr Barnes made his submissions under three main heads (1) agency, (2) trusteeship, and (3) subsistence of contract, and I propose to consider each of them in turn.

Agency

The main submission here was that Bronze was no more than the agent or tool of DHN and that everything it did was on account of its parent company. In support of this submission he first referred to *Smith, Stone & Knight Ltd v Birmingham Corporation* [1939] 4 All ER 116, which, he said, although accepting that legal entities cannot be blurred, established that the facts may show that a subsidiary company may occupy premises as an agent or tool of the parent company. He said that if a subsidiary company can occupy premises or carry on a business on an agency basis, there was no reason why it could not also have a legal title on an agency basis, the test being not a legal control but in reality what went on. I find that case no more than an application of a well-established principle that the actions of one company may be as agent for another and whether they are so is a question of fact. There is nothing in that case to support the proposition that one company may hold a legal title in the land as an agent for another company. He also referred to *Pegler v Craven* [1952] 2 QB 69, but in my view that takes the matter no further.

It seems to me that Mr Barnes' proposition is impracticable and without authority, save that the case of *Tunstall v Steigmann* [1962] 2 QB 593 is fatal to it. Certainly it is not enough that the alleged principal has complete control of his alleged agent; see p 602. I should perhaps mention that it was admitted in evidence by Mr J Field, a director of all the claimant companies, that as had been stated in an annual report of the directors the principal activity of Bronze was that of property investment and that in turn carried certain tax advantages. Moreover, the transaction for the acquisition of shares was carried out in that way in order to avoid stamp duty. I have no doubt that Bronze was carried on and intentionally carried on as a separate company and that there was never any intention that it should hold the freehold interest as agent for DHN; indeed it could not have done so consistent with the advantages it took by reason of being a separate company.

Trusteeship

Mr Barnes' submission under this heading was an extension of the rule that where a conveyance is made to one person but the purchase money is provided by another there is a resulting trust in favour of the person providing the purchase money, and that presumption can be rebutted only by evidence that the purchaser was intended to benefit. Mr Barnes referred to the fact that the freehold interest was transferred to Bronze in 1964, the money being provided by the bank. He then went on to say that when in 1966 DHN freed Bronze from claims by the bank, Bronze then held the property on a resulting trust in favour of DHN. He referred to an agreed fact that a deed of mortgage was entered into on February 8 1966, whereby Credit for Industry Ltd loaned £110,000 to DHN, and Bronze and DHN charged the reference properties and their interest therein in favour of the mortgagees. Thus, DHN raised the money to buy from the Israel-British Bank the shares in Bronze.

I was referred to *Dyer v Dyer* (1788) 2 Cox 92, *Diwell v Farnes* [1959] 1 WLR 624, *Gravesend Corporation v Kent County Council* [1935] 1 KB 339, and *Isaac v Hotel de Paris* [1960] 1 WLR 239.

I find it impossible to conclude from the facts that there was any intention that when DHN acquired the shares in Bronze from the bank it was intended that thereafter Bronze should hold the legal interest in the properties as trustees for DHN. The facts point the other way. Nothing could have been simpler in 1966, when DHN had raised the funds necessary to purchase the premises, to have done so. Instead they chose to purchase the whole of the issued shares of Bronze, with a declaration by the bank that it would have no further claims. Thereafter DHN accounts showed rent payable for the premises as a liability and note that the loan advanced to Bronze is repayable by quarterly instalments. There is no evidence that it was intended that Bronze should hold the premises as trustees, but on the contrary it was all along the intention that Bronze should hold the properties as a separate entity in order to gain various advantages, for example the avoidance of stamp duty, as was conceded by Mr Field. Similarly the Transport Company was established as a separate entity in anticipation of a Selective Employment Tax advantage, although that advantage did not materialise. I cannot reconcile the form of the 1966 transaction, nor the liability for payment of rent, nor the liability to repay the loan, with the relationship of trustee and beneficiary.

Subsisting Contract

The contract relied on was that dated May 27 1964 under which DHN were to purchase the properties from Bronze at the price of £120,000, the completion date as subsequently extended being on or before January 6 1966. Mr Barnes argued that as from the date of the contract DHN obtained an immediate equitable interest in the properties, and since the contract has not been rescinded, that interest subsists. Mr Eyre, on the other hand, says that the parties abandoned the contract in 1966 when the method of taking a transfer of shares was formulated. He referred to *Halsbury's Laws of England* (3rd ed), vol 8, para 296:

> "*Implied Rescission*
> A rescission of the original contract may be implied where the parties have subsequently entered into an agreement which is inconsistent with it: and where neither party has insisted upon the performance of the contract for a long period after it was made, or where the parties have acted inconsistently with its continuance, an intention to abandon the contract may be inferred. In that case the contract will be deemed to be rescinded although no express agreement to such effect has been made."

In a letter dated December 17 1965, DHN's solicitors said to DHN's accountants:

> "Bronze Investments Limited bought the property in January 1964 for the sum of £115,000. DHN entered into a contract to buy the property from Bronze Investments Limited for £120,000. It is now intended that, in order to obviate stamp duties so far as possible, DHN should buy the issued share capital of Bronze Investments Limited and the shareholders of Bronze Investments Limited are agreeable in principle."

The letter then went on to deal with the matters which culminated in the sale of the shares. On February 22 1966, a caution in favour of DHN in respect of the reference properties was removed from the Register of Titles. Mr Field in cross-examination said that it was recognised by all that the sale of shares was to be instead of the sale of property and that a similar amount of money was involved and that there was no prospect of or purpose in DHN buying the land, as it would involve paying £100,000 for what in effect they already had paid for. In the annual accounts of DHN and Bronze following the sale of the shares there are shown amounts for rent payable and receivable one to the other; that was inconsistent with one of the conditions of the contract. All those facts lead me to the conclusion that the parties impliedly rescinded the contract when the sale of the Bronze shares took place and that the contract thereafter no longer subsisted.

I now turn to the second question for my determination. Mr Barnes pointed out that section 20 of the Act of 1965 deals with two categories of tenancy, namely one for a year and the other from year to year. He argued that the words "compensation for the value of his unexpired term" are not apt for the second category, there being no fixed term. Hence, although it is necessary to assume for the purpose of valuing the interest in the land that the minimum notice would be given, that assumption did not fall to be made in assessing any loss or injury, as one is then not dealing with the value of land, and so regard should be had to the realities, namely that possession would last indefinitely. He said that the authority were bound to compensate the owner for all loss caused by the expulsion; *Horn v Sunderland Corporation* [1941] 2 KB 26 and *Harvey v Crawley Development Corporation* [1957] 1 QB 485.

Mr Barnes' contentions were advanced and argued before the tribunal in *Greenwoods Tyre Services Ltd v Manchester Corporation* (1971) 221 EG 1579 and were rejected. I, too, find myself unable to accept the argument. In addition to the reasons given by the tribunal in the Manchester case, I would add that if Mr Barnes' argument were valid then compensation would be excessive, in that the acquiring authority could have entered and served notice to quit without running the risk of paying any compensation for disturbance.

It follows from all the above that I determine the answers to the two questions of law are "yes" in both cases, with the effect that:

1. DHN and the Transport Company are confined to such compensation for disturbance as can be allowed under section 20 of the Compulsory Purchase Act 1965 and
2. Their compensation for loss or injury under section 20 has to be assessed on the basis that they could expect to remain in occupation of the subject land for only such period as would extend from March 17 1971 to the earliest date on which their tenancy could have been determined by a notice to quit served on that date.

The claimants will pay the acquiring authority's costs of this reference, such costs if not agreed to be taxed by the Registrar of the Lands Tribunal on the High Court Scale of costs.

DAVIES AND ANOTHER v WELSH NATIONAL WATER DEVELOPMENT AUTHORITY
(Ref/5/1974)
(J H EMLYN JONES FRICS)
February 13 1975

Estates Gazette March 22 1975
(1975) 233 EG 1113

Hill farmland in South Wales acquired for reservoir—Value of sheep-grazing and woodland—Timber of "little interest" to commercial contractor because of access and extraction difficulties—Loss of fishing rights—Injurious affection—Effect of presence of new road and access of visitors

Giving his decision, MR EMLYN JONES said: The claimants in this case, Mr Cyril Roger Davies and Mr John Philip Davies, have referred to the Lands Tribunal the assessment of compensation payable to them by the acquiring authority, the Welsh National Water Development Authority (formerly the West Glamorgan Water Board), following the compulsory acquisition by the authority of some 102 acres of land situated near Llanwrtyd in the County of Powys.

The land was acquired in pursuance of powers contained in the West Glamorgan Water Board (Llyn Brianne) Order 1968 and was required for the construction of a reservoir, now known as Llyn Brianne. Notice to treat was served on the claimants on July 26 1968, and shortly afterwards, on August 10 1968, the authority entered on to the land. It is agreed that this latter date is the relevant date of valuation.

The reference land formerly formed part of an upland hill farm, known as Trawsnant and Clynglas, in the upper valley of the river Towy. The main parcel of land taken, extending to 80 acres or thereabouts, was situated in the valley bottom and on the lower slopes on the left bank of the river, and has now been submerged beneath the waters of the reservoir. There was, in addition, a further detached strip of land, extending to 22.25 acres, which was higher up the hill, and required as the site of a new road.

"Very Remote" District

The farm, which now extends to some 1,500 acres, is used mainly for sheep-grazing and formerly carried also some registered hill ponies. The claimants, who are brothers, are owners as tenants in common, and the land has been held in the family for the last 25 years or so. The land at river level was at or about the 650-ft contour, and the main parcel of land taken went up to 915 ft above sea level, the normal water level of the reservoir being estimated at 905 ft above sea level. The highest point on the farm is just below the 1,700-ft contour and the strip of land taken for the road runs mainly between the 1,000-ft and the 1,100-ft contours. Most of the grazing was on the higher slopes, but there were small areas of land near the river amounting in aggregate to about 20 acres, which also provided grazing and shelter for sheep. The rest of the main parcel of land taken was woodland, covered mainly with native oak with some ash and alder. The farm as it exists today has no land below 915 ft above sea level, and the belt of woodland extends upwards from the new lower boundary of the farm, covering much of the land which lies between the 900-ft and 1,000-ft contours. Before the making of the reservoir the whole district was very remote and access was provided by a poor, unclassified county road, mainly unfenced. On the other side of the valley were large areas of land devoted to forestry and served by forestry roads. Since the reservoir has become established the area has become attractive to visitors and tourists, and the new road built on part of the land acquired carries a considerable amount of such traffic.

Mr Glyn Morgan of counsel [instructed by Sydney G Thomas & Co, of Builth Wells] appeared for the claimants and called evidence from the two brothers and also from Mr Victor Gordon Jones, a chartered surveyor practising in Brecon, and Mr P J Hackett BSc, who is an expert in forestry. For the authority Mr Trevor Davies of counsel [instructed by the solicitor to the authority] called evidence from Mr Leslie Donald Jones, a timber expert, and from Mr David Gareth Jones ARICS, the area estates officer of the authority.

In the course of his closing speech Mr Morgan said that no particular question of law arose in the case and that the compensation fell to be assessed in accordance with the provisions contained in section 5 of the Land Compensation Act 1961. I understood him to say, however, that in his submission the appropriate rule was rule (5), that is to say on the basis of equivalent reinstatement. Mr Davies did not demur, but he had at that stage completed his case. It may be that there has been some misunderstanding, but quite clearly this is not a case in which valuation under rule (5) is appropriate. In the first place the land taken cannot be said to be devoted to a purpose of such a nature that there is no general demand or market for land for that purpose (I shall deal later with the question of the timber). But in any case no evidence was before me about reinstatement in some other place, although it was stated by Mr C R Davies that a further 20 acres of land had been taken by him and his brother to be worked in conjunction with 270 acres which the claimants jointly farmed elsewhere. Finally I heard no evidence as to cost of any such reinstatement. In my judgment this is a clear case where the value of the land taken shall be the open-market value with a willing seller in accordance with rule (2) of section 5, and where the special effects of the acquisition on the claimants, in so far as it is appropriate to take them into account, are to be reflected under the headings of injurious affection or severance. It is certainly on this basis that the evidence as a whole was put forward by both sides.

Heads of Claim

It will be convenient if I set out at this stage the various headings of claim and for convenience I give in addition to the compensation claimed by the claimants the figures put forward on behalf of the authority.

Land taken	*Acquiring authority*		*Claimants*	
20 acres of grazing land in the valley bottom	at £30 per acre	£600	at £60 per acre	£1,200
60 acres of sloping land with standing oak wood	at £40 per acre	£2,400	at £80 per acre	£4,800
22.25 acres of mountain land taken for construction of new road	at £30 per acre	£670	at £40 per acre	£890
Fishing				
(a) Approx 1,400 yds on the East bank of the River Towy at 45p per yd ..		£630		£2,000
(b) Approx 2,120 yds on the Nant Gwrach at 10p per yd		£212		
(c) The Trawsnant stream		—		
Injurious affection				
(a) Loss of water supply to stock on 2 miles of rivers and streams		£300	No dispute	£300
(b) Loss of shelter		£120	No dispute	£120
(c) Loss of value to the remainder of the farm—nuisance, inconvenience, increased traffic, new road running through upper grazing land, etc		£1,100		£1,500
	Total:	£6,032		£10,810

Value of Grazing Land

I will deal first with the value of the grazing land, that is to say the 20 acres which was formerly grazed in the bottom of the valley. For the claimants, a value of £60 per acre is put forward, which is derived from the evidence of Mr V G Jones who put it forward as his valuation. He did not support this figure with any details of actual transactions and was unable to comment on any of the comparables put forward on behalf of the authority, since he had not inspected them and had no knowledge of them.

Both Mr Jones, however, and Mr C R Davies took the view that this land not only provided the best grazing on the farm but was better in quality than any of the other land in the vicinity which had been acquired for the reservoir. As to the strip of upper mountain land, Mr Jones said that this had a value " to the claimants " of £40 per acre.

For the authority, Mr D G Jones valued the land at £30 per acre. He supported this figure by putting in a schedule of five other transactions in the immediate locality which had taken place between January 1969 and October 1972. Three of these transactions related to acquisitions for the purposes of the reservoir under compulsory purchase powers, two of which show an agreed price of £30 per acre, paid in 1972 for 40 acres and 18.7 acres respectively. The third case was purchased at the original asking price of £27 per acre on 54.4 acres. A fourth comparable was a transaction in which the authority had purchased by agreement 344 acres of forestry land from the Forestry Commission at £27.50 per acre and a further 10 acres of oakwood at £25 per acre.

This transaction dated from January 1969. Finally, Mr Jones referred to the sale of 98 acres of upland grazing land by private treaty in May 1971. The price realised showed £22 per acre, but in the conveyance the purchaser covenanted to put up 1¾ miles of fencing and to construct a track with water-bound surface on adjoining land of the vendor for about half a mile. These comparables were criticised by the claimants on various grounds. For example, some of the land consisted of long narrow strips of river bank much of which was covered with young conifers. Some of the figures of compensation quoted represented only parts of larger figures of compensation including other items—for example, injurious affection and loss of fishing—and, it was suggested, the total figures of compensation represented the real measure of agreement. But as I have already indicated, no alternative transactions were referred to by or on behalf of the claimants.

I have visited the locality and inspected all the land referred to except that which has now been submerged. I have also studied the photographs of the area as it existed before the advent of the reservoir. I accept the opinion put forward on behalf of the claimants that the grazing land taken was of higher quality and a higher value than land which was the subject of the comparable transactions put forward by Mr D G Jones. At the same time I observed that these transactions were later (in two cases three years later) than the relevant date and I accept what Mr D G Jones says, that there was an increase in the value of agricultural land in the area during that period. I am left with the impression that the authority's figures do not adequately reflect the value of the valley land and I have come to the conclusion that the appropriate figure to adopt for this part of the claim is £35 per acre.

I do not consider that the same can be said of the 22.25 acres of hill land. I think that £40 per acre, suggested by Mr V G Jones, is too high, reflecting, as it appears to do, some element of injurious affection which is properly to be considered under that heading. I therefore adopt a value of £30 per acre as put forward by Mr D G Jones for this item.

Value of Woodland

A great deal of the evidence and of the argument related to the value to be applied to the 60 acres of sloping land with standing oakwood. There was substantial agreement between the expert witnesses who were called to give evidence on timber values. From their evidence I am satisfied that although the timber lost on the land taken was of value to the claimants, in that it provided materials for stakes, straining posts and other miscellaneous purposes on the farm, it was of little interest to any outside commercial contractor because of the difficulties of access to the land and the difficulties of extraction. This being so it seems to me that much of the technical evidence regarding the volume of timber and of its quality was of doubtful relevance. The purchaser in the open market for the reference land would clearly not be anyone interested in forestry. More probably it would be someone locally engaged in agriculture, who would make use of the timber for the sort of purposes for which it was in fact used by the claimants. I observe that both parties value this part of the claim at a third more than the valley grazing land, but having taken £35 per acre for that item I do not think I would be right in adding one-third in order to arrive at a value for the woodland. Mr D G Jones arrived at his figure of £40 by including an addition of £10 per acre to reflect the presence of the timber and I prefer to adopt this approach. I therefore determine a value of £45 per acre for the woodland area.

" One of the Great Salmon Rivers "

I turn now to the next main item of claim, the loss of fishing rights, which extended over 1,400 yds on one bank of the river Towy, over 2,120 yds on both banks of the Nant Gwrach, a tributary of the Towy, and also, according to the claimants, to a small stream, the Trawsnant. It is common knowledge, of course, that the Towy is one of the great salmon rivers of Wales, and the evidence of Mr J P Davies was that salmon, sewin and brown trout were to be caught not only in the Towy but also in the Gwrach and in the lowest 150 yds of the Trawsnant. I was referred to evidence given at the public inquiry which preceded the decision to build the reservoir at Llyn Brianne, where reference was made to the value of fishing rights lower down the Towy at £5,000 per mile per bank. Nevertheless, Mr D G Jones had agreed a value of 45p per yd for fishing rights on the west bank of the Towy, opposite the subject land, by agreement with the Forestry Commission. It also emerged during the hearing, however, that a further stretch higher up the Towy on the left bank had been valued at approximately 62p per yd in another settlement agreed by Mr Jones. The claimants put forward an overall figure of £2,000, which represents an average of about £1,000 per mile. It is clear, therefore, that the values on this stretch of the Towy are appreciably lower than those which may have been realised lower down the river. It is also clear from the evidence that there was a further fall in values higher up the river above the Fanog falls, which mark the upper limit of the salmon run. This is perhaps borne out by the fact that the claimants were not in the habit of letting the fishing rights on their strip of the river, although they occasionally fished the water themselves and invited their friends to do so. I was supplied with no evidence of the catch. It was pointed out, however, that the access to the bank of the river on the other bank was extremely difficult, owing to the forestry plantation, and I accept that if 45p per yd is the correct figure there, then something appreciably higher is to be expected on the subject land. I consider the amount to be attributed to the loss of fishing as the sum of £1,600, which is made up as to 75p per yd on the Towy, 25p per yd on the Gwrach, with a small rounding-up to reflect what I regard as very limited value on the Trawsnant.

" Considerable Disturbance of Quiet Enjoyment "

I turn lastly to the question of injurious affection and this I can deal with quite briefly. There is clearly considerable disturbance of the quiet enjoyment of the remaining land, owing to the presence of the new road and the access of visitors. This is quite clearly an admissible item. I also accept what is said on behalf of the claimants, that the disappearance of the lower land by the construction of the reservoir interferes with the movement of sheep to shelter from one side of the Gwrach valley to the other. Under this heading I consider that the figure of £1,500 put forward by the claimants is a reasonable one and I accept it. The compensation which I award amounts, therefore, to the sum of £7,590, made up as follows:

	£
Land taken	
20 acres of grazing land in the valley bottom at £35 per acre	700
60 acres of sloping land with standing oak wood at £45 per acre	2,700
22.25 acres of mountain land at £30 per acre, say	670
Loss of fishing	1,600
Injurious affection	1,500
Agreed items for loss of water and loss of shelter	420
Total:	£7,590

The acquiring authority will accordingly pay to the claimants the sum of £7,590, to which will be added a surveyor's fee, based on scale 5A of the scales of professional charges of the Royal Institution of Chartered Surveyors, also any legal costs properly incurred from the date of the notice to treat, up to the date of the notice of reference. The total sum thus awarded will carry interest from the date of entry at such rates as may be prescribed by regulations made by the Treasury from time to time.

The acquiring authority will pay to the claimants their costs of this reference such costs, unless agreed, to be taxed by the Registrar of the Lands Tribunal on the High Court scale.

LANDS TRIBUNAL FOR SCOTLAND

DONALD McKELLAR & SONS LTD v GLASGOW CORPORATION

(LTS/APP/3/40)

(W A ELLIOTT QC, President, and WM HALL FRICS)

February 3 1975

Estates Gazette May 3 1975

(1975) 234 EG 377

Butcher's shop in Glasgow—Owners a private limited company of retail butchers with several branches, head office and cooked meat factory—Continued occupancy for eight years after vesting date—Value of property—Tribunal's general comments on settlements—Disturbance—Total extinguishment of business—Forced sale of equipment—Loss of profit—Deduction of director's remuneration—Claim for loss of central profit because of increased overheads at factory not allowed

Mr J M S Horsburgh (instructed by Alexander Jubb & Taylor, of Glasgow) appeared for the claimants; Mr W D Cullen QC (instructed by the Town Clerk, Glasgow) for the acquiring authority.

Giving their decision, the tribunal said: The claimants were owners and occupiers of 257 Buchanan Street, Glasgow, which was compulsorily acquired by Glasgow Corporation under the Cowcaddens Compulsory Purchase Order No 2 1963 with vesting in the corporation on November 11 1965. This is agreed by parties to be the valuation date: *Trustees of Mrs Renfrew v Glasgow Corporation* [1971] RVR 478. The claimants, however, were allowed to continue trading at 257 Buchanan Street until they were displaced in 1973. Although no terms appear to have been agreed between the acquiring authority and the claimants about the basis of the continued occupancy, it appears to be the corporation's practice in these cases to offset statutory interest against what might have been charged as rent.

The claimants are a private limited company still carrying on the business of retail butchers. No 257 Buchanan Street was only one among many of their branches. Mr Thomas McKellar, a director of the company since 1948, gave evidence that there were four full-time working directors operating seven branch shops as at the valuation date in November 1965. Since that date, however, two of the branches have been compulsorily acquired, one closed, and another opened up, so that there were five branches operating at the date of the actual closure of the premises in 1973.

The directors' duties involved going to the market to make purchases, deciding on prices, supervising branches and generally running the business. When occasion demanded, however, they did actual butchering as well. The office administration was carried out at 70 Royston Road, Glasgow. There was also a factory at Drumchapel, where cooked meats were prepared for the various shops. In 1965 this factory employed two men, four youths and one woman and contained substantial equipment. According to Mr McKellar the cooked meats comprised about one-tenth of the total turnover from the shops.

It is not in dispute that the claim is correctly based on a total extinguishment of the branch of the business at 257 Buchanan Street. It is convenient to deal with the claim under separate headings.

Value of Property

A description of the shop, which is now demolished, was produced and agreed by parties. The shop had a ground-floor area of 320 sq ft and a basement of 277 sq ft. It had been extensively fitted out in 1957 and at the date of valuation was adapted as a butcher's shop with tiled walls, terrazzo floor, modern shop-front, hanging rails and cold store. The toilet used by the shop was at the rear of the building and did not appear from the titles to lie within the shop's curtilage; but there was no evidence that this in practice caused difficulties. The shop was in a location which appeared suitable for a butcher's business, being centrally situated and adjacent to a busy bus station. The nearest butcher's shop was at 44 Dundas Street adjacent to Queen Street Station.

Evidence was given by Mr Harry J Crone FRICS [of the firm of Henry J Crone & Partners, of Glasgow] in support of a claim for £16,900. He had taken the alleged price paid by Messrs McKellar in 1957 of £6,500 and applied a growth factor of 3 per cent for the period of eight years to 1965. This figure of £6,500 was arrived at by taking the total purchase price of £8,200 in 1957 and deducting therefrom a figure of £1,700 apportioned to the fittings. This implied that there had been no payment for goodwill. In addition he took the alleged cost of £8,100 spent on the property in 1957 and, allowing for monetary changes and the rise in building costs, applied again a 3 per cent per annum increase over the period. After adjustments the figures at 1965 proposed by Mr Crone were £8,200 for the basic property and £8,700 for improvements, giving a total value of £16,900. He conceded that not all the expenditure necessarily increased the value. This total figure represents, relative to a reduced area of 377 sq ft, a capital rate of £44 per sq ft of ground floor, ie £21 per sq ft for the basic shop and £23 per sq ft for improvements.

Mr Crone was of the opinion that in valuing a small shop a relatively higher rate per sq ft was appropriate. He did not, however, attempt to justify his opinion by comparable transactions. Those he did produce analysed at a maximum rate of £15 per sq ft. He considered, however, that the location of these comparables was such that they were of little assistance in arriving at the value of this particular butcher's shop.

The basic information on costs was given to Mr Crone by the company's accountant, Mr David S Edgar. The latter gave evidence that in 1957 the shop and business had been purchased for £8,200 and apportioned (apparently at the request of the sellers) as follows:

Property	£5,000
Goodwill	£1,500
Fittings	£1,700

There were no accounts available showing the business being done by the previous owners. £5,000 was agreed to be the consideration for the disposition of the heritage mentioned in the recorded disposition. The fittings had been scrapped and substantial alterations carried out to the shop at an alleged cost of just over £8,000. No receipts were available to show how this sum was made up, but Mr Edgar said that these costs were included in the accounts—some against capital but the remainder against revenue. The cost of the new floor, amounting to £666.14, was a capital item and is shown in the balance sheet.

In the revenue account the repairs and replacements over a number of years were as follows:

1955	£1,892	1958	£1,745
1956	£2,983	1959	£2,650
1957	£1,880	1960	£2,229

These figures, however, applied to all branches of the business and no details as to how they were made up are at this date available. The building is now demolished. The tribunal are unable to hold that the above figures substantiate in any way the indirect evidence of Mr Edgar and Mr Crone that improvements at the costs stated were carried out.

Mr Robert Drennan ARICS, district valuer of Glasgow 2, stated that in his opinion the value of the shop as at November 1965 was £8,000. This valuation was represented by the following capital rates per sq ft:

Ground-floor 320 sq ft @ £20	£6,400
Basement 227 sq ft @ £7	£1,589
	£7,989
say	£8,000

Mr Drennan also showed that the analysis of £8,000 as the value of the shop could be illustrated thus:

Ground floor 320 sq ft @ £12	£3,850
Basement 227 sq ft @ £3.50	£800
	£4,650
With addition for heritable improvements	£2,550
Chill room	£800
	£8,000

Counsel for the acquiring authority was anxious, however, that we should not take this as a valuation but only as a way of showing how it could be built up in a similar way to what Mr Crone had done. His figure for heritable improvements was broken down in evidence into the following items:

	£
Floor	670
Wall tiling	208
Shop front	750
Work in basement	250
Rear window	100
Removing pillar	250
Hanging rails	250
	£2,478

This, however, was the estimated cost in 1957 and no figures were included for plumbing and electrical work.

Settlements

In arriving at the alternative figures of £12 and £20 per sq ft Mr Drennan had regard to 19 compensation settlements in the area, mostly with chartered surveyors, including Mr Crone's firm. These settlements analysed to show in the same part of Buchanan Street rates from £10 per sq ft for unimproved premises to £24 per sq ft for a snack bar which had been modernised. In adjacent streets rates varied from £7 per sq ft to £21 per sq ft.

These settlements were all involved in the Cowcaddens Compulsory Purchase No 2 Order, and the figures for the actual properties—as opposed to disturbance—were apportionments made by his office of the total compensation settlements and not necessarily those apportionments agreed in all cases with the claimant's surveyors. Mr Crone nevertheless accepted that the apportionments involved in cases with which his firm was concerned were correct.

Mr Drennan explained, however, that in dealing with the initial settlements formed following the 1963 CPO regard was had by him to sales in some cases as far back as 1957 to assist in determining values as at 1965. Evidence of such sales, however, was not given in evidence. Mr Drennan accepted that the values of shops in this area would have increased by about 25 per cent from 1957 if there had been no scheme.

The recorded purchase price of £5,000 for the subjects had been taken into account by Mr Drennan—although he regarded it as on the high side compared with other sales. But we are unimpressed with this argument, as no evidence of such other sales was placed before the tribunal.

In terms of rule (2) of section 12 of the Land Compensation (Scotland) Act 1963 we have to determine the amount which the shop if sold in the open market by a willing seller might be expected to realise—disregarding any depreciation in value due to the corporation's redevelopment proposals. The tribunal are reluctant to accept settlement figures for compensation for properties acquired under a comprehensive redevelopment scheme where there exists evidence of comparable sales in the vicinity which are not affected by the scheme. The latter are the best evidence of open market value. We adopt, however, as a guide what was said by the Lands Tribunal (Sir Michael Rowe QC) in *Xerri and Shanks v Enfield London Borough Council* [1966] RVR 848, 850:

"The claimant's valuer distrusted the district valuer's settlements. In his view an owner anxious to sell had agreed to the first purchase, and once the district valuer had that precedent in his bag the pass had been sold and even surveyors, acting for owners or leaseholders not in a financial position to litigate, could not hope to do much better for their clients. This tribunal has said on more than one occasion that the danger to which the claimant's valuer drew attention is a real one, though it seldom carries with it any stigma of unfair conduct by a district valuer or his officers; but the tribunal has also said that if settlements of this kind are to be disregarded, there must be evidence clear and positive to prove that they were on too low a basis. If there have been sales of truly comparable properties in very close proximity to the subject land effected in the open market and not, so far as can be ascertained, affected by any special factors, those sales should be regarded as better evidence of market value than settlements effected by the district valuer even if surveyors were acting for the claimant."

In that case the learned President, however, declined to prefer two open-market transactions relating to property some distance away. In the more recent case of *Shaw v Hackney London Borough Council* (1974) 231 EG 1300 the tribunal did prefer evidence of two adjacent sales in preference to the acquiring authority's schedule of settlements.

We have been given no evidence of sales of modernised butcher's shops, and therefore in determining the value we have to do our best with the evidence and information which was put before us. This may be summarised as follows:

(a) A price of £8,200 was paid in 1957 for the going concern carried on at 257 Buchanan Street.

(b) The consideration for the property in the 1957 disposition was £5,000;

(c) Although the exact costs were not proved, considerable alterations were carried out to the shop after its purchase at a cost which was estimated by the district valuer at not less than £2,500 plus £800 for a chill room (ie £3,300) and by the claimants at £8,100;

(d) The district valuer has settled compensation for modernised shops within the same compulsory purchase order area at figures which analysed for the ground floor up to approximately £24 per sq ft.

A We consider that a purchaser wanting to buy a butcher's shop in this trading position would have some regard to the costs involved in buying a shop and reconstructing it to provide a modern shop, fitted and equipped for a butcher's business. If the purchaser had been considering these costs at 1965 they would be considerably higher than those of 1957.

In view of the lack of positive evidence before the tribunal we have experienced difficulty in this matter, but we eventually determine the value of the shop at £9,700 computed as follows:

Ground floor 320 sq ft @ £24	£7,680
Basement 227 sq ft @ £9	£2,043
B	£9,723
	say £9,700

The £9 per sq ft includes for the new chill room installed in the basement. As an overall check we also consider that the above figures are not unreasonable having regard to the figure of £5,000 paid for the heritage in 1957.

Disturbance

The claim under this heading is separated into three parts and we shall deal with it in the same way.

C (a) *Claim for loss on forced sale of equipment*

Equipment	*Cost* £
Scales (two)	850
Cash register	395
Refrigerated counters ...	1,200
Deep-freeze cabinet ...	600
Slicer	150
Mincers (two)	550
	£3,745
Less 20% depreciation ...	758
D	say £3,000

These figures were supplied by Mr Edgar, who had taken firstly the cost of replacing with new equipment at dispossession. The depreciation figure of 20 per cent, as we understand it, is to reflect the wear and tear of the equipment since it was purchased in 1957 or 1958. Mr McKellar stated that no sum had been received for the equipment, most of it being left on the premises. Two new scales had been taken away—the other two scales which had been converted for decimalisation were left. Mr McKellar explained that there E was no market for second-hand butcher's equipment. It appears that as a corollary there is no supply at any particular time. In consequence new equipment is usually purchased when fitting out a butcher's shop.

Mr Drennan proposed a sum of £756 for the loss on forced sale of equipment. This figure was based on a value of £840 for movable equipment which appears to have been agreed for estate duty purposes in 1968 as the value of the movable assets in the going concern at that time. Mr Drennan then made a deduction of 10 per cent in respect of further depreciation of the equipment for the five years up to 1973, when the claimants moved from the premises. But he F explained that this deduction was really to take into account the benefit the owner had in using the equipment during the period from the valuation date until the premises were vacated. His proposal accepts that there is no salvage value.

No Realisable Value

We have to determine the owners' loss. Both parties computed the value of the equipment to the owners at the date of dispossession. At the date of dispossession in 1973 the owners had equipment which was of value to them in the continuing business if it had not been compulsorily acquired. They contended, however, that the equipment had no realisable value if sold on the open market. This was accepted G by the acquiring authority. The factual loss would be incurred at the date when the equipment ceased to be used, ie in 1973. If there had been a realisable value, then one would have taken that as the figure actually received when it was sold. In view of the fact that it is agreed in this case that there is no realisable market value, the dispute between the parties is the value to the owner of the equipment at the date of the realisation or, in this case, of non-realisation when the equipment required to be scrapped. The claimants valued this scrapped equipment at £3,000 and the acquiring authority at £756—both at 1973.

We do not accept that the basis of value for estate duty H purposes at 1968 was necessarily on the basis we have now to take for the disturbance claim following a compulsory acquisition and which reflects value to the owner. The market value at 1968, when the relevant death occurred, would have regard to the fact that the business was being carried on in the premises at the whim of the acquiring authority and could be terminated at any time.

The costs to replace the equipment with new, as submitted by the claimants, were not challenged by the acquiring authority. The deduction of 20 per cent, however, is not convincing and the basis for that deduction is obscure. As we have stated previously we consider that the acquiring J authority's basis of value is suspect. The additional deduction for depreciation from 1968 to 1973 does not appear to take into account the presumably rising cost of new equipment during that period. Photographs of most of the relevant equipment in situ were taken just before dispossession and produced at the hearing. From these photographs the equipment appeared to be modern and attractive.

We consider that the value proposed by the claimants is too high and that the acquiring authority's value is too low. In these circumstances, we have to make a judgment on the information before us. We do so at the sum of £1,500, which is 50 per cent less than the cost of new equipment. It may be K noted that the apportionment of the price of £8,200 paid for the business in 1957 included £1,700 in respect of fittings. As there was no salvage value the loss under this item is determined at £1,500.

We have not made any discount back to the valuation date to allow for statutory interest as from that date, as the acquiring authority did not ask us to do so.

(b) *Claim for loss of profit due to cessation of branch*

Mr Crone supported the claim under this heading at £11,694 calculated as follows:

Net profit 1965 ...	£1,688 L
Add director's salary	2,210
	£3,898
Years' purchase ...	3
	£11,694

Mr Drennan, after hearing the claimant's evidence, proposed a sum for this item of £2,156, which represents two years' purchase of an adjusted net profit of £1,078. Both parties were agreed that we should have regard to 1965 profit levels for calculating loss of profits.

Both valuers derived their profit from a statement prepared M by Mr Edgar on behalf of the claimants showing the apportionment of profit and outgoings to the branch at 257 Buchanan Street. The gross profit of £11,242 and the outgoings of wages, rates, insurance, heating and lighting for the branch of £4,444 for the year ended October 31 1965 were agreed. The actual net profit from the branch as a contribution to the main company was therefore £6,796. Deductions were also made in respect of costs and expenses incurred by the company in respect of all their branches and these are the subject of a different approach by the two parties.

A In Mr Edgar's statement a deduction was made of one-fifth of the company's expenses for the year ended October 31 1965 and a sum of £2,210 for the salary of one full-time working director. The deduction of one-fifth appears to arise from the fact that there were five branches in 1973—just before Buchanan Street was closed. The net profit shown in the statement is £1,688, which is the figure adopted by Mr Crone as the base for his computation.

Mr Drennan's deduction for company expenses was done on the basis of the proportion of turnover of the branch to the whole business in the appropriate year. This produces a factor of 0.176 which means a lower deduction than the 0.20
B proposed by Mr Edgar. Mr Drennan has also applied the 0.176 to the total directors' remuneration. This gives a figure of £1,556, which again is a lower figure than Mr Edgar's.

Further adjustments, however, were made by Mr Drennan as follows:

(i) Evidence was given that in 1973 the costs of the head office and the running of the factory at Drumchapel were £10,500. This was apportioned as to £2,000 for head office expenses; £5,000 for the factory and £3,500 for transport and carriage costs. Mr Drennan said that the £3,500 for transport had already been taken into
C account in Mr Edgar's deduction, so that only the £7,000 had still to be set against the branch profits. As the £7,000 was a 1973 figure Mr Drennan reduced that to £4,900 after a study of the costs between 1965 and 1972. He then applied his ratio of 0.176 to that, giving a deduction of £860.

(ii) Mr Drennan also deducted £640 from the profits, being the notional rent of the shop and representing 8 per cent of his value of £8,000.

(iii) He also made a deduction for interest on capital of £200, which represented 8 per cent of £2,500—his allocation to the branch of the capital involved in the
D branch and the business so far as the branch was concerned. Although not specifically stated, this would appear to include the value of the equipment at £756.

After making these deductions Mr Drennan's calculation of the net profit was thus £1,078. Both parties are agreed that at least some of the company's general expenses should be deducted to arrive at the loss of profit. So far as the proportion of these expenses is concerned, Mr Drennan's ratio is more beneficial to the claimant, so we accept his ratio of 0.176.

Head Office and Factory Expenses

E The expenses of the head office and the factory are a part of the company's outgoings, and on the evidence these were necessary to maintain the profitability of the branches. The question of whether head office expenses should be deducted from the profits of a branch depends on the ability of the company either to absorb this expenditure or reduce it or to replace the turnover by opening another branch.

No evidence was led to show that the claimants in this case were not able to contain their costs following the closure of the branch. Over the years branches had been closed and opened by the company. It also appears difficult for the claimants in this case to concede that a substantial part of the
F company's outgoings should be deducted as a proportion—but that other expenses should not. In the circumstances of this case we therefore accept that the deduction of £860 proposed by Mr Drennan should be made.

So far as the rent deduction is concerned, this was not seriously opposed by the claimants and therefore we consider that a deduction is also appropriate. The deduction proposed by Mr Drennan was 8 per cent applied to his opinion of value. We have determined the value at £9,700; so applying 8 per cent we arrive at a notional rent of £776.

The deduction for interest on capital was not challenged by the claimants and we accept that this should also be made. As we have increased the value of the equipment we have G
made an adjustment in this item. Mr Crone in his computation added back the allocation made by Mr Edgar for director's salary. His reason appeared to be that the loss of the branch would cause a loss of remuneration to a director. But in the tribunal's view the proper question is whether director's salary should fairly be deducted in arriving at the claimant's net profits.

Director's Remuneration

The argument in connection with director's remuneration is linked to the argument about the deduction for company expenses. The question of whether such a deduction should H
be made to arrive at net profits has been dealt with in a number of Lands Tribunal cases in England. It has been held that in the case of a limited company with branches a deduction is appropriate: see *Cliffords (Dover) Ltd v Dover Corporation* (1965) 195 EG 821 and *Freeman Hardy & Willis v Bradford Corporation* (1967) 203 EG 1099. In the case of *Widden & Co v Kensington* (1970) 213 EG 1442, however, it was agreed by both the claimants and the acquiring authority witnesses that no deduction should be made for director's fees; but in that case the business was run by the director himself and his wife. The matter was also considered in the J
recent case of *Longbottom v Bingley* (1974) 230 EG 371, where no deduction was made for the director's remuneration in the case of a partnership.

This is a larger business involving a limited company with several branches where we consider that a deduction is appropriate. In the case of *Shulman (Tailors) Ltd v Greater London Council* (1965) 196 EG 453 at 458 the tribunal said:

> "I accept evidence that in the great majority of small family businesses profitability is more reliably measured by ignoring the directors' fees actually charged and substituting instead a reasonable figure being the value of the directors' services had they been employees." K

Mr McKellar gave evidence that the directors spent considerable time in running the business and that they were also actively and physically involved as butchers. The branch profits therefore were dependent on the directors' work. In this case the figure of £1,556 proposed by Mr Drennan appears to us to be reasonable and is less than the figure proposed by the claimants themselves. The adjusted net profit, allowing for the increased rental, is therefore £882.

Mr Crone had proposed a multiplier of three years' purchase to the net profit whereas Mr Drennan had applied two years' purchase. There was no direct evidence to establish what the correct multiplier should be. Mr Drennan indicated L
some transactions in connection with butchers' businesses; but no details were given of these and we were unaware whether these were one-man businesses or branches with head office expenses set against them. We consider in the circumstances of this case—where we have a shop well modernised and making a substantial contribution to the company organisation—that a multiplier of three years' purchase is more appropriate and we so determine. This gives a value of £2,646.

(c) *Claim for loss of central profit because of increased overheads at Drumchapel* M

The claimants propose a sum of £6,300 under this heading, which is computed as follows:

Distribution charges at Drumchapel	£10,500
There are five branches, therefore the loss in Buchanan Street branch	£2,100
At three years' purchase	£6,300

The acquiring authority rejected this claim on the basis that no loss had been proved.

We consider that the question of the overheads at Drumchapel and head office expenses has been dealt with under the previous heading. If the costs at Drumchapel had

A continued and no economies could be made and it was not possible to obtain another branch in which to absorb these costs, then this claim might have been valid. We consider, however, that this part of the claim is not established and we reject it.

We therefore determine the compensation at £14,019, calculated as follows:

Value of property	£9,700
Loss on forced sale of equipment	1,500
Loss of profit	2,646
Surveyors' fees under Scale 5 (a)	173
B	£14,019

We consider that there was some justification in the acquiring authority's complaints that they had been supplied with insufficient information to enable the disturbance claim to be properly assessed.

In the whole circumstances we make no finding as to expenses.

C

ROBERTS v GREATER LONDON COUNCIL
(REF/174/1972)
(J STUART DANIEL QC)
March 14 1975

Estates Gazette May 10 1975
(1975) 234 EG 443

Acquisition of Islington house—Resumed hearing to determine disturbance claim—Other compensation determined by tribunal 15 months earlier—Items allowed include interest on bank loan for deposit on new house and cost of school uniform and equipment for children at new D **school—Items unsupported by accounts or receipts disallowed or reduced—No order as to costs of the reference**

Giving his decision, MR DANIEL said: In the interim decision in this case given on December 11 1973, 229 EG 975, I left over for final decision the removal costs and the "Crawley" costs.

At the resumed hearing Mr J C Harper (instructed by the Solicitor of the Greater London Council) appeared for the council and Mr Roberts appeared in person.

Mr Roberts told me that he had sent to the council a copy of his particulars of claim for the above outstanding items on August 30 1974. The council, he said, had asked E for receipts and evidence in support of his claim and he had taken the view that it would involve him in an expenditure of £400 in photographs if he complied with the council's request. After some correspondence he therefore referred the matter to the Lands Tribunal.

I took the view, after a few minutes, that it would be foolish for me to go through all the items if some of them were evidenced by documents and were capable of agreement. Accordingly I invited the parties to sit round a table and see whether some agreement could be reached and I granted a short adjournment for this purpose. Some items were then agreed, in particular those where there was no F dispute in principle and where receipts were now furnished. I will now set out the whole of Mr Roberts' particulars of claim noting the agreed items and giving my decision on those still in dispute:

Item 1 *Expenses in relation to the old house: 112 St John's Way, London N19.*

(a) *legal fees* (already paid by the council) £93.50

(b) *insurance* (December 12 1973-July 12 1974) £9.42

Mr Roberts lived in the house during this period and had the benefit of the insurance cover. As Mr Harper said, Mr Roberts was no doubt well advised to keep this insurance alive. I do not see why the council should pay this item and G I disallow it.

(c) *loss of earnings* by Mr Roberts (four days for inspection by the council, two days for interviews by the council, two days for interviews for rehousing by Islington Council, a day at County Hall, a day for removal, a day for rates, three days for loan arrangements in respect of purchase of 81 Sotheby Road, N5 (abortive), 52 Ferme Road, N4 (abortive), 58 Victoria Road, N4, and a day to sign papers for gas, electricity and telephone at both the old house and the new house) (15 at £6.81 a day) £95.34

No 58 Victoria Road is Mr Roberts' new house. It is clear that he lost a certain amount of time making arrangements H in and about the change of home and in investigating abortively other houses. Mr Roberts claims at the rate of £6.81 per day which is his present rate of wages; he agreed that his wages had varied from time to time, and indeed that till June 1971 he had been a student. Mr Harper questioned the number of visits and the necessity of losing whole days on these occasions. He suggested that some of these matters could have been dealt with by post. Mr Roberts said that all these things required personal attention and that it was difficult to obtain a part only of a day's wage. I find that this claim is so imprecise that it is very difficult to assess. Mr Harper accepts that a reasonable figure should be allowed J and, doing the best I can, I fix the amount at £30.

Item 2. *Costs of reference*
I will leave this item to the end.

Item 3. *Expenses in relation to new house: 58 Victoria Road.*

(a) *legal fees, etc.*

(i) land registration	now agreed	£31.50
(ii) solicitor	now agreed	£93.93
(iii) surveyor	now agreed	£44.00 K

(iv) interest on loan and late completion—(£25 + £6.54) £31.54

Mr Roberts said that he claimed £25 interest in respect of the deposit of £300 which he had had to pay on the new house and which he had had to borrow from the bank during the second half of 1974. The £6.54 was interest caused by late completion. I am quite satisfied that completion was not delayed by the council and I disallow the £6.54. But judging from what Mr Roberts told me and from some figures he showed me and from some letters produced by the council it does seem that there was some interest payable on a L deposit. I therefore allow £25.00

(b) *abortive expenses*

(i) *surveyor's fee for structural survey of:*

(a) 14 and 14a Rathbourne Road	£15.00
(b) 17 Burgoyne Road	£15.00
(c) 81 Sotheby Road	£15.00
(d) 52 Ferme Road	£44.00

As regards the first three of these properties, Mr Roberts said he had paid for the surveys in cash and had no receipts. As regards the fourth, he produced a receipt for £44 and this item was agreed. This is the kind of item which needs to be M vouched for. I disallow the first three items and allow the fourth at: £44.00

(ii) *solicitor's fees*

(a) 14 and 14a Rathbourne Road, SW12 (£20.20).

Of this item £17.20 was agreed. Mr Roberts produced three cheques from himself to his solicitors which in all totted up to £22.20. But, he said, £2 was in respect of another matter. I allow £20.00

(b) 52 Ferme Park Road, N4: this was agreed at £ 7.58

A (iii) *travelling expenses to agent, etc*

(a) agents	£ 8.50
(b) solicitors	£ 5.40
(c) viewing hours	£12.60

Mr Roberts said that this expenditure had been incurred between January 1972 and May 1974; all this travelling had been by bus; he had 53 times travelled to view houses at a fare of 16p to 20p; he had visited different agents and different solicitors; it had been impossible to jot down the actual amounts spent.

Mr Harper accepted that something should be allowed
B here but suggested that without any proper evidence or records the claim as it stood was exorbitant. I agree. I allow £10.00

(iv) *telephone calls, correspondence and postage to:*

(a) agents	£20.77
(b) solicitor	£ 8.10
(c) Greater London Council	£ 1.00
(d) vendors and agents of:	
(i) 14 and 14a Rathbourne Road, SW12	£ 2.00
(ii) 17 Burgoyne Road, N4	£ 0.52
(iii) 81 Sotheby Road, N5	£ 0.50
C (iv) 52 Ferme Road, N4	£ 0.80
(v) 58 Victoria Road, N4	£ 0.60

This is the same story. Nothing really is proved, though obviously something was incurred. I allow £10.00

(v) *loss of earnings*

(a) 3 days at Rathbourne Road at £6.81 a day	£20.43
(b) 2 days at Burgoyne Road at £6.81 a day	£13.62
(c) A day at Sotheby Road at £6.81 a day	£ 6.81
(d) 3 days with solicitor to sign the contract and interviews, etc, at £6.81 a day	£20.43

D This item is very similar to that at item 1 (c). I allow £25.00

Item 4. *Expenses in relation to moving to new house:*

(a) removal costs, these are agreed at £63.80

(b) insurance of goods, £2.50 agreed; tips (on moving goods) £5.00

I allow this, making £ 7.50

(c) cost of 8 electric plugs, etc: agreed at £ 7.40

(d) disconnection of 6 gas points at £3.09 each: agreed at £18.54

(e) disconnection of electricity, agreed £ 1.06

E (f) reconnection of 4 gas points at £5.09: £20.32 (amended to £52.13)

Mr Roberts showed me some documents which mentioned, among other things, some auto-check meters. He said he did not want these for tenants, but simply for security. But these documents, as Mr Harper pointed out, were only "specification and estimates." I am afraid they prove nothing and this is the kind of item which I do not think I can allow without receipts. I disallow this item.

(g) cost of floor and stair coverings: £250.89

I found this item of claim difficult to follow. But, as far
F as I could understand him, Mr Roberts said that when he measured the floor coverings at the old house some of these were fitted carpets, others were linoleum with loose squares of carpet on top of the linoleum; then on the stairs and hall there was felt-backed linoleum (Rossella); all these items were in good condition; if he replaced these items it would originally have cost £93.50, but some were unobtainable and the prices went up and up; he had been using the floor coverings for 11 years but they would have lasted another five years; he thought the claim of £250 was fair. Mr Roberts produced receipts for some 60 yds of Rossella which he had bought for the new house; he had spent £79 on this.

In connection with this item Mr Harper called Mr D R G
Hunt, of the supplies department of the council. Mr Hunt said that he had visited Mr Roberts at 112 St John's Way in February 1974; he had inspected the floor coverings in detail; generally they were of poor quality, though they were better in the hall and on the stairs; he agreed that there were three fitted carpets and some loose squares of carpet over linoleum; he valued all the coverings *in situ* from his experience at the figure of £44.20. I accept Mr Hunt's valuation at £44.20

(h) reconnection of telephone was agreed at £27.50

(i) redirection of mail for one year was agreed at £ 5.00 H

(j) cost of postage of notification of change of address to relatives, friends, etc, was agreed at £ 1.52

(k) school uniform and equipment for two children at new school £64.95

With regard to (k) Mr Roberts said that when he moved house it had been necessary to move the children to a new school at Stroud Green; he and his wife had called on the headmistress who had told them that it would be necessary to provide certain items of clothing; the headmistress furnished a list of necessary items and also the name of the shop where the things could be bought. Mr Roberts produced J
receipts for £64.95.

I asked Mr Harper if the council could investigate this matter and the council's representative did this during the adjournment. Mr Harper then told me that parents were at any rate submitted to a strong persuasion to provide these things, which made it reasonable to accept this claim in principle, though he queried whether some of the items were not standard, or could not be more cheaply bought.

I think this claim is made out and I allow it at: £64.95

Item 5. *Accountant's fee for preparing this claim*: £20.00

This item was not supported by an account or receipt. I K
disallow it.

This means that the total of the figures now agreed or decided by me in respect of the outstanding items is £676.18. As the council has already paid £93.50 (item 1 as above) there remains due to the claimant the sum of £582.68.

Before I come to the question of costs I should mention that Mr Roberts referred me to section 37 of the Land Compensation Act 1973. This section is concerned with disturbance payments to persons without compensatable interests. I do not think it has any relevance to these proceedings.

Costs of Reference L

This only leaves the costs of the reference. Here the situation is rather unusual. On November 11 1973, that is two or three days before the first hearing of this case started, the council made an unconditional offer to the claimant to pay him £16,000 for the freehold and for any other compensation to which he might be entitled, and to pay a surveyor's fee of £195.30, and all proper legal costs incurred from the date of the notice to treat up to the date of the notice of reference.

After I had read my interim decision in court this sealed offer was opened by agreement of the parties and put before me and it was submitted to me that the costs so far could M
be dealt with there and then. Quite apart from the figures, reliance was placed on behalf of the council on section 4 (1) (a) and (b) of the Land Compensation Act 1961, where it is provided:

4 (1) Where either—

(a) the acquiring authority have made an unconditional offer in writing of any sum as compensation to any claimant and the sum awarded by the Lands Tribunal to that claimant does not exceed the sum offered; or

(b) the Lands Tribunal is satisfied that a claimant has failed to deliver to the acquiring authority, in time to enable them to make a proper offer, a notice in writing of the amount claimed by him, containing the particulars mentioned in subsection (2) of this section;

the Lands Tribunal shall, unless for special reasons it thinks proper not to do so, order the claimant to bear his own costs and to pay the costs of the acquiring authority so far as they were incurred after the offer was made, or as the case may be after the time when in the opinion of the Lands Tribunal the notice should have been delivered.

(2) the notice mentioned in subsection (1) of this section must state the exact nature of the interest in respect of which compensation is claimed, and give details of the compensation claimed, distinguishing the amounts under separate heads and showing how the amount claimed under each head is calculated.

It was submitted for the council (after I had read my interim decision on December 11 1973, in which I awarded £15,500 and surveyor's fees of £190) that having regard to the amount of the offer and the fact that the claimant had not particularised his claim for disturbance the council should have their costs after the date of the offer. Mr Roberts said (indeed he said it during the first hearing in November 1973) that he could not give details of his disturbance claim because the disturbance was not yet complete, and as appears from my interim decision I decided to leave the whole question of costs to abide the final outcome of the case.

The outcome is that the total amount of compensation now agreed and determined is as follows:

Interim Decision (December 11 1973)	£15,500
add surveyor's fee	190
Already awarded	£15,690
This Decision (March 14 1975) re: outstanding matters	£ 676.18
Total:	£16,366.18

The unconditional sealed offer of compensation made by the Greater London Council on November 12 1973 was £16,000 plus a surveyor's fee of £195.30, giving a total of £16,195.30. Thus the offer is about £170 below my award. Under these circumstances section 4 (1) (a) of the Land Compensation Act 1961 does not therefore come into play. It remains to consider whether section 4 (1) (b) and subsection (2) are applicable. The sealed offer of November 1973 was a package deal; it purported to make an offer for all the compensation to which the claimant might be entitled. After the interim decision the parties knew that there was £500 (£16,000 less £15,500) in hand to meet the reserved matters. It was open to the council, as I see it, to make a further offer protecting themselves in costs, more particularly in respect of the second hearing (when it became apparent that this would be necessary). I am satisfied, however, that the claimant's particulars of August 1974 as to the outstanding matters were not a notice in writing in sufficient detail in accordance with section 4 (1) (b) such as to enable the council to make a further offer, for although the particulars distinguished the amounts under separate heads they did not as regards several of the heads (and these the more debatable ones) show how the amounts were calculated. It is arguable, therefore, that in accordance with section 4 (1) (b) the claimant should pay the council's costs after the time when I think further particulars should have been delivered. But no request for these was made, I think, and it is in any case difficult to apply the strict provisions of section 4 in all the special circumstances of this case, where the interim decision dealing with the claim for the freehold and some other matters was followed by this decision on the remaining matters.

I think on the whole justice will be done if I make no order as to the costs of the reference between the parties.

MARZELL v GREATER LONDON COUNCIL

(REF/104/1974)

(J D RUSSELL-DAVIS FRICS)

March 19 1975

Estates Gazette May 24 1975

(1975) 234 EG 621

Shop premises in Bethnal Green, London, included in clearance area as unfit—Occupied as tobacconists/confectioners by private limited company of which freeholder claimant and his wife the only shareholders—Company held not the "alter ego" of the claimant—No entitlement to business supplement under Housing Act 1957, Schedule 2, para 6—Point not previously tested before tribunal

Giving his decision, MR RUSSELL-DAVIS said: This reference was initiated by Mr Morris Marzell, the owner of a shop and premises known as 51 Cheshire Street, Bethnal Green, London E2, which is the subject of compulsory acquisition under the provisions of the Housing Act 1957. It concerns the compensation to be paid under section 59 and Part II of the Second Schedule to that Act, and in particular, a fairly narrow point of construction of, and application to the claimant of, paragraph 6 in relation to his entitlement to a supplementary payment which, I am informed, has not previously been tested before the Lands Tribunal.

Mr Bernard Marder (instructed by Howard Kennedy & Rossi) appeared for the claimant and called evidence from Mr Stewart David Marzell FCA, chartered accountant, the claimant's son and financial and business adviser. Mr Michael Rich (instructed by the Solicitor and Director of Legal Services to the Greater London Council) appeared for the acquiring authority but called no evidence.

Agreed Facts

A number of facts are agreed which I will summarise:

1. The subject property comprised shop, parlour and kitchenette on the ground floor with storage basement, yard and external wc, and in the upper part, to which there was a separate entrance, there were four rooms on two floors, which to varying degrees were used as stock rooms and for storage.

2. It was included in the London County Council (Fuller Street, Bethnal Green) Compulsory Purchase Order made on November 27 1964 under Part III of the Housing Act 1957. The premises were included in the clearance area as unfit and were confirmed as such by the Minister of Housing and Local Government on April 4 1966.

3. Notice to treat was served on September 7 1966 and notice of entry on January 4 1972, possession being taken by the acquiring authority on February 5 1973, which in consequence becomes the date of valuation for this purpose.

4. The claimant first held the premises in 1938 as a tenant. In 1948 he purchased the freehold. He carried on business as a tobacconist and confectioner in his own name until October 1963 and his own name remained on the facia until demolition. The family occupied the upper part as a dwelling until 1958-59.

5. On October 16 1963 M Marzell (Confectionery) Ltd was incorporated as a private limited company with an authorised share capital of 100 ordinary shares of £1 each. Only two shares were actually issued, one to the claimant and one to his wife. This company took over the business. The claimant retained the freehold interest in 51 Cheshire Street; and the company paid no sum for goodwill established at that address; it was not granted a lease or any form of tenancy agreement and no rent was charged against profits.

6. The claimant also had lock-up premises in High Road, Wood Green, London N22, of which he held the lease from March 25 1954 and at which he conducted a similar business until the premises were compulsorily acquired and possession

A given in June 1973, following which he removed to 143 High Road close by. This business was also included in the transfer to the limited company.

7. The Wood Green business was conducted by the claimant without any staff, the shop being open from 9.30 am to 5.30 pm daily except Sundays. He opened the Cheshire Street business at 7.30 am and after the Wood Green shop was closed he returned to Cheshire Street until that shop closed at 7.15 to 7.30 pm. On Sundays he worked there from 8.30 am to 3 pm. The shop was therefore open for 66½ hours a week, and the business was conducted by Mrs Marzell, his wife, with a part-time help from a female assistant at a low
B wage. Her children occasionally helped out.

8. One set of trading accounts was prepared annually for the company by the firm of Ash, Marzell & Co, chartered accountants, in which firm Mr S D Marzell is a partner.

9. The value of the cleared site of the subject premises as at the date of the valuation and in accordance with the provisions of section 59 is agreed at the sum of £850. The full compulsory purchase value of the freehold interest on the assumption that it had not been represented and confirmed as unfit is agreed to have been the sum of £3,850.

10. If the claimant is held to be entitled to the supple-
C mentary payment of £3,000, it is agreed that the disturbance claim shall be settled at £2,950 as follows:

loss of goodwill	£2,600	
loss on forced sale of stock and fixtures and fittings	300	
loss on transfer of stock	25	
accountant's fees for preparing separate accounts	25	
	£2,950	plus surveyor's fees

11. If the claimant is unsuccessful in establishing that he is entitled to a supplementary payment it will be open to the
D company to make a fresh claim for disturbance which the authority will consider on its merits.

Relevant Statutory Provisions

At this stage it is convenient to set out the relevant parts of sections 59 and 61 of, and paragraph 6 of the Second Schedule to, the Housing Act 1957 (as amended) as follows:

"59 (1) Where land is purchased compulsorily by the local authority under this part of the Act, the compensation payable in respect thereof shall be assessed in accordance with the Acquisition of Land (Assessment of Compensation) Act 1919, subject to the following provisions of this section.

E (2) The compensation to be paid for land, including any buildings thereon, purchased as being land comprised in a clearance area shall be the value at the time the valuation is made of the land as a site cleared of buildings and available for development in accordance with the requirements of the building by-laws for the time being in force in the district:
Provided . . . (3) . . . (4). . . ."

"61. The provisions of Part II of the Second Schedule to this Act shall have effect as respects payments to be made in certain cases in respect of houses comprised in a clearance area or redevelopment area."

Paragraph 6 in Part II of the Second Schedule has the head
F note "Business Premises" and the relevant part reads as follows:

"(1) This paragraph shall have effect where on or after the 13th day of December 1955 a house has been purchased at site value in pursuance of a compulsory purchase order made by virtue of Part II . . . of this Act or vacated in pursuance of (a) . . . (b) A clearance order.

(2) If at the date of the making of the order the house was occupied wholly or partly for the purposes of a business and the person entitled to the receipts of the business held an interest in the house, the local authority by whom the order was made shall make in respect of that interest a payment of the amount hereinafter specified:

Provided that no payment shall be made under this paragraph in G respect of any interest in a house unless the house was occupied wholly or partly for business purposes, and a person entitled to the receipts of a business carried on wholly or partly therein held an interest in the house, either on the 13th day of December, 1955, or at all times during the two years preceding the date of the making of the order.

(3) The amount of the payment shall be the full compulsory purchase value of the interest less the compensation which was or would have been payable in respect of the interest in connection with the compulsory purchase of the house at site value:
Provided . . . (4) . . . (5). . . ."

The question before the Lands Tribunal for determination is: Did a person entitled to the receipts of the business H carried on therein hold an interest in the house at all times during the two years preceding the date of the making of the compulsory purchase order (that is to say, since before November 27 1962)? If so, the claimant will be entitled to the sum of £6,800 by way of compensation calculated as follows:

Site value	£850
Supplementary payment under paragraph 6 of the Second Schedule	3,000
Disturbance	2,950
	£6,800 J

If he is not so entitled he will receive only the site value of £850, it then being open to the company to establish a claim for compensation for disturbance.

From the evidence of Mr S D Marzell, I find the following additional facts:

(a) The transfer by the claimant to the company in October 1963 provided for no cash payment for fixtures and fittings, stock and a motor car, etc, but the directors' loan account was credited with the value thereof.

(b) There was no service contract with the directors. Mrs Marzell acted as secretary. Mr Marzell kept the day-to-day books of K the business. Mr S D Marzell provided accountancy services and advised his father throughout.

(c) Annual accounts were prepared in the form required by the Companies Acts and consisted of a directors' report, balance sheet, trading and profit-and-loss accounts.

(d) No dividends were distributed, the directors taking their remuneration in the form of "directors' remuneration" by periodical drawings, any overdrawing being debited to the loan account. The directors' remuneration was unallocated, but Mr Marzell took about 80 per cent and Mrs Marzell about 20 per cent. A formal annual general meeting was held.

(e) The claimant personally ran the two businesses and arranged for the buying of stock. He was solely responsible for policy and L all matters affecting the businesses.

Under cross-examination Mr Marzell said that the reason for the formation of the company was two-fold, firstly, tax advantages and, secondly, enhancement of status when dealing with suppliers. The company had no other activity or asset. Mr Marder contended that the whole circumstances made the company artificial, and the creature and the *alter ego* of Mr Marzell. Only two shares were issued. No dividends were paid. There were no outside directors. There is a long line of cases on the legal identity of a company in the context of disturbance claims, but none has ever arisen on this particular aspect of a supplementary payment, and although there was M a general rule that the "corporate veil" cannot be lifted to reveal the individual underneath, it was permissible to examine what degrees of control were exercised in certain of the cases.

Smith, Stone & Knight's Case

Although it was a "stray" case, he thought the judgment in *Smith, Stone & Knight Ltd v Birmingham Corporation* [1939] 4 All ER 116, which had the greatest relevance to the present case, should be followed. That was a compulsory purchase case in which a parent company carried on business

A by a subsidiary company in the subject premises, all the shares of the latter, except five, being held by the parent, those five being held by the directors in their respective names. The profits of the subsidiary were treated as the profits of the parent, who appointed the persons who conducted the business, and were in constant and effective control; the books of account were kept by the parent and none by the subsidiary. It was held that possession by a separate legal entity was not conclusive as a right to claim, and as the subsidiary was not operating on its own behalf, but on behalf of the parent, the latter was the party to claim compensation.

This was a question of fact analogous to the present case
B in which Mr Marzell would be in the shoes of the parent. Six questions were posed by Atkinson J (as he then was), all of which could be answered favourably for the present claimant.

1. Were the profits treated as profits of the parent?
2. Were the persons controlling the business appointed by the parent?
3. Was the parent the head and brain of the trading venture?
4. Did the parent govern the adventure, decide what should be done, and what capital should be employed in the
C venture?
5. Did the parent make the profits by its skill and direction?
6. Was the parent in effectual and constant control?

He added " Indeed, if ever one company can be said to be the agent or employee, or tool or simulacrum of another, I think the [subsidiary] was in this case a legal entity, because that is all it was." There was nothing to prevent the claimants at any moment saying " We will carry on this business in our own name."

D

Other Decisions Distinguished

Mr Marder distinguished the following:

Taylor v Greater London Council (1973) 226 EG 123, LT. In this recent case T owned the freehold and the limited company, a travel agency, was in occupation of the subject premises without lease or tenancy agreement, and no rent passed. The company had an authorised share capital of 6,000 ordinary shares of £1, of which 3,900 were issued. There were four directors in addition to the claimant and his wife, who each held 1,300 shares; two of these outside directors (one class B and one class C) each held 650 shares. It was provided that a director could be removed only by a
E majority of shareholders in his own class, so that the claimant (class A) could not remove those two in classes B and C. It was held that the company was a separate legal *persona,* and the claimant was not entitled to claim for disturbance.

Lander Equipment Co v Glasgow Corporation (Lands Tribunal for Scotland) (1972) 25 P & CR 460. A claim for disturbance. The company, Bothwell, was in occupation, the shares in which were owned by Mr and Mrs Lander as trustees for the claimant firm. There was no lease by the claimants to Bothwell and no rent passing. Bothwell kept its own books, retained its profits and paid tax thereon. The tribunal considered that this case had certain similarities to
F *Smith, Stone & Knight,* but that there were serious differences and it could be distinguished, as was done by the Lands Tribunal in the case of *Brain and Drive-Yourself-Hire Co (London) Ltd v London County Council* (1957) 170 EG 210, in which it was said, " I do not find that Mr Brain could deal with his companies as easily as [in *Smith, Stone & Knight*]. From the evidence I infer that each of them was an active separate entity, and that rather than the companies being servants to Mr Brain he in fact is the servant of the companies."

Roberts v Bristol Corporation (1960) 175 EG 679 LT. The company in occupation had one outside director as well as the freeholder claimant; the member said " I consider the G
circumstances in this case to differ from those in *Smith, Stone & Knight.*"

Melias Ltd v Manchester Corporation (1972) 222 EG 303 LT. The landlords were a subsidiary property company, owning a number of properties besides the subject premises, and a wholly-owned private company having the same directors as the parent company, the tenants. The member found that the landlords were an active company with other interests, and were a separate legal entity, adopting the judgment in *Tunstall v Steigmann* [1962] 2 QB 593 CA, a landlord and tenant case.

On the construction of paragraph 6, said Mr Marder, line H
3 of sub-paragraph (2) does not read " the *only* person," and a singular would connote the plural; neither does it read " *solely* entitled," nor " *all* the receipts." He distinguished " *the* " person entitled in line 2 and " *a* " person entitled in the proviso at line 9. The company was Mr Marzell and Mr Marzell was the company, and Mr Marzell had all the qualities defined in paragraph 6.

He did not consider *Tunstall* to have any application to the present circumstances, since not only is the context of landlord and tenant different but paragraphs (f) and (g) of subsection (1) of section 30 of the Landlord and Tenant Act 1954 used the subjective expression " intends to demolish or J
reconstruct . . . " and " intends to occupy . . . for the purposes . . . of a business. . . . "

Mr Rich contended that the company and Mr Marzell were distinct entities. The transfer of the business, whether or not in the circumstances it was wisely made, was for advantage, and it was not permissible to lift the " corporate veil."

The company acquired the business lock, stock and barrel, and it was immaterial that nothing was paid for goodwill, and nothing was paid in cash for the fixtures, fittings, etc. The date of the compulsory purchase order was November 27 1964, and Mr Marzell had been in business on his own K
account up to October 1963, the date of incorporation of the limited company, which was a changed status; he therefore lacked 11 months of the two-year qualifying period. No dividend was paid, and the directors were in effect paid a salary in the form of " directors' remuneration."

Mr Marzell was not " entitled to the receipts." He and his wife were entitled as shareholders to profits in the form of a dividend, if one had been declared, which was different. The issue of only two shares gave Mrs Marzell an equal interest with her husband, but she had no interest in the property. This case differs from *Smith, Stone & Knight. Tunstall* was to be relied on in that in that case the landlord's company, L
formed since the originating application, was held to be different from the previous landlord, who was an individual.

I find this case to have its difficulties and it would seem logical for it to follow the same rules as for entitlement to make a claim for disturbance, which have been the subject of a number of decisions in recent years; indeed, the acquiring authority has conceded that if Mr Marzell's claim succeeds under paragraph 6, he will be entitled to the agreed figure for disturbance, such not being for determination in this reference.

The Nub of the Case

But the nub of this decision is the identity of the company. M
Each case has to be distinguished on its own facts, and here these appear to be very close to *Smith, Stone & Knight.* But *Tunstall,* albeit in the context of landlord and tenant, is persuasive in the other direction: it would seem to be as close, or even closer, to the *alter ego* argument, since the landlord held all the shares in the company except for two which were issued to her nominees.

The *alter ego* argument owes its origin to an *obiter dictum* of Lord Evershed MR in *Pegler v Craven* [1952] 1 All ER 685 at 690, but was not endorsed by the Court of Appeal in *Tunstall,* having been previously held to be contrary to the

A House of Lords' decision in *Salomon v Salomon & Co Ltd* [1897] AC 22.

The other disturbance cases cited by Mr Marder all follow the same direction, although all of them could possibly be regarded as being further away still from the *alter ego* or agency argument than the present case. I have also looked at a very recent decision (January 1975) by the learned President of the Lands Tribunal in *DHN Food Distributors Ltd v Tower Hamlets London Borough* (1975) 233 EG 1109 on disturbance, which on the agency argument follows the same lines, although it is true to say that the company structure in that case was much more complex.

B Even if the *alter ego* argument were entitled to succeed, in the present case the company could be said to be the *alter ego* of *both* the claimant and his wife, and the claimant *alone* was not entitled to the receipts of the business: he and his wife were so entitled; but he and she together did not have an interest in the house, only he.

I am aware that this seems to run counter to Mr Marder's construction of paragraph 6, by which, as I understand him, he makes the proposition that the last word in line 2 "*the*" person, should be read as "*a*" person to conform with "*a*" person in line 3 of the proviso, at first sight an apparent example of unfortunate drafting. But the proviso goes on to
C say "entitled to *the* receipts" of a business, not "entitled to receipts"; and I think this disposes of his argument that "the receipts" can mean the profits. "Receipts" might be so interpreted, but "the receipts" can only mean "*all* the receipts."

The claimant and his wife were shareholders, and it is significant that profits were taken as "directors' remuneration" by periodical drawings before the end of the trading year, with the obligation to refund out of the loan account any overdrawing. No dividend was in fact declared. Only the company was entitled to the receipts (ie gross takings), to be applied in buying further stocks, defraying expenses, etc.

D Mr Rich acknowledged the acquiring authority's sympathy with the claimant in the situation in which he finds himself, and I, too, believe he is the victim of hardship on account of an unfortunate decision to form his business as a company, the consequences of which could not have been appreciated at the time, which was 13 months before the compulsory purchase order was made. Reluctantly, however, I have to find against the claimant that he is not entitled to claim the business supplement provided for in Part II of the Second Schedule to the Housing Act 1957. Accordingly the acquiring authority will pay to the claimant compensation in the sum
E of £850 to which will be added a surveyor's fee based on scale 5A of the Scales of Professional Charges of the Royal Institution of Chartered Surveyors, and any legal charges properly incurred from the date of the notice to treat up to the date of the notice of reference.

The claimant will pay to the acquiring authority the costs of this reference, such costs unless agreed to be taxed by the Registrar of the Lands Tribunal on scale 3 of the County Court Scales of Costs.

F J DAVY LTD v. HAMMERSMITH LONDON BOROUGH

(Ref/160/1974)

(J STUART DANIEL QC and J R LAIRD FRICS)

March 19 1975

Estates Gazette May 31 1975 and June 7 1975

(1975) 234 EG 685 and 753

Leasehold interest in Fulham, London, land and buildings held from British Rail on 21-year term from 1954—Several years' negotiations for new lease of 99 years suddenly broken off by lessors in 1960 because of "possibility of the overall development of the area"—Planning permission for petrol filling station granted three months later, but not implemented and would have lapsed shortly before expiry of lease—Tribunal dismisses claim that prospective purchaser at date of entry in 1970 would have paid additional sum on account of planning permission—Value of permission "too speculative" G

Giving their decision, the TRIBUNAL said: This reference is concerned with the assessment of the compensation (or part thereof) for the compulsory acquisition by the London Borough of Hammersmith (the "corporation") of a leasehold
interest in some 3,260 sq yds of land and buildings (the H
"leased premises") on the east side of North End Road, Fulham, SW6, in the County of London. The acquisition was under the "Hammersmith (North End Road Railway Sidings) Compulsory Purchase Order 1967" (the "1967 CPO") made by the corporation on March 18 1968 under section 97 of the Housing Act 1957 for the purpose of providing housing accommodation. The CPO was confirmed without modification by the Minister of Housing and Local Government on September 5 1969. Notice to treat was served on October 30 1969 and possession was taken on July 15 1970 as at which date it is agreed that the compensation must be assessed. J

The Sole Issue

The parties came to agreement on all heads of compensation except one. The sole surviving issue is this, that on June 11 1960 the claimants obtained from the then London County Council (the "council") planning permission for the erection of a petrol filling station on part of the land in their lease. This permission was never acted on, but the claimants contend that a prospective purchaser of the remainder of the lease would at July 1970 have given £25,550 on account of the permission. (The value of the lease without the planning
permission was agreed to be worth £13,500.) The corporation K
say that in all the circumstances of the case the tribunal should award nothing in respect of the value of the planning permission; alternatively some very much lesser sum.

Although the issue was described as a short one, the evidence and arguments took some five days to be heard.

The claimants were represented by Mr Leolin Price QC and Mr D Ritchie (instructed by Hewitt Woollacott & Chown, of Redhill), who called three witnesses, Mr J J Davy, a director of the claimant company, Mr P F Jordan, a chartered accountant who was for 17 years (1954-1971) with the company, for the last seven years until 1971 its managing director, L
and Mr H K Marsh BSc (Est Man) ARICS, a partner of the firm of Henry Berney Meadow & Co, surveyors of 87 Regent Street, London W1. The corporation was represented by Mr J Sullivan (instructed by the London Borough of Hammersmith), who called one witness, Mr E L Parison BSc (Est Man) ARICS, senior valuer to the corporation for the last eight years.

Before we deal with the case as presented by the parties we propose to set out the facts as agreed or found by us relating not only to the leased premises but also to a much larger area of surrounding adjacent land and buildings and
also to the time-table of events relevant to the issue now M
before us.

At one time, and certainly in the 1950's, the British Transport Commission (hereafter called the "commission" or "British Rail") owned property known as the "West Kensington Coal and Goods Depot" together with certain other property totalling some 16.711 acres (hereafter called the "railway property"). This property lay between North End Road to the west and the Lillie Bridge Depot to the east; away to the north was West Kensington Underground Station and to the south, separated by a row of small old shops and then small old terraced houses, was Lillie Road. The railway pro-

A perty fronted on to the east side of North End Road. Somewhere about the middle of this frontage, in 1954, was a " piece or parcel of land containing 3,260 sq yds or thereabouts . . . together with a building known as 241 North end Road . . . and other workshops, sheds, lock-up garages, timber racks and other erections thereon."

On February 3 1954 the commission leased these premises to Ram & Austin Ltd, of 241 North End Road, Fulham, for a term of 21 years commencing on June 24 1953 and expiring on June 24 1974.

There Ram & Austin carried on the business of " joinery B manufacturers, and shopfitters and general woodworkers " until early 1955, when this company was taken over by J Davy Ltd (sometimes " Davys " and later the " claimants "), now of 9 Logan Place, Kensington, London W8 6QW.

By clause 2 (7) of the lease the lessees covenanted:

> " To use (a) that part of the demised premises which comprises the building known as number 241 North End Road aforesaid as offices and for storage purposes in connection with the lessees' business as joinery manufacturers and shopfitters and general woodworkers (b) that part of the demised premises which comprises the said lock-up garages as private lock-up garages, and (c) the remainder of the demised premises for purposes, in connection with the lessees' business of joinery manufacturers and C shopfitters and general woodworkers and for no other purpose. . . ."

And by clause 2 (19) they covenanted:

> " Not to keep or to permit or suffer to be kept on the demised premises any combustible or inflammable material or substance except materials and goods used by the lessees in connection with their trades and businesses aforesaid and such motor spirit as shall be in the tanks of motor vehicles upon the demised premises and (subject to all statutory provisions and all provisions contained in any regulation made by any duly constituted authority relating to the storage and use of petrol being observed by the lessees and their tenants) two-gallon cans of motor spirit in each of the said D lock-up garages."

And by clause 2 (22) they covenanted:

> " Not to make any alterations or additions to the demised premises or use the same otherwise than as, herein before, specified, without the previous approval in writing of the commission."

Early in 1955 Mr J Davy and his brother purchased the share capital of Ram & Austin and the lease was informally transferred to J Davy Ltd, whose business then was the " garaging and repairing of motor vehicles." Written consent to this change of use, but only so far, was given by the commission on June 27 1955 to Wheeler & Atkins, the claimants' E surveyors.

The claimants wished to obtain an extension of the lease beyond June 24 1974, or else the freehold, and negotiations proceeded accordingly over some years. The commission apparently vacillated somewhat in its policy until October 20 1959, when its London Midland Region surveyor wrote to Wheeler & Atkins with heads of agreement on most points. The arrangement was that the existing lease would be surrendered and the claimants would be granted a new 99-year lease. It was a part of this arrangement that the claimants should apply for planning permission for the installation and use of a petrol filling station on a small (front) part of the leased F premises.

On December 7 1959 a firm of chartered architects and surveyors acting for Davys applied to the then London County Council (the " council "), for planning permission for development of a more-or-less rectangular site with frontage of some 116 ft to North End Road and depth from 40 ft to 60 ft; we will call this the " petrol station site."

A " Volte Face "

Then on March 7 1960 the commission performed a *volte face* when its estate and rating department (London Midland Region) wrote to the claimants' surveyors as follows: G

> " With reference to your letter of December 18; your proposal has been further considered and in view of the possibility of the overall redevelopment of this area, it is not desired to enter into any further lease arrangements for the present.
>
> I regret having to convey this decision to you, particularly in view of our lengthy negotiations, but you may be assured it has not been made lightly and should the position change in the future, your clients would, of course, be given the opportunity of reopening the matter."

Davys wrote to Sir Reginald Wilson, chairman of the London Midland Area Board of the commission, on March 28 1960 protesting at this change of mind, but Sir Reginald H replied on May 3 1960 as follows:

> " I have now examined the course of the negotiations to which you refer in your letter of March 28, and though I understand your disappointment and irritation that they should have come to nothing at the last minute, I am satisfied that in the circumstances our Estates Department were right to act as they did. Moreover, their letter to your agents explained the situation fully and courteously—a copy is attached. I must regret the consequences to you of our decision not to complete—as indeed does the Estates Department itself—but these things happen in commercial life.
>
> We have already promised that you will receive particular consideration if the position changes again in future, but so far as the present position is concerned I see no reason to intervene." J

Conditional Planning Permission

Then on June 9 1960 the London County Council granted (conditionally) the planning permission we have mentioned. The permitted development was described as follows:

> " The development of part of the front portion of the premises 241-245 (odd) North End Road, Fulham, as a public petrol filling station together with the erection of a single-storey sales and car hire office, the formation of one new and one altered access to North End Road and alterations to the front elevation of the K existing single-storey workshop building upon the rear boundary of the proposed petrol filling station."

Then there were conditions:

> " (1) No part of the petrol filling station premises shall be used for the display and sale of new or secondhand motor vehicles.
>
> (2) The width of crossovers being not less than 16 ft or more than 18 ft."
>
> The reasons for these conditions were:
>
> " (1) In order to safeguard visual amenities and to preserve a satisfactory appearance.
>
> (2) To safeguard the free flow of traffic in North End Road. L
>
> I have to inform you:
>
> (1) That the Borough Council should be consulted regarding (a) the construction of the crossover on the public way, and (b) under the Town and Country Planning (Control of Advertisements) Regulations 1948 to 1951, in respect of any proposed signs or advertisements, and
>
> (2) that this decision is in no way a decision upon the ' future showroom with offices over ' shown on the drawings, which should form the subject of a separate application."

We will call this the " planning permission," and the works proposed in connection therewith the " proposed development." This proposed development involved and envisaged:

(1) the demolition of parts of the buildings on the proposed M petrol-station site (eg, part of the offices), demolition of part of the car-showroom (outside the petrol-station site but part of the leased premises), and the giving-up of part of the main frontage of the open yard used for car-display purposes.

(2) user of the premises for purposes not permitted by the lessees' covenants in the said lease.

The planning permission was related to two (architects') plans. The first (No 2381/10A) was drawn to a scale of 1/32 in to 1 ft and was described as a " site plan "; it showed that the proposed petrol filling station was to have a frontage

A to North End Road of some 116 or 117 ft out of the total frontage of the leased land of some 180 ft and a depth in the main of some 44 ft 6 in behind which was to be a new single-storey office block measuring, say, 48 ft long by 16 ft wide (to be used for car-hire business and for petrol sales). The council had prescribed some years previously an "improvement line" some 10 ft back from the inner pavement line of North End Road.

The other drawing (No 2381/12A), to a scale of ⅛ in to 1 ft, was described as a "sketch scheme"; it was in greater detail and showed *inter alia*:

B

(a) the improvement line

(b) those parts of existing buildings that were to be demolished to make way for the said proposed petrol filling station and the necessary forecourt

(c) that the site for which planning permission had been granted was to be enclosed at either end (that is to say the north and south ends) by "new timber screens"; the frontage to North End Road was to be enclosed by 3 ft-high railings and inset flower boxes with new gates at either end at the new access or crossover points from the main road; such gates would, presumably, have been opened when the station was C selling petrol. The rear or western boundary was to consist partly of a new brick gable-end to the foreshortened existing workshop building, also the new office block described above, and two sets of new gates on either side of this block giving access to the rear part of the leased premises. On this petrol-station site were shown two "islands," each with three petrol pumps, making a total of six pumps with six underground petrol tanks each of 3,000 gallons.

This plan also showed, though in less detail, a future showroom with offices over at the north end of the site fronting North End Road. In the planning permission it was stated:

D "(2) that this decision is in no way a decision upon the 'future showroom with offices over' shown on the drawings, which should form the subject of a separate application."

It will be clear from this description that the proposed petrol filling station was intended to be virtually self-contained save for the two entrance/exits off the North End Road and the two entrance/exits to the rear of the major part of the leased premises.

Thus at June 1960 the situation was that development of a petrol filling station on the site was legally possible so far as planning control was concerned, but could not be done E in accordance with the lessees' covenants in the existing lease without a licence from the landlords, the British Transport Commission. This licence was never granted or even requested and there followed a period of about five years during which the claimants simply held as tenants under the existing lease with the existing buildings without having any firm information about the future of the site. Thus, at the time of entry by the corporation on July 15 1970, the material date in this reference for the valuation, no action had been taken by the claimants to implement the aforesaid planning permission. Mr Davy told us that at the end of 1965 and in 1966 there was a great deal of conjecture as to F what was to happen and who was to acquire the railway property; there had been rumours that the land would be acquired by the Greater London Council or by the Hammersmith London Borough Council, but it had been difficult to get hard news; he did not know what was the form of the "overall development" referred to in the letter of March 7 1960 or who would carry it out. He and Mr Jordan had approached the architects' department of the then London County Council in January and March 1965 in an attempt to ascertain something definite, and on April 13 1965 Mr A I Williams of the valuation department of the GLC (the GLC having taken over from the LCC on April 1 1965) wrote to G J Davy Ltd as follows:

"*Town and Country Planning Act*
239/245 North End Road, W14

With reference to your letter of March 25, the above-mentioned properties are zoned as to the frontage land for light industry and programmed in the second period (1960-1972) as being suitable for redevelopment and in regard to the back land which is, I understand, railway freehold, the zoning is currently for goods yard and free from programming.

The railway goods yard is, however, to some extent surplus to railway operational requirements and consideration is being given towards the use of the area, including possibly the North End H Road frontage at this point, for residential development for municipal purposes. It is not yet clear, however, to what extent this will be carried out in the near future but it is not likely that major redevelopment of the premises you occupy would be advisable at the present time.

If there are any particular points on which you wish to raise further queries at the present time perhaps you would telephone me."

Mr Jordan gave similar evidence. He said that he had not been directly concerned with the negotiations for a new lease, but he had seen the letters; when the commission broke off negotiations he believed that a local authority would take J over, for there was much in the press about releasing railway land such as the Marylebone Goods Yards (sold by British Rail in 1964), but at 1960 this was really all conjecture; it was quite possible that British Rail themselves might redevelop the railway property including the leased property; he did not know at all what the form of the redevelopment would be.

Meetings and Discussions

After Mr Williams' letter of April 13 1965, there followed various letters, meetings and telephone conversations. We were shown some of the letters and notes of the meetings K and conversations as made by Mr Jordan. We do not think it necessary to set these out in full. A typical note was that from Mr Jordan to other members of J Davy Ltd on May 3 1965 recording a conversation with Mr Williams:

"Mr Williams phoned to say that he was in the process of answering my letters dated January 26 and March 25, 1965, but unfortunately could not be specific so he thought a conversation would be more helpful and would amplify his letter.

There had been extreme agitation to use all surplus railway land for residential purposes. There were 120 locations currently under consideration, including the West Kensington Goods Yard. So far nothing had been agreed, although, following the Milner L Holland report, there was great emphasis being placed on residential development of surplus railway land and one of the reasons railway land was so desirable was that there was no rehousing problem involved.

Dealing with the location on which we were situated, he gave as his view that the northern part, ie, Cromwell Road end, would be developed in 4/5 years. How far down the development would extend was uncertain because the southern part of the location, which possibly includes us, may be retained by the railways as a fuel depot. If this was the case it is possible that we would be allowed to remain with our frontage to North End Road, although we would probably lose our car park.

M If the railways gave up the whole of the location, the machinery would be put into motion to rezone the railway land and the light industrial land fronting North End Road, to residential. In this event efforts would be made by the GLC to conveniently resite us, in view of our local connections and having been established a number of years.

If no decision is reached by the Ministers concerned within one year he suggested that we make a town planning application, at the minimum cost, in order to precipitate a clarification of the position.

He did not think his letter to us would be very helpful and suggested that we telephone him at three-monthly intervals to ascertain progress."

Another conversation with Mr Williams was recorded on June 14 1965 in which Mr Jordan noted:

"As a consequence of Mr H O R Davy's memo dated June 3, I telephoned Mr Williams to ascertain the current position. He said that no definite decision had been taken with regard to the West Kensington Goods Yard, although it was fairly certain that a large chunk of it would be taken over by the GLC for residential development—possibly to be carried out by the local Borough Council. He would be in a position to give more information after June 28, on which date the GLC were having a meeting with British Railways to discuss the above-mentioned location, together with others. BR's are being awkward about some of the area under consideration, which they claim they wish to retain.

If it is finally decided that the GLC will take over substantially the whole of the site, they will eventually consider making a Compulsory Purchase Order which implies compensation and/or will endeavour to resite. No movement along these lines is expected for at least three years."

And so on. At one stage meetings were held between Davys and Mr R W Maydwell FRICS, of Folkard & Hayward (consultant surveyors to British Rail). A meeting was held on February 25 1966 to discuss the "sale of West Kensington Goods Yard to GLC." One matter broached by Mr Maydwell, but without any mandate from his clients, had been the possibility of Davys acquiring the freehold of the leased land. On March 16 1966 in a telephone conversation between Mr Jordan and Mr Maydwell, the latter is noted by Mr Jordan as having told him that his clients were "quite happy" to sell the freehold to Davys but that he had not been able to obtain the agreement of the GLC.

Finally, the compulsory purchase order was made on March 18 1968 by the London Borough of Hammersmith, who by then had acquired the freehold of the leased premises. About this time, judging from the map referred to in the 1967 CPO, the locality was in a somewhat neglected and rundown state as the plan dated November 22 1967 shows; the following descriptions were noted on three sides of the leased property, "disused railway tracks," "derelict stables," and "derelict workshops"; in addition to Davys' garage there was a large "vehicle park" (tarmac surface), a "Metropolitan Police car pound" (ash surface), various "works," a "warehouse," the "Seven Stars" public house, and a large advertisement hoarding about 120 ft long fronting North End Road.

The leased premises were virtually square in shape, having a frontage of approximately 180 ft to the road and a depth of some 180 ft or thereabouts. Davys were occupying this frontage to varying depth as offices, car-showrooms, car display area (in an open yard), and for a car-wash. The remainder, or major part, of the leased premises lying behind this frontage was used by Davys for various purposes connected with motor-cars, such as workshops, garages, spray-booths, offices, stores, etc. As we have mentioned, possession was taken on July 15 1970. At that date the lease had just under four years to run, for it would have expired at midsummer 1974, whereas the planning permission would have lapsed at the end of March 1974, because, although the permission purported to be unlimited in time, the effect of section 65 of the Town and Country Planning Act 1968 was that the permission was deemed to have been granted subject to a condition that the development must be begun not later than the expiration of five years beginning with the commencement of the section (April 1 1969).

As we said at the outset, the parties came to an agreement on all items of compensation except one. The matters agreed between Leopold Farmer & Sons, who lodged the original claim dated November 26 1969 on behalf of Davys, with the corporation may be summarised under the following heads:

Leasehold interest

Expiring June 24 1975 ... ... (excludes any item referrable to the planning permission)		£13,500

Disturbance

Disposal of plant and equipment, etc. ...	£4,240	
Removals, general terminal costs and redundancy payments	£4,367	
Costs re alternative premises including directors and staff time and preparing claim	£4,207	
Loss of profit—paint and panel workshops	£22,000	
Loss of car-sale franchise	£21,497	
	——	£56,311
Total agreed		£69,812

(90% has been paid, pending a decision in this reference)

So far we have summarised an historical account of what happened.

It will be convenient, before we refer to some of the evidence given by the expert witnesses, to summarise the main submissions made by counsel on behalf of the parties.

The parties are agreed that the question is whether, at July 1970, a prospective purchaser of the leasehold interest would have paid anything, and if so what, on account of the planning permission. Mr Price emphasised that at that date the permission had a limited life and would expire before the date on which the reversion would fall in; therefore, he said, the freeholders could get no value from the permission unless they made terms with the tenant; they would therefore have been under considerable pressure to do so.

The "Trocette" Case

Mr Price introduced his argument by reference to the case of *Trocette Property Co Ltd v Greater London Council and Southwark London Borough Council* (1974) 28 P & CR 408. In that case the claimants had a lease from the LCC of a disused cinema. The lease would expire in 1984. Negotiations for a new long lease took place and in 1968 a valuer of the council (now the GLC) recommended that a new lease should be granted. But in 1970 a plan was produced by the highways department of the GLC for a new road crossing the site, and the council's valuer said that a new lease could no longer be considered. The claimants failed to obtain planning permission for developing the site and a purchase notice followed. The Lands Tribunal held that it was to be assumed that, apart from the road scheme, one form or another of planning permission could reasonably be expected for the reasonable development of the site; and that such permission would reasonably attract the interest and concurrence of the landlord to the extent that a new long-term lease would be granted to allow development for those reasonable purposes to proceed; and that any effect upwards or downwards which the road scheme might have had on the valuation of the claimants' interest was to be excluded. The tribunal awarded compensation of £39,000. The council appealed, contending that since the council would in the circumstances have been unwilling to grant the claimants a new lease, the claimants' interest was worth only £2,000. The Court of Appeal dismissed the appeal, holding (by a majority) that the knowledge by a purchaser of the lease that the freeholder would not come to terms eliminated any question of "marriage value," but that the resulting depreciation was caused by "indications" within the terms of section 9 of the Land Compensation Act 1961 and therefore fell to be ignored.

Section 9 provides as follows:

"No account shall be taken of any depreciation of the value of the relevant interest which is attributable to the fact that (whether by way of designation, allocation or other particulars contained in the current development plan, or by any other means) an indication has been given that the relevant land is, or is likely, to be acquired by an authority possessing compulsory purchase powers."

The facts of that case, said Mr Price, were in some respects very near to the facts of the present case, but there were also significant differences; in that case there was only an assumption that planning permission was reasonably to be expected—here there was no speculation but an actual permission; in the *Trocette* case the council made it clear that there was no question of proceeding with a surrender of the lease (and the exploitation of a permission depended on a surrender), but in the present case it was not necessary to have a surrender; this was only an alternative, for Davys could simply have developed under the existing lease with their rights under Part II of the Landlord and Tenant Act 1954 thereafter, continuing to hold at a market rent unless some special reason prevented them; moreover, reverting to the possibility of a surrender and a new lease, British Rail (unlike the council in the *Trocette* case) had not been at all firm in their refusal to come to terms; the negotiations were broken off by British Rail " for the present," because of the possibility of overall redevelopment of the area; and, as the commission had written on March 7 1960: ". . . should the position change in the future, your clients would, of course, be given the opportunity of reopening the matter."

Claimants' Reliance on Section 9

Secondly, Mr Price said, the claimants relied on section 9 of the Act. He referred to the case of *Mountview Estates Ltd v London Borough of Enfield* (1968) 20 P & CR 729 and to the case of *London Borough of Hackney v MacFarlane and Another* (1970) 21 P & CR 342, and in particular to the judgment of Lord Denning in that case as showing that the word " indication " in section 9 should be read in a broad sense. In the present case it seemed probable that the " overall development " mentioned in British Rail's letter of March 7 1960 referred to the likelihood of compulsory acquisition; it seemed most unlikely that British Rail would have broken off negotiations for any other reason; from 1960 onwards there was a planning permission in existence for a petrol filling station, the exploitation of which for both landlord and tenant was dependent on one matter only—the landlord's consent, but here (again unlike the *Trocette* case) there was nothing to evince an intention not to exploit the permission; there was no bar to commercial commonsense, more particularly once the Act of 1968 had created a time limit; by the time of the compulsory purchase order in the instant case the Hammersmith Borough Council had become the landlords; then, of course, the landlords, being a local authority, would probably have done the same thing as the landlord in the *Trocette* case, but then section 9 came into play. Mr Price then reviewed the correspondence and records of conversations to which we have already referred, and submitted that these were clear indications that the subject site was likely to be compulsorily acquired.

Acquiring Authority's Argument

We now turn to Mr Sullivan's argument on the main issue. Countering Mr Price's proposition that it was commercially sensible to exploit the permission, Mr Sullivan said that a purchaser would not have given anything for the permission because it was clear that its value was too speculative; the landlord would not have been interested in the permission because he would have been more interested in his own more valuable redevelopment; he had to wait for only four years till the lease fell in and then he could do what he liked; in March 1960 British Rail had decided not to grant a new lease at the present because of the possibility of overall development; if in that situation the planning permission for a petrol station was so valuable, it was strange indeed that neither British Rail nor Davys reopened the matter and that there was no movement at all between 1960 and 1965. Mr Sullivan quoted from the claimants' notice of reference in this case, a passage in paragraph 6 which, having referred to British Rail's letter of March 7 1960, continued:

> " The possibility of overall development referred to arose because at this time it had become known that the London Borough of Hammersmith were considering arrangements for comprehensive redevelopment of an area which included the premises; and this made it likely that land in that area, including the premises, would be acquired by an authority possessing compulsory purchase powers. That likelihood continued to affect the premises at all times until the CPO was made and the premises were acquired by the London Borough of Hammersmith. . . . "

But, said Mr Sullivan, referring to the earlier part of this passage, there was no evidence at all to connect the British Rail scheme (the overall development) with the corporation's compulsory order. It was quite clear from the evidence of Mr Jordan and Mr Davy that they did not know at 1960 what was the " overall development," or who might carry it out; indeed, Mr Jordan had said that he thought that at that time British Rail might have been contemplating the scheme themselves.

Landlord in " Very Strong Position " ?

Mr Sullivan accepted that since the corporation acquired the freehold, and thus became the landlord, the effects of its own scheme must be ignored. But the question was: would a hypothetical purchaser have given anything for the permission in 1970? He referred to the question raised in the *Trocette* case whether the council would necessarily have granted a new lease but for the road scheme (see pp 416 and 417 of the report in the Court of Appeal); in the instant case, he said, the question was not really whether the landlord would have refused to come to terms, but rather whether he would have been interested in coming to terms, for he might well think that he had bigger fish to fry and would not want to endanger these by the tenant's project for a petrol filling station; if the landlord had wanted commercial development on the frontage, no doubt there were alternative possibilities, and indeed the land was still zoned for light industrial use in 1970; at that date the petrol planning permission still had four years to run, so the landlord was not pressed; indeed he was in a very strong position, for between 1970 and 1974 he could plan and start a phased development; the existing permission might not be what the landlord really wanted, but he could revise his own scheme and let the tenant go, for it was clear that under section 30 (1) (f) of the Landlord and Tenant Act 1954 a new lease could have been refused. As to section 9, neither British Rail's letter of March 7 1960 nor the oral evidence showed that there had been a depreciation in the value of the planning permission caused by any indication that the subject site was likely to be acquired compulsorily.

Mr Sullivan in his turn reviewed the documents and submitted that there was no indication which qualified under the section between 1960 and 1966; later, of course, there was the compulsory purchase order itself; but if a depreciation in the value of the permission was suggested, this had occurred long since; no other explanation would give an answer to the question why, during all the detailed notes of meetings and conversations about the future of the land, the planning permission was never once mentioned.

Meaning of " Likely "

In reply on the main issue Mr Price submitted that it was not right to say that the landlord (British Rail) was not under pressure, for it was clear that in 1959 and early 1960 British Rail had been ready to come to terms and had been baulked only by the possibility of overall development; later the 1968 Act had injected against that background a new factor, a factor of urgency, in that the planning permission would expire within the lease and could be exploited only if the

A landlord came to terms with the tenant; there could be no certainty that the permission would be renewed and there, was, therefore, very strong pressure to exploit it before it expired; it was clear from the letter of March 7 1970 that British Rail had suspended negotiations " for the present " and had kept the position open; the suspense had been prolonged (for corporations moved slowly), but meanwhile events such as the Marylebone Goods Yard sale were occurring; Mr Williams' (GLC) letter of April 13 1965 was clearly an indication that compulsory acquisition was likely. Mr Price referred to *Stroud's Judicial Dictionary* for the meaning of the word " likely "; of the various meanings B there listed he suggested the most suitable in this context was that " of a tendency or real possibility," not " more likely than not." Judged by this test, the letter of April 13 1965 and the various notes of conversations constituted a series of " indications " within the meaning of section 9; it was clear that the whole situation was clouded by municipal interest from 1960 onwards, for British Rail did not want the site for their own use and the planning permission for a petrol station would have been exploited but for one thing—the likelihood of a local authority acquiring the site.

C

Valuation Evidence for Claimants

We now turn to notice the effect of the evidence of the expert witnesses. Mr Marsh had prepared and lodged two valuations; the first was dated September 1974 and the second was dated January 23 1975; each contained the same opening paragraph thus: " Once it is accepted in principle that in the particular circumstances the planning consent has a value, it is necessary to quantify the amount."

In them, Mr Marsh took the following steps designed to:

1. Apportion the agreed leasehold compensation of £13,500 to exclude petrol forecourt area.

D 1a. Calculate the value of freehold subject to lease including forecourt area on existing-use basis.

1b. Calculate the value of freehold subject to lease but excluding forecourt area on existing-use basis.

2. Value the petrol site as an unencumbered freehold.

2a. Value the freehold petrol site—subject to lease.

2b. Value the leasehold interest in the petrol site.

From the above Mr Marsh then sought to adduce:

3a. Value of planning consent to freeholder subject to lease.

E 3b. Value of planning consent to lessee.

4. Marriage value of the unencumbered freehold in the petrol station.

5. His claim in respect of the planning consent, which was as follows:

> "An addition of the existing use value of the remainder of the site and the value of the petrol forecourt consent plus 50 per cent of the 'marriage' value from value 4.

Lease excluding petrol station site (from 2 above)	£10,264
F Petrol forecourt (from 2 (b) above) ...	£22,000
Share of marriage value at 50% (from 4 above)	£6,800
Value of entire leasehold interest ...	£39,064
Less agreed value of lease excluding planning consent	£13,500
	£25,564
Claim for planning consent (ie, additional value to leaseholder): say	£25,550 "

G Mr Marsh, in support of his five foolscap pages of calculations covering the above headings, then added two further pages consisting of check calculations:

6. Loss to freeholder if planning consent expires.

7. Lessee's claim to share of value to freeholder.

Basically, Mr Marsh's calculations proceeded on his assumption that the throughput of petrol would have been at the rate of 800,000 gallons per annum (if " merchandised "), but he did not adduce any evidence whatever in support other than his opinion. His second lodged valuation contained, *inter alia*, the following " note 6 " among his reasons for submitting it: H

> " My original valuation was of course made in the knowledge of compulsory acquisition which in my experience produces in most types of case settlements lower than would be achieved in the actual open market. Therefore whilst bringing my figures into line with those of Hammersmith Borough Council I have adjusted the value of the unencumbered freehold petrol site slightly upwards."

[*Tribunal Note:* From £105,000 to £113,500 achieved, in the main, by dropping his years' purchase rate per cent from 8 per cent in his first valuation to 7 per cent as applied in the said second valuation.] J

Mr Marsh's second valuation was as follows:

" 2. *Value of Petrol Site as unencumbered Freehold*

Forecourt 800,000 gallons at 1½d rental value	£5,000	
Less Green Shield Stamps at 25% ...	£1,250	
		£3,750
Add Special value at 1½d per gallon (expected " throughput overbid " by oil company operating site themselves)		£5,833
Sales building		£741 K
		£10,324
Y.P. at 7% in perpetuity ...		14.3
		£147,633
Less development costs (inc tanks)		£35,000
		£112,633
Add parking area along front boundary (between petrol pumps and pavement)		£1,630
		£114,263 L
Say		£113,500 "

We return to Mr Marsh's " note 6 " later on.

It is to be noted that Mr Marsh's calculations depended on the acceptance in principle that in all the circumstances the planning permission had a value (the rest was simply quantification). The acceptance of this proposition in principle was explained by Mr Marsh in oral evidence by the following main considerations. He accepted that the project of a filling station required the landlord's licence; but the landlord had M already varied the lease considerably in his view; the permission had become temporary under the 1968 Act and would run out before the lease; a filling station was, he thought, more valuable than Davy's other motor-car uses; the tenant would expect to get a new lease under the Act of 1954 and the court would grant a new lease on the same terms as before (except for a shorter term and at a market rent); British Rail had ordinary commercial instincts and would be likely to negotiate with the tenant; it would suit the landlord and the tenant to keep the permission alive and therefore enter into negotiations before it expired.

A

Council's Valuer's Evidence

Turning now to the evidence of Mr Parison, who has been with the corporation some eight years, he said very frankly that he had not valued a petrol filling station before; but he thought that the planning permission was in all the circumstances valueless as at July 1970, for a purchaser of the tail-end of the lease other than the corporation would have known that in 1960 British Rail had refused to grant a new lease because of "the possibility of the overall redevelopment." He excluded, he said, the question of redevelopment by the corporation because he accepted absolutely that the effect on value of the compulsory purchase order, or on the B corporation's "scheme" which led to it, must be ignored; but the fact remained that the landlord had only to wait four years, successfully oppose a new lease (section 30 (1) (f) of the Landlord and Tenant Act 1954), and then the old lease had gone. Assuming that redevelopment by the corporation had not occurred, Mr Parison considered that the 14.205 acres of railway property (16.711 acres less 2.02 acres earmarked for school purposes and 0.486-acre for extension to telephone exchange) in the west part of London, would almost certainly have been used for private housing, so that the prospective purchaser would have clearly contemplated redevelopment over this large area as soon as the lease fell in on June 24 C 1974. The intervening four years from July 1970 would give him time in which to apply for planning permission, negotiate, and arrange contracts for a phased redevelopment of the whole area. Thus the freeholder was in an extremely strong position and certainly would not have been prepared to pay the lessee £25,550, as Mr Marsh proposed, as the value of the planning permission for the last four years of the lease. Indeed the freeholder might well think that such an estate would be depreciated by the advent of a petrol filling station.

We think it perhaps not insignificant that a petrol filling D station has not been incorporated in the corporation's redevelopment of the 14 acres. In a nutshell, Mr Parison believed that the project of a petrol filling station would probably have depreciated the value of the very much larger area and that the freeholder, and for that matter a prospective purchaser of the tail-end of the lease, would have regarded that project as far too speculative and risky in 1970 to attach any value to it.

We have dealt very shortly with the expert evidence, because we think that, in truth, the issue turns on the broader considerations rather than on the detailed investigation of figures. Mr Parison, as well as Mr Marsh, put forward E different calculations and valuations, some parts of which were much criticised and were subjected to very close scrutiny in cross-examination; some palpable hits were registered, but we do not think that the outcome of this reference depends on this sort of detailed analysis.

"Somewhat Depressed" Locality

We think we should say a few words about the North End Road, which runs more or less due north and south, linking the east-west Talgarth Road at its northern end by West Kensington underground station with the east-west Lillie Road. The latter crosses the former at right-angles at about F its centre, and then continues southwards until its junction with the east-west Fulham Road and Fulham Broadway. We were given very little evidence about North End Road as a traffic route, but our recent inspection, admittedly some four-and-a-half years after the material date in this case, shows that the locality is somewhat depressed in the vicinity of what had been the leased premises; there are small shops fronting the western side of the road, behind which lie streets of old, dilapidated, two-storey terraced houses. Opposite, on the east side of the road, there remains the Seven Stars public house and surrounding it, on three sides, is the new West Kensington Housing Estate of the Hammersmith Corporation, consisting of blocks of local authority flats built on the aforesaid railway G property. When the corporation was developing this estate no attempt was made to incorporate any shops at ground-floor level fronting North End Road or, indeed, a petrol filling station; instead, the flats are at pavement level up to the improvement line with small areas of grass in front. This would seem to indicate to us that there was little, if any, commercial value in this lengthy frontage to North End Road. To the south of the Lillie Road crossing, North End Road has what appears to be a flourishing kerb-side street market (the stalls and barrows are wired up to main electric light), which considerably restricts the free flow of vehicles along the 30-ft wide carriageway when the market is operating H most days of the week.

This has been a puzzling case. The planning permission which was the bone of contention was never exploited. This non-event must be investigated against a background which consists partly of historical events and partly of statutory assumptions. The parties have wisely, and we think rightly, agreed to state the issue in these terms:

> As at 1970, would a prospective purchaser of the tail-end of the lease have given anything and, if so, how much on account of the permission? If the value of the permission was depreciated by the fact or the likelihood of compulsory acquisition, that depreciation must be discounted in accord- J ance with section 9 of the 1961 Act.

Was there indeed such a depreciation? The claimants' case in its simplest form says "yes," because clearly the exploitation of the permission foundered on first the likelihood, and then on the certainty, of compulsory acquisition; and the hypothetical purchaser, freed from either the certainty or even the likelihood in question would have regarded the permission as a valuable asset; the likelihood was there by 1965 or 1966 at the latest, perhaps even very much earlier, and the certainty was there in 1968.

We agree that in terms of section 9 the word "likely" may K very well mean "reasonably possible," and we think that British Rail's letter of April 13 1965 made it clearly "likely" that the subject site would be compulsorily acquired. But we are not satisfied that the likelihood, or later the fact, caused the non-exploitation of the permission.

The Genesis of the Project

The links of causation are difficult to disentangle, but we think Mr Sullivan was right when he said that, if depreciation there was, it resulted from events which happened much earlier. It is clear beyond doubt that the whole genesis of the project for a petrol filling station was part and parcel of L the arrangement that the existing lease should be surrendered and that a new lease for 99 years should be granted. The application for planning permission was made in accordance with that proposal and the planning permission was not acted upon, as we see it, because the proposed arrangement was in the meanwhile broken off by British Rail. The actual conduct of the parties all points the same way, for, as soon as the new lease failed to materialise, all thought of exploiting the planning permission seems to have been ignored. If, indeed, a purchaser of the lease would at 1970 have regarded the permission as valuable, why then had not Davys (long since) made the attempt to exploit it by embarking on the necessary M first step of seeking British Rail's licence under the lease to the change of use?

In the claimants' notice of reference, and indeed in Mr Price's opening, it was clearly suggested that negotiations were broken off in March 1960 because it was thought that a local authority would acquire the site; but there is no evidence of this. Indeed there is the very frank evidence of Mr Davy and of Mr Jordan that they did not know the form of the "overall development" which did cause the break, and that this might well have been development by British Rail. Of course it can be said that as time went on, probably from

A 1965 onwards and certainly in 1970, it would have been true to say that the exploitation of the permission was impossible because a local authority was going to acquire compulsorily. But we think it would be putting the cart before the horse to say that the obstacle which eventually materialised was the proximate or significant cause of the non-exploitation. The proximate cause had occurred earlier. "Overall redevelopment" can mean almost anything.

It is significant in a negative way that no evidence was available from British Rail or from either of the local authorities explaining either orally or in the form of official minutes or resolutions or other documents what indeed was B the immediate occasion, in the spring of 1960, of the break in the negotiations for a long lease. We note that Mr A J Williams, who was in the valuer's department of the GLC in June 1965, had transferred to the London Borough of Hammersmith by October 1965; he was not called. No doubt the claimants' advisers explored the possibility of finding such evidence, but it was not forthcoming. British Rail had apparently vacillated before. When they decided not to grant a new lease, all the parties, it seems to us, treated the petrol filling station as pigeon-holed with the new lease. *Quaere* whether without a new lease British Rail would have granted the necessary licence for use as a filling station.

C However that may be, we feel unable to be satisfied on the facts that the decision not to grant a new lease was caused by the likelihood of compulsory acquisition. We think the planning permission was simply one part of an abortive negotiation which happened to emerge from the local authority machine. Incidentally, it seems most unlikely that the LCC would have given the permission in question on June 11 1960 if at that time they had had the slightest intention of acquiring the land for municipal purposes. In connection with the notion of a prospective purchaser bidding for the lease in the market in 1970, there was much discussion about the respective position of the landlord and tenant and about D the respective "pressures" each would have been under to come to terms. It was said on behalf of the claimants that the pressure on the landlord would have been very great in this sense—that he would have been most anxious to ensure that the permission did not lapse, and that he would have done everything to that end, and that the purchaser would realise this and bid high accordingly, in order to be in the shoes of a tenant in such a strong position. We do not agree. We think it much more likely, as Mr Sullivan put it, that the landlord would not have been interested in the permission. For the landlord owned a much larger area of land and might well have had other plans of his own. He had only to E sit and let the reversion fall in and the purchaser might very well have thought the prospects of renewal were slim or non-existent.

At the end of the day we have not been satisfied that a prospective purchaser of the tail-end of the lease would have paid anything in the summer of 1970 for the planning permission. For the reasons which have just been given we are not at all persuaded that the prospective purchaser would have been inhibited in the market by matters which should have been ignored in accordance with section 9, and we do not think that the potential, but never exploited, value of this permission can be, as it were, resuscitated by the provisions F of that section.

If that is so, then we are left with the effect of the evidence of the two expert witnesses.

We have already referred to "note 6" to Mr Marsh's second valuation. That note 6 statement leaves us with the feeling that his complicated and lengthy calculations do not reflect open market values at June 1970 as they should have done. Our general conclusions are that his calculations proceeded on an incorrect basis and contained too many variables, imponderables and unsupported assumptions, some of which we consider were over-optimistic. We accept Mr Sullivan's criticism that, in certain respects, Mr Marsh was endeavouring to have his cake and to eat it. Furthermore, G some of the assumptions underlying his calculations do not seem to us to have been made good, for we do not think it at all certain that British Rail would have granted the necessary licence or would have been so anxious as he suggested to negotiate.

We prefer Mr Parison's simple proposition, which was primarily that the value of the planning permission was too speculative in 1970 for a purchaser to attach any value to it.

In the final analysis we have not been satisfied either in law or by valuation that the claimant company has established any claim to compensation in respect of the planning permission in this reference. The claim is therefore dismissed. H

The claimant company will pay the corporation's costs of this reference on the High Court scale; these costs, if not agreed, to be taxed by the Registrar of the Lands Tribunal.

HOBBS (QUARRIES) LTD v SOMERSET COUNTY COUNCIL

(Ref/9-10/1974)

(DOUGLAS FRANK QC, President, and J D RUSSELL-DAVIS FRICS)

April 23 1975

Estates Gazette June 14 1975

(1975) 234 EG 829 J

Revocation of planning permission and making of discontinuance order—Limestone quarry in Mendip Hills, Somerset—Probability of being awarded subcontract for supply of rockfill and base material for construction of section of M5—Loss of profit held proper basis for compensation—Allowance for deferment—Interest not payable from date of revocation order—Tribunal unwilling to accept Inland Revenue statement on tax deduction made in "Bostock Chater" case—Accepts submissions of revenue counsel appearing as "amicus curiae" on parties' joint instructions

Mr David Widdicombe QC and Mr Anthony Anderson (instructed by Hallinan, Blackburn, Gittings & Co, of Cardiff) appeared for the claimants; Mr Alan Fletcher (instructed by the solicitor to Somerset County Council) for the compensating authority. K

Giving their decision, the TRIBUNAL said: These are references to determine the amount of compensation payable consequent upon the revocation of a planning permission and the making of a discontinuance order. We find the following facts:

1. The land concerned, which we shall call "the quarry," has an area of about six acres, is known as North Quarry and is situated at the western end of the Mendip Hills, at Webbington, Somerset. It overlooks and is within quarter-of-a-mile of the M5 motorway. L

2. The quarry had been used for very many years for the working and winning of limestone, and in 1947 planning permission for that purpose was granted.

3. The quarry was occupied as a depot by the Somerset County Council under a lease for 15 years dated June 8 1959 but surrendered at the end of 1965.

4. On July 21 1965 the claimants purchased 56.85 acres of land, including the six acres comprising the quarry, for £50,000, but they were unaware of the intended revocation and discontinuance orders. M

5. On November 19 1965 the county council made an order under section 27 of the Town and Country Planning Act 1962 revoking the planning permission, and on December 3 1965 another order under section 28 of that Act ordering the discontinuance of the use of the quarry for the winning and working of minerals and the removal of all buildings and works. The claimants objected to those orders and accordingly a public inquiry was held on November 24 1966, but by a letter dated June 1 1967 the Minister of Housing and Local Government confirmed those orders.

Motorway Subcontract

6. At the time of the confirmation of the orders the reserves of limestone within the quarry amounted to 2m tons, all of which could have been extracted by conventional quarrying methods within the time and at the rate required for the construction of the Easton-in-Gordano to East Brent section of the M5 motorway. For those reasons and because of its proximity to that section of the motorway in relation to other quarries able to supply the material, it is almost certain that the claimants would have been awarded the subcontract for the supply of rockfill and base material for that section of the motorway. The probability of being awarded that subcontract was the claimants' main motive in purchasing the quarry.

7. The total amount of rockfill and subbase material required for the construction of that section of the motorway was about 2,161,000 tons. Having allowed for the buying-in of 161,000 tons of material, the claimants would have made a profit of £295,805. If, however, it is correct to assume that the valuation date for this purpose is June 1 1967, and that it is correct to allow for deferment over the period of the contract, the amount becomes reduced to £250,100.

8. If the claimants had failed to win the motorway contract they would have been able to sell the limestone to the successful subcontractor at a profit of £200,000, less the remaining value of the land, £300, that is, £199,700. This figure is derived from the sale of other stone-bearing land for the purpose of the motorway.

9. If there had been no motorway and the claimants had merely bought the quarry and installed plant for supplying the general market and their own ready-mixed concrete plant at Weston-super-Mare, the quarry would have had a life of 11½ years and with the capital value arising from the annual profit allowing for interest on and redemption of capital, ie sinking fund, they would have made a profit as at November 1969, after allowing for the value of the land after the revocation order, of £107,650. If, however, this amount is discounted back to June 1 1967 it becomes £84,000.

10. The parties are agreed that the depreciation of the market value of the interest in the quarry as at June 1 1967 due to the revocation order was £72,000, excluding the costs of preparing the claim and any interest payable on that amount.

11. The cost of complying with the discontinuance order, excluding any interest thereon, is agreed at £1,250.

Basis of Compensation

The main issue before us was whether the claimants are entitled to compensation on the basis of loss of anticipated profits or whether they are restricted to the depreciation in the value of their interest in the land. Before turning to the arguments and submissions it is convenient to set out the statutory right to compensation which is provided in section 118 (1) of the Town and Country Planning Act 1962 in these terms:

"118 (1) Where planning permission is revoked or modified by an order under section 27 of this Act, then if, on a claim made to the local planning authority within the time and in the manner prescribed by regulations under this Act, it is shown that a person interested in the land:

(a) has incurred expenditure in carrying out work which is rendered abortive by the revocation or modification, or

(b) has otherwise sustained loss or damage which is directly attributable to the revocation or modification,

the local planning authority shall pay to that person compensation in respect of that expenditure, loss or damage.

(4) In calculating, for the purposes of this section, the amount of any loss or damage consisting of depreciation of the value of an interest in land, it shall be assumed that planning permission would be granted for development of the land of any class specified in the Third Schedule to this Act."

For the claimants Mr Widdicombe's general submission was that the words "loss or damage" are wide and intended to cover loss of all kinds suffered and are not limited to loss in the value of the land. They do not necessarily exclude the latter loss, but if compensation is assessed on that basis, then, and only then, section 127 of the Act of 1962 and, therefore, the rules set out in section 5 of the Land Compensation Act 1961 apply. In the present case there are no grounds for reading the words as giving the claimants less than their true loss, as would the county council's construction of the section. He pointed out that it was commonplace for compensation for loss of profits to be awarded in compulsory purchase cases, subject to exclusion on the ground of remoteness, and that there was no logic in the county council's argument that compensation for loss of profits could be awarded only where the claimants' business on the site was an existing one. In the case of a quarrying company that factor must be particularly irrelevant, because any mineral-bearing land was part of the totality of the stock-in-trade of the company. Where there is a revocation order, there is no purchase of the land or any interest therein. The revocation order effects a restriction on what can be done on the land comparable in a mineral case with the restriction imposed by a counter-notice under the mining code.

He referred to the well-known case of *Bwllfa & Merthyr Dare Steam Collieries v Pontypridd Waterworks* [1903] AC 426, saying that everything in the judgments applied to the present case. They are both mineral cases; in neither case was there a purchase, and in both cases compensation was for the prevention of the working of minerals and was to be made "as in other cases of disputed compensation." He stressed that Lord Halsbury insisted that "the true inquiry here is not what is the value of the coalfield or of the coal, but what would the colliery company, if they had not been prohibited, have made out of the coal during the time it would have taken them to get it," and that Lord Robertson said "it follows that what is due to the appellants is not the price on a transaction of sale, but compensation for a continuing embargo of working."

Three Relevant Questions

Mr Widdicombe went on to say that in a revocation case the question is that stated in the statute, namely whether the loss or damage is directly attributable to the revocation. This is not the same as the question where a claimant has been disturbed in his occupation of the land for the purposes of the law of compensation for compulsory purchase; and that cases decided under that law are not relevant, but in any event they were decided upon their facts and not on any principle of law. He said there are three relevant questions, namely:

(1) Would the company have made profits from the quarry?

(2) If so, is the loss of profits directly attributable to the revocation?

(3) If so, what is the quantum?

This accords with the decision in the case of *Cottrill v Steyning & Littlehampton Building Society* [1966] 1 WLR 753. He went on to say:

"In the present case the claimant company is an established business, and an experienced quarry operator, and the subject land is merely a prospective extension of its business premises. The company owned the subject land, which had the benefit of an existing planning permission for quarrying, and was the site of an existing quarry. It owned some 21 other quarries, including four within a radius of 18 miles of which two were within 2½ miles. There is no reason to doubt that but for the revocation the company would have worked the quarry for a profit. The

A county council clearly thought this was so, and that revocation and discontinuance orders were necessary to prevent it."

For the Somerset County Council, Mr Fletcher summarised his submissions as follows:

1. The grant of a permission to develop land (except in so far as the permission otherwise provides) enures for the benefit of the land and of all persons for the time being interested therein, and the revocation of permission affects the land and all interests in it.
2. The relevant date for the assessment of compensation under a revocation order is the date when the order is confirmed, for it is on that date that the interests then B subsisting in the land suffer damage.
3. The depreciation in the open market value of the land is to be assessed under Rule (2) and compensation for disturbance or any other matter not directly based on the value of the land is to be ascertained under Rule (6).
4. The potentiality of the land for the making of profits by a purchaser is reflected in the open-market value of the land, and if the owner is not already carrying on a business on the land the value to him is normally the open-market value. Only in a case where he has a better chance than a purchaser of making profits is he entitled C to a payment for disturbance representing the additional value of the land to him.
5. In this case the relevant date for the assessment of compensation is June 1 1967. There is no evidence that at that date the claimants had a better chance than a purchaser of making profits out of the land and there is no valuation put forward by the claimants on this basis.

Mr Fletcher also said that the principle of compulsory purchase applies; that is where business is about to commence and the imminence and degree of probability of profit are relevant factors in the valuation of the land. If a particular owner has a better chance of profit he gets something extra. D But he also said that if the claimants' business had been a going concern at the quarry, compensation would have been on the basis of the capitalised loss of profits for the loss of the motorway contract, and so it would if the claimants, instead of buying the quarry, had bought the shares in a company already operating the quarry.

We find this case very difficult because we see some force in the argument that the loss to the claimants cannot be more than the depreciation in the market value. That argument, however, presupposes that there was another quarry available to be bought at market value which could serve the claimants' E purposes and provide them with an equal profit; but there was no evidence to that effect. Thus, it is clear that the claimants have been deprived of the profits they would have earned and that much is not disputed. We find the cases decided on claims under the heading of disturbance by compulsory purchase of little help, because for the most part they were decided on the grounds of remoteness or because they offended the rule against double compensation as in *D McEwing & Sons Ltd v Renfrew County Council* (1960) 11 P & CR 306, or because no reasons for the decision were given. On the other hand we find no material difference in the compensation provisions in the *Bwllfa* case appeal or in F *Eden v North-Eastern Railway Co* [1907] AC 400.

So we think, adopting the words of Phillimore J in *Thomas McGhie & Sons Ltd v British Transport Commission* [1963] 1 QB 125 at 148, "The applicant is to be compensated, not for the value of the minerals or for the sterilisation of a capital asset, but for what he would, if not prohibited, have made out of the minerals during the time it would have taken to work them and he is to be compensated for the loss caused by the specified minerals being left unworked." Further, we cannot see any justification for drawing a distinction between a company which has already started operations and one which has not done so, for in both cases the loss is identical.

Nor do we think that there is necessarily any difference G between a revocation order case and a compulsory purchase order case, provided that the rules against double compensation and remoteness of damage are observed.

We have had regard to those rules and we find that the damage resulting from the loss of the motorway contract is not too remote: indeed, it was conceded on behalf of the county council that the claimants had been in a strong position to win it and, of course, as there is no claim for the value of the land the question of double compensation does not arise. Accordingly, we find that the proper basis for compensation is by reference to the loss of profits from the motorway contract. H

Deferment

Mr Fletcher submitted that, if it were right to assess compensation by reference to future profits rather than the value of the land on the date when the order was confirmed, then allowance should be made for the fact that there would be a delay of about two years before any profits at all could be earned, and then that the earning of those profits would be spread over another two years. We did not hear much argument as to the appropriate date for valuation, presumably because there is no alternative to the date of confirmation of J the revocation order. If, therefore, compensation had been paid immediately after that date on the basis of estimated loss of profits, the claimants would have the benefit of the use of the whole of that money for two years, when, but for the revocation, they would not have been earning any profits, and then for a further two years to profits which they would only have had in part until the exploitation was completed. Accordingly, it seems to us that Mr Fletcher's argument is well founded and we reduce the principal sum on that account to the agreed sum of £250,100, that figure being the amount should deferment be appropriate.

K

Payment of Interest from Date of Revocation

It is common ground that, if allowed, interest on the amount of our award should run from the date of the award and we so order.

The dispute between the parties is, however, whether interest is payable from the date of the revocation order, having regard to the fact that the statute is silent on the question, whereas express provision is made for the payment of interest on the money under the Land Compensation Act 1961 and the Compulsory Purchase Act 1965. Mr L Widdicombe argued that express provision is necessary in those cases, having regard to the basis of compensation which is restricted to the value of the land taken and is therefore not apt to comprehend the payment of interest. He contrasted the basis of compensation under those Acts with that under section 118 (1) (b) of the Act of 1962, the terms of which are sufficient to cover not only the loss of profits or other loss but also the loss of the use of the money from the date of revocation. He referred to the case of *Jefford v Gee* (CA) [1970] 2 QB 130, which, he said, established that if this were a common-law case the claimants would get loss of profits plus interest and it would be wrong if the compen- M sation law were different. He likened the position of the Lands Tribunal to that of the Admiralty Courts as explained in that case.

Mr Fletcher, on the other hand, said that there is a well-established principle that until compensation is quantified no interest is payable and he referred to the *London, Chatham & Dover Railway Co v South-Eastern Railway Co* [1893] AC 429, *The Caledonian Railway Co v Sir William Carmichael Bart* (1870) LR 2 HL (Sc) 56, *Re Richard and Great Western Railway* [1905] 1 KB 68, and *Swift v Board of Trade* [1926] AC 520.

A In our view we have to assess the amount necessary fully to compensate the claimants as at the date of the revocation and discontinuance orders, and having done that we are *functus officio*. Such compensation is limited to the loss arising from the service of the notice and order. It follows from those considerations that we are entitled neither to compensate for a loss which arose not from the making of the order and notice but from the fact that the compensation moneys were not paid on or immediately following the date of service, nor to compensate for a loss which in truth is not part of the damage to the interest in the land but arises from the fact that the claimants had not had the use of the
B money. We think that sufficient authority for our conclusion is to be found in *Richard v Great Western Railway* (*supra*), and we also get some support from the recent case of *Wallerstiner v Moir, The Times*, January 29 1975. In the latter case it was said by the Master of the Rolls that interest was awarded not under the statute but under the equitable jurisdiction of the court. The Lands Tribunal, however, is a creature of statute and has no equitable jurisdiction. Accordingly we find that no sum is payable in respect of interest.

C

Deductions for Tax

During the course of the hearing the question arose as to what deduction in the calculation of compensation should be made for tax having regard to the cases of *British Transport Commission v Gourley* [1956] AC 185, and *West Suffolk County Council v W Rought Ltd* [1957] AC 403. We were referred to a written statement made by the Board of Inland Revenue handed in to the tribunal during the hearing of the case of *Bostock, Chater & Sons Ltd v Chelmsford Borough Council* (1973) 227 EG 141, setting out the advice which the board had given regarding reduction of compensation in respect of tax. The tribunal in that case had little alternative
D but to accept that statement, but we were unwilling to do so. Accordingly we requested the solicitor to the board to instruct the board's standing counsel to appear *amicus curiae*, but for reasons which we accept the board's solicitor declined our request. However, the parties agreed jointly to instruct revenue counsel, namely, Mr Michael Musgrave, to appear *amicus curiae* on the understanding that neither of them would be bound by his submission. We wish to acknowledge the helpfulness of that action taken by the parties and our gratitude to Mr Musgrave for dealing with this matter so thoroughly over a weekend. In the event neither of the
E parties disputed Mr Musgrave's submission, but we must record that they reserve their right to do so elsewhere.

Mr Musgrave referred to the cases of *British Transport Commission v Gourley, West Suffolk County Council v W Rought Ltd, Thomas McGhie & Sons Ltd v British Transport Commission, Re Associated Portland Cement Manufacturers Ltd's Application* [1966] Ch 308, *Glenboig Union Firclay Co v Commissioners of Inland Revenue* (1921) 12 TC 427, 463 and 465, *Parsons v BNM Laboratories* [1964] 1 QB 95, *Taylor v O'Connor* [1971] AC 115; and to the Finance Act 1965, sections 42 and 45 (5) and submitted:

F (1) Where the compensation is assessed by reference to loss of anticipated profits it must be reduced where tax would be paid on those profits in the hands of a recipient. The sum reduced should be grossed up by the amount of the extra tax payable.

(2) No deduction for tax should be made where compensation is assessed by reference to a sale of the quarry.

(3) If interest is awarded it should be awarded gross and the tax deducted and accounted for by the acquiring authority.

We accept Mr Musgrave's submissions and his method of putting them into effect, which will be seen in our award.

Award G

We find that the compensation payable in respect of the discontinuance order is the sum of £1,250 and in respect of the revocation order £192,168, the latter being made up as follows:

Principal sum after deferrer back to 1/6/67 applied		£250,100
Deduct for corporation tax on lost profits in 1969/70 and 1970/71, adjusted to deferred value ...		101,332
Profits net of corporation tax		£148,768 H
Add for capital gains tax, compensation net of corporation tax as above	£148,768	
Less cost of acquisition of quarry, six acres apportioned	47,500	
	£101,268	
Capital gains tax thereon @ 30%	£30,380	
Grossed up to allow for tax on award 10/7		43,400
Amount of award		£192,168 J

The total award therefore is £193,418.

In addition to the above sums the acquiring authority will pay to the claimants in each case a surveyor's fee based on scale 5A of the scales of professional charges of the Royal Institution of Chartered Surveyors and any legal costs properly incurred from the dates of confirmation respectively of the discontinuance order and the revocation order up to the date of the notices of reference.

Having caused the sealed offer submitted on behalf of the county council to be opened and read, we find that our award exceeds the amount thereof. We have been addressed by the K solicitor for the claimants and counsel for the county council, and we order that the county council do pay two-thirds of the claimants' cost of these references, which, if not agreed, shall be taxed on the High Court scale by the Registrar of the Lands Tribunal.

HULL TRINITY HOUSE CORPORATION v HUMBERSIDE COUNTY COUNCIL

(Ref/69/1974) L

(DOUGLAS FRANK QC, President, and E C STRATHON FRICS)

April 23 1975

Estates Gazette June 21 1975

(1975) 234 EG 915

Purchase notice—Land at North Ferriby, Hull, required for highway purposes—Planning permission likely for residential development—Abortive agreement on price in 1969—Agreement reached by parties' valuers in December 1973 repudiated by council and unenforceable—Question of appropriate date of assessment—Tribunal rejects "with some reluctance" claimants' submission for date of negotiations and, although it M considers "commonsense and convenience point overwhelmingly to date of notice of reference," it adopts date of hearing, since parties agreed on that as the alternative—Council's argument that value of subject land fell between December 1973 and February 1975 fails

Mr I D L Glidewell QC and Mr D M W Barnes (instructed by Rollit, Farrell & Bladon, of Hull) appeared for the claimants, and Mr A P Fletcher (instructed by the director of administration of Humberside County Council) represented the acquiring authority.

Giving their decision, THE TRIBUNAL said: This is a reference to determine the amount of compensation payable for the deemed compulsory acquisition of the freehold interest of land having an area of 1.68 acres fronting Melton Road, North Ferriby. The village of North Ferriby straddles, but is predominantly to the south of, the A63 Leeds/Kingston-upon-Hull trunk road. It is approximately 8 miles to the west of the centre of the city of Kingston-upon-Hull. North Ferriby is a commuter residential village within the sphere of influence of Kingston-upon-Hull. There are reasonable local shops, schools and social facilities. North Ferriby lies on the Selby/Goole/Kingston-upon-Hull railway line, and there is a passenger station within the village and reasonable services to Hull and Selby. The site is unused, but it is common ground that if the land had not been required for highway purposes, it was likely that planning permission would have been granted for 12 detached dwellings, each plot to have approximately a 70-ft. frontage to Melton Road. The first issue between the parties is the appropriate date at which compensation should be assessed. The claimants say that it is December 1973, when a price was agreed by the parties' respective valuers. The respondents say that it is the date of our hearing, or in August 1969 when there was an abortive agreement. There are also issues on quantum.

History

On December 20 1966, the Beverley Rural District Council refused an outline planning application for the residential development of the subject land. The reason for the refusal was " the land is required for a possible road improvement at the junction of Melton Road and the Ferriby By-Pass." On December 28 1966, the claimants served on the Rural District Council a purchase notice under section 129 of the Town and Country Planning Act 1962, which notice was confirmed by the Minister of Housing and Local Government on July 24 1967, subject to the substitution of the East Riding of Yorkshire County Council as purchasing authority.

In a letter of July 31 1967, the clerk to the East Riding of Yorkshire County Council told the claimants' solicitors that the county land agent had been "instructed to negotiate settlement of compensation on the basis that the [subject] land would attract a certificate of appropriate alternative development for residential purposes."

On May 17 1968 the county planning officer informally told the county land agent that if the land had not been required for highway purposes planning permission as already described would have been granted.

In December 1968 and January 1969 the claimants' surveyors and the county land agent agreed a figure of compensation amounting to £23,050 plus fees and costs. An abstract of title was delivered and conveyancing procedures were put in hand, but a dispute arose as to the area of the subject land, which dispute was not resolved. However, on April 28 1969, the county land agent in a letter to the claimants' surveyor, having stated that a number of unexpected difficulties had to be overcome, went on to say: " the county council will soon be in a position to pay interest on the amount of the agreed settlement." Nevertheless the council did not proceed to implement that agreement but, on the contrary, on May 2 1969 the clerk wrote to the claimants' solicitors a letter in which he said " I am now asking the Ministry of Transport to accept responsibility under the purchase notice and will write to you again on hearing from the Minister." The solicitors replied as follows:

> " We thank you for your letter of the 2nd instant and note that you are now asking the Ministry of Transport to accept responsibility under the purchase notice in this matter, but with respect we do not see how this concerns our clients in any way. As you are aware, our clients have been concerned with the delay in settling this matter, negotiations having commenced in September 1967, and the compensation having been settled in August 1968, subject to adjustment in respect of compensation repayable under the Town and Country Planning Act 1954. As the precise amount of the compensation has now been agreed and you received a report of the district valuer on or about the 25th ultimo, and you are also in possession of the abstract of title, it would appear that completion should take place immediately."

In reply to that letter the clerk wrote: " it has been obvious throughout these proceedings that your clients' land ought to be purchased by the Minister of Transport, and I am instructed to attempt once again to clarify the Minister's intentions." In reply to that the solicitors reminded the clerk of the terms of the purchase notice as confirmed and thenceforth continued to press the clerk for a settlement. On August 1 1969 the clerk said that he was " now in a position to proceed with the conveyancing respecting the above land " and then went on to deal with conveyancing matters. But again the dispute as to the precise area arose and culminated in an acceptance by the claimants that the area to be acquired was 1·68 acres and not 1·85 acres as stated on the purchase notice plan. Accordingly, on October 1 1971 the parties agreed that the surveyors and the county land agent should agree a price for the reduced area. Further negotiations proceeded and in the summer of 1973 oral agreement was reached, and on November 21 1973 the county land agent wrote to the claimants' surveyors thus:

> " Further to previous correspondence and discussions I am now in a position to make certain recommendations, as follows:
>
> 1. *Purchase Notice—High Street*
>
> (a) The county council to purchase the freehold interest in the area of approximately 1.68 acres of land shown on the County Surveyor's Plan No 3052/L/B for the sum of £84,000. From this has to be deducted the Registered Charge amounting to £402.17 thereby making a net purchase price of £83,597.83, say £83,600 (eighty three thousand, six hundred pounds).
>
> (b) Vacant possession to be given on completion.
>
> (c) The land to be transferred free of any easements, restrictions or outgoings, and any interests of the owners in mines and minerals to be included.
>
> (d) No accommodation works to be specified.
>
> (e) The county council to pay the vendor's proper legal costs regarding the conveyance to the land.
>
> (f) The county council to pay the vendor's surveyors fees of £905.10 (nine hundred and five pounds and ten pence) plus 10% VAT on the assumption that the vendors do not have a registered business for VAT."

The claimants' surveyors replied on December 7 1973, saying " We have now taken our clients' instructions regarding this matter and wish to advise you that they are prepared to accept the offer of £83,600, together with the other terms set out in your letter," and went on to give the name and address of the claimants' solicitors. On January 4 1974 those solicitors asked the clerk of the county council for a draft conveyance for approval as quickly as possible, but the clerk replied that the county council were unwilling to shoulder the financial burden involved and would need a clear assurance of reimbursement by the Secretary of State as the highway authority actually concerned. Not surprisingly the claimants' solicitors wrote a long letter protesting at the council's attitude and failure to proceed with the agreement, but the clerk did little more than reiterate what he had said in his previous letter, as he did in a further letter in which he suggested that the claimants should drop their claim for compensation and apply again for planning permission. On April 5 1974 the present respondents assumed the functions of the former county council, and not surprisingly Mr Alan Fletcher, who appeared for the Humberside County Council, did not seek to justify the actions of his clients' predecessors.

On March 4 1974 the claimants applied for an appropriate alternative development certificate under section 17 of the Land Compensation Act 1961. On July 12 that certificate was issued and specified frontage residential development of not more than 12 houses with paired accesses and the provision of turning spaces within the curtilage of each dwelling and the planting of a tree belt. That application was made because it had been said by the clerk of the former county council that the Lands Tribunal would require such a certificate. On March 5 1974 the claimants, through their solicitors, referred the determination of the assessment of the compensation to the Lands Tribunal.

A

Date of Assessment

Mr. Glidewell conceded that no enforceable agreement was concluded in 1973, but as that was the time when the parties agreed to assess compensation it should be adopted. He referred to *Birmingham Corporation v West Midland Baptist (Trust) Association (Inc.)* [1970] AC 874 for the proposition that where there is no contract and where possession has not been taken, their Lordships' speeches point to the date of assessment as being appropriate; see, for example, Lord Morris of Borth-y-Gest at p904. He said that in the peculiar circumstances of this case agreement had been reached and it was only necessary to make the reference to the Lands Tribunal in order to force the council B to complete it.

Mr. Fletcher invoked the decision in *W & S (Long Eaton) Ltd v Derbyshire County Council* (1974) 233 EG 403 for his submission that where entry has not occurred and there is no enforceable contract, compensation is to be assessed at the date of the hearing. That case, he said, was rightly decided. Further, he maintained that Mr. Glidewell's test was imprecise, because, for example, when did the valuers agree the assessment? Was it in July 1973, when they reached oral informal agreement, or was it in November, when the agreement was confirmed in writing by the county surveyor, or was it in December, when the amount was accepted. C If, he said, the rules of contract were not to be followed, then it would be appropriate to take the early agreement, even though it transpired that the parties were not *ad idem* as to the area of the land to be acquired.

In the *Derbyshire* case the tribunal said:

> "It seems to us that we should conclude from the *Birmingham* case that there are at least three different events, the date of occurrence of which may determine the date of assessment. They are: (i) entry by the acquiring authority; (ii) the making of an enforceable contract; and (iii) the hearing by the Lands Tribunal."

The tribunal went on to find that a contract which was not enforceable did not qualify. However, we may say in passing that D we understand that an appeal is pending in that case. A somewhat different issue is raised in this case, namely, whether where the parties' valuers are negotiating the notional market price as at the date of their negotiations, then that date is or should be held to be the date of assessment, so that there is yet another category of events. We can see that in the present case justice and convenience supports that view. Indeed, there might be much to be said for it in other cases because, for example, it would prevent a party resiling from his agreement because of a change in the market between the agreement and completion. In this case the council repudiated the agreement by their predecessors' valuer E because they thought that by doing so they would be able to pay less for the land. It could also be said that both parties had committed themselves, not to the values agreed by their respective valuers but to having compensation assessed at the date of their agreement. We are attracted by the theory and it is with some reluctance that we find against it. We do so because it must be necessary to extract from it a date certain and that cannot be done, particularly where negotiations are protracted as they were in this case. We do not think the references in the *Birmingham* case to the date of assessment mean the date of negotiations.

The important question now arises: which is the appropriate F date? The *Birmingham* case requires us to disregard the date of the deemed notice to treat. There is no contract. A notice of entry has not been served. It seems to us that where there is a reference to the Lands Tribunal three dates fall for consideration, namely, the date of reference, the date or dates of the hearing, and the date of the decision. It could be argued that the date of the decision is the appropriate date because thereafter either party can call for specific performance, but in practice it would be impossible to assess values at the date of decision save in those cases where the decision was given on the day of or within a few days of the hearing. There are also disadvantages in adopting the date of the hearing, because at the time of the hearing values may have changed and changed quite substantially between the G lodging of documents pursuant to rule 42 and that date. They can also change during the course of a protracted hearing, particularly if there are long adjournments; in the *Derbyshire* case the council's valuer contended that there had been a fall in values of 20 per cent over a fortnight, although on the evidence that was not accepted by the tribunal. The other alternative, namely the date of the reference, has certain obvious advantages. It enables either party to fix the date of assessment; on the contrary argument, that can only be done at present by the acquiring authority giving notice of entry, thus enabling it to select a date most advantageous to it. It also fixes a single date for the date of assessment of values without either party standing H to benefit or disbenefit by the shifting sands of the market. The choice of that date as the valuation date has one advantage, that valuations and particulars of comparable properties do not have to be altered right up to the time of, or during the course of, the hearing.

Our feeling is that in the interest of justice and convenience there is much to be said for the date of the reference. The date of the reference is, no doubt, in practice the date at which valuations submitted under rule 42 of the Lands Tribunal Rules 1963 are assessed. We do not find, at least in a rule (2) case, that there is anything in the *Birmingham* case to the contrary. However, we have not heard argument in favour of it, and indeed when asked J to deal with the suggestion both Mr Fletcher and Mr Barnes (in the absence of Mr Glidewell) argued in favour of the date of our hearing, although it must be said that they had very little time to consider the alternative we put to them. We think that commonsense and convenience point overwhelmingly to the date of the notice of reference. Nevertheless, as the parties seem to be agreed that the alternative to Mr Glidewell's submission is the date of the hearing, we adopt that date as a matter *ex concessis* and we will now consider valuations on that basis, without prejudice to the possibility that in some later case the date of the notice of reference may be put forward as being the right date. K

Assessment of Value

The valuation of Mr Seaton FRICS, the claimants' valuer, was £83,600. Mr J M Bell BSc (Est Man) FRICS, now county estates and valuation officer, produced two valuations, one of £49,200 and the other £49,700. The main issue is whether the value of the subject land has fallen between the December 1973 provisional agreement and February 1975, the date of the hearing. It is clear that both Mr Seaton and Mr R Gregory ARICS, then an assistant to the county land agent, reached agreement on values in 1973 by direct comparison with the sale (in lots comprising one L or more plots) in March 1973 of the Aston Hall Estate, an area of 4.5 acres lying about 250 yds from the subject land. In January 1975 one of the plots was resold for £7,000 by comparison with its 1973 price of £6,500, a sale which Mr Seaton regarded as supporting his opinion that the value of the subject land had not fallen since 1973. Mr Bell had concluded at the outset that the values of land, including the subject land, had fallen since 1973; he had therefore rejected the 1973 Aston Hall sales and for his primary valuation had relied on a sale by auction in September 1974 of three plots in a village six miles west of the subject land.

We do not accept that the evidence adduced (in so far as it was admissible) as to a fall in the values of land was sufficiently M dependable as to cause the rejection of the Aston Hall sales, which we consider to be the better comparables. Indeed, Mr Bell himself did not regard his comparables as adequate evidence on which to rely for his valuation.

We now turn to Mr Bell's other valuation, in which he adopted the 1973 agreement as the starting point and reduced it by a percentage which was derived directly from the difference in the sale prices of two areas of land in August 1973 and October 1974 respectively. In our view the characteristics of the two sites are so disparate that a reliable judgment could not be made as to the amount, if any, of a fall in value between those dates.

A We therefore accept Mr Seaton's basis of valuation and his basic valuation of £83,600. But there are two subsidiary matters for decision. On the point of paired accesses, specified in the section 17 certificate, none of the witnesses had had experience as to whether such a condition had any effect on the selling price of building plots. We decide that no deduction falls to be made pursuant to such a condition. As to drainage, it was clear that neither Mr Seaton nor Mr Gregory was aware that the main sewer was not laid along the frontage of the subject land. The issue was argued and various estimates of the costs of connection were given in evidence. We have concluded that in the market a purchaser would reduce his bid on this account and he would B round down the basic price from £83,600 to £80,000.

Accordingly we determine the compensation payable at £80,000.

The acquiring authority will pay the claimants the sum of £80,000, to which will be added a surveyor's fee based on scale 5 (a) of the scales of professional charges of the Royal Institution of Chartered Surveyors, also any legal costs properly incurred from the date of the confirmation of the purchase notice up to the date of the notice of reference.

Having read the decision in this matter and having then opened the sealed offer of compensation lodged by the acquiring authority, we find that the amount thereof is less than the C amount of our award; accordingly the acquiring authority will pay to the claimant its costs of this reference, such costs, unless agreed, to be taxed by the Registrar of the Lands Tribunal on the High Court Scale.

RANDELL v WEST GLAMORGAN COUNTY COUNCIL

(Ref/96/1974)

(J H EMLYN JONES FRICS)

May 14 1975

D Estates Gazette June 28 1975

(1975) 234 EG 999

Blight notice—Detached house in ¾-acre grounds near Swansea—Strip of land about 20 ft wide, less than one-tenth of site area, to be acquired by council—Case under section 202 (2) of Town and Country Planning Act 1971—Tribunal holds that strip cannot be taken without material detriment to house or without seriously affecting its amenity—Council must purchase whole property

Mr D M W Barnes (instructed by T B Walker-Jones & Gardner, of Swansea) appeared for the claimant; Mr M J E Clare, solicitor, for the acquiring authority.

Giving his decision, MR EMLYN JONES said: The claimant, Mrs M B Randell, is the owner in fee simple of a dwelling-house known as Cefn Bryn, Caswell Road, Caswell, Swansea. Cefn Bryn is a substantial detached dwelling-house with five bedrooms and two bathrooms standing in three-quarters of an acre of land. On or about January 15 1974 Mrs Randell served on the highway authority (at that time the Swansea City Council) a blight notice pursuant to the provisions of section 193 (1) of the Town and Country Planning Act 1971 requiring the authority to purchase her interest in the property on the grounds:

F (1) That she was the freeholder;

(2) That part of the hereditament had been included in land falling within paragraph (e) of section 192 (1) of the Act, that is to say that it was:

> "Land shown on plans approved by a resolution of a local highway authority as land comprised in the site of a highway as proposed to be constructed, improved or altered by that authority";

(3) That she had made reasonable endeavours to sell her interest but had been unable to find a purchaser at a price which might reasonably have been expected in the absence of the proposed highway improvement; and

(4) That her interest qualified for protection in that she G was a resident owner-occupier.

The authority on February 20 1974 lodged a counternotice objecting to the blight notice. In its original form the notice challenged Mrs Randell's entitlement under the grounds which I have set out in paragraph (3) above, but this objection was subsequently not proceeded with. The authority, who were in due course succeeded by the West Glamorgan County Council, continued to rely on the grounds set out in section 194 (2) (c) of the Act:

> "That the appropriate authority propose in the exercise of relevant powers to acquire a part of the hereditament . . . but H (unless compelled to do so by virtue of these provisions) do not propose to acquire any other part of that hereditament . . . in the exercise of any such powers."

There is no dispute as to the facts. The authority intend to acquire a strip of the garden of Cefn Bryn for road-widening purposes. Of the total site area of about 3,450 sq yds, rather less than one-tenth, 315 sq yds, is required. Before the acquisition the house at its nearest point is set back some 54 ft 3 in from the boundary. After the strip has been taken the distance will be 34 ft 8 in. The claimant wishes the authority to acquire the whole of the property. The authority unless required to do so propose to acquire only the strip. J

Section 202 (2) of the Act provides:

> "(2) The said provisions [including the provisions of section 194 (2) (c)] shall not affect the right of a claimant under section 8 of the Compulsory Purchase Act 1965 to sell (unless the Lands Tribunal otherwise determines) the whole of the hereditament, . . . which he has required the authority to purchase; and accordingly in determining whether or not to uphold an objection relating to a hereditament on the grounds mentioned in section 194 (2) (c) of this Act the tribunal shall consider (in addition to the other matters which they are required to consider) whether:
>
> (a) in the case of a house, building or manufactory, the part proposed to be acquired can be taken without material detriment to the house, building or manufactory; or K
>
> (b) in the case of a park or garden belonging to a house, the part proposed to be acquired can be taken without seriously affecting the amenity or convenience of the house."

These two matters which the tribunal is required to consider are expressed in identical terms to those used in section 8 (1) of the Compulsory Purchase Act 1965, which the Lands Tribunal would be required to consider in applying the provisions of that section.

Mr Barnes, for the claimant, called evidence from Mr Leslie Wynne Davies FRICS, senior partner in the Swansea L firm of chartered surveyors, Astley Samuel, Leeder & Son. Mr Wynne Davies had been engaged in various unsuccessful attempts to dispose of the property on behalf of the claimant. Prospective purchasers had lost interest in the property when told about the road-widening proposals. In the opinion of Mr Wynne Davies the removal of the strip of garden, involving as it would the destruction of a large part of the existing screen of trees and shrubs, would have the effect of changing entirely the character of the house and would largely destroy its existing amenities and the seclusion which it presently enjoys.

Mr Clare, for the authority, called the assistant county M estates and valuation officer, Mr John Alan Howells ARICS, who took the opposite view. He thought that by the carrying out of the proposed accommodation works, involving the planting of trees and shrubs, there would be no serious effect on the house as a result of the road-widening. Although the strip to be taken was about 20 ft wide, the traffic would not in fact be brought that much nearer to the house, since provision was being made for a footpath at the side of the carriageway where none existed at present. In so far as the character of the house was changing, Mr Howells thought that this was primarily caused by the separate sale of two

A parcels of the original plot totalling in aggregate about half an acre, on one of which a bungalow had been erected.

I heard argument as to whether paragraph (a) or (b) of section 202 (2) was applicable, and reference was made to the decision of the Lands Tribunal in *Ravenseft Properties Ltd v London Borough of Hillingdon* (1968) 206 EG 1255, which was concerned with the interpretation of the wording of section 8 (1) of the Compulsory Purchase Act 1965. In Mr Barnes' submission the present case came under both paragraphs. The house under (a) should be taken to include its curtilage, for as Mr Barnes pointed out, if the house under (a) referred only to the building, then, since no part
B of the building was being acquired, the section would not operate, and although paragraph (b) would come into force if part of a garden were being acquired it could not apply in cases where the whole of a garden was being acquired. Such an anomaly would, in Mr Barnes' submission, be absurd. It is true that section 202 of the 1971 Act refers to the hereditament, but no such word is used in section 8 of the 1965 Act from which the wording of paragraphs (a) and (b) is derived.

Section 202 clearly presents difficulties of interpretation, since if Mr Barnes is correct then the word "house" is to be
C interpreted in two different ways in the two paragraphs—as house with curtilage in paragraph (a) and as a house excluding its garden in paragraph (b). In the present case I do not find it necessary to reach a final conclusion. It seems to me that the case falls squarely within paragraph (b), but on the facts which I have found from the evidence and from my inspection I arrive at the same conclusion under both headings. In my judgment the strip of land in the present case cannot be taken without material detriment to the house nor can it be taken without seriously affecting the amenity of the house. In other words I accept the evidence of Mr Wynne Davies that the removal of the strip of land takes the house out of one category and puts it into another of a D
quite different character.

Finally, I should say that I have reached this conclusion on the evidence. It also seems to me that this is the conclusion which best accords with the rights of the citizen which Parliament has been at some pains to protect. The claimant, Mrs Randell, is quietly enjoying her property when the county council, for what are no doubt perfectly proper public reasons, come along and require to take some of that property away. It is true that under the compensation code she is to be compensated not only for the value of the land taken but for injurious affection on the part which the authority do not require. She is also entitled to a claim for disturbance. E
But she no longer wishes to live in the house in its changed surroundings, and she may well ask in these circumstances why she should be expected to suffer the uncertainties of putting her house on the market and anticipating what the depreciation in value may be when agreeing to a determination of injurious affection. The blight notice provisions of the 1971 Act provide a remedy whereby the authority can be required to take over the whole property and thereby minimise the trouble and inconvenience caused to the claimant. I am glad to be able to find in the present case that the claimant is protected in this way. I therefore do not
uphold the objection of the authority and declare that the F
purchase notice dated January 15 1974 is a valid notice in relation to the whole of the hereditament to which it relates. I direct that a notice to treat in respect of the whole hereditament is deemed to have been served by the authority on the claimant on March 15 1974.

The authority will pay the costs of the claimant, such costs failing agreement to be taxed by the Registrar of the Lands Tribunal on scale 4 of the County Court Scale of Costs, with discretion as to counsel's fees, and certificate for expert witness fit to qualify.

LANDS TRIBUNAL
LEASEHOLD REFORM

NASH v CASTELL-Y-MYNACH ESTATE
(LR/49/1974)
(R C WALMSLEY FRICS)
March 4 1975

Estates Gazette April 26 1975
(1975) 234 EG 293

Price of freehold of detached house in open country in South Glamorgan on site of over ¼ acre—1½ years' unexpired term—No mains water—Tribunal determines site value by "a considered apportionment" between site and buildings—"Dulwich" case considered—An established "land market" in Cardiff based directly on tribunal decisions

Giving his decision, MR WALMSLEY said: This is a reference for the determination under section 9 of the Leasehold Reform Act 1967, as amended, of the price which should be paid for the freehold interest of Hill Cottage, Groesfaen, near Pontyclun, in the county of South Glamorgan, by the resident lessees, Mr and Mrs C H Nash, to the freeholders, Castell-y-Mynach Estate. The lessees' interest in the property is for the residue of a 99-year term from December 24 1875 and at a ground rent of £2 per annum. The relevant date for the assessment of the price is June 13 1973, being the date on which the lessees served notice of their desire to have the freehold, when the lease had about 1½ years unexpired.

The subject property is a two-storey, stone-built detached dwelling nearly 100 years old, standing on a site of over a quarter of an acre fronting to the main (A4119) road from Cardiff to Llantrisant. About half a mile further along this road is the village of Groesfaen, and about half a mile to the north is the village of Creigiau.

At the hearing Mr Philip Price (instructed by Ivor W Shea & Co, of Cardiff) appeared for the tenants and called evidence from Mr W K R Ricketts ARICS, who is in practice on his own account under the name of the William Ricketts Partnership. Mr Vernon Pugh (instructed by L G Williams & Pritchard, of Cardiff) appeared for the landlords and called evidence from Mr T R W Davies BSc ARICS, who is a partner in the firm of W H Cooke & Arkwright.

Parties' Valuations

The respective valuations, slightly adjusted for ease of comparison, may be summarised thus:

Valuation for the tenants by Mr W K R Ricketts ARICS

Ground rent reserved		£2	
YP for 1½ years at 7 per cent ...		1.37	
			£3
Reversion to section 15 rent:			
Site value £2,250 at 12 per cent ...	£270		
YP in perpetuity deferred 1½ years at 12 per cent		7.04	
			1,901
			£1,904
		Say	£1,905

Valuation for the landlords by Mr T R W Davies ARICS

Ground rent reserved	£2		
YP for 1½ years at 7 per cent ...	1.37		
		£3	
Reversion to section 15 rent:			
Site value £8,361 at 7 per cent ...	£585		
YP in perpetuity deferred 1½ years at 7 per cent	12.92		
		7,558	
		£7,561	
	Say	£7,500	

Mr Ricketts and Mr Davies, while not disagreeing on factual matters, were at variance on practically everything else. On the more important issues I shall be summarising their respective opinions before coming to my findings, but for the rest I record simply my findings based on the evidence given and in the light of my own inspection.

Despite its name the subject property has the characteristics not of a cottage but of a fairly substantial house. It stands facing south in attractive open country, but at the rear it has the detraction of a small area of waste ground (at a lower level) on which there are some abandoned and derelict cars. The house-site falls to the rear rather steeply, but the house itself is on level ground and there is ample level space for any extensions that might be required. The main driveway to the house is from the A4119 road at a potentially dangerous bend, but an alternative access has been formed off a side road (involving a detour of about 200 yds), and this affords a safe ingress-egress point on the main road.

As to services: mains electricity is connected, and drainage is to a cess-pit. Up to about 1972 water had always been drawn through pipes from a spring about 123 yds away. For some time prior to 1972, however, this water had required boiling, because the collecting chamber at the spring had fallen in, and debris and surface water were introducing contamination at the source; furthermore the pipeline itself required renewal. The use of this private supply was accordingly discontinued, and since 1972 water has been delivered fortnightly by the local authority. This is kept in a large storage tank at the back of the house. Mains water is some distance away—566 yds to the east, or 700 yds to the west.

Turning now to the valuations, it will be seen that there is disagreement as to:

(a) The site value,

(b) The rate of return to arrive at the section 15 rent,

(c) The rate of return for capitalising the section 15 rent.

Site Value

Mr Ricketts's site value figure of £2,250 represents 30 per cent of an assumed entirety value of £7,500. To arrive at his entirety value Mr. Ricketts had looked at all the sales of country properties which had taken place during 1973-74 within the radius of about 8 miles from Hill Cottage.

As it happened these were only three in number. On the basis of these comparables he had formed the opinion that the value of the subject property could be put at £7,500. His proportion of 30 per cent site value had been derived from previous decisions of the tribunal, and in particular from the "standing house" approach as described in *Farr v Millersons Investments Ltd* (1971) 218 EG 1177.

Mr Davies's site value figure of £8,361 represents an assumed value of £9,000 for the site with water available, less £639 being his estimate of the cost of renovating the private water supply. "I looked at the property," said Mr Davies, "and assumed the house was not there. I notionally took the house away." The comparables cited by Mr Davies comprised transactions during 1972-73 relating to various single-dwelling building plots within a distance of about 8 miles from Hill Cottage.

With due respect to Mr Davies I think the approach he has adopted is incorrect. Although the section 15 formula of a "value of the site (without including anything for the value of buildings on the site)" does not specifically require the "value of the site" to be taken as being a component part of the value of the whole premises, it is only when the subject house is nearing the end of its economic life that the required "site value" is helpfully to be ascertained by reference to the prices of sites sold for development or redevelopment.

This being so (and also because none of his building plots was in open country), I do not find Mr Davies's comparables helpful. Nor, I may mention, do I get any assistance from details which were produced by Mr Ricketts of prices realised for certain *other* houses (all lying within the village of Groesfaen) additional to his three comparables, which are described hereunder.

Tenants' Comparables

Mr Ricketts adopted, I think, the more dependable approach, namely the ascertainment, as a first step, of an entirety value derived from a consideration of the prices that were being realised for other properties in or near open country at or about the relevant date. Brief details of Mr Ricketts's comparables—including certain findings of my own thereon—are as follows:

A. *Elm Cottage, Ystradowen.* Sold by auction in November 1973 for £7,000. Site area 769 sq yds. The house in a dilapidated condition, requiring major works to render it habitable. Accommodation: two sitting-rooms (about 13 ft x 13 ft and 12 ft x 11 ft) and four bedrooms. Mains electricity. No mains water. Cesspit drainage. A fairly isolated position but on a main road—the house wall actually abutting the road. About 11 or 12 miles from Cardiff.

B. *Bryndafod, Prisk.* Sold by auction in April 1974 for £4,200. Site area 1,742 sq yds. The house so dilapidated as to be fit only for demolition. No services at all. A completely isolated position at the bottom of a small valley. About 11 or 12 miles from Cardiff.

C. *Llidiard Mawr, St Mary Hill.* Sold by auction in July 1973 for £8,200. Site area 800 sq yds. The house in fair condition but unmodernised. Accommodation: sitting-room (14 ft 6 in x 10 ft 8 in), living-room/kitchen (11 ft 8 in x 9 ft 6 in), bathroom/wc and two bedrooms. Mains electricity. Mains water. Cesspit drainage. A cottage-type dwelling not far from a group of about 50 houses built by the Land Settlement Board. About 15 or 16 miles from Cardiff.

For convenience of comparison I append the following details of the subject property:

Hill Cottage, Groesfaen. Valuation date June 1973. Site area 1,296 sq yds. The house in fair condition. Accommodation: two sitting-rooms (each about 15 ft x 15 ft), kitchen, bathroom/wc, and four bedrooms. Mains electricity. No mains water. Cesspit drainage. A country position, half a mile from a village. About 8 miles from Cardiff.

Mr Davies, on a consideration of comparables A, B and C above, expressed the opinion that these sales pointed to a value for Hill Cottage not of £7,500, as put forward by Mr Ricketts, but of £15,000. The comparables provide only meagre evidence, but they suffice in my opinion to show that £7,500 must be rejected as much too low a figure. I am prepared to accept the opinion of Mr Davies, based as much on his experience as on the comparables, that Hill Cottage would have fetched £15,000 in June 1973—but only, I think, with a satisfactory water supply. I have described the circumstances regarding water: a purchaser would in my opinion have reduced his £15,000 bid by, say, £1,000 to cover the likely costs of renovating the private supply and also to cover the contingency that this particular supply might prove to be inadequate in quantity and/or in quality.

Value on Reconstruction and/or Extension

Having found the entirety value of Hill Cottage, at the relevant date and in its then condition, to be £14,000, the next step is to estimate the "value of the site (without including anything for buildings on the site)." In the circumstances of the present case the "standing house" approach adopted by Mr Ricketts, with its conventional assumption of up to one-third of the entirety value being attributed to the site, is no more reliable than the "cleared site" approach adopted by Mr Davies. The reason for this is that there is no dependable evidence as to what the entirety value of Hill Cottage would be, not only if it were put into good condition, but also if it were to be reconstructed and/or extended so as to develop the potential of the site up to the limit of any likely planning permission. Mr Ricketts regarded the existing house as already fully developing the potential of the site, but I agree with Mr Davies, who visualised a purchaser demolishing part of the present structure and incorporating the remainder in a house some 10 per cent larger than now exists. Mr Davies suggested £21,000 as the then likely value of the property, but this was no more than a spot figure hazarded by him in answer to a question from myself.

In order to arrive at a figure for "value of the site" in these circumstances, there is no alternative but to make a considered apportionment, between "site" and "buildings on the site," of the figure of £14,000, being the value of Hill Cottage unimproved. For this purpose it is necessary first to identify the most likely type of purchaser, and then to identify those attributes for which such a purchaser would be looking. The location of the subject property, only about 15 minutes' distance by road from the centre of Cardiff, points to the most likely purchaser being someone who works in that city but prefers to live in the country. I took note in this connection that the nearby villages of Groesfaen and Creigiau are regarded as being dormitory areas of Cardiff. To such a purchaser the attributes of a site in open country would be at least as important as the bricks and mortar standing thereon. Looking at the subject property in particular, I consider that such a purchaser would be paying not less than one-half of his £14,000 for the "site" and not more than one-half for "buildings on the site." This finding, of a site value of, say, £7,000 for Hill Cottage, seems to me congruous with the price of £4,200 paid for comparable B, where the asset obtained by the purchaser comprised, in effect, simply the site for the erection of a house, that site being in much less attractive surroundings than Hill Cottage, with no services at all, and not so near to Cardiff. £7,000 happens to be also, coincidentally, one-third of Mr Davies' estimated "standing house" entirety value of £21,000.

Rate of Return to Arrive at Section 15 Rent

Mr Ricketts uses the 12 per cent table, and Mr Davies the 7 per cent table, to arrive at the section 15 rent. The substantial reason for Mr Davies using 7 per cent is that this is apparently the rate of return customarily used by surveyors in Cardiff at this stage of a valuation, when agreeing prices under the 1967 Act. Having now studied the decision of the tribunal in the *Dulwich* case (*Carthew v Estates Governors of Alleyn's College of God's Gift* (1974) 231 EG 809) Mr Davies considered that his 7 per cent in respect of this house near Cardiff was wholly consistent with the 6 per cent that had been proposed by the landlords and adopted by the tribunal in respect of houses at Dulwich.

Mr Ricketts, for his part, agreed that he had himself made settlements which incorporated 7 per cent at this stage of the calculation. "But," said Mr Ricketts, "7 per cent was the result of many negotiations. I have gone along with this, but I have never *accepted* it." He agreed also that if 6 per cent were right in the *Dulwich* case, then he "would not expect much difference here." "But," said Mr Ricketts, "I profoundly disagree with the evidence given for the landlords in *Dulwich*. I would myself have advised the Dulwich Governors to press for much more than 6 per cent—say 10 per cent." Mr Ricketts told me he derived his "strength" for a 12 per cent return, in calculating the section 15 rent, from what the landlord would have to accept for that rent when it had been created: the landlord would be aware that he would have to accept a price for his section 15 rent which showed the purchaser a return of, say, 12 per cent, and accordingly the landlord would require 12 per cent on his site value when he created that rent.

With due respect to Mr Ricketts he is putting the cart before the horse. The *first* step is to decide what section 15 rent could be reasonably expected: it would be of no avail for the landlord to "require" 12 per cent on his site value if, from the tenant's point of view, that calculation produced a rent that was unacceptably high. The implications of section 15 were exhaustively explored in the *Dulwich* case; during the course of that five-day hearing the tenants advanced many arguments and adduced much evidence directed to establishing that a section 15 rent could not be as high a figure as emerged by taking even 6 per cent on site value. The tribunal found on the evidence in that case that 6 per cent was right for houses at Dulwich. And on the evidence in the present case I find that 7 per cent is right for Hill Cottage.

This finding gives a section 15 rent of £490 per annum.

Rate of Return for Capitalising the Section 15 Rent

The final stage of the valuation is to capitalise the section 15 rent. For this purpose Mr Ricketts uses the 12 per cent table, and Mr Davies the 7 per cent table.

Mr Ricketts supported his rate of return by reference to open market sales of freehold ground rents in Newport and Cardiff showing prices which devalue on average to about 13.5 per cent. The reason Mr Ricketts had adopted only 12 per cent was an assumption that the purchaser of a section 15 rent would be satisfied with some lower return than 13.5 per cent because the amount of that rent could be reviewed after 25 years.

The reason for Mr Davies using 7 per cent was that this is apparently the rate of return customarily used in Cardiff at the final stage of the valuation. Mr Davies very fairly agreed, however, that if he were acting for an intending investor-purchaser of a section 15 rent secured on Hill Cottage his valuation would be at a figure showing a return to the purchaser of 10 per cent or even more.

In the light of the evidence I find that any purchaser of the ascertained section 15 rent of £490 pa, other than the tenants of the property, would have required a return of, say, 11 per cent.

Should the valuation then be completed by capitalising the section 15 rent on the 11 per cent table? Or on the 7 per cent table? Or on some intermediate table? For the following reasons I have decided that the price properly payable under section 9 is more appropriately arrived at by using the 7 per cent table.

1. The disparity between the rates of return above is nothing unusual: it has been a recognised characteristic of Leasehold Reform Act valuations subsequent to the amendment of the 1967 Act in 1969.

As the tribunal noted in *Chilcott v Clinton Devon Estates* (1973) 229 EG 39:

> "... the main flow of evidence has been all one way; that an investor-purchaser would expect a greater return on his money than the freeholder would obtain on the granting of a lease under the terms of section 15. For convenience a name 'adverse differential' has become attached to this difference between the two rates of return. But ... this 'adverse differential' does not rest on any theory or principle of valuation; nor does it involve any question of law; it is indeed no more than a by-product of the cumulative factual and opinion evidence which the tribunal has received in the cases heard under this branch of its jurisdiction."

The general valuation practice, up to the time of the Court of Appeal's judgment in *Official Custodian of Charities v Goldridge* (1973) 227 EG 1467, had been to reflect this particular characteristic by adopting differential rates of return. But since that judgment, in cases coming before the tribunal (including the present case), valuers have discontinued the use of differential rates.

Practice Since "Goldridge"

2. The evidence given by Mr Davies in the present case was that he had handled personally about 400 cases under the 1967 Act, and that the number of cases handled by his firm now totalled over 6,500. Prior to *Goldridge* his practice, in accordance with a line of decisions by the tribunal, had been to adopt differential rates of return, normally 6 per cent and 8 per cent. Since *Goldridge* his invariable practice (and, to his knowledge, the general practice of other surveyors in Cardiff) was to assume a 7 per cent rate of return at both stages of the calculation. "Since that Court of Appeal decision," said Mr Davies, "it is no longer permissible to adopt a higher rate at the second stage."

3. The Court of Appeal (per the Master of the Rolls and Sir Eric Sachs) in *Gallagher Estates Ltd v Walker* (1974) 230 EG 359 directed the tribunal in cases under this jurisdiction to look, for dependable evidence, primarily to the "land market," and to the "money market" only as a factor. The "land market" approach, of determining values on the evidence of comparables, is one to which the tribunal is well accustomed.

"Land Market" Approach

In compulsory acquisition cases the best evidence of the "land market" is that provided by open market transactions, in default of which, however, it is common practice to look at the evidence of settlements—but cautiously. In Leasehold Reform Act cases the assumptions postulated under section 9 of the 1967 Act, coupled with the exclusion of the tenant's bid by section 82 of the Housing Act 1969, means that, in practice, directly comparable open market transactions are unavailable, and the "land market" comparables cited to the tribunal have taken the form of prices which have been agreed for other freeholds under the 1967 Act. "Land market" evidence of this nature was accepted by the tribunal in the early case of *Delaforce v Evans* (1970) 215 EG 315, and the same approach has been applied in a number of cases since *Gallagher*.

4. The valuation exercise under section 9 of the 1967 Act is not only hypothetical, it is also artificial: this was brought out in the *Dulwich* case, where it was common ground that no landlord in the open market would contemplate granting a lease under the terms postulated by section 15, nor would any well-advised tenant wish to take such a lease. With a jurisdiction of this nature, consistency of application of the statutory formula must be of quite exceptional importance. Over the years the tribunal has, to the best of its ability, set out an evolving line of acceptable valuation techniques, and these have come under review by the Court of Appeal. The tribunal's decisions, as so reviewed, will have been used as the basis of settling many thousands of cases which have never reached the tribunal. On the evidence in the present case I am satisfied that there is an established "land market" in Cardiff, based directly on tribunal decisions (as reviewed). It would in my opinion be inapt now to reject the valuation technique of that "land market" on the grounds of evidence drawn solely from the "money market."

Purchase Price

The outcome of these considerations is to produce a figure of £6,325 calculated thus:

Ground rent reserved	£2	
YP for 1½ years at 7 per cent ...	1.37	£3
Reversion to section 15 rent:		
Site value £7,000 at 7 per cent ...	£490	
YP in perpetuity deferred		
1½ years at 7 per cent	12.92	6,330
		£6,333
	say	£6,325

I determine the price to be paid by the tenants of Hill Cottage for the freehold interest to be £6,325.

Having opened the sealed offer of purchase price lodged by the tenants I find that the amount thereof is less than the amount of my award. The tenants will pay to the landlords their costs of this reference, such costs unless agreed to be taxed by the Registrar of the Lands Tribunal on the High Court Scale.

ADDISON v HACKNEY LONDON BOROUGH

(LR/68/1974)

(E C STRATHON FRICS)

April 15 1975

Estates Gazette May 10 1975

(1975) 234 EG 446

House at Clapton, London—Four days before hearing, landlord council accepts tenants' unconditional offer made some nine weeks earlier—Amount £10 less than "without prejudice" offer made by council 12 months previously—Council awarded costs to date of unconditional offer and to pay tenants' costs thereafter

Giving his decision, MR STRATHON said: This case concerns the price which Mr and Mrs D Addison, the tenants, should pay for the freehold interest in the house 7 Millfields Road, Clapton, London E5, under the provisions of the Leasehold Reform Act 1967 as amended.

The tenants served notice of their desire to have the freehold on October 16 1973 and the landlords, the London Borough of Hackney, admitted the right of the tenants to purchase on December 20 1973.

I sat in London on April 15 1975. When this case was called I was told that I should not be troubled with the question of the amount of the purchase price, since the landlords had on April 11 1975 accepted an unconditional offer of £750 made by the tenants on February 4 1975.

Mr David Mole, of counsel (instructed by M Wernick & Co), for the tenants, submitted that the landlords should pay the costs of the tenants after February 4 1975. But Mr D Martineau, of counsel, for the landlords, submitted that had the tenants accepted an offer of £760 made by the landlords on February 14 1974, the reference to the Lands Tribunal in October 1974 would have been unnecessary. Mr Martineau, accepting that the offer was in a "without prejudice" letter by the borough valuer in the form that he would be prepared to recommend such a purchase price to the council, submitted that the balance of probabilities was that the council would have accepted such a recommendation.

In all the circumstances of the present case I decide, in the exercise of my discretion, that I should make an order as to costs in the following terms: the tenants will pay the landlords their costs of the reference up to and including February 4 1975 and the landlords will pay the costs of the tenants as from February 4 1975, such costs not to include the presence of the tenants' surveyor at the hearing. The costs unless agreed are to be taxed by the Registrar of the Lands Tribunal on Scale 4 of the County Court Scales of Costs.

GUIVER v FRANCINE PROPERTIES LTD

(LR/63/1974)

(R C WALMSLEY FRICS)

May 5 1975

Estates Gazette June 7 1975

(1975) 234 EG 741

Freehold price of Barnet semi-detached house with 59 years unexpired—Well-established procedure for operating Act in this part of London—Average going price £350—Deduction for "Delaforce" effect the sole issue—Evidence in "Finkel" case distinguished—Section 9 (4) costs

Giving his decision MR WALMSLEY said: This is a reference for the determination under the Leasehold Reform Act 1967, as amended, of the price which should be paid for the freehold interest of 98 Sherrards Way, Barnet, Hertfordshire, by the resident lessee Mr Peter Ernest Guiver to the freeholders Francine Properties Ltd. The lessee's interest in the property is for the residue of a 99-year term from March 25 1933, and at a ground rent of £9.45 per annum. The relevant date for the assessment of the price is March 21 1973, being the date on which the lessee served notice of his desire to have the freehold, when the lease had about 59 years to run.

The subject property is a two-storey, three-bedroom, semi-detached house, typical of many such houses in this part of London.

At the hearing Mr P A S Mulready, of P A S Mulready & Co, solicitors, appeared for the tenant and Mr J E D Macvie, of Elliott & Macvie, solicitors, appeared for the landlords. Mr Mulready gave factual evidence himself and called evidence from Mr R J Maunder Taylor FRICS, who is a partner in the firm of J Maunder Taylor & Son. Mr Macvie (who happens to be a director of Francine Properties Ltd as well as being their solicitor) gave factual evidence himself but did not call any opinion evidence.

The valuation produced by Mr Taylor for the tenant was in the sum of £225 as follows:

Average of sale prices of similar properties	£350
Allowance for *Delaforce* effect	£125
	£225

The case for the landlord, in brief, is that the price properly payable is £350 without any deduction.

Costs a Bargaining Factor

By the close of the hearing a great deal of common ground had been reached. It appears that in this part of London, for houses of this type built about 40 years ago on 99-year leases, a procedure for operating the 1967 Act has become well established. A qualified lessee, I was told, will start by serving on his landlord a notice of leaseholder's claim in proper form and will either accompany this notice with the offer of a purchase price or will inquire of the landlord what price is expected for the freehold interest. After this originating formality, negotiations apparently proceed informally without further specific reference to the requirements of the Act. Should the parties reach agreement on a price, this is regarded as still leaving the question of costs as a bargaining factor. Although both higher and lower prices have been known, the average going price for the enfranchisement of houses such as 98 Sherrards Way is agreed to be £350.

As has been seen, both parties in the present reference adopt the "land market" approach as described in *Finkel v Simon and Others* (1975) 231 EG 329, and the sole issue between them is the question of what deduction, if any, falls to be made for the "*Delaforce*" effect. This effect, as fully explained in *Delaforce v Evans* (1970) 215 EG 315, may be succinctly described as an adjustment, from a price ascertained on "land market" evidence, to bring into account a deemed anxiety by purchasing tenants to settle rather than face the expense and/or delay and/or uncertainty involved in coming to the tribunal. Mr Taylor's sole support for his deduction of £125, for the *Delaforce* effect, was the adoption by the tribunal of that figure in the *Finkel* case. Mr Taylor pointed to the closely parallel circumstances: the subject houses were very similar in the two cases; the ground rents were identical; the lengths of unexpired term were nearly the same; both valuation dates were 1973; and the "land market" evidence in *Finkel* pointed to much the same kind of price, *viz* £375 as compared with £350.

I recognise and accept that if the practice of the tribunal were blindly to follow its previous decisions, then it might be logical to assume that an allowance of £125 for the *Delaforce* effect would follow in the present case. The tribunal's practice, however, has always been to reach each decision on the evidence and arguments presented in the particular case before it, but—especially with any new jurisdiction—considering each fresh case against the background of earlier decisions in order to obviate potential inconsistencies developing.

The evidence presented to the tribunal in the *Finkel* case included the factors:

(a) that an enfranchising tenant of a house in the same road as the subject house, who purchased the freehold, at a cost of (plus or minus) £375, would, on the evidence, immediately increase the value of his ownership by about £1,000, from about £14,000 leasehold in possession to about £15,000 freehold in possession.

(b) that the tenants' valuer had been aware for some time that the landlord would have been willing to accept a price of £225 instead of £375; this reduction in price being the equivalent of a quantification, by the landlord, of the *Delaforce* effect in the sum of £150.

It was against the above background that the tribunal decided on a deduction of £125 for the *Delaforce* effect in the *Finkel* case, and determined the price to be paid for the freehold interest in the sum of £250.

Ownership Value Might Not be Increased

In the present case the evidence is different. Mr Taylor offered the opinion, which I accept, that an enfranchising tenant in Sherrards Way might not increase the value of his ownership by even as much as the cost of enfranchisement: the value of a typical house in the locality could be put at between £12,250 and £12,500 leasehold in March 1973, and its value freehold at, say, £12,750. Thus a tenant who purchased his freehold at £350 might increase the value of his ownership by as much as £500, but on the other hand he might find it increased by only £250. Mr Taylor told me that although a tenant might be expected to offer a higher price for the freehold if he were intending to sell his house, he (Mr Taylor) knew of a case where, on a sale of the leasehold interest, the purchase of the freehold had not been proceeded with either by the vendor or by the purchaser.

Mr Macvie's submission was that the factual evidence disclosed no anxiety by tenants to settle; the landlord company had never made any approach to tenants; it was always the tenant who made the approach. Three hundred and fifty pounds, said Mr Macvie, was the average figure freely negotiated in the real market with tenants who were professionally advised, and the tribunal should recognise this as being the price properly payable under the Act.

Looking at the facts of the present case I perceive little, if any, evidence of anxiety on the part of the tenants to settle. The operation of the local "land market" as it was described to me, gives the impression of being a relaxed one, made up of landlords under no pressure to accept *less* than they regard as a proper price, and tenants under no pressure to offer *more* than they regard as a proper price. The *Delaforce* effect, however, may be the reason underlying a differing treatment of the costs of enfranchisement: in some of the cases the landlord has been obtaining his £350 with the tenant also paying such costs as are provided for in section 9 (4), but in a number of other cases the landlord has had to find any such costs out of the purchase price. As to *quantum*, Mr Mulready and Mr Macvie agreed that the amount of these costs might be taken as £25. In the present case section 9 (4) applies, requiring the tenant to bear the costs of enfranchisement in addition to paying the purchase price. My finding is for a purchase price of £325, which may be regarded as:

Land market price (agreed)	£350
Adjustment for *Delaforce* effect	£25
	£325

I determine the price to be paid by the tenant of 98 Sherrards Way for the freehold interest to be £325.

Having opened the sealed offer of purchase price lodged by the tenant I find that the amount thereof is less than the amound of my award. The tenant will pay to the landlords their costs of this reference, such costs unless agreed to be taxed by the Registrar of the Lands Tribunal on Scale 3 of the County Court Scales of Costs.

BRIDDON v FIELD

(LR/69/1974)

(E C STRATHON FRICS)

April 30 1975

Estates Gazette June 14 1975

(1975) 234 EG 840

Freehold price of Lenton, Nottingham, three-storey, semi-detached house—Lease 23 years unexpired—Fundamentally different approaches by valuers in arriving at section 15 rent, yet only £8 apart—Tribunal's own figure some £55 lower—Sales of long-distant reversions held not reliable as evidence for short unexpired terms—Basis of value per sq yd derived from settlements "fraught with uncertainty"

Mr Pell, of Wells & Hind, of Nottingham, appeared for the tenant; Mr C C Hodson, of Rotheras, of Nottingham, for the landlord.

Giving his decision, MR STRATHON said: This is a reference for the determination under section 9 of the Leasehold Reform Act 1967, as amended, of the price which should be paid for the freehold interest of 31 Albert Grove, Lenton, Nottingham, by the resident lessee, Mrs Beatrice Briddon, to the freeholder, Mr John Edmund Field. The lessee's interest in the property is held for the residue of a term of 99 years from March 25 1898 at a ground rent of £7.50 per annum. The lessee served notice of her desire to have the freehold on June 14 1974, which is therefore the relevant date for assessment of the price and when the lease had 23 years unexpired.

The subject house is a three-storey, semi-detached dwelling about 50 years old. It stands on a double plot of land. The accommodation comprises: on the ground floor two living rooms and kitchen, on the first floor two bedrooms, bathroom and wc, on the second floor two attic bedrooms, and outside a wc, garage and a stable.

At the hearing, valuation evidence for the tenant was given by Mr G T Harlow FRICS, who is a partner in the firm of Harlow Shelton & Co, and for the landlord by Mr R J Freeston ARICS, a partner in the firm of Walker Walton & Hanson.

Parties' Valuations

The respective valuations are reproduced as follows:

Valuation for the tenants by G T Harlow FRICS

Existing ground rent	£7.50	
YP for 23 years at 12%	£7.718	£58
Modern ground rent	£180.00	
YP in perpetuity at 12% deferred 23 years	0.61490	£110
		£168
	say	£170

Valuation for the landlord by R J Freeston ARICS

Present ground rent	£7.50	
YP for 23 years at 6%	£12.30	£92
Modern ground rent	£172.00	
YP in perpetuity at 8% deferred 23 years	£2.13	£366
		£458
	say	£450

It can be seen that while Mr Harlow and Mr Freeston differ on rates of interest they are but £8 apart in their estimates of the reversion to a modern ground rent (the section 15 rent). But hidden by this apparent agreement are differences of approach by the surveyors in arriving at the section 15 rent. I will first summarise these respective approaches.

Mr Harlow said that he had adopted the "land market" approach in accord with the judgment in *Gallagher Estates Ltd v Walker* (1973) 230 EG 359. He had therefore looked for sales of freehold ground rents and had found that at four such sales by auction in June and July 1974 ground rents, with reversions long distant, had realised yields of 27.8 per cent, 14.3 per cent, 14.1 per cent and 13 per cent respectively.

Ignoring the 27.8 per cent yield as being out of line, Mr Harlow had concluded that 14 per cent could be taken as the average yield at a time when the yield on undated War Loan was 15¼ per cent. Since the reversion in the reference property was 23 years distant, he had adopted a 12 per cent yield for capitalising the existing ground rent. To arrive at the section 15 rent Mr Harlow had first formed an opinion that the value of the site was £1,500. He had then again taken the view that if for an investment in a distant reversion the "land market" appeared to expect a return of 14 per cent, it would accept a return of, say, 12 per cent where a rent review would take place in 25 years. Finally, as it had appeared to him that in its recent decisions the Lands Tribunal had determined that the "adverse differential rate" should no longer be used, he had adopted a rate of 12 per cent for capitalising the section 15 rent.

Mr Freeston relied for support for the various stages of his valuation on settlement of prices which he had agreed for other freeholds under the 1967 Act. He produced details of six such settlements in all of which the rate used for capitalising the existing ground rent was 6 per cent and for capitalising the section 15 rent 8 per cent; he had adopted these rates in his valuation for the reference property. He had arrived at a section 15 rent of £172 by applying to the site area of 223 sq yds a value of 75p per sq yd. He relied for this value on the values per sq yd comprised in the settlements, five of which, on behalf of the Church Commissioners, had been negotiated with a firm of surveyors; Mr Freeston had also used this basis of valuation in reaching settlements in 20 to 25 other cases.

I turn now to the details of evidence and to my findings:

Capitalising the Existing Ground Rent

Mr Harlow's concept of the "land market" had directed him to reported sales of ground rents with long-distant reversions. I recognise that distant reversions were the subject of consideration in *Gallagher*, where the subject property of that reference itself comprised a lease with a long period of years unexpired. But in the instant case the lease has only 23 years unexpired, with a reversion to a lease on the terms of section 15 and subject to review at the end of 25 years. Reliable evidence of the "land market" for this type of property is not to be sought in the sales of long-distant reversions, but is more probably to be found in settlements of comparable transactions under the provisions of the 1967 Act. Mr Freeston's experience had been that throughout all his negotiations a rate of 6 per cent for capitalising the existing ground rent had been found to be acceptable: such a percentage, as Mr Freeston said, has been accepted by the tribunal in numerous other references, and I accept that in the instant case it is more appropriate than a rate of 12 per cent.

The Section 15 Rent

Mr Harlow for the tenant is of the opinion that the section 15 rent should be £180, while Mr Freeston, for the landlord, supports a rent of £172. It might be thought that it is therefore unnecessary to consider the differing bases from which the respective values are derived. But the differences which distinguish the two approaches are so fundamental that they must be examined. In support of a site value of £1,500 Mr Harlow relied on (a) sales of undeveloped land comprising areas varying between 471 sq yds and 34 acres, (b) 11 sales of new houses, and (c) seven sales of older houses. By comparison with these data his valuation of the entirety of the reference property was £6,900 of which he attributed a value of £1,500 to the site, which is equivalent to a proportion of 21.8 per cent of the value. He had applied to this £1,500 a rate of 12 per cent to produce a section 15 rent of £180.

Mr Freeston had arrived at a rent of 75p per sq yd in the following manner:

(1) In an area of poor quality property, a settlement in 1974 at 65p per sq yd by comparison with three settlements in 1970 at 45p per sq yd was equivalent to an increase of 44%.

(2) A settlement in 1969 made direct with a resident lessee in respect of 50 Albert Road (in the same road as the reference property) had been based on a section 15 rent of 65p per sq yd which, increased by the above 44% produces in respect of the reference property a 1974 value of	93p per sq yd
(3) Less 20% on account of larger area of reference property	18p
Section 15 rent	75p per sq yd

A Mr Freeston also produced a valuation by reference to the entirety value of the reference property. By comparison with (unspecified) modern houses, and with the recent sale of 8 Albert Grove at £5,800, Mr Freeston's valuation of the entirety of the subject property was £7,500. He had estimated the value of the site at £2,250, which was equivalent to 30 per cent of the whole value, a percentage which had appeared in certain decisions of the Lands Tribunal. By applying a percentage rate of 7 per cent Mr Freeston had found a section 15 rent of £157, but this had been rejected in favour of his first valuation of £172.

Mr Freeston's first approach is fraught with so much
B uncertainty that I do not accept it as a reliable basis on which to found the section 15 rent. Nor is there sufficient evidence to support Mr Freeston's other section 15 valuation of £157 which he himself rejected. On the other hand, Mr Harlow's evidence on comparability was more dependable and I accept his valuation of the site value of the reference property at £1,500.

But, as I understand the evidence, it would not follow that Mr Harlow would have applied to this site value a rate of 12 per cent to arrive at section 15 rent of £180 if the rate was not also to be applied for capitalising that rent. For the reasons given in this decision I conclude that 12 per cent is
C not an appropriate rate of return to apply in this case. I take the view that Mr Freeston's "land market" of settlements points to the percentage that has proved to be acceptable generally, namely 8 per cent, and on the evidence I am satisfied that this rate should be applied both to the site value, in order to arrive at the section 15 rent, and also to capitalising that rent. D

Purchase Price

Summarising my conclusions a purchase price of £350 results as follows:

Existing ground rent	£7.50	
YP for 23 years at 6%	£12.30	£92
Reversion to section 15 rent site value £1,500 at 8%	£120.00	
YP in perpetuity deferred 23 years at 8%	£2.13	£255
		£347
		say £350

E

I determine the price to be paid by the tenant of 31 Albert Grove, for the freehold interest to be £350.

Having opened the sealed offer submitted by the tenant, I find the amount thereof is less than the amount of my award. I also received submissions on costs by both parties and in particular that the contents of the sealed offer were not known to those representing the landlord until the case for the tenant had been concluded at the hearing before me. F

I determine that the tenant will pay the landlord his costs of this reference, which costs if not agreed, are to be taxed by the Registrar of the Lands Tribunal on Scale 3 of the County Court Scales of Costs.

LANDS TRIBUNAL
RATING

ARSENAL FOOTBALL CLUB LTD AND SMITH (VO) v ENDE
(LVC/33-34/1974 (OL))
(DOUGLAS FRANK QC, President, and V G WELLINGS QC)
February 5 1975

Estates Gazette April 5 1975
(1975) 234 EG 54

Proposal to increase rateable value of Arsenal Football Stadium by person who is manager of block of flats in same local authority area but resides in neighbouring area —Held not a ratepayer in relevant rating area nor a "person aggrieved"—"Person who receives the rents" in section 24 of the 1967 Act is person who receives them beneficially not as agent—Proposal invalid

The Hon David Trustram Eve QC and Mr Guy Roots (instructed by Lawrence Jones & Co) appeared for the Arsenal Football Club Ltd; and Mr Alan Fletcher (instructed by the Solicitor of Inland Revenue) for the valuation officer. The respondent appeared in person.

In their decision Mr Frank and Mr Wellings stated: This is a preliminary issue to determine whether the proposal dated March 23 1972 made by the respondent, Mr T A Ende, was bad in law and void in that he was not "a person aggrieved" within the meaning of section 69 (1) of the General Rate Act 1967. In that proposal he claimed to be aggrieved by the value ascribed in the valuation list for the rating area of the London Borough of Islington to the hereditament, the Arsenal Football Stadium, Avenell Road, occupied by the first appellant, the Arsenal Football Club Ltd (whom we will call "Arsenal") and by the description therein contained of that hereditament.

The hereditament was described in the above-mentioned valuation list as "Football grounds, stands and offices." At the date of the making of Mr Ende's proposal the hereditament was shown in that valuation list as having a rateable value of £9,250.

Mr Ende does not live in the London Borough of Islington: he lives in the London Borough of Hackney. However, he is, and at all material times was, the manager of a property, 22 Beaumont Rise, within the London Borough of Islington. That property consists of 18 self-contained flats and 10 garages. Mr Ende's name at all material times appeared and appears in the rate books of the council of the London Borough of Islington as the person liable to pay rates on that property.

Grounds of Proposal

Mr Ende's notice of proposal was received by the second appellant, the valuation officer, on March 28 1972 and by the rating authority, the council of the London Borough of Islington, on March 29 1972. The alteration proposed in the valuation list was: "A more comprehensive description of the hereditament in the valuation list available to the public and increase of the rateable value to £60,000." The grounds of the proposal were stated to be: "The original rateable value of £8,600 made for the 1963 valuation list disregarded altogether the rental value bestowed upon the hereditament by free motor-car and coach-parking facilities and free police services and continues to disregard these facilities and services and the increase in them since then; undervalued the rental value of the hereditament at that time; and the increase of £650 since then undervalues the rental value of the improvements and additions made to the hereditament since the original valuation list was compiled."

The notice of proposal also stated: "Capacity in which signed (eg as Occupier, Owner, Agent for occupier, Agent for owner, Clerk to Rating Authority, etc): Aggrieved person." In the proposal Mr Ende gave as his address his home address in Hackney.

By letter dated May 22 1972 Arsenal's solicitors asked Mr Ende to state the grounds upon which he claimed to be an aggrieved person in respect of his proposal. Mr Ende replied on May 23 1972: "It must be self-evident that a person who is a ratepayer can be aggrieved by any value ascribed in the valuation list to a hereditament or by any other statement made or omitted to be made in the valuation list with respect to a hereditament (a) in his rating area and (b) in his precepting area. It must also be self-evident that a person contributing into the general fund of taxation must be an aggrieved person in respect of the same matters in any rating or precepting area if the rating or precepting authority is drawing upon the general fund of taxation for support of its activities in local government. Finally, if a person manages real property from which rents of rated occupiers within a rating or precepted area issue, he evidently qualifies as an aggrieved person."

Local Valuation Court's Decision

Arsenal on May 22 1972 and the valuation officer on June 21 1972 objected to the proposal on the ground that it was bad in law and that the proposed assessment was incorrect, excessive and bad in law. The matter was heard by the local valuation court on October 23 and 24 and November 1 1973. In a fully and carefully reasoned reserved decision dated January 16 1974 the local valuation court decided that the entry in the valuation list in respect of the appeal hereditament should be amended to an assessment of rateable value £13,900 and stated:

"Having heard all the facts and submissions we have come to the conclusion that Mr Ende should be treated as a person aggrieved for the following reasons:

(a) His name appears on the rate books of the Islington Borough Council as the rated occupier for a property for which he collects rents on behalf of the owner;

(b) Although resident in the London Borough of Hackney he lives near enough to the appeal premises to be affected by the parking of vehicles and other disturbances created during home matches.

(c) He is a resident within the same precepting area."

We pause to observe that the second reason was not advanced by Mr Ende before us. We need say no more about it than to remark that it appears to have no relevance whatsoever to any question arising out of the preliminary issue,

A although it might have some relevance on the question of valuation.

The appellants in their statements of case and Mr Ende in his reply duly set forth their respective contentions.

Finally, in reviewing the history of the matter, we should mention that on July 28 1972 Arsenal took out an originating summons in the Queen's Bench Division seeking a declaration that Mr Ende was not a person aggrieved within the meaning of section 69 of the Act of 1967 and that the originating summons was struck out on July 5 1973 on the ground that the claim was an abuse of the process of the court in that the local valuation court had jurisdiction to resolve the
B matter. An appeal to the judge in chambers was subsequently dismissed.

" Unfairly Low Assessments "

Mr Ende argued his case well. He said that the appeals represented one of a number of cases arising out of his contention that the valuation lists in all the Greater London Boroughs are consistently undervalued and this undervaluation obscures undervaluation of certain classes of hereditament, of which local-authority-owned dwellings form one
C class and football stadiums where professional football is played are another. The effect of this was that those ratepayers whose assessments were correct because they are in tone with the generality of assessments are compelled to contribute a sum of money each rating year into the domestic budget of occupiers of hereditaments which have been favoured by unfairly low assessments. In support of this broad contention he referred to four previous decisions of the Lands Tribunal in three of which he had been successful and in not one of which was the point taken against him that he was not " a person aggrieved." He also referred to the decision of a local valuation court in which it had been held
D specifically that he was " a person aggrieved." Accordingly, he said, counting the decision of the local valuation court in the present appeals, there had been two decisions of local valuation courts that he was " a person aggrieved " and four decisions of the Lands Tribunal in which he was tacitly treated as such.

In his argument before us he submitted that his sources of grievance were as follows:

(1) That in his capacity as a ratepayer of Islington (by virtue of his liability to pay rates under section 24 of the Act of 1967) and in his capacity as a precept-payer to the Greater London Council (by virtue of his being a ratepayer in Hackney and/or
E Islington) he is prejudicially affected if the appeal hereditament be insufficiently assessed, because in that event Arsenal will not be contributing a fair sum into the funds of the rating authority or the Greater London Council and he and his fellow ratepayers/precept-payers will be in effect making a contribution to the domestic budget of Arsenal;

(2) That in his capacity as a taxpayer, underassessment of the appeal hereditament affects him adversely if the result is to increase the rate support grant provided by the central government;

(3) That the valuation officer has taken upon himself the duties of the rating authority by determining without adequate informa-
F tion the means of Arsenal to pay rates;

(4) That the description of the appeal hereditament in the relevant valuation list, in that it was inadequate, prevented ratepayers, like himself, who were entitled to inspect the valuation list under section 108 of the Act of 1967, from taking full advantage of the provisions of that section.

Mr Ende conceded that the third submission was inapplicable in relation to the preliminary issue, and as to the fourth submission he conceded that he had in no way been inhibited by the description of the appeal hereditament in the valuation list from making a valuation of it. Accordingly we are in effect concerned only with Mr Ende's first two submissions.

VO and Proposer Agree G

The first submission involves the question whether Mr Ende is to be regarded as a ratepayer of Islington. Mr Ende and the valuation officer agree that the answer to that question is in the affirmative. If they are right, the question (raised by the valuation officer but eschewed by Mr Ende) as to whether the fact that he is a ratepayer suffices to make him an aggrieved person if a hereditament in the area is underassessed will arise.

Mr Ende gave no evidence before us, but for the purposes of enabling us to give a decision on the preliminary issue we assume that the local valuation court was right in assessing H
the appeal hereditament at rateable value £13,900.

Is Mr Ende a ratepayer in the rating area of the London Borough of Islington?

It is alleged by Mr Ende and by the valuation officer that because the valuation officer has exercised his power under section 24 of the Act of 1967 to treat the whole of 22 Beaumont Rise as a single hereditament and because Mr Ende receives the rents of the parts of that property, the whole is deemed by the section for the purposes of rating to be in Mr Ende's occupation. If that is right he is liable for the rates payable in respect of the single hereditament and is thus a ratepayer of Islington. The section provides: J

" Where a building which was constructed or has been adapted:

(a) for the purposes of a single dwelling; or

(b) as to part thereof for such purposes and as to the remainder thereof for any purpose other than that of a dwelling,

is occupied in parts, the valuation officer, in preparing a new valuation list or in altering a current valuation list, may, if he thinks fit, having regard to all the circumstances of the case, including the extent, if any, to which the parts separately occupied have been severed by structural alterations, treat the building or any portion thereof as a single hereditament, and a building or portion of a building so treated as a single hereditament shall, for the purposes of rating, be deemed to be a single hereditament K
in the occupation of the person who receives the rents payable in respect of the parts."

The argument as put by Mr Alan Fletcher of counsel on behalf of the valuation officer and supported, so we assume, by Mr Ende, is that the words " the person who receives the rents payable in respect of the parts " means, where the rents are collected by the owner's agent, that agent. It is common ground that at all material times Mr Ende had collected the rents of the parts of 22 Beaumont Rise on behalf of the owner thereof. We were referred to *Adams and Watts v Southall Rating Authority* [1943] 1 KB 359, where the proprietors of a block of flats agreed with the occupiers to L
pay the general rates on the flats and employed estate agents to collect the inclusive rents and to pay the rates to the rating authority on behalf of the proprietors. The estate agents were summoned by the rating authority for rates due. They were held by the Court of Appeal not to be liable therefor. The matter arose under section 118 of the Middlesex County Council (General Powers) Act 1938, which provided:

" Where the owner of any hereditament has agreed with the occupier thereof that the owner shall pay the general rate charged on such hereditament, the owner shall be liable to pay to the council of the district in which such hereditament is situate so much of any payment in respect of rent received by him from the M
occupier as shall represent the proportion of rate included in such payment."

By subsection (2) " owner " was defined as meaning the person who " is entitled to receive the rent payable in respect " of the hereditament. Tucker J (as he then was) said at p 361:

"*Prima facie* the definition of 'owner' in section 118, subsection (2), of the Act of 1938 means a person who is entitled to receive rent in his own right. That that is the meaning to be attributed to 'owner' in that section appears from the fact that when the legislature wishes to render agents who have collected rents liable for the payment of rates it uses very clear language to achieve that result. An example is to be found in section 11, subsection (4), of the Rating and Valuation Act 1925. In my opinion, it would want words as clear as those of that subsection to render persons in the position of the appellants liable to pay the rates in respect of premises the rent of which they have merely collected as rent collectors on behalf of the owners."

Section 11 (4) of the Act of 1925 (now to be found in section 57 (1) of the Act of 1967, which, unlike its predecessor, applies to London) was in these words:

> "Where in pursuance of this section the owner is rated or has undertaken to pay or collect the rates charged in respect of any hereditament, the amount due from him in respect of the rates shall be recoverable by the rating authority from him or, where the rates are collected by an agent of his, either from him or from that agent, in the same manner and subject to the same conditions in and subject to which rates are recoverable from occupiers of rated hereditaments."

"Owner" had the same definition as that to be found now in the second part of the definition of that word in section 115 (1) of the Act of 1967.

Mr Fletcher said that the contrast between "entitled to receive" and "receives" in section 24 of the Act of 1967 means that where an agent receives the rents of the parts section 24 is speaking of him. Where the principal is not disclosed, he said, the agent to all intents and purposes does receive the rent. He conceded, however, that in statutes intended references to agents are usually express.

Agent's Liability for Rates

Mr David Eve QC, on behalf of Arsenal, submitted that the reference to the person who receives the rents of the parts in section 24 is to the person who receives them beneficially and not as agent. There were a number of express references to agents in the Act of 1967 and the Act of 1925: see Act of 1967, sections 36 (5), 57 (1), 60, 108 (3) and the first part of the definition of "owner" in section 115 (1). If Parliament had intended the owner's agent to be liable for rates under section 24 it would have said so expressly.

On this part of the case we agree with Mr Eve. In our judgment, it is inconceivable that section 24 can, without express words (contrast section 36 (5)), intend an agent to be liable in place of his principal or to be entitled to exercise, to the exclusion of his principal, the rights given to ratepayers by the Act (see, for example, section 108). Even in the case of express references to agents in the Act of 1967, recourse to the agent is, apart from the circumstances mentioned in section 36 (5), permitted only as an alternative to recourse to the owner. We decide that the reference in section 24 to the person who receives the rents of the parts is a reference to the person who receives them beneficially, that is to say the "owner" or indeed the landlord, and that Mr Ende was at no time a ratepayer of the London Borough of Islington.

Does the fact that a person is a ratepayer in itself suffice, without regard to other considerations, to constitute him a person aggrieved if there is any error, unfairness or lack of equivalence in the valuation list? Strictly speaking, because of our decision on the first question, this question does not call for a decision by us, but, in case we should have been wrong, we will nevertheless give a decision on the matter.

Mr Fletcher and his instructing solicitor following him (with our permission in his absence on the second day of the hearing) submitted that any ratepayer is a person aggrieved if hereditaments in his own rating area are underassessed. They referred to the observation of Lord Hewart CJ in *Sevenoaks Urban District Council v Twynam* [1929] 2 KB 440 at 443, where he said:

> "Now undoubtedly those words, 'a person aggrieved,' have very often been considered, and, if one looked at the mere terms apart from their context and apart from the particular circumstances, it would have been quite easy to marshal decisions of contradictory import. But as has been said again and again there is often little utility in seeking to interpret particular expressions in one statute by reference to decisions given upon similar expressions in different statutes which have been enacted *alio intuitu.* The problem with which we are concerned is not, what is the meaning of the expression 'aggrieved' in any one of a dozen other statutes, but what is its meaning in this part of this statute?"

They also referred to the provisions relating to appeals of the Poor Relief Act 1743, the Poor Rate Act 1801, the Union Assessment Committee Act 1862 and the Rating and Valuation Act 1925. Valuation lists were introduced by the Act of 1862 and the provisions of statutes prior to that Act relating to appeals were concerned with appeals against the rate only. We were referred also to the decisions of the Queen's Bench Divisional Court in *Double v Southampton Assessment Committee* and *Hunter v Swindon and Highworth Union Assessment Committee* and of the Court of Appeal in *Pointer v Norwich Assessment Committee,* all reported in [1922] 2 KB at 213, 630 and 471 respectively.

The conclusion which Mr Fletcher and his instructing solicitor sought to draw was that if the present appeals had arisen before 1925 Mr Ende would have been entitled to object to the assessment of Arsenal's hereditament as of right. We think that there is some force in that suggestion. However, section 69 (1) is not in quite the same words as those of the provisions relating to appeals of the pre-1925 Act or even that Act itself. The Act of 1967 is a consolidating Act, and the current provision in section 69 (1) derives from section 40 of the Local Government Act 1948, which Act removed the assessment of hereditaments from the control of rating authorities and transferred it to valuation officers of the Inland Revenue. Section 69 (1) of the Act of 1967 provides:

> "Subject to subsection (6) of this section, any person (including a rating authority) who is aggrieved:
>
> (a) by the inclusion of any hereditament in the valuation list; or
>
> (b) by any value ascribed in the list to a hereditament or by any other statement made or omitted to be made in the list with respect to a hereditament; or
>
> (c) in the case of a building or portion of a building occupied in parts, by the valuation in the list of that building or portion of a building as a single hereditament,
>
> may at any time make a proposal for the alteration of the list so far as it relates to that hereditament."

Mr Fletcher and his instructing solicitor place their main reliance in support of their submission on the decisions:

(1) Of the Divisional Court in *Grainger v Liverpool Corporation* [1954] 1 QB 351, where it was held that the Liverpool Corporation, a rating authority, had power to employ a surveyor to review on its behalf the assessments of offices in Liverpool with a view to the authority making proposals under section 40 of the Local Government Act 1948 to alter the valuation list and where Lord Goddard CJ said at p 357:

> "Further, the rating authority, which in this case is the Corporation of Liverpool, is itself a ratepayer, and can be a person aggrieved. The corporation can be a person aggrieved under section 40 because it may be able to say that its own property, such as the town hall and other property, is rated too high, or that other persons in the area of rating are rated too low. It seems to me, therefore, that it can make a proposal to the local valuation officer either as the rating authority or as a ratepayer in the city, and it matters not whether it describes itself as one or the other. The question is whether the corporation has a *locus standi* to be heard, and I do not think that it is even faintly suggested in this case that it has not. What is suggested is that it has not the power to employ and pay a valuer."

The "Peachey" Cure

(2) Of the Court of Appeal in *R v Paddington Valuation Officer, ex parte Peachey Property Corporation* [1966] 1 QB

A 380, a case in which the company sought to attack, by way of prerogative orders in the High Court, the valuation list on the ground that it was invalid and in which Lord Denning MR said:

(a) with reference to the statutory remedy, at p 399:

"Summarised, the remedy is this: the Peachey Property Corporation, if they are aggrieved, are entitled to make proposals for the alteration of the valuation list. They can propose alterations, either downwards or upwards. On the one hand, they can ask that the values on their own Park West flats be reduced. On the other hand, they can ask that the values on other people's flats in converted houses be increased."

B (b) With reference to the prerogative remedy, at p 400:

"The second question is whether the Peachey Property Corporation are persons aggrieved so as to be entitled to ask for *certiorari* or *mandamus*. Mr Blain contended that they are not persons aggrieved because, even if they succeeded in increasing all the gross values of other people in the Paddington area, it would not make a pennyworth of difference to them. Strange as it may seem, owing to the way expenses are borne in the County of London, the rate poundage of Paddington would remain the same even if the assessments of the flats in converted houses were greatly increased. If the assessments were increased by £1,000,000, the rate poundage
C would be reduced by a penny. But I do not think grievances are to be measured in pounds, shillings and pence. If a ratepayer or other person finds his name included in a valuation list which is invalid, he is entitled to come to the court and apply to have it quashed. He is not to be put off by the plea that he has suffered no damage, any more than the voters were in *Ashby v White* (1704) 2 Lord Raymond 938. The court would not listen, of course, to a mere busybody who was interfering in things which did not concern him. But it will listen to anyone whose interests are affected by what has been done, just as it did in *Greenbaum's* case [*R v Thames Magistrates' Court, ex parte Greenbaum* (1957) 55 LGR 129] and in *Attorney-General of the Gambia v N'Jie*
D [1961] AC 617. So here it will listen to any ratepayer who complains that the list is invalid."

"Busybody"

Mr Eve's submissions on this question were:

(1) The observations of the Court of Appeal in the *Peachey Property Corporation* case did not mean that any ratepayer, a mere busybody, such as Mr Ende undoubtedly was, could, without regard to the circumstances, come forward and allege that he was a person aggrieved;

(2) Some demonstrable effect on the pocket, rights or interests of a proposer, particularly singled out, must be shown to exist, although "grievance" was not
E limited to financial loss or injury;

(3) Mr. Fletcher's submission was illogical: if a person liable to pay a rate was merely by virtue of his status as a ratepayer entitled to make a proposal, many other persons, such as tenants or lodgers paying inclusive rents were excluded. Why should this be so? If not excluded, a different test must be applied in their case: why should this be so?

(4) The question whether a person is aggrieved is one of fact: beyond that he would not attempt a definition of the expression.

F Mr Eve had not intended to refer to the decision of Salmon J (as he then was) in *Buxton v Minister of Housing and Local Government* [1961] 1 QB 278, but at our request cited that decision, in which it was held, following the well-known decision of James LJ in *re Sidebotham* (1880) 14 Ch D 458 at 465, that "person aggrieved" in section 31 of the Town and Country Planning Act 1959 meant a person with a legal grievance. See also *Turner v Secretary of State for the Environment* (1972) LGR 380; 228 EG 335.

The use of the words "person aggrieved" in statutes has more than once been criticised. They are, however, not words which on their face suggest that they bear a technical meaning and may cause little difficulty if they be regarded as G raising a question of fact (a point on which Mr Ende and Mr Eve both agree). It appears to us that Mr Fletcher's submission puts a gloss on the meaning of the words just as much as does the dictum of James LJ. Again we accept the argument of Mr Eve and accordingly, even if Mr Ende were to be regarded as a ratepayer of Islington, we decide this second question in the negative. We may add that if the expression "a person aggrieved" means "any ratepayer" without more ado, it is surprising that the legislature, which used the last-mentioned expression in section 108 of the Act of 1967, did not use it in section 69 (1).

Is Mr Ende, if he is a ratepayer, on the facts aggrieved by the underassessment of Arsenal's hereditament? H

There is no doubt that Mr Ende was not authorised or required by his principal to make a proposal and that his principal had not made a proposal. Indeed Mr Ende does not claim to have made his proposal on behalf of his principal. He does claim that, except in so far as he made it in his capacity as a precept-payer or taxpayer he made it in his personal capacity arising out of his alleged liability under section 24 of the Act of 1967. He said that although in making his proposal he did not in fact represent any one else, he in effect represented the occupiers of 22 Beaumont Rise and other persons in the neighbourhood. He happily accepts J the appellation of "busybody" which Mr Eve applied to him.

He relied particularly on the words of Lord Goddard CJ in *R v Surrey (Mid-Eastern Area) Assessment Committee, ex parte Merton and Morden Urban District Council* [1948] 1 All ER 856 at 858:

"A 'person aggrieved' for the purpose of this Act, that is to say, a person who is entitled to make a proposal either for increasing or decreasing the amount of the valuation, must mean a person who considers that he is aggrieved and may be able to show that he is aggrieved because until the decision is given K either by the Assessment Committee or Quarter Sessions, if there is an appeal, no one can say whether the person is aggrieved. It may turn out that he has no case for grievance at all."

He also relied on *Grainger v Liverpool Corporation, supra; Ryde* (12th ed), p 657. He said that the question whether he is a person aggrieved is one of fact. We have already expressed our agreement with that view.

In response to the expert evidence called by Arsenal that whether or not the assessment of its hereditament be increased to £13,900 or £60,000 the increase could have had no effect whatsoever on the rate poundage in the London Borough of Islington for the years 1971-72 and 1972-73 L and that the effect of his proposal in financial terms upon himself would have been nil or *de minimis* (which evidence we accept), Mr Ende argued that the proposal would have meant nevertheless an increase, not estimated for, in the funds of the rating authority and precepting authority of about £4,000 in each of the years in question.

In our judgment Mr Ende has signally failed to justify himself as a person aggrieved, even if, as the result of section 24 of the Act of 1967, he is to be regarded as a ratepayer of the London Borough of Islington. We cannot see, on the assumption that the Arsenal hereditament has been underassessed, that he has suffered any grievance personally or M otherwise. It is true, as he says, that if he does not pay the rates for which he is said to be liable he may suffer distress on his own effects under section 96 of the Act of 1967, but, of course, it is to be assumed that so long as he receives the rents which he is deputed to collect he will pay the rates out of those rents and that if he ceases to receive the rents his liability under section 24 will cease also. Finally we mention that it was agreed between Mr Fletcher and Mr Eve that if Mr Ende was not aggrieved by virtue of his status as a ratepayer merely, he was not aggrieved at all. For all these reasons we decide this question against Mr Ende.

Is Mr Ende, considered as a precept-payer or taxpayer, a person aggrieved?

In argument Mr Fletcher expressed the view that a ratepayer in a rating area alien to that in question cannot be "a person aggrieved," because the rating area is the relevant unit: see section 68 of the Act of 1967. We agree with that view, which is sufficient to dispose of this question. We however also think that a grievance suffered by a precept-payer or contributor to the general fund of revenue from taxation is too remote. We are not satisfied that Mr Ende has suffered any injury or prejudice in his capacity as a precept-payer or taxpayer. We therefore answer this last question in the negative also.

In the result we decide the preliminary issue in favour of Arsenal and against Mr Ende and, in so far as the valuation officer supported Mr Ende, against him also. The effect of this decision is that Mr Ende's proposal was invalid, that the local valuation court had no jurisdiction in the matter and that Arsenal is aggrieved by the local valuation court's decision. For this reason it appears to us that our decision on the preliminary issue substantially disposes of the proceedings on the appeals and that under rule 48 of the Lands Tribunal Rules 1963 we should allow the appeals and confirm the assessment of £9,250 originally entered in the valuation list.

Finally we should say that we have not mentioned all of the many authorities cited to us and have not dealt with one point, of a technical nature, taken by Mr Eve.

We award the Arsenal Football Club their costs to be taxed by the Registrar of the Lands Tribunal on the High Court Scale and to be paid by the Valuation Officer and by Mr Ende in the proportions of 40 per cent and 60 per cent respectively.

STANBRIDGE v SPITTLE (VO)
(LVC/489/1974)

(J H EMLYN JONES FRICS)

March 17 1975

Estates Gazette May 17 1975
(1975) 234 EG 552

Land used as mooring—Ratepayer owns and occupies house-boat using mooring under licence from National Trust on annual payment of £83.20—No services connected—Comparison with assessments of nearby bungalows and dwelling-houses rejected—Annual payment best evidence available of annual value—Assessment of £80 NAV upheld

The appellant and the respondent valuation officer appeared in person.

Giving his decision, MR EMLYN JONES said: This appeal concerns the rateable value to be attributed to a hereditament described as "land used as mooring," which has the postal address, c/o New Haw Lock, New Haw, Weybridge, Surrey. The hereditament was entered in the 1973 valuation list with an assessment of rateable value £80. On May 24 1973 the ratepayer, Mr D W Stanbridge, made a proposal seeking a substantial reduction in the assessment. Mr Stanbridge described himself as occupier/owner, but this is in fact incorrect. He owns and occupies a house-boat, "Oleine," which makes use of the mooring, but the land itself is owned by the National Trust.

When the matter came before the local valuation court on October 4 1974, the court confirmed the assessment in the list, as it did in 11 other similar cases, one of which was also described as "land used as mooring." The remaining 10 hereditaments concerned were described simply as "mooring."

Mr Stanbridge appeals to the Lands Tribunal. The respondent valuation officer, Mr G Spittle, contends that the assessment of net annual value £80, rateable value £80, is not incorrect or excessive.

The appeal hereditament is a plot of land about 40 ft square with a frontage to the River Wey Navigation at New Haw, near Weybridge. The water is navigable to the north into the River Thames at Weybridge and as far as Godalming in the south, the total navigable distance being about 22 miles. The land is situated on the western bank of the river, which is at this point an artificial cut, and is approached from the southern end of a cul-de-sac known as Common Lane and thence by way of a footpath some 130 yds long, which leads to the western bank of the river and gives access to a number of separate plots of land used as moorings for house-boats and other river craft. There is no electricity, gas, mains water, or drainage connected to the land, and refuse has to be carried about 50 yds to a refuse collection point.

Mr Stanbridge's submission, put quite simply, is that it cannot be right to assess at rateable value £80 a parcel of land with a 40 ft frontage to the river with no services, when a pair of bungalows in Common Lane are each assessed at gross value £85, rateable £50, and four semi-detached houses have assessments varying between gross value £125, rateable value £78, and gross value £150, rateable value £98. It is true, says Mr Stanbridge, that he has chosen to live on a house-boat, but in doing so he does not expect to have to pay rates on the assumption that he makes use of various services which are not available.

The respondent valuation officer, Mr Spittle, put in what he described as "a schedule (and analysis) of rent returns." This schedule gives details of the amounts paid by the occupiers of other house-boats on this stretch of the river, from which it is clear that the payment made to the National Trust for the mooring rights was calculated at the date of this proposal at the rate of 4p per week per foot of frontage. Mr Stanbridge agrees that he also pays on this basis, the total coming to £83.20 per annum, which the valuation officer relies on as supporting his assessment of rateable value £80. The valuation officer dismisses the relevance of the comparables put forward by Mr Stanbridge on the grounds that they are not the same sort of hereditament. He further indicated that if one of those properties, by extending its plot boundaries, acquired a frontage to the river, then it would, in his opinion, become necessary to revalue such a hereditament in order to reflect the added value to be attributed to a mooring facility; and the measure of that added value would be the amount paid for that facility.

The valuation officer seeks to rate only the hereditament described in the valuation list, that is to say "the land used as mooring." He does not seek to include the house-boat as part of the hereditament, nor that part of the bed of the river covered by the house-boat. He treats other house-boats along this stretch of the river in the same way but does not seek to assess as "land used as mooring" other plots where the moored vessels are mobile river craft.

It is clear that the mere right to a mooring, being an incorporeal hereditament, is not rateable *per se* unless it is clearly linked with the occupation of land. But as Lord Buckmaster said in *Port of London Authority v Orsett Union* [1920] AC 273 at 305:

> "The actual hereditament of which the hypothetical tenant is to be determined must be the particular hereditament as it stands, with all its privileges, opportunities and disabilities created or imposed either by its natural position or by the artificial conditions of an Act of Parliament."

In the present case the hereditament which the valuation officer seeks to value is the plot of land occupied by the appellant with the advantages created by its natural position on the bank of the River Wey Navigation and the right to use it as a mooring for craft on the river. Having identified the hereditament, there then comes the question of valuation. In *Robinson Bros (Brewers) Ltd v Houghton and Chester-le-*

A *Street Assessment Committee* [1937] 2 KB 445, Scott LJ said at 469:

> "It is the duty of the valuer to take into consideration every intrinsic quality and every intrinsic circumstance which tends to push the rental value up or down just because it is relevant to the value and ought therefore to be cast into the scales of the balance."

It is not disputed that the hereditament is one which is to be valued direct to net annual value in accordance with the provisions of section 19 (3) of the General Rate Act 1967, which provides as follows:

B
> "(3) The net annual value of any other [ie than as defined in the previous sub-section] hereditament shall be an amount equal to the rent at which it is estimated the hereditament might reasonably be expected to let from year to year if the tenant undertook to pay all usual tenants' rates and taxes and to bear the cost of the repairs and insurance and other expenses, if any, necessary to maintain the hereditament in a state to command that rent."

The valuation officer regards the annual payment made by the appellant to the National Trust as the best evidence of the net annual value as defined in the Act. The appellant produced a copy of the licence originally granted to him by the National Trust in 1969 and which was accepted as covering
C the terms of his occupation at the relevant date, although the amount paid by him to the trust had been increased. The licence was granted to the appellant "to moor a house-boat called 'Oleine' on the River Wey Navigation at New Haw, Weybridge, Surrey, and to use and keep the same as and for a house-boat." Under the licence, the licensee agreed to pay an annual sum by way of equal quarterly payments, to observe certain regulations with regard to the mooring and maintenance of the house-boat, to pay and discharge all lock tolls in connection with the passage of the said house-boat through or along the navigation. He further agreed not in any way to dispose of possession of the house-boat without
D the previous consent in writing of the National Trust nor to use the bank of the navigation "except for the purpose of mooring thereto and as a means of access to the said craft or vessel and will not place or permit or suffer to be placed any material or rubbish on the bank of the navigation" but "will keep the said bank adjacent to the said craft or vessel in a clean and tidy condition."

Whatever the terms of the licence may be, it is clear that the ratepayer is in rateable occupation of the land notwithstanding the fact that certain other people have the right to pass across it in order to reach other mooring places further
E downstream. The real question for determination is whether the annual sum paid under the licence is equivalent to the rent which might reasonably be expected in accordance with the statutory definition of net annual value. The licence is silent on the question of the repairs and maintenance of the bank. If the National Trust were to undertake to keep the bank in F repair this would certainly justify some reduction in the annual payment in order to convert it to net annual value, where the maintenance of the bank is to be deemed to be undertaken by the hypothetical tenant. The valuation officer admitted that he had not considered this point, but he considered that any such adjustment would be so small that it would be sufficiently covered by his rounding off at net annual value £80 from the annual "rent" of £83.20. Mr Stanbridge, however, indicated that the National Trust had never in his experience done anything to the bank, although he had from time to time done a certain amount of work on it. In these circumstances I cannot find that any further G adjustment is properly to be made on account of the liability for repairs.

In considering what rent a tenant might reasonably be expected to pay for the "land used as mooring," the fact that Mr Stanbridge is prepared to pay £83.20 per annum for the licence to moor his house-boat to the bank at this point, even though he has not specifically undertaken to carry out repairs and maintenance of the land, must represent a significant pointer. The annual payment under the licence may not be the same thing as the rent for the hereditament. Nevertheless, it represents the best evidence of value put before me. H I have heard no evidence from the appellant of any adjustments which should be made to that annual sum, and all that can be said on his behalf is that a figure of net annual value £80 appears at first sight to be inconsistent with the assessments on bungalows and dwelling-houses in Common Lane. This discrepancy is clearly one which surprises the appellant but, as the valuation officer rightly suggests, the hypothetical tenant who occupies the appeal hereditament is someone who has special tastes and preferences which are not those of people who choose the more traditional types of habitation. The proper test to be applied is one of annual value; and in the absence of evidence to the contrary I must J assume that the assessments applied to the bungalows and dwelling-houses are properly arrived at in accordance with rental evidence available. If that be right, then it must follow that the annual payments made by the appellant in respect of the appeal hereditament are higher than, for example, the rents to be attributed to the bungalows; the annual value of the appeal hereditament is, in my opinion, clearly established by those payments, which not only the appellant but also the occupiers of other house-boats are clearly prepared to pay.

Mr Stanbridge's dissatisfaction can be well understood, but the statutory basis of valuation for rating is that of annual value of property and is not related to the value K of services received by the ratepayer.

I therefore dismiss the appeal and confirm the assessment determined by the local valuation court. The appellant will pay £10 costs to the valuation officer.

LANDS TRIBUNAL

Law of Property Act 1925 — Section 84

MEMVALE SECURITIES LTD'S APPLICATION

(LP/37/1973)

(DOUGLAS FRANK QC, President)

November 5 1974

Estates Gazette February 15 1975

(1974) 233 EG 689

Modification of covenant in lease sought to enable ground and first floors of Wimpole Street premises in " medical enclave " on Howard de Walden Estate to be used as administrative offices of institutions and cultural or learned societies—Restriction originally to use and occupation as dwelling-house only—Both floors converted to consulting rooms with planning permission conditional on use being restricted to medical and dental professions—Estate licenses use by five named medical practitioners, but floors used as administrative offices of medical institute unbeknown to estate—On discovery of breach reasonable steps taken by estate to terminate or regularise it—Laches held not to arise—Application refused —Covenant secures to estate practical benefits of substantial value or advantage

Mr G H Newsom QC and Mr D Levy (instructed by Berwin Leighton) appeared for the applicants; Mr Paul Baker QC and Mr J S Colyer (instructed by Baileys, Shaw & Gillett) for Howard de Walden Estates Ltd, the objectors.

Giving his decision MR FRANKS said: This is an application under section 84 of the Law of Property Act 1925, as amended by section 28 of the Law of Property Act 1969, for the modification of a covenant contained in a lease of 35 Wimpole Street, St Marylebone, London W1:

> " that (without the previous written consent which may be temporary or permanent revocable or irrevocable or otherwise howsoever framed or qualified of the lessor) there shall not be carried on or exercised in or upon any part of the said premises any profession trade or business whatsoever but the said messuage shall be used and occupied as and for a private dwelling-house only. . . ."

The covenant was set out in a demise on July 29 1918 by the then Lord Howard de Walden to one of the applicants' predecessors in title. The modification sought by the application is to enable the ground and first floors of the property to be used as offices, but the application was subsequently amended to the effect that such use should be confined to administrative offices of institutions, cultural or learned societies which are associated with the medical and dental professions and other cultural or learned societies.

Long History of Property and Area

There is a long history attached to the property itself and the area in which it is situated, and as that history has some bearing upon the issues before me, I must outline it. In the latter part of the 18th century the then Lord Howard de Walden started to lay out for development what is commonly referred to as the Howard de Walden Estate. The boundary of the estate, which has not changed materially over the years, commences at Cavendish Square then runs along Wigmore Street, Marylebone Lane and then irregularly to the rear of properties fronting and then along Marylebone High Street, then along Marylebone Road and down Harley Street as far as No 122 and thence to Hallam Street, Queen Anne's Street and the southern part of Harley Street.

Since 1918 there have been at least two successors in title to the freehold interest in the estate, including a company called General Real Estates Investment and Trust Ltd (" Great "), who were succeeded by the present objectors. In practice, but not in law, the estate has at all times been retained by the Howard de Walden family as shareholders, but that is a factor which I shall ignore. It has been the policy of the successive owners of the estate to preserve amenities, secure observance of covenants, improve income and effect good estate management. Thirty-seven per cent of the properties are let on 999-year leases, 20 per cent on 99-year leases, and the remainder are in hand or let on occupational leases. The covenant with which this application is concerned is common to all the long leases.

All, or nearly all, the properties were constructed as dwelling-houses and were, as stated, subject to covenants restricting any other use, but in 1881 medical practitioners first came to Harley Street from the City. By 1900 there were 150, by 1920 over 500 and there are now nearly 1,700. There thus evolved what is called the " medical enclave." This is bounded by part of Marylebone Road, the properties on the west side of Devonshire Place, Wimpole Street and Upper Wimpole Street, the east side of Harley Street, and at the rear of premises fronting Queen Anne's Street. Within that area, the ground and first floors (but not above, save where there is a lift) of most of the houses are used as consulting rooms by doctors and dentists. A few houses (in addition to the well-known London Clinic) are used for, or in connection with, clinics and the National Heart Hospital, a few more by institutions mainly medical, and there are a few non-medical uses. Many of the houses are listed statutorily as being of special architectural or historic interest and the area forms part of an area designated as a conservation area under section 277 of the Town and Country Planning Act 1971.

Conditional Planning Permission

I turn now to refer specifically to 35 Wimpole Street. The indenture made on July 29 1918 was for a demise for 999-years of the dwelling-house at 35 Wimpole Street and of 9 Wimpole Mews. Until 1949 the property had been used only as a dwelling-house, but on June 2 1949 planning permission was granted for the conversion of the basement into a caretaker's flat, for consulting rooms on the ground and first floor and a self-contained flat on the second floor and a maisonnette on the third and fourth floors. The permission was subject to a condition that the use of the consulting rooms should be restricted to the medical and dental professions; the reason given for that condition was " in order to safeguard the character of the street." The application was made on behalf of Hilstone Trust Ltd, who on June 12 1949 took an underlease of the property for the unexpired term less 10 days. That underlease contained a covenant by the underlessee not to carry on any profession, trade or business and to use the messuage as a private dwelling only until the conversion work had been carried out and thereafter, subject to the licence of the superior lessor being obtained in writing by the lessor, to use only for medical suites and/or residential premises and/or any other form of occupational use approved by the superior lessor and the lessor and the appropriate town planning authorities.

On February 15 1950 " Great " granted a licence for the works of conversion, but it was expressly declared " that the marking of the said plan as consulting rooms shall not be deemed to con-

A stitute licence as to user, but written consent in respect of each practitioner using any of such consulting rooms must first be obtained." The plan showed consulting rooms on the ground and first floors.

On October 23 1950 the London County Council refused planning permission for the office use of the ground and first floors for the reason that "such use does not accord with the council's planning requirements for the area and is considered inappropriate."

On February 20 1951 Hilstone's solicitors wrote to the estate's solicitors and said:

> "in connection with the licence of February 15 1950, our clients B have requested us to ask you for permission to lease the ground and the first floors to the Institute of Cardiology, whose registered office is at the National Heart Hospital, Westmoreland Street, W1. We believe that these people are already tenants of other premises belonging to you on this estate and they require to use the premises for consulting rooms and offices."

Letting as "Consulting Rooms and Offices" Not Agreed

That letter was sent to the estate's surveyor, who replied that the estate would not agree to the ground and first floors being let for consulting rooms and offices and went on to say that they could only be used as consulting rooms by surgeons, doctors and dentists together with their secretaries. On February 25 the C estate's solicitors told the estate's office that they had informed Hilstone's solicitors of the decision. Although there is no record of that decision having been conveyed, I have no doubt that it was, as it will be seen from the next letter that the application for office use was dropped. On March 9 Hilstone wrote to the estate confirming a telephone conversation that it was proposed to lease the ground and the first floors as consulting rooms in the following order. There were then set out the names of five doctors and the rooms it was intended they should occupy. The letter then went on "I shall be glad to hear that these occupiers are acceptable to the estate. I might add that all these doctors are associated with the National Heart Hospital." D On March 14 the estate's surveyor, in a letter to Hilstone, said that he had now certified to the estate's solicitors for the necessary licences to be prepared permitting the five medical practitioners enumerated in the letter to practise their profession in the ground- and first-floor consulting rooms and to have a nameplate each.

By a lease made on May 25 1951 Hilstone demised to the Institute of Cardiology the ground and first floors of 35 Wimpole Street for 21 years from February 8 1951. That lease was subject to a lessee's covenant "not to use the demised premises for any other purpose than as a doctor's consulting room."

On June 15 1951 Hilstone's chairman wrote to the estate office E in which he said: "some weeks ago I spoke to you about the nameplates at this property in respect of those doctors who are occupying the premises and are connected with the Institute of Cardiology and who are our main lessees." He then went on to submit a nameplate for approval. That nameplate, of which Mr Newsom makes much, was shown by a sketch to measure 4 in by 2 in. It was headed "Institute of Cardiology"; on the left-hand side there is the word "Director" and against that "Dr Paul Wood"; again on the left-hand side is the word "Dean" and against that "Dr Graham Heywood." The estate approved of that nameplate. Those doctors were two of those who had been named in the letter of March 9 1951. But on July 5 1951 licences signed on behalf of both the estate and F Hilstone were issued in respect of the five named medical practitioners.

On July 12 1955 Hilstone wrote to the estate to the effect that the Institute of Cardiology would like to redecorate their rooms, but the estate replied to the effect that the application should be made through the head lessees.

On December 12 1962 Hilstone wrote to the Institute of Cardiology referring to a recent meeting and stating that they were prepared to accept the suggestion put forward in a letter that the institute would have two secretaries working in the board room and there would be an additional rent, but that there would be no need for the lease to be amended. There was no evidence that the estate were in any way aware of those nego- G tiations and arrangements.

On October 28 1966 the property was inspected by one of the estate's surveyors, who recorded that the ground and first floors were being used by the Institute of Cardiology as offices and no other use is recorded. Indeed, it was observed that no consent had been given for office use and that the licence authorised five practitioners to occupy and it was noted at the time that they were connected with the institute but they had now all left and the two floors were used entirely as administrative offices. There had subsequently been added against these observations in manuscript "no action yet—reinspect autumn 1969. In the meantime negotiate—rental terms." H

"Concern" Over Institute's Occupation

On February 14 1969 the estate's manager in a letter to the head lessees, namely the trustees of the Watney Mann Pension Scheme, said that the estate had for some time been concerned over the occupation by the Institute of Cardiology and pointed out that no formal consent had ever been given for the administrative offices of the institute and that the estate had refused the original request for that occupation in 1951. He said that the estate had at that time licensed doctors, who belonged to the institute, to conduct medical practices at that house but that it seemed that they carried on the institute's business. He acknowledged the excellent work done by the institute but J said that the estate could not allow it to continue indefinitely, but without prejudice would not object to the institute's continued use during their tenancy provided that it terminated within the next three years. The trustees took up the matter with Hilstone but without getting any satisfaction, Hilstone asserting that the estate had waived their right to object. In a further letter on October 29 1969 the estate manager wrote again to the trustees and said that he understood that it was still the institute's intention to move to other premises the following year and that in these circumstances the estate proposed to take no action. That was stated to be without prejudice to the terms of their head lease and did not in any way condone the breach K of covenant.

On April 14 1972 the underlease became merged by an assignment. On November 12 1973 Hilstone Trust Ltd by its liquidator transferred its interests in the property to the present applicants.

On November 29 1972 the local planning authority in response to an application gave a determination under section 53 of the Town and Country Planning Act 1971 that the use of the ground and first floors as offices did not constitute redevelopment and that an application for permission in respect thereof was not required. Clearly the word "redevelopment" should be read as "development." It is common ground that that decision rested upon the fact that the use of offices had been carried on for L more than four years, and as the law stood at that time the planning authority thereby were precluded from taking any enforcement action.

On December 7 1973 the application to the Lands Tribunal was made, but some discussions took place between the applicants and the objectors. Following the applicants' modification of their application, the objectors offered a deed of variation of the covenant whereby the ground and first floors could be used as dental or medical consulting rooms for five doctors or dentists to be approved in writing by the lessor or as administrative offices or headquarters of any society or institution associated with the medical and/or dental profession (other than M any society or institution concerned with abortion), provided that any such body should first have been approved by the lessors whose approval should not unreasonably be withheld. The applicants made no response to that offer and so on September 30 1974 it was withdrawn.

The facts I have so far related are either agreed or not challenged and I now turn to the grounds relied upon by the applicants.

Applicants' Case

Mr Newsom said that his case rested primarily upon ground (b) in section 84 (1) of the Act of 1925; that is to say that the objectors had agreed by implication, by their acts or omissions, to the covenant being modified. He conceded that there had been no express agreement. He said that he also relied to a lesser extent upon ground (c) and upon (a) and (aa) in section 84 (1).

In the particulars set out in the application it was alleged that the ground and first floors had been used as office premises by the Institute of Cardiology continuously from May 25 1951 to March 25 1972 without complaint on the part of the respondent, except in the year 1970, and that since the complaint was made in that year, rent had been tendered and accepted and no proceedings had been commenced. It was further said that during the greater part of that period the nameplate of the institute was affixed to the exterior door of the said property.

In their objection the estate said that if the premises had been used as offices since May 25 1951 then they had been deliberately deceived by Hilstone and to such an extent as to amount to fraud. I should have not the slightest difficulty in finding, should it be necessary, that right up to, and indeed beyond, that date everything that Hilstone did was calculated to lead the estate into believing that the premises would be used for consulting rooms and for no other purpose. However, Mr Newsom conceded that the use began after May 25 1951 and not as pleaded, which enabled Mr Paul Baker to withdraw the charge of fraud.

Mr Newsom then contended that the office use started some time in 1951, that it was the predominant use by 1953, and that by 1955 there was almost total office user. As evidence in support of the earlier date, he relied upon the nameplate to which I have referred. For my part, however, I do not find that nameplate inconsistent with the use of the rooms concerned as consulting rooms only. On February 20 the estate had been told of the nexus between the institute and the National Heart Hospital and on March 9 they had been told the names of the doctors who were to occupy the floors as consulting rooms and at the same time it was said that they were associated with the National Heart Hospital. Accordingly, when the estate received the letter of June 18 referring to the doctors occupying the premises and stating that they were connected with the Institute of Cardiology who were their main lessees, the natural construction to put on that letter was that although these doctors were connected with the Institute of Cardiology, as was stated in the letter and clear from the proposed nameplate, they nevertheless intended to occupy the rooms solely as consulting rooms.

I find that the estate did not and could not reasonably have been led to believe from that letter and the sketch of the proposed nameplate that the rooms were to be used other than as consulting rooms. Further, having regard to the covenant which the institute themselves had given, I have considerable doubt whether at that time they had any intention of using the premises in breach of that covenant. The year 1953 advanced by Mr Newsom is derived from the evidence of Mrs Annie Miller, the present housekeeper at the premises. She succeeded her aunt in that post in 1968, but for many years had visited her from time to time. She said that in 1953 she noticed on the premises that there were two secretaries, a museum, a boardroom with two desks in it, all on the ground floor, and on the first floor there were doctors' rooms and a secretaries' room with two secretaries. She did not then notice nameplates. She did not recall that there were couches in the doctors' rooms. In 1955 she again visited her aunt and recalls that there were then nameplates by the door and that there were girls working in all the rooms except one doctor's room at the front. In cross-examination she said that the front windows were always curtained.

As against that Mr I Bloomfield, a director of the applicant company and one of the companies from which the applicants' derived title, stated that the rooms were always used as consulting rooms and that the office use was as medical offices, that is ancillary to the consulting rooms. However, Mr Bloomfield was concerned to answer the charge of deliberate fraud, and he said that he did not know " what was going on in the place." His assertion was based on hearsay and I found him totally unreliable as a witness. Certainly his evidence did nothing to help his company's case, rather the reverse.

Probability of Gradual Change of User

For my part I find it impossible on the evidence to conclude when the office user started. By that I mean office user as the main use and not merely ancillary to a consulting room use. The probability on the evidence is that there was a gradual change, which may well have been completed by the autumn of 1957 when an inspection was carried out by one of the directors of a company associated with the applicant company. The applicants did not call any representative of the institute, who I should have thought could have proved the date they regard as so important. However, one thing certain is that there is no positive evidence whatsoever that the estate were aware of any change of use to offices before their surveyor's inspection on October 28 1966

It was conceded by Mr L F Hall FRICS, a director of the objectors, and the estate's surveyor, that the estate always have been very vigilant to ensure that no uses are carried on without their consent, and that nameplates are clues which they look out for. However, he had been with the estate only since 1959, but he said the estate's records did not show the office use before 1966 and that it was very unlikely that the property was inspected. He had always assumed that the use at No 35 had been as consulting rooms.

Mr Newsom also submitted that the letter of June 15 1951 was a notification that the institute were their lessees. That may well have been the case, as indeed the estate had previously been informed that the institute were to be the lessees, but what that letter says is that it was the doctors who were occupying the premises and who were the main lessees. Even if the letter is capable of being construed to the effect that the institute were to be their lessees, having regard to what had gone before, the only sensible conclusion was that nevertheless the doctors would be occupying the premises and occupying them as consulting rooms.

I am satisfied, and I so find, that the objectors were not aware of any breach of covenant before the inspection in 1966 and thereafter that they took such steps as were reasonable to terminate or regularise the breach. It follows from all the foregoing that in so far as the claim under paragraph (b) is based on acquiescence, as in part it was, it fails.

Laches

However, Mr Newsom said that his case under paragraph (b) rested primarily on laches, for which he said his *terminus a quo* was 1955, and contended that the objectors, being *sub-silentio* for over 20 years, must be taken to have agreed to the change of use and that proof of knowledge was not necessary. He contended that the purpose of paragraph (b) was to enable the Lands Tribunal to achieve by positive act that which previously could have been achieved only by an action in equity and equitable relief. On that footing he invited me to consider whether in 1966 an injunction would have been granted restraining the office use. That is an invitation I decline to accept, having regard to the discretionary nature of the powers to grant an injunction. Thus, for example, an injunction might have been granted, but suspended to the end of the institute's term. In any event, in my assessment it is not a useful test to apply in deciding whether the objectors have agreed to the modification of the covenant, but I would like to remark in passing that Mr Newsom in his well-known and valuable work on restrictive covenants [*Restrictive Covenants Affecting Freehold Land*, 5th ed] remarks at page 148 that a plaintiff is not bound to take any steps to discover breaches, for example by going to inspect the premises, and he gives an authority for that proposition.

It seems to me, on a proper construction of paragraph (b), that a person cannot have been deemed to have agreed to a modification unless he had knowledge of the subject of it. I can find no evidence that they had that knowledge. Moreover laches is only that which the court will infer from something having gone on for a long period unquestioned in the absence of some explanation other than acquiescence. I find that there is ample evidence to show that the estate's conduct was perfectly consistent and reasonable and has not caused any injustice.

A From the beginning they made it clear to the applicants' predecessors that they were strongly opposed to office use, and as soon as the knowledge came to them they reiterated that objection and so the question on laches does not arise.

It follows that I have concluded that the applicants' case under paragraph (b) fails, but I would add this, that the most that could be said against the objectors is that they have agreed to modify the covenant only to the extent of its breach; *ergo* they have not agreed to an office use other than by the Institute of Cardiology, and certainly not non-medical uses. Mr Newsom seemed to accept that relief under paragraph (b) would only operate *qua* medical office use, but he submitted that given
B that situation the other uses would fall under ground (c).

Benefit of the Covenant

I now turn to deal with the application under ground (c). Here arises a preliminary question of law, namely whether the estate are to be looked at only as the reversioners of 35 Wimpole Street or alternatively whether regard may be had to all other properties on the estate and particularly those in the medical enclave. Mr Newsom submitted as a general proposition that section 84 was passed to ensure that the landlord could not use his position of strength under ancient covenants other than to protect his legitimate interests. He could not, for example, use the covenant as a means of extracting a premium. However,
C Mr Baker readily conceded that the modification could cause no injury to the reversion of 35 Wimpole Street and if that were the only test under paragraph (c) then the application must succeed, save for the question of discretion. Mr Newsom invoked the rules in *Miles v Easter* [1933] Ch 611, which rules are summarised in the headnote to the report as follows:

> "Where on a sale otherwise than under a building scheme a restrictive covenant is taken, the benefit of which is not on the sale annexed to the land retained by the covenantee so as to run with it, an assign of the covenantee's retained land cannot enforce the covenant against an assign (taking with notice) of the covenantor unless he can show (i) that the covenant was taken for the benefit of
> D ascertainable land of the covenantee capable of being benefited by the covenant, and (ii) that he (the covenantee's assign) is an express assign of the benefit of the covenant."

He conceded that it is permissible to ascertain the land benefited dehors the deed on the authority of *Marten v Flight Refuelling Ltd* [1962] Ch 115, and he did not contend that there would be any difficulty in that respect in this case, but he argued that the covenant in the lease of 1918 was not stated to have been taken for the benefit of other land nor was there any evidence of the benefit of the covenant having been assigned to the objectors.

Mr Baker conceded that the benefit of the covenant is not
E expressly annexed to any land, but he went on to say that one would not expect it to be in a lease, and he derived support for that proposition from Mr Newsom's book, where at page 13 it is said:

> "There are three, and only three, ways in which a plaintiff, not being the original covenantee, can become entitled to the benefit of a restrictive covenant:
>
> A. He may be an assign of land to which the benefit of the covenant is annexed.
>
> B. He may be an express assign of the benefit of the covenant, and of some or all of the land for the protection of which it was taken.
>
> C. Both he and the defendant may own lands subject to a scheme
> F of reciprocal rights and obligations.
>
> There is no fourth class."

Mr Baker said that A does not arise because B operates, in that merely by taking an assignment of the reversion subject of the lease the objectors have the benefit of the covenants and do not have to show a chain of assignments as they would in freehold cases. It seems to me that no express assignment of the benefit of the covenant was necessary, because assignment occurred by statute, namely section 141 of the Law of Property Act 1925, and that that is conclusive. Certainly in the case of *Ridley v Taylor* [1965] 1 WLR 611, which was another lease-hold case, apparently without express assignments of the benefit G
of the covenant and without any specific annexation, it was not even argued that the landlords had not the benefit of the covenant. Russell LJ clearly based his judgment on the assumption that it was proper to take into account damage to property of the landlords other than that the subject of the covenant. Mr Baker asserted that section 10 (4) of the Leasehold Reform Act 1967 assumes that leases may be capable of benefiting other property of the landlord. That may be so, but as Parliament is not deemed to know the law I do not take that into account. I am satisfied that the test is met if, at the date of the covenant, the covenantee had ascertainable land to which the covenant could be relevant, and that in this case there was such land and that it can be ascertained with reasonable precision. It is that H
land which I have described as the Howard de Walden Estate.

Injury to Estate

Having concluded that the objectors are entitled to the benefit of the restriction, regarding it as applicable to the whole estate, there remains the question whether any injury to the estate would result from the proposed discharge or modification. In *Ridley v Taylor* (*supra*) Russell LJ expressed the view that paragraph (c) is, so to speak, a long-stop against vexatious objections to extended user, and that it would be a strong thing to find that in an estate of the experience of that in the case [Grosvenor J
Estate]. Certainly it would also be a strong thing to find that in the case in the experience of the objectors. However, for reasons which will appear when I consider the application under ground (aa) I am satisfied that the discharge would injure the estate as a whole.

I now turn to ground (aa). The first question which arises is whether the proposed use is reasonable, and I find this question virtually impossible to answer in relation to paragraph (a) of subsection (1A), because no test is prescribed as to what is reasonable. I suspect that that part of the subsection really is applicable only in considering whether the covenant is contrary to the public interest. Mr Baker referred me to the current
development plan, which provides that it is the local planning K
authority's intention that professional and cultural uses should as far as possible be maintained in the area of Portland Place, but in my view that is not sufficient to show that the proposed use is not reasonable. Development plans are subject to review from time to time and, moreover, one could not say with confidence that an application for planning permission for the uses proposed by the applicants necessarily would be refused either by the local planning authority or by the Secretary of State on appeal.

Accordingly, I propose to assume that the proposed user would be reasonable, and turn to consider the vital question, namely whether the restriction secures to the objectors practical L
benefits of substantial value or advantage. It was conceded by the applicants that the objectors exercise strict control over the estate, and Mr J P Stacey FRICS, who gave evidence for them, approved of that policy. However, he said the fact of the office use for 20 years not even having been noticed shows that that use is harmless. He accepted that the grant of the modification might generate other applications, but he thought that they would be very severely limited in number because of the difficulty in obtaining planning permission. Further he pointed out that a use cannot now be established for planning purposes merely by its having been carried on for a certain period.

Mr Newsom contended that the applicants were asking for
no more than the estate had granted to the Medical Defence M
Union for their lease of 1-5 Devonshire Place, but it was pointed out that the estate was anxious to attract that body, which had rebuilt the houses concerned at a cost of £½m. It was therefore a special case. Mr Hall had stated (as had been made clear by the estate's offer) that the estate accepted medical and dental institutions in principle, subject to their approval of the proposed occupier, although he preferred them to be located in the cross streets rather than in Harley Street and Wimpole Street. On the other hand it had been the policy of the estate to avoid having other office uses within the medical enclave.

Risk of " Domino " Effect

I have come to the conclusion that the covenant does secure to the objectors practical benefits of substantial value or advantage to them. It has always been the policy of the objectors and their predecessors in managing this estate to impose and, save in very exceptional cases, to enforce this covenant as part of their management of the estate. It is not, in my view, for the Lands Tribunal to interfere with that policy of estate management if the covenant still serves or may serve a useful purpose. I think the covenant does serve a useful purpose, for it enables the objectors to keep control over the use to which the houses are put and to keep the area as one predominantly used for residential and medical purposes and it also enables them to keep out other kinds of occupants who, they may think, are undesirable. That seems to me to be a control valuable not only to the objectors, and here I bear in mind that some of these houses are on short leases or in hand, but also to their lessees. I also consider that it is proper to have some regard to the large number of listed buildings and the designation of the conservation area, because, to some extent at least, those factors are a consequence of estate management. I am further of the opinion that there would be a real risk of a " domino " effect resulting if this restriction were modified.

Clearly, a precedent would be set, because there are such a large number of comparable houses. The fact that this one house now has an established office use under planning law is not very persuasive. Today there is some force in the argument that planning permission for office use would be difficult to obtain, but, as I have already pointed out, planning policies can change and I see no reason why the estate should be called upon in effect to transfer their rights of control to another body. If, as it seems, office use is more valuable than use as a private residence or as a medical institute, then it is probable that many other lessees would want to exploit their leases in the same way as the present applicants. There was some argument whether the loss of bargaining powers resulting from the modification of the covenant is a legitimate consideration. Mr Newsom pointed out that, having regard to section 19 (2) (3) of the Landlord and Tenant Act 1927, there was no bargaining position to be lost, but on the other hand Mr Baker said that that provision merely made an increase in rent irrecoverable, not unlawful, and he reminded me of the decision in *Ridley v Taylor* (*supra*) and also referred to *Comber v Fleet Electrics* [1955] 1 WLR 566. I think Mr Newsom is right and have not assumed that, but for the order of a modification, the objectors might obtain an improved rent.

However, as I have said, I am satisfied that the restriction is of substantial value and advantage to the objectors. It follows that this application is refused.

The applicants will pay the costs of the objectors, such costs if not agreed to be taxed by the Registrar of the Lands Tribunal on the High Court Scale, but there shall be deducted therefrom a sum of £100.

COLLINS' AND OTHERS' APPLICATIONS

(LP/2-4/1974)

(DOUGLAS FRANK QC, President)

December 16 1974

Estates Gazette March 8 1975

(1974) 233 EG 948

Application to permit erection of detached houses at density of 3.2 houses an acre at North Fleet, Hants, refused—Proposal alien to purposes of original building scheme—Certain published planning documents ruled inadmissible —" Inappropriate for a hearing under this jurisdiction to become something in the nature of a town planning inquiry "

Mr William Glover QC and Mr A Anderson (instructed by Withers) appeared for the applicants; Mr G H Newsom QC and Mr V Hallett (instructed by W Bradly Trimmer & Son, of Alton; Ellis & Fairbairns, of Reading; and Stewart Green & Co, of Winchester) for the objectors.

Giving his decision Mr Frank said: These are three applications, (1) by Mr A J R Collins, Mr T J Davies and Mr O B Harris; (2) by National Westminster Bank Ltd; and (3) by Col E L Fanshawe, for the discharge of restrictive covenants affecting land at Elvetham Road and Stockton Avenue, formerly known as the Stockton House Estate, North Fleet, Hampshire, to enable 30 detached dwellings to be erected in accordance with a planning permission granted on December 21 1972.

On February 3 1933 the Stockton House Estate, comprising about 32 acres, was conveyed to one Herbert Pool, who over the years sold off the estate in plots with mutual covenants. The main covenants standing in the way of the applicants' proposals are those prohibiting the erection of more than one dwelling-house on each plot sold and against the making of any road or path upon any part of the plot for use in connection with any land or premises other than the land conveyed. The effect of the applicants' proposals if carried out would be to have within the areas concerned 32 houses instead of the four houses allowable under the terms of the covenants. In order to achieve that, the applicants require the covenants to be modified in one case so that building can take place where none is permitted and in other cases to enable more than one house per plot to be erected, and so that roads can be made across various plots and so that fences may be erected less than 40 ft from Waverley Avenue. The scheme for 30 houses extends beyond the Stockton House area, and the effect of it within the covenanted area would result in 20 houses, including one existing house within that area, and produce a density of 3.2 houses per acre gross in that area.

The Existing Development

The development which has been carried out is mainly along one side of Stockton Avenue and along both sides of Waverley Avenue, which is a crescent off Stockton Avenue. The houses are substantial, generally having five or six bedrooms with three reception rooms and mainly having plot areas varying from about a half to one acre. Part of Waverley Avenue has a metalled carriageway, but some of it is completely unmade and none of it has been adopted. Along most of its length it is bordered by many attractive trees with shrub undergrowth. The gardens are well maintained, giving a spacious and rather lush appearance, and have within them many fine, mature trees. Most of the application area has within it many mature trees and, it is understood, is subject to a tree preservation order.

The owners of 15 of the houses entitled to the benefit of the covenants and who objected to the applications were all represented by Mr Newsom and Mr Hallett. One objector did not appear and was not represented. Several of the objectors gave evidence, and evidence also was given by Mr J E Montague FRICS. It was conceded for the applicants that all the objectors had the benefit of the covenant.

Mr Glover said that he relied entirely upon paragraph (aa) of Section 84 (1), that is to say that the proposed development would be a reasonable user of the land and that the restriction in impeding that user does not secure to the persons entitled to the benefit of it any practical benefits of substantial value or advantage to them; and further that in impeding that user the restriction is contrary to the public interest. He said that only two houses would be affected by the development and then only to a minor degree and that the payment of compensation, £250 in each case, would be adequate for any loss or disadvantage suffered by the owners of those properties.

I now proceed to consider the questions arising in these applications by reference to the questions adopted for the purpose of paragraph (aa) in *re Bass Ltd's Application* (1973) 228 EG 927, 1123, 1321.

A *Question 1: Is the proposed use reasonable?*

As I see it, I have to answer this question on the hypothesis that the covenants do not exist. I think that having regard to the residential character of the area and the type of housing proposed, the answer must be "Yes."

Question 2: Do the covenants impede the proposed user? Clearly they do.

Question 3: Does impeding the proposed user secure to the objectors practical benefits?

Question 4: So are those benefits of substantial value or advantage?

B In the context of this case it is convenient, and I think sensible, to consider questions 3 and 4 together.

The houses proposed are to be built fronting three new cul-de-sacs with hammer heads, each having a junction with Waverley Avenue, and with one having another cul-de-sac leading off it. Within the Stockton Estate there would be 20 houses on a gross area of 6.2 acres, giving a gross density of 3.2 houses per acre and plot sizes of about $\frac{1}{4}$-acre.

"Not Merely Arithmetic of Density which Matters"

This density is low compared with most modern development. It is nevertheless a modern form of development, with
C a density much greater than that existing and that permitted under the covenants for the application area, viz about one per acre. However, it is not merely the arithmetic of the density which matters, but the general effect on the amenity of the area. In my assessment, and I have viewed the area, the applicants' proposal will introduce a form of development quite out of character with that now existing. Mr Glover described the area as now being "suburban" in the true sense of the word, that is to say neither urban nor rural. I adopt that description but would describe the applicants' proposal, because of the density and the numbers and types of roads, as being urban in character and alien to the pur-
D poses of the building scheme brought about by the covenants. I further think that the objectors are right when they say that the development would, by introducing more people into the area, increase traffic and increase noise of various kinds, and involve loss of privacy. It would also necessitate the clearing and felling of more trees and shrubs than would be necessary under the very limited development permitted under the covenants.

Two Tests of Practical Benefits

Mr. Glover posed two questions as a test for deciding
E whether practical benefits are secured to the objectors. They were: do the covenants maintain the ethos of the estate and, if so, would the development destroy it? In my judgment both questions must be answered affirmatively. I further think that the covenants are of substantial advantage to the objectors, though as Mr Montague, the objectors' expert witness, conceded, the development would not result in any depreciation in market value of the objectors' houses. However, it is clear from those who gave evidence that they regard the covenants as being of substantial advantage to them, and indeed some of them bought their houses in reliance upon the covenants. It seems to me that nothing
F material has changed since the building scheme first started and it is fair to assume that it was effected for the purpose of giving substantial advantages to the purchasers by reason of the restrictive covenants.

Question 5: Is impeding the proposed use contrary to the public interest?

Mr Glover relied upon the following published documents, viz:

(a) *The South-East Study,* published by the Ministry of Housing and Local Government in February 1964, being a report of a study of the problems that may be expected to arise in the South-East of England over the 20 years from 1961 to 1981.

(b) *A Strategy for the South-East,* published by the South- G
East Economic Planning Council in August 1967.

(c) A number of Government White Papers and Circulars.

(d) Development Control Policy Note No 2, published by the Ministry of Housing and Local Government in 1969.

(e) The Aldershot, Farnborough and Fleet Town Map, approved by the Minister of Housing and Local Government in 1965 as part of the Hampshire County Development Plan.

(f) The North-East Hampshire Interim Plan, published in July 1974, being a summary of the then current policies of the county planning authority until their structure plan for North-East Hampshire is approved.

(g) The North Fleet Local Plan 1971. H

Paragraph 24 of the North Fleet Local Plan, referring to an area which includes the application sites, says:

"These areas contain substantial groups of trees, dwellings of semi-formal arrangements and some roads of special character. It is proposed that new development shall conform to this pattern and dwellings on plots of less than one quarter acre (approximately 0.1 of a hectare) will not normally be permitted."

The purpose of producing these documents is to show:

(i) That there is a shortage of land available for building houses in South-East England.

(ii) That published plans provide for some of the J
deficiency to be met in the outer metropolitan area in which Fleet is situated.

(iii) That the sites the subject of these applications are allocated in the approved development plan, that is, the town map, for residential development.

(iv) That it is county council policy that the sites should be made available for development pending a strategic plan and local plan.

(v) That it is Government policy to encourage development at high densities.

I have considerable doubt whether any of these documents, K
other than the approved development plan, are admissible. Subsection (1B) of Section 84 provides that:

"the Lands Tribunal shall take into account the development plan and any declared or ascertainable pattern for the grant or refusal of planning permissions in the relevant areas, as well as the period at which and context in which the restriction was created or imposed and any other material circumstances."

"Highly Undesirable, Impracticable and Unfair"

Mr Newsom argued that "any other material circum- L
stances" must be construed *ejusdem generis* with the words immediately preceding it as far back only as the comma. I think that is right, because had Parliament intended to embrace every plan that could be produced by central or local government, I think it would have said so. Moreover, I think it highly undesirable, impracticable and unfair to expect persons entitled to the benefit of a covenant to take part in what could be a long and wide-ranging discussion of planning policies for the south east or indeed for the whole of England. The only document which has been subjected to all the statutory tests and has any statutory standing is
the approved development plan. That was published, was M
open to objections which in turn were considered at a public inquiry, and finally it was subject to the Minister's approval or otherwise. None of the other documents have been submitted to any such tests and it would seem most inappropriate for a hearing under this jurisdiction before the Lands Tribunal to become something in the nature of a town planning inquiry. Accordingly, I hold that all those documents, other than the approved town map, are inadmissible, but even if I am wrong as to that, I do not think that they advance the applicants' case very much.

Nobody is objecting to the residential development of the

A application sites; the only issue is how many houses. Although successive governments have pressed for higher densities as a general proposition, as the Ministry stated in the Policy Note No 2 "whether or not a particular density is appropriate in any given case will, in practice, depend on local conditions and on the design and layout of the scheme, rather than on some predetermined scale of densities for the area." Further, the extract from the local plan which I have quoted shows that the local planning authority's policy is that the application area should have not less than one quarter acre plots; therefore, it does not rule out a lower density.

B

In my view, for an application to succeed on the ground of public interest it must be shown that that interest is so important and immediate as to justify the serious interference with private rights and the sanctity of contract. In my judgment this case comes nowhere near satisfying that test. C

Questions 6 and 7, which ask whether money would be an adequate compensation, do not under the circumstances arise.

It follows from all of the above that these applications are dismissed.

The applicants will pay the objectors' costs, such costs, if not agreed, to be taxed by the Registrar of the Lands Tribunal on the High Court Scale.

D